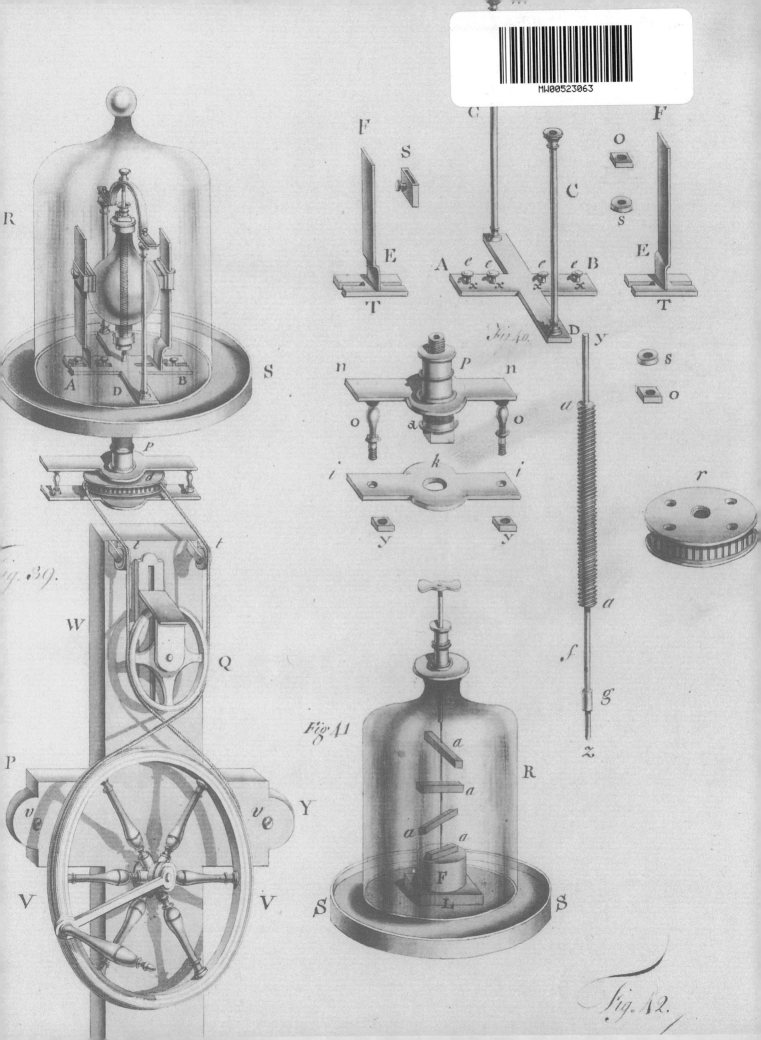

Fig. 39.

Fig. 40.

Fig. 41.

Fig. 42.

PUBLIC & PRIVATE SCIENCE

Public & Private Science
THE KING GEORGE III COLLECTION

Alan Q. Morton & Jane A. Wess

OXFORD UNIVERSITY PRESS
in association with the
SCIENCE MUSEUM
1993

Oxford University Press, Walton Street, Oxford OX2 6DP

Oxford New York Toronto
Delhi Bombay Calcutta Madras Karachi
Kuala Lumpur Singapore Hong Kong Tokyo
Nairobi Dar es Salaam Cape Town
Melbourne Auckland Madrid
and associated companies in
Berlin Ibadan

Oxford is a trade mark of Oxford University Press

Published in the United States
by Oxford University Press Inc., New York

A catalogue record for this book is available from the British Library

Library of Congress Cataloging in Publication Data
(Data available)

ISBN 0-19-856392-2

Typeset by Cotswold Typesetting Ltd, Gloucester
Printed in Hong Kong

FOREWORD

Neil Cossons

Director, National Museum of Science & Industry

The Science Museum's *King George III Collection* is an assemblage of early scientific instruments without parallel. The items in it are varied, often visually attractive, and richly informative about eighteenth-century science. The publication of this first definitive catalogue of the whole collection coincides with the opening in the Museum of a gallery in which many of the instruments are newly displayed. The book and the gallery together make available to scholars, and to a wider public, not only the instruments themselves, but much new information about their origin and the way they were used.

Within the collection, two main groups of instruments are associated with George III: the apparatus which the King commissioned for his own collection, and the equipment used by a public lecturer who later became superintendent of the King's own observatory. Both groups take us back to a period when science was still an innocent pastime, an accessible and fashionable pursuit for people in many different walks of life. In the second group in particular, we glimpse the beginnings of popular science, scientific ideas made available to a wide audience for the first time, as they are today by the successors of these lectures—books, television, museums, and the education system.

George III was a man of great intellectual curiosity, whose strengths have only grudgingly been acknowledged by history. His interests ranged widely across the arts and sciences. The astronomer William Herschel, one recipient of the King's patronage, described him as 'the best of kings, who is the liberal protector of every

art and science',[1] remarking on another occasion that 'the King has very good eyes, and enjoys observations with the telescopes exceedingly.'[2] Herschel proposed that the planet he had recently discovered should be dedicated to the King, and suggested the name *Georgium Sidus* or George's star; not surprisingly the name was soon supplanted by the classical one the planet bears today, Uranus.

George III's enthusiasm for science had been kindled when he attended a course of lectures on 'natural philosophy' in his youth. These were given by Stephen Demainbray, an itinerant lecturer whose work of explaining and demonstrating scientific ideas to public audiences is described in the introductory chapters of this book.

Much later, in 1769, the King's imagination was caught by the prediction of a rare astronomical event. This was the transit of Venus when, for a few minutes, the planet could be seen in silhouette as it passed across the face of the sun. George III gave the Royal Society £4000 to support Captain Cook's first expedition to the South Seas to observe the event and use it to make an accurate measurement of the size of the solar system. At the same time the King arranged for the building of his own observatory at Richmond (now Kew Observatory), from which he himself observed the transit of Venus.

When Stephen Demainbray was appointed superintendent of the King's new observatory, he brought with him the workaday demonstration apparatus he had used as a travelling lecturer. A few years earlier, the

---

[1] Lubbock (1933), p. 124.

[2] Lubbock (1933), p. 118.

King had commissioned from the leading instrument-maker, George Adams, some splendid pieces of apparatus to be used to entertain and educate his family and friends. Equipment from two widely differing sources thus came together under one roof, and later passed into the care of the Science Museum.

In 1993, more than two centuries after the *King George III Collection* first began to take shape, it is a source of pride to the Science Museum that through this catalogue and the accompanying gallery we are able to restore this remarkable collection to the prominence it so fully deserves.

ACKNOWLEDGEMENTS

While working on this book we have benefited from the help of many people. In particular we would like to thank D. J. Bryden and V. K. Chew who provided detailed comments on the manuscript at several stages, and R. G. W. Anderson, Geoffrey Cantor, J. R. Millburn, and Simon Schaffer for their helpful discussions. G. L'E. Turner was particularly generous in giving us the benefit of both his great knowledge of scientific instruments and his wide practical experience of preparing catalogues.

In writing and producing this book at the Science Museum we have had the immense advantage of the assistance of our colleagues, past and present, throughout the museum. We are most grateful for their generous and unstinting help. Kevin Johnson, who has assisted us over the years, has done more to ensure the smooth running of our work than we could ever have reasonably expected. David Exton, John Lepine, and Adrian Whicher, who photographed the collection, went to immense trouble on our behalf. We are also grateful to John Ward who provided the entries on photographic items.

Manuscripts

Information from a manuscript in the Royal Archives is quoted by gracious permission of Her Majesty the Queen. We are also grateful to the following for permission to quote from manuscripts in their possession: Aberdeen University Library, Bibliothéque Municipale de Bordeaux, British Library, Linnean Society of London, The Marquess of Bute, Trustees of the Natural History Museum.

Illustrations

Illustrations taken from items in the Royal Library, Windsor Castle and Royal Collection, St James's Palace are reproduced by gracious permission of Her Majesty the Queen. These are p. 7 (lower), p. 10, p. 98, p. 272 (P32), p. 294 (M6), p. 295 (M7), p. 298 (M11), p. 304 (M19), p. 310 (M30), p. 313 (M34), p. 314 (M35), p. 317 (M38), p. 322 (M45), p. 323 (M46), p. 329 (M50), p. 331 (M53), p. 336 (M60, M61), p. 349 (M76), p. 356 (M85), p. 362 (M94).

We are also grateful to the following organizations for allowing us to reproduce illustrations in their collections.

Aberdeen University Library: p. 95, p. 103.

Bibliothéque Municipale de Bordeaux: p. 135 (D1), p. 137 (D3), p. 144 (D11), p. 147 (D14), p. 152 (D19), p. 153 (D20), p. 154 (D21), p. 157 (D25), p. 158 (D26), p. 164 (D29), p. 189 (D62).

Bodleian Library, University of Oxford, Ms.Rawl. D.871, fols.141r and 142r: p. 53.

British Library: p. 23 (lower), p. 42 (upper left), p. 82, p. 169 (D37), p. 170 (D39), p. 487 (E159).

British Museum: p. 60.

Christ's Hospital: p. 45.

Governor and Company of the Bank of England: p. 110.

Guildhall Library: p. 47.

John Rylands University Library, Manchester: p. 97.

King's College, London: p. 28, p. 35.

Library of Congress: p. 73.

Microform (Wakefield) Ltd: p. 94 (upper).

Museum of the History of Science, Oxford: p. 68, p. 115, pp. 128–33, p. 210 (D105).

National Maritime Museum: p. 27 (lower).

National Portrait Gallery: p. 16, p. 33, p. 44.

Public Record Office: p. 20 (lower).

Royal Botanic Gardens: p. 9.

Royal Geographical Society: p. 101.

Royal Society: p. 26, p. 461 (E102).

Victoria and Albert Museum: p. 8, p. 58, p. 198 (D72).

Worshipful Company of Clockmakers: p. 30.

Yale Center for British Art, Paul Mellon Collection: p. 27 (upper).

CONTENTS

PUBLIC AND PRIVATE SCIENCE IN MID-18th CENTURY LONDON
The King George III Collection at the Science Museum

Introduction

THE KING GEORGE III COLLECTION at the Science Museum is one of the most remarkable collections of eighteenth century scientific apparatus to survive to the present day. It contains around 1000 items used for the demonstrations that were a standard feature of courses on natural philosophy by 1750. These courses covered such topics as mechanics, pneumatics, optics, electricity, magnetism and hydrostatics.[1] The historical interest of the collection is twofold; the design of the apparatus is typical of its period, yet we have also established a detailed provenance for many items, something it is rarely possible to do for material of this kind.

The range of historical enquiry opened up by the collection is vividly illustrated by comparing two groups of material within it. On the one hand there is the apparatus King George III commissioned from the instrument-maker George Adams shortly after coming to the throne in 1760. This apparatus is of high quality,

made from expensive materials, or ornate, and was used for entertaining or instructing members of the royal family. In marked contrast, there is another set of items made for the same type of demonstrations envisaged by Adams, but simpler and more utilitarian. This is the apparatus assembled by Stephen Demainbray, who made his living by lecturing on natural philosophy in the 1750s. When an observatory was built for the King at Richmond in 1769 (now Kew Observatory) and Demainbray was appointed its superintendent, the building came to house both the apparatus of the royal family and Demainbray's own equipment. Consequently the two collections, illustrating both the private and public aspects of science, became amalgamated.

Considered together, these items from different sources span the range of social groups interested in natural philosophy in the mid-eighteenth century: from an aristocratic élite who owned their apparatus to a wider audience who paid to attend lectures. In outline this book explores some of the issues that arise from

---

[1] A catalogue covering many of the items in the collection was published in 1951. See Chaldecott (1951). We would like to record our appreciation of Dr Chaldecott's pioneering work on this collection.

the collection, starting with its history and going on to give an account of natural philosophy on the public stage in mid-eighteenth century London. The history of the collection is plotted in Chapter I, first as a collection of collections assembled at Kew Observatory, but later as a construct shaped by institutions, after it was removed to King's College, London in the mid-nineteenth century and finally to the Science Museum in 1927.

A common influence throughout the different sections of the collection is that of lecturing on what was called natural or experimental philosophy. (Terms such as science or physics were rarely used in the eighteenth century—many of the subject divisions which are familiar today such as physics or chemistry, and indeed the word 'scientist', date from the nineteenth century.) Consequently, the origins of public lecturing on natural philosophy in London in the early eighteenth century and its later development are covered in the chapters that follow. Some items in the collection provided clues about the business of lecturing and here the items used by Demainbray are of particular interest because it is very unusual to be able to identify the actual apparatus used by an itinerant lecturer on natural philosophy of the mid-eighteenth century.

Lecturing

Through the process of lecturing, the subject of natural philosophy was defined for a wider audience and represented to that audience. Though the work of the lecturers drew on the earlier achievements of the 'virtuosi' of the Royal Society, the approach they took to natural philosophy was also influenced by other factors connected to broader cultural developments such as the coming of newspapers, or the productions of the hacks of Grub Street. One result was that the subject of natural philosophy became better known with much of this knowledge being spread by itinerant lecturers.

The activities of those who lectured were sustained by networks of individuals and in turn the lecturers played a part in extending and sustaining these same networks—a system of mutual benefit. For example, the transits of Venus over the Sun's disc in the 1760s attracted a great deal of interest, in part fostered by the lecturers who had seized on the opportunities to lecture on such a novel experience. But the transits, which provided a rare opportunity to estimate the size of the solar system, were also the occasion for more complex alliances of interests. Many international expeditions were mounted to observe the transit, including Captain Cook's first voyage to the South Seas, and the lecturers helped to gather support for these efforts. Thus the lecturers helped in the formation of the rash of societies that became a feature of both national and provincial life in Britain in the later eighteenth and early nineteenth centuries. In these respects, the activity of lecturing was an important connection between the development of natural philosophy in the seventeenth century and physics in the nineteenth.

The first public lectures on natural philosophy in London began in the early years of the century and Chapters II and III give an account of these lectures from their origins until 1770. The lectures were one important way in which knowledge of natural philosophy was formulated for an audience. As the development of the audience is one of the most intriguing aspects of the history of lecturing in the early years of the eighteenth century, several factors that contributed to its development are considered. The lectures on natural philosophy were just one aspect of the growth of the business of public lecturing in general, and so lectures on other subjects such as anatomy or chemistry are also considered. In the period 1745–70, a new generation, Demainbray among them, now tried to make a living from lecturing. For a few years in the late 1750s there was a modest boom for the lecturers on natural philosophy but otherwise there was a gradual decline. Some explanations are advanced for this pattern such as the influence of new institutions, particularly the Society of Arts founded in 1754, and the fact that lecture courses on natural philosophy, unlike those on medical subjects, were not vocational. Chapter IV covers Demainbray's life and work and provides much of the background information about the items he collected for his lectures.

A full assessment of the significance of the development of public lecturing on natural philosophy in the eighteenth century falls outside the scope of this work; nevertheless, an attempt has been made, using the apparatus in the collection and other sources, to give a partial account. Though developments in London are the only ones considered in detail, what happened there set the pattern for what happened elsewhere in Britain in that period.

Through studies such as this one we can come to a deeper understanding of the uses of natural philosophy in the eighteenth century. An understanding of how the lecturers were involved in laying the foundation of the boundaries around science and establishing these demarcations in the public mind, may also illuminate more modern concerns such as the nature of science and its relation to other bodies of knowledge. Similarly, as the lecturers both demonstrated expertise, while acknowledging the authority of others (principally a few individuals like Newton or Boyle) and built up public support for aspects of scientific authority, a study of their work focuses on the nature of expertise in the field of natural philosophy.

I

THE
KING GEORGE III COLLECTION
A collection of collections

Introduction

THE KING GEORGE III COLLECTION consists of scientific apparatus brought together from a variety of sources over much of the eighteenth and nineteenth centuries. This chapter gives an outline of the events leading to the formation of the collection and its subsequent development. The changing content of this collection and its transformation from a selection of apparatus for demonstrations in natural philosophy into a collection in a museum is an involved story—a story best approached by seeing the whole as a collection of collections. The collection has four main parts, each significant in different ways: the apparatus provided by George Adams for the King in 1761–2, equipment used by a lecturer on natural philosophy, Stephen Demainbray (1710–82), other miscellaneous eighteenth century items from the royal family, and lastly a range of nineteenth century material, mainly added while the collection was housed in King's College, London. Despite the apparent diversity of the sources of material in the collection, the different elements have a common use in that the items were designed for lectures on natural philosophy. The collection as a whole was shaped by the business of lecturing on natural philoso-

phy, and this explains much of the background and provenance of the items in it.

Overall the collection, the items and their provenance, provides a view of the wider arena of the development of lecturing in natural philosophy; clearly, many aspects of eighteenth century natural philosophy bear on the history of the collection. Natural philosophy was then an area of interest for several reasons. In broad terms, the ruling and new professional élites continued to develop a 'high' culture which diverged from 'popular' culture with ideas from natural philosophy as part of this process of differentiation. For example, new ideas of rationality and experimental science helped to counteract the magic or superstition of popular culture. Furthermore, these new attitudes were debated amongst the different sections of the élite who stood to gain or lose by the outcome; the ideas of Newton and his followers on natural philosophy were important for the latitudinarian wing of the Anglican Church as it became more influential in the early part of the century.[1] Contemporary with these large-scale changes was the development of public lecturing on natural philosophy; the activity of lecturing was an important part of the process through which these new ideas and attitudes were disseminated to and adopted by various social groups.

---

[1] See Gascoigne (1989).

In England, this type of lecturing was an innovation of the early eighteenth century. The novel features of the lectures were twofold: they were open to the public, that is anyone willing to pay the fee to attend, and they were based on working demonstrations carried out with apparatus similar to that in the King George III Collection. The development of public lecturing on natural philosophy came about for several reasons. While natural philosophy as a field of study had a long history before the beginning of the eighteenth century, there was by then a growing audience with an interest in the subject. Though it is difficult to be precise about the numbers, this audience was large enough for a few individuals to make a living by lecturing or writing on natural philosophy and related topics at that time. However, these individuals are not easily categorized for they were few and did not make a cohesive or distinct social group.

An intriguing aspect of this early eighteenth century lecturing was its novelty in a period that was otherwise fallow in scientific terms by comparison with what had gone before and what came after.[2] However, the growth of popular forms of natural philosophy, such as lecturing or the collecting of instruments, is important precisely because it links the scientific developments of the seventeenth century to those of the later eighteenth century, or in other terms, the work of Newton to that of Priestley, to mention two individuals who lived in London. Though this link is indirect, it is nevertheless important, for it helped establish networks of patronage and interest that promoted scientific activity later in the century.[3]

The activity of lecturing obviously helped create and sustain an audience with at least a passing interest in scientific matters. By examining these developments we can arrive at a broader assessment of scientific activity, one which encompasses the appearance of scientific ideas in a variety of forms, such as lectures or popular books, and which explains how these ideas became transformed into part of a wider culture. Such an assessment directs our attention outwards from groups such as the Fellows of the Royal Society or skilled instrument-makers towards a much wider section of the population, the audience who attended the lectures, and it provokes questions about the techniques used to reach that new audience and how they reacted to the lectures.

The Origins of the Collection

Many items of the collection, particularly those dating from the mid-century, provide evidence about the activity of collecting and the practice of patronage and how these changed. Only a few items were made between 1690 and 1740 and were likely to have come into possession of the royal family then. This was probably due to the situation of the monarch, newly installed and not yet firmly established on the throne. In addition, the equipment of the type used for lectures was not very common in this period and collections of this material were even rarer, a situation that was to change by the mid-century. The earliest item in the collection, a sundial by Hager with the monogram of William and Mary, serves to emphasize the paucity of other early material.[4] However, the lack of early material does not indicate that members of the royal family were not interested in the new philosophy. Like others in London society, they were intrigued by the innovations of the early eighteenth century such as lectures on natural philosophy. Members of the royal family heard lectures from J. T. Desaguliers who showed his experiments to George I in 1716, just a few years after the start of such lectures in London.[5]

Perhaps Caroline, then Princess of Wales, was most active in pursuing an interest in the fashionable subject of natural philosophy. Her interest arose from her close

---

[2] See Cantor (1983). To summarize a complex series of events, often called the Scientific Revolution, a new way of describing nature emerged in this earlier period, a way which involved a greater use of mathematics, as in Newton's work on gravity, a new model of the solar system, and novel kinds of instruments such as the barometer, thermometer, and air pump. At the same time societies, and the Royal Society is a prime example, were set up to bring together the practitioners of the new knowledge. Events of this magnitude did not occur in the first part of the eighteenth century. The later eighteenth century was more eventful, however, with developments in chemistry and celestial mechanics coinciding with the appearance of specialist discipline-based societies in a pattern that is familiar today.

[3] Cf. Jacob (1988) who argues strongly that the diffusion of this knowledge was very important for the process of industrialization in Britain.

[4] See catalogue entry E46.

[5] See Chapter II. Desaguliers (1744) Vol. 1, pp. IV, 281.

Hager dial, catalogue number E46.

Queen Caroline of Ansbach, by Joseph Highmore, c. 1735.

The exterior of the Hermitage at Kew.

acquaintance with Leibniz whom she knew before she came to England. At that time Leibniz was caught up in a priority dispute with Newton over the invention of calculus, an acrimonious argument which had broken out in 1705 and had been carried on by Newton's acolytes John Keill and Roger Cotes. This dispute entered a new phase in 1714 on the death of Queen Anne when the succession was settled on the Elector of Hanover who became King George I. Leibniz, as librarian and political secretary to the Elector, had hopes of preferment when George I ascended the throne of England (and expectations of receiving his arrears of pay as well). However, Newton and his followers had similar aspirations and so questions of patronage

became bound up with issues about natural philosophy. In 1715, to strengthen his claim, Leibniz wrote to Caroline, by then Princess of Wales, suggesting that Newton's works had contributed to a decline of natural religion in England. On Newton's behalf, Samuel Clarke replied and Caroline passed on his reply to Leibniz. This correspondence continued until Leibniz's death in November 1716. The following year it was published by Clarke. During the course of this dispute, in February 1716, Newton and Clarke visited Caroline to discuss 'Sir Isaac's System of Philosophy'.[6]

Caroline's regard for these 'savants' continued after she became queen in 1727 when George II came to the throne. She embarked on a remodelling of the royal

[6]   See Cowper (1865) p. 74. See Alexander (1956), Westfall (1980), Manuel (1968), and DNB on Caroline for details of the

dispute with Leibniz.

Terracotta by Rysbrack for Newton's Monument in Westminster Abbey.

estates at Kew which by then included the Richmond Lodge Estate bought in 1721. These plans featured a Grotto or Hermitage designed by William Kent around 1731.[7] The Hermitage was an expression of royal recognition and a monument to those who had influenced Caroline's views on natural knowledge for it contained

> . . . *a curious Busto of the Honourable and justly celebrated Robert Boyle Esq., incompassed with Rays of Gold. And on each side of him below, are placed Sir Isaac Newton; Mr*

> *Locke; Dr Clarke and Mr Wollaston, author of the Religion of Nature delineated.*[8]

The sculptor of some of these busts was Michael Rysbrack. At about the same time he also collaborated with Kent on another project which paid homage to Newton but in a different setting. After Newton's death in 1727, Kent and Rysbrack had been commissioned by Conduitt (the husband of Newton's niece and his successor as Master of the Mint) to design the monu-

---

[7] See Wilson (1984) pp. 143–4 and Sicca (1986). The Secretary to the Prince of Wales was Samuel Molyneux who was interested in astronomy and he, together with James Bradley, later Savilian Professor of Astronomy at Oxford, investigated problems resulting in the discovery of the aberration of light.

[8] See Curll (1736). Clarke was Samuel Clarke who had taken Newton's part in the dispute with Leibniz. Wollaston's work sold 10 000 copies. Franklin worked for a time printing it—and wrote a refutation of it as a result.

The busts of Newton (left) and Boyle (right), by Rysbrack, 1733, from the Hermitage.

ment for Newton in Westminster Abbey which was completed in 1731.[9]

The Hermitage in its carefully contrived rustic setting signalled an attitude to nature in which the work of human beings artfully blended with that of the Creator. The busts themselves were homage to individuals who, in Caroline's view, were in the forefront of the work of interpreting nature in ways consistent with the scriptures.[10] However, within a generation or so attitudes and tastes had changed so that, when Capability Brown remodelled the gardens, the Hermitage with its busts of natural philosophers in a rustic setting was swept away to be replaced by an observatory with its instruments and staff.

The completion of the Hermitage around 1732–3 provided an opportunity for some sport by the readers of a new periodical, the *Gentleman's Magazine*, who were

able to give their own views of royal largesse. The magazine published verses on the Hermitage and its busts, eventually running a competition for the best composition. Jonathan Swift was one contributor who took the opportunity to allude to the disappointments of ungranted patronage, something he had experienced at first hand, and to draw a very pointed comparison with the more enlightened policy of the King of France.

> *Lewis the living genious fed,*
> *And raised the Scientific Head;*
> *Our Q--, more frugal of her Meat*
> *Raises those heads which cannot eat.*[11]

In the eighteenth century questions of patronage were as important for natural philosophers as they were for religious and literary figures such as Swift. A few years before, Desaguliers had also used verse to attract

9   For details of his monument see Webb (1954) pp. 82–4, Wilson (1984) p. 139.

10 See Colton (1974).

11 See *Gentleman's Magazine* **3** (1733) 207. For details of the competition see ibid., p. 208.

William Hogarth 'A performance of "The Indian Emperor"'. This picture shows a bust of Newton in the home of the Conduitt family. Desaguliers is the figure in the background prompting the children.

Queen Caroline's attention through flattery rather than wit. Desaguliers's poem was entitled 'The Newtonian System of the World, the best model of Government'. His expressed intention was to '. . . divert her most Gracious Majesty with my first Poetical Experiment, as I have had the great Honour of entertaining Her with Philosophical ones.'[12] He based his approach to political issues on his understanding of natural philosophy for

. . . among my Philosophical Enquiries, I have consider'd Government as a Phaenomenon, and looked upon that Form of it to be most perfect, which did most nearly resemble the Natural Government of our System, according to the Laws settled by the All-wise *and* Almighty Architect *of the Universe.*[13]

Indeed, his poem made an interesting comparison

[12] Desaguliers (1728) Preface p. [v].

[13] Desaguliers (1728) Dedication pp. iii–iv.

Plate from Desaguliers (1744) showing his planetarium.

between the difficulties of doing experiments on mechanics and the political realities of the recently installed House of Hanover.

> *Bodies rightly plac'd still rolling on,*
> *Will represent our fix'd Succession . . .*[14]

In 1728, the year of his poem's publication, the question of succession was particularly significant for Desaguliers. With George II on the throne and Caroline queen, Frederick became Prince of Wales. That year Frederick came to England, leased a house at Kew, and

[14] Desaguliers (1728) pp. 31–2.

Orrery by Thomas Wright, 1733, catalogue number E34.

Benjamin Cole's trade card showing a grand orrery c. 1745.

Armillary sphere made by Sisson, 1731, catalogue number E35.

John Stuart, Third Earl of Bute, by Sir Joshua Reynolds, 1773.

appointed Desaguliers his Chaplain. In a further twist to the issues about patronage and Desaguliers's part in it, Frederick became a Freemason, initiated in 1737 at a ceremony at which Desaguliers presided.[15]

In the same year Desaguliers arranged courses for Frederick on more material matters. One member of the household wrote:

The Prince lives retired, seeing no company. We have a new amusement here which is both very entertaining and instructive. Dr Desaguliers has a large room fitted up at the top of the house, where he has all his mathematical and mechanical instruments at one end and a Planetarium at the other; which is an instrument he has invented which is very much superior to the Orrery, and shows the motions of the heavenly bodies in a plainer and better manner. The Doctor reads lectures every day which the Prince attends diligently. I have gained some credit by the little knowledge I have in Astronomy . . .[16]

A few years before, at the end of 1733, Thomas Wright had been paid £336 'for making new all the Machinery and Wheel-Work to perform the Motions of all the planets & Co to the Great Orrery in Her Majesty's Gallery at Kensington. . .'.[17] However, despite the value of the 'Great' orrery for marking the status of its owner and users, it seems that Desaguliers preferred his Planetarium to what on occasion he had called 'pompous orreries'.[18]

The Next Generation

In 1751 Frederick died and George became Prince of Wales at the age of 12. George was considered a backward youth and responsibility for his education was given to Lord Bute in 1755. Bute already had considerable influence with Augusta, George's mother, a matter of great public speculation.[19] Earlier, Frederick had indicated his doubts about Bute, reportedly saying 'Bute is a fine showy man, and would make an excellent ambassador in a Court where there is no business.'[20] But whatever the overall effect of Bute's influence on members of the royal family, he helped prepare the foundations for the future king's collection of scientific instruments. For Bute arranged for George and his brother Edward to follow a course of lectures on natural philosophy by Stephen Demainbray early in 1755.[21] In the 1750s, too, the future king attended demonstrations of electrical phenomena given by William Watson.[22]

The connection with Demainbray was important because it stimulated George's interest in natural philosophy but also because it was one of the chain of events which led to Demainbray's apparatus becoming part of the King George III Collection.

King George III

In terms of natural philosophy Bute was influential in yet another respect. At his Luton estate Bute had a cabinet of mathematical and philosophical apparatus which, according to one traveller, '. . .may be reckoned the most complete of the kind in Europe'.[23] The young and impressionable George may have wanted to emulate this example, and his unrivalled opportunity to do so came when he ascended the throne in October

[15] See Hamill (1986) pp. 44–5, Hurst (1928) p. 18.
[16] King (1985) p. 55.
[17] See Public Record Office LC 5/19 f155. The Wright Orrery is catalogue entry E34. Interestingly, one of the few other instruments that can be dated to this period is another astronomical device, an armillary sphere by Sisson made for Frederick, Prince of Wales in 1731 (catalogue number E35).
[18] See *Daily Advertiser* Monday 2 November 1741 No. 3365 p. 2. See Desaguliers (1744) Vol. 1, pp. 448–66 for a description of his planetarium.
[19] See Brooke (1972) pp. 42–3. Watkins (1831) p. 19 'The King, though his own education was far from being well conducted, possessed more general knowledge, particularly in the practical sciences, than the world gave him credit for.' It was in this period that Demainbray introduced his young brother-in-law,

John Horne, to the Prince (see Chapter IV). A few years later Horne published a scurrilous pamphlet alluding to a liaison between Bute and Augusta. It contained a view of Lord Bute's 'erections' at Kew, a picture which drew attention to a door between the gardens of Bute's house and that of Augusta at Kew. See Horne (1765).
[20] Jesse (1867) Vol. I, pp. 55–6.
[21] See Chapter IV for the background.
[22] See Pulteney (1790) Vol. 2, p. 315.
[23] Dutens continued: 'It is there that, since the year 1766, the time when he declared in the House of Peers that he no longer saw the King, and that he took no further part in public affairs, Lord Bute has lived, more like a philosopher than a man of the world: . . .' Dutens (1806) Vol. 2, p. 415. Bute's collection was sold in 1793. See Turner (1967a).

1760 on the death of his grandfather, George II. King George III appointed one of the best known London intrument-makers, George Adams, to be his Mathematical Instrument Maker on 15 December 1760.[24] By the end of the year it seems that he had commissioned Adams to make the Pneumatic and Mechanics apparatus that form a large part of the Collection today, for a sketch of the Philosophical Table by Adams is dated 'Jan. 5 1761'.[25]

A further indication of the interest that the new king took in collecting scientific instruments comes in a letter from Demainbray to Bute on 5 November 1760.

His Majesty's Collection of philosophical instruments increases daily, and as it is finely chosen, requires one conversant to keep it in Order; would it be adviseable, My lord, to apply, on the Grounds abovementioned Considerations, to be nominated as Keeper of the same, for to spend the Wane of Life, and to approach his Majesty's sacred Person, tho' as the lowest of his Servants, is my whole Ambition.[26]

It seems likely that Demainbray was referring to the apparatus commissioned from George Adams which he hoped to turn to his own advantage through Bute's patronage. Though Demainbray was not granted his request on this occasion, it was a step in a long campaign that was ultimately successful.

George III's interest in experimental philosophy continued throughout the first years of his reign although the evidence for this is scanty and anecdotal. In 1761, for example, James Ferguson, who wrote and gave courses on astronomy, was awarded a pension of £50 a year by the king.[27] Perhaps the attraction of natural philosophy for the king was that it was an interest he could indulge by collecting things, an activity which extended in his case to clocks, watches, and books as well. However, gratifying an interest in natural philosophy and collecting the requisite instruments were fashionable pursuits in this period, and so George III's interests were exceptional only in the size and quality of his collection. Another example is

The first page of the Pneumatics *manuscript, Adams (1761a).*

provided by a German princess for whom Euler wrote a popular account of natural philosophy in 1760.[28]

In November 1762 a contemporary wrote of George III's activities:

His Majesty is extremely busy in a course of experimental philosophy, but I doubt whether under the direction of any real philosopher.

The writer, Dr Birch, continued:

... Mr. Champion, a famous writing-master, has been

---

[24] See Public Record Office LC/3/67 p. 30. Mr Dolland was appointed Optician and Dr Robert Smith, Master of the Mechanicks at the same time. LC/3/67 pp. 31–2.
[25] Science Museum MS 203, Pl. 7.
[26] Demainbray to Bute 5 November 1760 Brentford. Bute Papers.
[27] See Henderson (1870) p. 54.
[28] See Euler (1795).

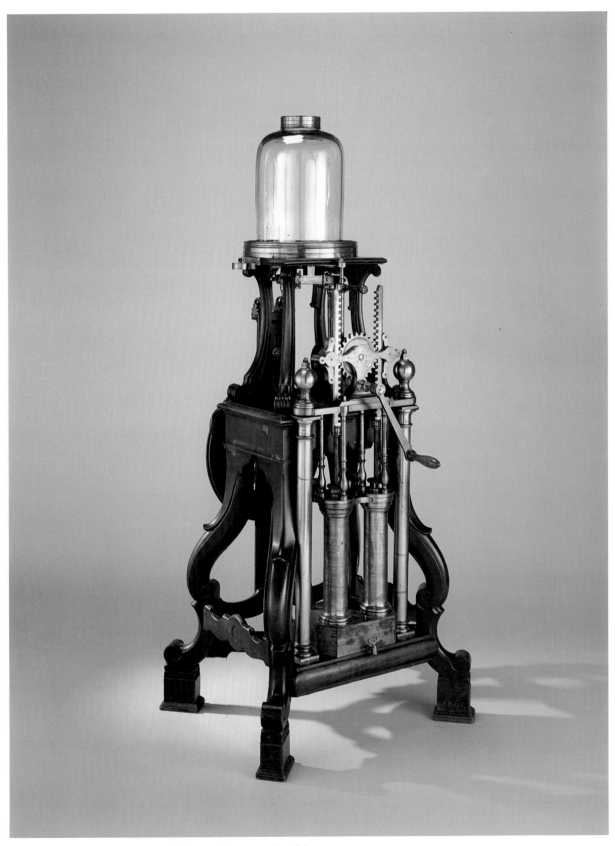

Air pump by George Adams, 1761, catalogue number P1.

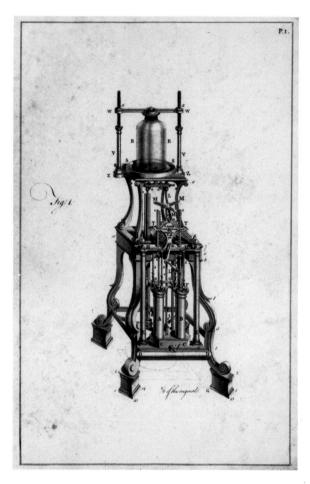

Plate of pump by George Adams from the Pneumatics *manuscript, Adams (1761a).*

The record of Adams's appointment as Mathematical Instrument-maker to George III in 1760.

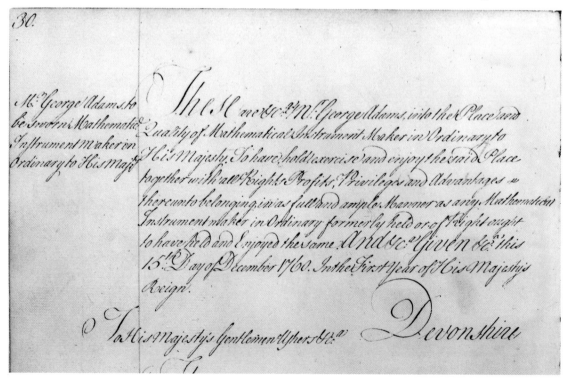

Silver microscope by George Adams, catalogue number E157.

lately taken out of his business of teaching to transcribe in a fine hand the discourses drawn for the royal use on the several branches of philosophy. This latter fact was told me by a friend of Mr. Champion, whose new employment was called by his friend, writing out the King's exercises; and he was in hopes of procuring it to be turned into a patent place. [29]

These 'discourses' written out by Champion are the two manuscripts, on Mechanics and Pneumatics, that survive today, one copy with the collection at the Science Museum and another in the Royal Library at Windsor Castle. These manuscripts were compiled by George Adams to accompany the apparatus he provided for the King and they consist for the most part of descriptions of demonstrations taken from a number of sources, principally 's Gravesande's *Mathematical Elements of Natural Philosophy*. At that time the most recent edition had been translated from the Latin original by Desaguliers and published in 1744–5 by his son, Thomas Desaguliers. However, it seems likely that Adams used his own translation.

Sometimes the King's apparatus would be demonstrated to visitors and on one occasion Adams was called upon to demonstrate the equipment himself. During a visit to England in 1763, the French astronomer Lalande recorded in his diary that Adams operated the large pump for him at the King's command. On that trip to London Lalande also saw a very elaborate silver microscope made by Adams for the King. A similar elaborate microscope was made for George III's own son George, the Prince of Wales. [30]

Patronage

The relationship between George III and George Adams illustrates aspects of patronage in the eighteenth century. As the King's mathematical instrument-maker, Adams supplied a large quantity of expensive apparatus, for which he was probably well paid. In turn the reputation Adams enjoyed as a result of his royal connection may have helped him to sell other instruments to Government bodies, such as the Office of Ordnance, and to well-off individuals. [31]

However, royal patronage had limits. One of the demonstrations taken from 's Gravesande by Adams used a guinea and feather and showed that they fell at the same rate if there was no air to hinder them. On one occasion John Miller, who worked for Adams, showed this apparatus to the King.

Mr Miller . . . used to tell that he was desired to explain the airpump experiment of the guinea and feather to Geo: III. In performing the experiment the young optician provided the feather the King supplied the guinea and at the conclusion the King complimented the young man on his skill as an experimenter but frugally returned the guinea to his waistcoat pocket. [32]

A more substantial example of patronage involving Adams came a few years later, in 1766, when he published his book on *The Use of the Globes*. There he illustrated a pair of 18 inch globes on elaborate stands, probably the globes now in the King George III Collection. Adams presented a copy of this book to the King, one account of this event being as follows:

When Adams, the optician and mathematical instrument maker, presented his 'Treatise on the Use of the Globes' to the King, who had given him permission to bring the volume out under the royal sanction, his Majesty looked over the dedication, and said, 'This is not your writing.' 'No, Sire,' replied Adams,' it was composed for me by Dr Johnson.' 'I thought so,' answered the King; 'it is excellent—and the better for being void of flattery, which I hate.' [33]

In fact, this 'excellent' dedication did owe something

---

[29] See Harris (1847) Vol. 3, p. 291.

[30] In Lalande (1980) p. 72 the 'Adams' mentioned is identified as Robert Adam the architect. However, this seems very unlikely since he is mentioned in connection with a large air pump and a silver microscope, both items supplied by George Adams to George III. For the silver microscope belonging to King George III see Lalande (1980) p. 70. It is now in the Museum of the History of Science, Oxford. The other silver microscope, which belonged to George IV, is now in the Science Museum (catalogue number E157).

[31] See Brewer in McKendrick et al. (1982) for the problems that tradesmen had with aristocratic patrons. See Millburn (1988) for the case of the Office of Ordnance. For discussions of patronage see Turner (1976).

[32] See Clarke et al. (1989) p. 54, note 29 quoting a document in the Scottish Record Office SRO GD 76/464/3–4. John Miller later returned to Edinburgh to set up as an instrument-maker.

[33] Watkins (1831) p. 509. The globes are catalogue entries E38 and 39.

GEORGE ADAMS,

MATHEMATICAL INSTRUMENT-MAKER TO HIS MAJESTY,

At Tycho Brahe's Head, in Fleet-Street, London,

MAKES and SELLS all Sorts of the moſt curious MATHEMATICAL, PHILOSOPHICAL, and OPTICAL INSTRUMENTS, in Silver, Braſs, Ivory, or Wood, with the utmoſt Accuracy and Exactneſs, according to the lateſt and beſt Diſcoveries of the modern MATHEMATICIANS.

HADLEY's QUADRANTS, with the lateſt Improvements, in the moſt exact Method, with Glaſſes whoſe Planes are truly parallel.

Azimuth and Steering Compaſſes, invented by Dr. *Gowin Knight*, F.R.S. approved of, and uſed by his Majeſty's Royal Navy.

N.B. Theſe Compaſſes are all examined and certified by Dr. *Knight*.

Large ASTRONOMICAL QUADRANTS, TRANSIT and EQUAL ALTITUDE INSTRUMENTS, for obſerving the Tranſits of the Sun and Stars over the Meridian, &c.

Sun Dials HORIZONTAL, for Pedeſtals in any Latitude; with Variety of PORTABLE ones, either UNIVERSAL, or for ſeveral different Latitudes, with new Improvements.

Choice of curious Caſes of DRAWING INSTRUMENTS, in Silver, Braſs, &c. containing a Sector, Scales, Proportionable, and other Compaſſes; Drawing-Pens, a Protractor, Parallel Rules, &c.

A new-invented Portable MICROSCOPE for viewing all Kind of Minute Objects, as well Opake as Tranſparent, in ſo conſpicuous and conciſe a Manner, as to comprehend all the Uſes of all the other Sorts of Microſcopes in one Apparatus; and magnifies to ſo great a Degree, as to diſcover the Circulation of Blood in Animals, the Periſtaltic Motion of Inſects, the Farinæ of Vegetables, and many other ſurpriſing Phænomena, otherwiſe not perceptible.

The double Conſtructed MICROSCOPE; Mr. *Ellis's* ÆQUATIC MICROSCOPE; SOLAR MICROSCOPES; MAGELLESCOPES, &c.

The New ACROMATIC TELLESCOPE, with a compound Object Glaſs, approved by all the Curious in Optics; with all other Sorts of REFRACTING TELLESCOPES; NIGHT TELLESCOPES, &c.

REFLECTING TELLESCOPES, of the lateſt Improvement.

MICROMETERS of the neweſt Conſtruction, elegantly fitted to Refracting or Reflecting Telleſcopes.

ORRERIES and PLANETARIUMS, greatly improved.

Inſtruments proper for GUNNERY, FORTIFICATION, &c.

PANTOGRAPHERS, for reducing Drawings and Pictures of any Size, in the moſt complete Manner.

Inſtruments for taking the true Perſpective of any Landſcape, Building, Gardens, &c. and others for copying of Drawings.

New GLOBES; mounted in a peculiar Manner, whereby the Phænomena of the Sun, Earth, Moon, &c. are exhibited according to Nature.

AIR-PUMPS, or Engines, either for exhauſting or condenſing the Air, and this by turning one Cock only, with all their Appurtenances; whereby the Properties of that moſt uſeful Fluid are diſcovered and demonſtrated by undeniable Experiments; HYDROSTATICAL BALANCES, nicely adjuſted for determining the ſpecific Gravity of Fluids and Solids, &c.

Curious BAROMETERS, Diagonal, Wheel, Standard, or Portable, with or without Thermometers. Alſo the ſo much famed QUICKSILVER THERMOMETERS, made after any of the Forms.

THEODOLITES, of the lateſt Conſtruction; WATER LEVELS, which may be adjuſted at one Station; MEASURING WHEELS; Pocket and Coach WAY WIZERS, for meaſuring the Way, &c.

SPECTACLES ground on Braſs Tools, in the Manner approved of by the Royal Society, ſet in Variety of convenient Frames: Alſo READING GLASSES of all Sorts, ſet in Silver or other Metal, to turn into Caſes of various Kinds.

PRISMS, for demonſtrating the Theory of Light and Colours.

The CAMERA OBSCURA for drawing in Perſpective, in which all external Objects are repreſented in their proper Colours and exact Proportions.

CONCAVE, CONVEX, and CYLINDRICAL MIRRORS, OPERA GLASSES, MULTIPLYING GLASSES, SPECTACLES of the true *Venetian* Green Glaſs, MAGIC LANTHORNS, &c.

ZOGRASCOPES, for viewing Perſpective Prints.

N.B. Gentlemen may have any Model or Inſtrument made in Metal or Wood, with Expedition and Accuracy, and carefully packed up to be ſent to any Part of the World.

The trade card of George Adams, c. 1760.

The New TERRESTRIAL GLOBE, As Improved and Conſtructed by GEO: ADAMS In Fleet Street LONDON.

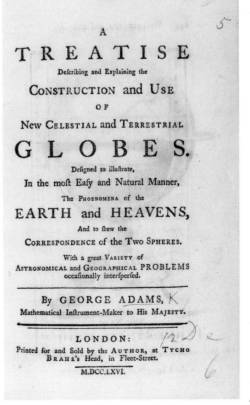

A TREATISE Deſcribing and Explaining the CONSTRUCTION and USE OF New CELESTIAL and TERRESTRIAL GLOBES. Deſigned to illuſtrate, In the moſt Eaſy and Natural Manner, The PHŒNOMENA of the EARTH and HEAVENS, And to ſhew the CORRESPONDENCE of the Two SPHERES. With a great VARIETY of ASTRONOMICAL and GEOGRAPHICAL PROBLEMS occaſionally interſperſed.

By GEORGE ADAMS, Mathematical Inſtrument-Maker to His MAJESTY.

LONDON: Printed for and Sold by the AUTHOR, at TYCHO BRAHE's Head, in Fleet-Street. M.DCC.LXVI.

Plate of the terrestrial globe and title page from Adams (1766).

Terrestrial globe by George Adams, catalogue number E38.

Celestial globe by George Adams, catalogue number E39.

Detail of the celestial globe showing illustration of an air pump.

to flattery, demonstrating that in the interview between Adams and the King what was perceived as flattery or truth depended on your vantage point. Johnson had written in the dedication,

Geography is in a peculiar manner the science of Princes. When a private student revolves the terraqueous globe, he beholds a succession of countries in which he has no more interest than in the imaginary regions of Jupiter and Saturn. But Your MAJESTY must contemplate the scientifick picture with other sentiments, and consider, as oceans and continents are rolling before You, how large a part of mankind is now waiting on Your determinations, and may receive benefits or suffer evils, as Your influence is extended or withdrawn.[34]

Here Johnson was following a tradition of 'global' politics. Earlier, for example, Coronelli had made globes 13 feet in diameter for the French King. These were accompanied by a dedication from Cardinal d'Estrees who

...Has consecrated this Terrestrial Globe, to render continual Homage to his Glory, and to his Heroick Virtues, by shewing the Countries in which a thousand great Actions were perform'd, by himself, or by his Orders, to the Astonishment of so many Nations whom he might have subdued to his Empire, if his Moderation had not stop'd the Course of his Conquests, and prescrib'd Bounds to his Valour, which is yet greater than his Fortune.[35]

Clearly, both the size of the globe and whether the king ruled in an absolutist state, or otherwise, determined the extravagance of the dedication.

While much of the King George III Collection can be seen as the relics of successful bids to gain royal patronage, particularly the equipment provided by Adams and by Demainbray, there are also traces of less successful bids. In 1761 David Lyle presented the King with a set of silver drawing instruments of a type he had designed, at the 'desire' of Lord Bute.[36] Bute's influence in the 1760s extended much further than collections of

Bust of King George III by Nollekens commissioned by the Royal Society using the surplus funds from the 1769 Transit Expedition.

scientific instruments, though it both waxed and waned in these years. By February 1761, a few months after George III came to the throne, Bute was a Secretary of State; about a year later he had risen to be first minister, but his increasing unpopularity meant that he had to resign soon afterwards.

Bute took a house at Kew around 1761 and played a part in planning the gardens there.[37] In 1764 Capability Brown produced a new plan and, as stated above, in the course of the work of remodelling the gardens the Grotto or Hermitage built by Kent for Caroline was dismantled and the busts of Boyle, Newton, and the others were

[34] Adams (1766) Dedication.

[35] *Daily Courant* Wednesday 20 December 1704 No. 837 p. 1. The dedication for the celestial globe ran '... all the Stars of the Firmament and the Planets are plac'd in the same Situation they were in at the Birth of that glorious Monarch; to preserve to Eternity a fix'd Image of that happy Position of the Heavens, under which France receiv'd the greatest Present that Heaven ever made to Mankind'.

[36] See entries E1 to 4. Lyle mentions his gift in the very lengthy dedication to Lord Bute of this 'Art of Shorthand Improved'. Lyle (1762) p. vi.

[37] A move which increased speculation about Bute's relationship with Augusta, George's mother. See Chapter IV. Bute had a great interest in natural history. See Miller (1988).

OBSERVATORY. RICHMOND GARDENS.

Observatory, Richmond Gardens, by
G. E. Papendick, c. 1820.

Captain Cook by Nathaniel Dance,
1776.

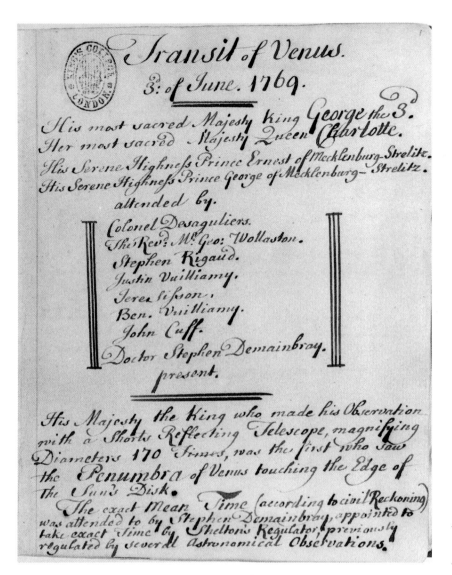

List of those observing the transit of Venus recorded at Kew Observatory, 1769.

removed.[38] Changing taste meant that Brown's plans for the gardens supplanted those of Kent's as one image of nature gave way to another. But it also meant that one form of royal interest and homage to men of science, the Hermitage and its busts, was replaced by an observatory complete with staff and instruments.

The Transit of Venus

The interest in nature that the Hermitage originally represented now manifested itself in new ways. In 1769

Venus moved across the Sun's disc when viewed from the Earth. Such a transit of Venus is a comparatively rare event but it occurred twice in the 1760s and became a subject of widespread interest.[39] The transit was significant because it provided an opportunity to measure the size of the solar system with reasonable accuracy; the ratio of the distances to the Sun of the Earth and Venus being known, it was possible to calculate the size of the Sun and the dimensions of the solar system by observing the transit from different places on the Earth at known distances apart. This calculation was relatively simple, but carrying out the

---

38   Some of these busts are now in Kensington Palace.

39   One group with a special interest in the transit was the lecturers on astronomy in the capital. Martin, for example, advertised

lectures on the subject several years in advance of the transit. See *Daily Advertiser* Monday 2 November 1767 No. 11495 p. 2.

measurements with sufficient accuracy in places far apart was not. However, the event had been predicted long in advance and so there was the opportunity to lay elaborate plans to observe it. The transit of Venus in 1769 was one of the first major collaborations between scientific societies (the earlier transit in 1761 had been something of a rehearsal for the later). The scale of the collaboration was great and the cost of the expeditions sent to observe the transit from different parts of the Earth was huge. For example, the Royal Society, after formulating its plans, had to petition the Crown for funds of £4000 to cover the cost of their proposal for sending observers to the South Seas.[40] King George III granted their request and in due course James Cook set sail in the *Endeavour* with Joseph Banks, amongst others, on board. The expedition set an unusual precedent in one respect—it cost less than the sum that had been allocated for it and so the Royal Society used the surplus to honour their benefactor, George III, by commissioning a bust from Nollekens.[41]

The King took a keen interest in the transit of Venus and had a scheme, more modest than the Royal Society's expedition in scientific terms but still conspicuously expensive, to build an observatory at Richmond (now Kew Observatory) where he could observe the transit himself. The new observatory was designed by Sir William Chambers and was ready by the time of the transit in June 1769. Stephen Demainbray, who had long sought royal patronage, was appointed the first superintendent.[42]

On the day of the transit the King came with Demainbray and several assistants to observe events. Jeremiah Sisson, the instrument-maker, George Wollaston, and Rigaud (Demainbray's son-in-law and father of the Oxford astronomer S. P. Rigaud) also took part.[43] Though they made their observations from a building specifically designed for the purpose, equipped with some of the best obtainable instruments including chronometers by Harrison, and the people there were familiar with that equipment, they did not publish their results. This omission is striking and contrasts with what was accomplished on the expedition organized by the Royal Society and the fact that the observations of 138 observers in 63 locations were eventually reported.[44] Cook's voyage to the South Seas had wide repercussions; Demainbray's sojourn at Kew did not. Episodes such as these illustrate some of the different ways that patronage could be used in eighteenth century English society.

Lightning Conductors

Another long-running issue underlay a dispute about a scientific matter that erupted in the 1770s. In 1772 the Board of Ordnance had requested the Royal Society's advice about the best way to protect the powder magazines at Purfleet from the effects of lightning. A committee which included Henry Cavendish, Benjamin Franklin, and William Watson had recommended installing pointed rods projecting above the roof of the buildings. Benjamin Wilson, a dissenting member of the committee, had argued instead for rounded rods below the roof. The Board of Ordnance duly fitted pointed conductors, but in 1777 the House of the Board at Purfleet was struck by lightning. The controversy about the best shape for the conductors erupted once more and on this occasion Wilson carried out an elaborate series of experiments at the Pantheon in London to prove his case. The King saw these experiments and, accepting Wilson's conclusions, ordered that the pointed conduc-

[40] See Woolf (1959) for a general discussion of issues to do with the transit of Venus. Woolf (1959) p. 166 gives the cost of the Royal Society's proposed expedition.

[41] The bust is in the entrance hall of the Royal Society.

[42] Rigaud (1882) p. 282. The King and Queen had visited the Royal Observatory at Greenwich for the first time in June 1768. See Howse (1989) p. 102.

[43] King's College archives K/MUS/1.

[44] Woolf (1959) p. 21. In fact one Fellow of the Royal Society, John Bevis, observed the transit from Kew, near to the observatory, and reported his results in *Philosophical Trans-*

actions. He made a passing reference to the new observatory, the sole reference in the vast bulk of reports on the transit. 'Mr Kirby's house [where his own observations were made AQM] is exactly 4″ 3/4 of time east of his Majesty's domestic observatory, and 1′ 14″ west of the Royal observatory at Greenwich.' Bevis (1769) p. 189. With no more than that passing remark, the work of the King's domestic observatory was ignored in 1769 and afterwards. When the time came to write the history of the transit observations, the work was mentioned only in a footnote. See Woolf (1959) p. 188.

Philos. Trans. Vol. LXVIII. Tab. III. p. 245.

A View of the Apparatus and part of the Great Cylinder in the Pantheon.

Engraved by J. Basire.

Electrical experiment at the Pantheon, Philosophical Transactions, 1778.

tors to Franklin's design should be taken down from Buckingham House.[45] By this time events in America, in particular the opposed views of Franklin and the King, had coloured perceptions of the situation. On hearing the news of the King's decision Franklin wrote to one correspondent,

The King's changing his pointed Conductors for blunt ones is . . . a Matter of small Importance to me. If I had a Wish about it, it would be that he had rejected them altogether as ineffectual, For it is only since he thought himself and Family safe from the Thunder of Heaven, that he dared use his own Thunder in destroying his innocent Subjects.[46]

[45] See Heilbron (1979) pp. 380–2. According to Heilbron, the supposed denouement involving Pringle, royal physician, friend of Franklin, and President of the Royal Society, is apocryphal: '. . . the king told Pringle to instruct the Fellows that lightning rods would henceforth end in knobs. "The prerogatives of the president of the Royal Society do not extend to altering the laws of nature," Pringle replied, and forthwith resigned.' See also Wilson (1778).

[46] Franklin to [Lebègue de Presle] 4 October 1777. Willcox (1986) p. 26.

Harrison's chronometer (H5) which was tested at Kew Observatory.

After 1770

In the 1770s Demainbray, and through him the King, became involved in a long-running dispute between John Harrison and the Board of Longitude about tests on his chronometer. The Board doubted whether Harrison's chronometer met the criteria which would qualify him for the prize offered for improved methods to find the longitude of ships at sea. A further trial was carried out at the King's new observatory and three

Cole theodolite with the arms of the Duke of Cumberland, catalogue number E25.

Duke of Cumberland, studio of David Morier, 1740–9.

Wheelcutting engine, possibly made by John Smith, catalogue number E65.

separate keys were needed to inspect the chronometer, one kept by the King, one by Demainbray and the last by John Harrison's son, William.[47]

Demainbray himself was now able to dispense patronage as a result of his position as superintendent of the observatory; he was able, for example, to employ John Smith 'in the instrument way'. Smith had been an apprentice of Henry Hindley in York, an instrument-maker who had met and impressed Demainbray over twenty years before. Hindley had also made a great impression on John Smeaton, the eminent civil engineer.[48] It seems likely that Smith made a few items that are now in the present-day collection. Material

from other sources also accumulated in the observatory. The Duke of Cumberland, 'Butcher' Cumberland, died in 1765, and his nephew, King George III, inherited his estate which included scientific instruments and surveying equipment, perhaps relics of his involvement with the survey of the Highlands of Scotland after the rebellion in 1745. It seems likely that some of this is now in the present-day King George III Collection. In May 1770, soon after the observatory opened, further material, allegedly belonging to Robert Boyle, was added to the collection. Unfortunately it has not been possible to confirm this connection, nor to suggest another possibility.[49]

47 See Horrins (1835) pp. 1–18. Howse (1989) p. 124 identifies this chronometer as H5.
48 See Smeaton (1786).
49 One suggestive occurrence is that a few months earlier, Mrs

Miles, the widow of Henry Miles, had deposited Boyle's papers at the Royal Society. However, these events may be unconnected.

Illustrated London News *engraving of the opening of the King George III Museum in 1843.*

When Demainbray died in February 1782, his successor had to be chosen and the selection process illustrates again the complex nature of patronage. Although Demainbray's son had been promised the post by the King, this idea disappointed Joseph Banks, by then President of the Royal Society. A few weeks later Banks heard from William Watson about an alternative plan to make William Herschel the astronomer at Kew. Watson wrote

It gives me likewise great pleasure to be able to inform you that since Dr Demainbray's death, the King has again

twice spoken to Mr Griesbach in relation to Mr Herschel, & told him that Mr Herschel was to come to him as soon as the Concerts at Bath were over. These are very encouraging circumstances, & make me still hope that the King has some notion of making him Demainbray's successor. . .[50]

Herschel had recently discovered the planet Uranus for which he had suggested the name Georgium Sidus to enourage royal interest. Despite these manoeuvrings by the President of the Royal Society, the King did not change his mind. However, Herschel did receive a pension from the King which helped to compensate to

[50]   British Museum—Natural History, Dawson Turner Collection Vol. 2, p. 108, Watson to Banks 27 March 1782. See also Lubbock (1933) pp. 121–2. A week later he continued, '. . . The King the first time he saw him at Windsor asked him [Mr Griesbach AQM] after his *uncles* at Bath, & how Mr H's Telescope went. To which Mr Griesbach answered that his Uncle was preparing them for the inspection of his Majesty. The King, you see, has very often made enquiries after Mr H

since Dr Demainbray's Death, & indeed, I find since, oftner than I have mentioned to you, which makes me hope he has him in his eye yet . . .' Vol. 2, pp. 118–19. At the end of June Watson was still optimistic, '. . . nothing remains now to be done in order to gain him the Post he so much covets, than to inform the King of these particulars, & of Mr Herschel's ardent wishes to serve his Majesty by succeeding the late Dr Demainbray.' Vol. 2, p. 144 Watson to Banks 29 June 1782.

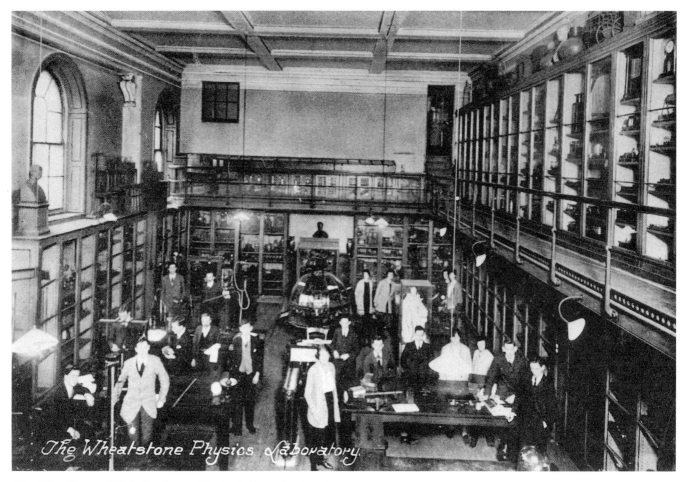

The King George III Collection at King's College, London, in the 1920s.

some extent for his own disappointment and that of others such as Banks.[51] The second Stephen Demainbray was superintendent of the observatory for 58 years, until his retirement in 1840.[52]

Conclusion

The material assembled at Kew Observatory by the 1770s became the core of what became known as the King George III Collection. As far as we can ascertain, little was subsequently added while the collection was kept at Kew during the later eighteenth and early nineteenth centuries except for a few photographic items deposited in the later 1830s. Subsequently the bulk of the collection went to the newly founded King's College, London, where it was housed in a King George III Museum opened in 1843 and the items were sometimes demonstrated to the students.

During this phase of the history of the collection items were donated to King's for the King George III

---

[51] See Herschel (1876) p. 50 who gives a comment made by Caroline Herschel, William's sister, 'The prospect of entering again on the toils of teaching, &c., which awaited my brother at home . . . appeared to him an intolerable waste of time, and by way of alternative he chose to be Royal Astronomer, with a salary of £200 a year. Sir William Watson was the only one to whom the sum was mentioned, and he exclaimed, 'Never bought [a] monarch honour so cheap!'''.

[52] The records of the weather are now at King's College, London. S. P. Rigaud was also an observer there on occasion. See Rigaud (1882) p. 283.

Museum, such as eighteenth century thermometers made by Six, a model steam engine, and examples of new processes such as electrotypes. More importantly, King's bought some new items for teaching purposes, and these mainly nineteenth century pieces, including some material belonging to Charles Wheatstone, one of the two professors of natural philosophy, have found their way into the collection.[53]

The observatory building itself was taken over by the newly founded British Association for the Advancement of Science.[54] Some telescopes and other astronomical apparatus were given to other observatories in Britain.[55] A few other items like some of the microscopes are now in museums, having taken roundabout routes to their present homes. Some anatomical specimens went to the Hunterian Museum in London but were destroyed during the Second World War. In 1927 the bulk of the early material in the King George III Museum came to the Science Museum when King's College decided that they no longer wanted the collection and would prefer to use the space taken up by the Museum for lecture theatres.

Study of the material in the King George III Collection illuminates broader issues about scientific activity in the eighteenth century and augments information from the more usual literary sources. But as well as the material in the collection being germane to these general arguments, it also provides specific insights because the collection is evidence—almost

archaeological evidence—of actual events and processes that occurred. For example, in 1761–2, when the new King received the Pneumatics and Mechanics apparatus he commissioned from the London instrument-maker, George Adams, this apparatus represented, at least in function, what was by then the standard equipment for lectures. At the same time it was a collection of elaborate and magnificent pieces, the latest in a long tradition of royal and aristocratic collecting.

The assembly of such collections of demonstration equipment, however, marked a new twist of fate for those who tried to make a living by lecturing. With this equipment and a textbook, the collector could carry out the demonstrations himself. Along with other factors this led to a decline in the number of public lectures in London, and lecturers such as Stephen Demainbray had to look for alternative employment. Fortunately for Demainbray he became superintendent of the King's new observatory at Kew, where the King's own apparatus was housed, with the consequence that Demainbray's own apparatus was amalgamated with that of the King. As a result the King George III Collection today has eighteenth century material illustrating both public lecturing on natural philosophy and also the private collection of apparatus which, to an extent, supplanted public lecturing. The nature of the collection has changed over time: in many cases the same objects have had multiple functions with different uses and meanings depending on context.

[53] See Hearnshaw (1929) pp. 148, 190, 420, 479 who recounts some of the problems created for King's College by the collection.

[54] See Scott (1885) pp. 45, 49. The Royal Society had originally solicited the use of the building but then rejected it.

[55] Some items went to Armagh Observatory. See Lindsay (1969), Butler and Hoskin (1987). See also Chaldecott (1951) pp. 82–4.

II

SCIENCE AS POLITE CULTURE
Early scientific lectures in London 1700-45

For now Philosophy being admitted into our Exchange, our Church, our Palaces, and our Court, has begun to keep the best Company, to refine its fashion and appearance, and to become the Employment of the Rich, and the Great, inste[a]d of being the Subject of their scorn: Whereas it was of old for the most part only the Study of the sullen, and the poor, who thought it the gravest part of Science to contemn the use of mankind, and to differ in habit and manners from all others, whom they slighted as madmen and fools.

From this arrogant sordidness of such Principles, there could not be expected any Magnificent Works, but only ill-natur'd and contentious Doctrines. Whatever the Poets say of the Moral Wisdom, that it thrives best in Poverty; it is certain the Natural cannot: for in such mean and narrow conditions men perhaps may learn to despise the World, but never to know it.[1]

Introduction

IN THE FIRST CHAPTER we explained that the background of much of the King George III Collection is provided by the public lectures on natural philosophy. In London these lectures began in the early years of the eighteenth century and in this chapter we shall look at this development in detail.

Thomas Sprat's *History of the Royal Society of London for the improving of Natural Knowledge* was published in 1667 when the Royal Society was a few years old. Sprat's purpose was to praise the virtues of the fledgling Society and to make grandiose claims for the usefulness of the new experimental philosophy. He also commended its Fellows for their dedication to the pursuit of these ends. By the date of the second edition, 1702, the Society was not altogether flourishing and many of Sprat's claims about its role in advancing natural philosophy must have seemed grossly inflated.[2] Nevertheless, natural

---

[1] Sprat (1959) pp. 403–4.

[2] Hunter (1982) p. 48. Johnson commented on Sprat's work, 'This is one of the few books which selection of sentiment and elegance of diction have been able to preserve, though written upon a subject flux and transitory. The history of the Royal Society is now read, not with the wish to know what they were then doing, but how their transactions are exhibited by Sprat.' Johnson (1925) Vol. 1, p. 304.

philosophy continued to extend its acquaintance amongst 'the best Company', as Sprat had put it, and significant innovations were still being made by people with connections to the Royal Society. One innovation was the start of public lectures on natural philosophy in London with the first course being given in 1705 by a Fellow of the Society, James Hodgson.

This new style of lecturing, with demonstrations using specially designed apparatus, became established in London within a few years and lasted for most of the century. It was a kind of polite rational entertainment with a serious educational purpose. But, as such, it was only one of several new 'rational' entertainments in this period. In addition there were lectures on other subjects, on anatomy, on chemistry, and on mathematics, for example, and newspapers and periodicals came from the presses in increasing numbers.[3] In this situation, lecturing on natural philosophy was influenced by a wide range of factors reaching far beyond the interests of the Fellows of the Royal Society. Amongst the most important of these factors were changing attitudes to education, particularly vocational education, and the continued growth of the instrument-making trades. Furthermore, in this period matters of great and lasting importance, such as techniques for finding longitude at sea and the reform of the calendar, helped enlarge the markets for both instruments and lectures.[4] Some of this interest in the use of instruments would benefit the lecturers on natural philosophy.

In contrast to similar lecture courses at the Universities of Oxford and Cambridge, the lectures in London were not, in general, given under the auspices of any particular institution.[5] However, the Court, the City, and the Royal Society all provided a potential audience for the lecturers. The success of the lectures depended on the patronage of a public which frequented coffee-houses and read newspapers and periodicals.[6] Because of the links between lecturing and these other activities, the growing market for lectures can be seen as one aspect of the growth of consumption and a particular example of what has been characterized as the birth of consumer society.[7] Some of the connections between the lectures and other cultural changes are traced below.

The Royal Society

It was the concerns of Royal Society circles in the late seventeenth century that set the pattern for the content of lecture courses on natural philosophy. The subject matter, and the procedures and apparatus used for the public lectures, for example, repeated much that was said and exhibited to the more exclusive audience at the meetings of the Royal Society. The air pump provides a good illustration of this connection between the concerns of the Society and the business of public lecturing. Many of the experiments using an air pump at the Royal Society were demonstrated by Francis Hauksbee, who then showed these demonstrations to a wider audience at his own lectures. Lectures on pneumatics became a standard feature of public courses.

However, when the air pump was used in the late seventeenth century at the Royal Society, it was not just a physical effect that was being demonstrated. Its use was part of a process of fixing the boundaries of the community of people competent to carry out experimental work and to show how that community should conduct its business.[8] In such ways the circle of 'virtuosi' around the Royal Society came to an understanding of both the appropriate language to use to describe their observations and a style of argument that they should adopt amongst themselves.[9] Through using the appropriate language and kinds of argument, disputes about

---

[3] There were also new types of opera.

[4] The question of whether the lectures were important, either in helping to find solutions to these problems or publicizing new discoveries or inventions, has puzzled many. See Musson and Robinson (1969), Redwood (1976).

[5] Though some of the lecturers, e.g. Desaguliers and Hauksbee, did have affiliations with the Royal Society. Lectures were given at one institution, the 'Academy' in Little Tower Street. Desaguliers himself had lectured at Oxford University before coming to London and Whiston had done the same in Cambridge. For Oxford see Simcock (1984) and G. L.'E.

Turner, 'The physical sciences' pp. 659–81, in Sutherland and Mitchell (1986). For Cambridge see Gascoigne (1984, 1989), Gowing (1983).

[6] In broad terms the clientele for lectures in mid-eighteenth century London consisted of the patrons of the salons and coffee houses and often came from the ranks of the developing professions. See Holmes (1982) and Schaffer in Porter et al. (1985).

[7] See Plumb in McKendrick et al. (1982) and Weatherill (1988).

[8] See Shapin and Schaffer (1985).

[9] See Shapin (1988a), Dear (1985).

the phenomena observed with the air pump could be settled and credible accounts of them prepared. In short, the Fellows developed new social conventions around the carefully contrived demonstrations shown in the rooms of the Royal Society.[10]

At the same time the Fellows were concerned to find ways to 'boost the prestige of the new philosophy'.[11] This also motivated the people who lectured. Obviously public lecturers were in a position to transmit the concerns of the virtuosi about natural philosophy, its purpose, and the usefulness of the enterprise to other sections of society. By using an air pump and demonstrating the findings of Boyle and his contemporaries, for example, the lecturers extended the process of defining the community of natural philosophers whilst bringing the work of that community to the attention of others. The lecturers demonstrated the knowledge and expertise of the community members and the authorities within it at the same time as they demonstrated the pump. While repeating the experiments and reiterating the conclusions accepted by the Fellows of the Royal Society, the lecturers inculcated the members of the audience with the values of the virtuosi (or at least reinforced the social messages for those viewers who already held them).[12] This enabled the Fellows of the Royal Society to command the respect, or the acquiescence, of others.[13]

If the Royal Society needed the lecturers in the early years of the eighteenth century, it was not just as broadcasters of its views. As the Royal Society was in a sorry state, some hoped that the lecturers might help revive its fortunes. This point was grasped by one contemporary, Abraham Sharp, in a letter referring to Hodgson's plans to lecture:

> Surely your Royal Society, or some generous members thereof or others, might contribute to so laudable a work, which seems destined and is likely to be largely instrumental in reviving that Society now, in appearance, sadly drooping

> and languishing, the products of their consultations and deliberations being of late very inconsiderable.[14]

For these reasons, amongst others, many of the lecturers in the early eighteenth century were (or became) Fellows of the Royal Society. But this situation gradually altered so that by the 1750s only a minority of lecturers were Fellows, one indication that the social position of the lecturers changed through the century.[15] At the same time as the social significance of the virtuosi of the Royal Society and their claims for knowledge were subtexts for the lectures, so too was the social position of the lecturers themselves. As a marginal occupational group they were yet to find a place and secure status in eighteenth century society.

A major obstacle hindering an assessment of the social position of those who gave courses on natural philosophy in the eighteenth century is that information about the audiences attending the lectures is virtually non-existent. As a result the significance of their activities has to be gauged by other means. One source of information is furnished by the histories of those who lectured: because they were involved in numerous schemes apart from lecturing, their biographies reveal the choices available to them. Furthermore, by comparing the careers of lecturers on natural philosophy with those of others, such as lecturers on medical topics or engineers, some of the complexities of life in eighteenth century London become clearer. In turn this illuminates the growth and change of professions, an important feature of the period.[16]

The First Lectures on Natural Philosophy in London

James Hodgson FRS (1672–1755) was the first to give a course of lectures of natural philosophy in London. It

---

[10] See Shapin (1988b).
[11] Golinski (1989).
[12] Cf. Shapin (1988b) pp. 373–4 on demonstration.
[13] Cf. Schaffer (1983).
[14] See Cudworth (1889) pp. 81–2 and Rowbottom (1968) p. 199. See Hunter (1982) p. 48.
[15] Of course, the social position of the Royal Society also changed. Cf. Schaffer (1983). Heilbron (1979) p. 158 has pointed out that there were several identifiable types of lecturer, 'At the top

were the public lecturers associated with learned societies. . .A second class of lecturer consisted of members of learned societies who set up independently of their institutions. A notch lower, perhaps came the unaffiliated entrepreneurs, who taught in rented rooms, and the itinerant lecturers, who performed in public houses. At the bottom of the heap were the hawkers of curiosities, the street entertainers, and the jugglers who held forth at the fairs of Saint Laurent and Saint Germain.'
[16] Cf. Holmes (1982) and George (1966).

Hodgson's lecture advertisement from the Daily Courant, *1705.*

began on Monday 15 January 1705 (NS). To reach his audience Hodgson advertised in a newspaper, the *Daily Courant*. A typical advertisement read:

> *For the Advancement of Natural Philosophy and Astronomy, as well as for the benefit of all such Curious and Inquisitive Gentlemen as are willing to lay the best and surest Foundation for all useful Knowledge. There is provided Engines for Rarafying and Condensing Air, with all their Appurtenances, (according to Mr Hauksbee's Improvements) Microscopes of the best Contrivance, Telescopes of a convenient length, with Micrometers adapted to them, Prisms[,] Barometers, Thermometers, and Utensils proper for Hydrostatical Experiments, in order to prove the Weight and Elasticity of the Air, its Usefulness in the Propagation of Sounds and Conservation of Life: The Pressure or Gravitation of Fluids upon each other: Also the new doctrine of Lights and Colours, and several other matters relating to the same Subjects, by James Hodgson Fellow of the Royal Society. All Gentlemen that are willing to encourage so great an Undertaking or are willing to be benefited by it, must*

Air pump by Francis Hauksbee the elder, c. 1708, inventory number 1970-24.

> *subscribe two Guineas, one at the time of Subscription, the other two months after the Course begins, which will be on Monday next, at Mr More's (formerly Coll Ayres's) at the Hand and Pen in St. Paul's Church-yard, where Subscriptions are taken in, likewise at Mr. Hauksbee's in Giltspur-street without Newgate, at Mr. Rowley's under St. Dunstan's Church Fleet-street, at Mr. Senex's Bookseller next the Fleece-Tavern in Cornhill, where Proposals at large may be seen.*[17]

[17] *Daily Courant* Thursday 11 January 1705 No. 855 p. 2. This course was first advertised in the *Daily Courant*, Saturday 9 December No. 828 p. 2 to begin on Monday 8 January but it does not seem to have started until the 15th. The *Daily Courant* for Wednesday 10 January carries the first advertisement in January and mentions the lecture on Monday next. This is confirmed by an advertisement on 15 January. Interestingly,

Ayres was one of the foremost teachers of book-keeping of his time. See Murray (1930) p. 256. The overlap between those who gave lectures on experimental philosophy and those who lectured on arithmetic and book-keeping needs investigation. See Stewart (1986*b*) p. 183 who mentions that Ayres was involved with Lowther and the building of steam engines.

Plate of Hauksbee's air pump from Hauksbee (1709).

It is hardly surprising that Hodgson's references to the 'Advancement of Natural Philosophy' and the 'Foundation of all useful Knowledge' echo the rhetoric used to describe the endeavours of the Royal Society, for he had been elected a Fellow the previous year.[18] In his turn he used this rhetoric to try and attract an audience of gentlemen whose social position and interests would overlap with that of the Fellows of the Royal Society. The close relationship between Hodgson's lectures and the Society's activities is further emphasized by the mention of Hauksbee's pump and the reference to the 'new doctrine of Lights and Colours'. Hauksbee was demonstrating just such a pump (probably the same one) before the Society, having begun to give demonstrations for them during 1704,[19] and the 'new doctrine of Lights and Colours' is a reference to the views that Newton, the President of the Royal Society, had advanced in his *Opticks*, published the same year.

In fact Newton had been elected President of the Royal Society in November 1703,[20] only shortly before Hodgson's first lectures. This coincidence is striking and it is tempting to see the lecture courses as one manifestation of Newton's growing influence. However, the lectures were not exclusively concerned with his views or those of his followers; while there were close links between Newton and some lecturers, the lectures were not confined to London nor indeed to a 'Newtonian' approach to natural philosophy—the lecturing phenomenon occurred elsewhere in Europe, carried on by individuals with Cartesian or other backgrounds.[21]

Irrespective of the precise role that Newton himself played, his influence and that of the Royal Society were not the only factors shaping the development of the lectures in London. An important feature of the course of lectures offered by Hodgson was that it brought together his talents as a teacher with those of an instrument-maker, Hauksbee. The contribution of each was vital to the success of the enterprise and, moreover, the combination of teacher and instrument-maker emphasizes two other developments in the early eighteenth century—a growing concern about the importance of education and the growth of the instrument-making trades.

A further significant feature was the use of newspaper advertisements to reach potential members of the audience. Again, this was a comparatively new possi-

---

[18] At the same meeting Newton became President.
[19] This pump is now in the Science Museum's Vacuum Technology collection, inventory No. 1970-24.
[20] Westfall (1980) p. 629.
[21] Cf. Rowbottom (1965, 1968). The question of whether 'Newtonian' means anything in this context is left on one side. See, for example, Cantor (1983). This point about lecturing elsewhere has been emphasized by Heilbron. 'The introduction of experimental philosophy had . . .begun before the advent of Newtonian experimental philosophy. For a time, it proceeded on the continent under Cartesian fellow travellers opposed to Newton's methods and bemused by his apologetics.' Heilbron (1979) pp. 140–1. See Hanna (1982) for the case of Polinière in Paris.

Sir Isaac Newton, by Sir Godfrey Kneller, 1702.

bility because the growth of newspapers was a result of the demise of the last of the Printing Acts at the end of the seventeenth century, Acts which were effective in restricting both printing and the number of printers. An example of this growth is the appearance in 1702 of the *Daily Courant*, the first daily paper to be published in London and the one favoured by Hodgson for announcing his lectures. At the time there were altogether some 25 periodicals in the capital, a figure which had quadrupled by 1760.[22]

[22] See Rudé (1971) p. 78.

Education

Courses of public lectures on natural philosophy were one manifestation of an interest in new educational possibilities around 1700. Other new courses also proliferated; there were courses on topics like anatomy and chemistry (considered below) besides existing courses on mathematics and related topics given during the seventeenth century.[23] However, the earlier courses had not used experimental demonstrations to the same

[23] See Taylor (1954).

King Charles II, by Marcellus Laroon, c. 1670. Charles II was patron of The Royal Society.

extent, if at all, and had been given to audiences with professional and vocational interests. Robert Hooke, Professor of Geometry at Gresham College and lecturer in mechanics from 1664, is a prominent example of someone who, early in his career, lectured to an audience with more professional and vocational interests but later went on to give public lectures on the history of art and nature for a number of years, despite the fact that Cutler, who was to pay for these lectures, did not do so for over 25 years.[24] Another London lecturer, John Harris, gave a series of mathematical lectures from 1698, first in Southwark and then at the Marine Coffee House in Birchin Lane, paid for by Charles Cox.[25] Both Hooke and Harris lectured to the public on mathematical topics, globes, and various useful devices, topics that were taken up and continued by later lecturers. However, one difference between these and subsequent lectures was that the earlier ones were paid for (though in Hooke's case, this was the intention, if not the reality) by individual patrons. The later lectures, including those at the Marine Coffee House, were not subsidized by Cox but were financed solely by the people attending them, indicating that by the early years of the eighteenth century a 'market' had been established for lectures, and a subsidy was no longer so necessary since there was now an audience both willing and able to pay for instruction.

The introduction at the end of the seventeenth century of a scheme for teaching mathematics and natural philosophy to certain boys at Christ's Hospital was another example of the growing concern about the educational utility of these subjects. The scheme was intended to ensure the supply of suitable candidates for posts in the Navy but it faced difficulties as there were few people who could teach these subjects.[26] For those who had a background in natural philosophy, however, the scheme meant that a number of new opportunities opened around 1700 in public lecturing or teaching. Hodgson, for example, went to teach at Christ's Hospital and gave up public lecturing. The opportunities available to those such as Hodgson illustrate the changes that took place as a result of the expanding educational possibilities.[27]

Instrument-making

The lecturers drew heavily on the resources provided by the instrument-making trades that had grown up in London. As some of the apparatus for the lectures (and the demonstrations at the Royal Society) was expensive and sometimes had to be developed specifically for lecturing, the lecturers needed the co-operation of instrument-makers.[28] Around 1700 the business of instrument-making, clockmaking, and similar occupations, was expanding for several reasons, such as an increasing demand for specialized instruments used by surveyors or seamen. But the market for items used in domestic settings had also grown significantly with, for example, seventeenth century inventions like the barometer becoming more common.[29]

In the early eighteenth century Edward Saul wrote

[22] See Rudé (1971) p. 78.

[23] See Taylor (1954).

[24] See Ward (1740), Espinasse (1956), Hunter (1989) Chapter IX.

[25] See Taylor (1954) p. 138, Stewart (1986a) p. 48. Cox was a brewer.

[26] See Newton Correspondence Vol. 4 for Newton's involvement in these 'Schemes' and Hans (1951).

[27] According to Hans (1951) p. 136, the Gresham Lectures were revived in 1706. Ward (1740) p. 188 mentions in the case of Hooke that 'He likewise often gave out, that he designed to dispose of the greatest part of his estate in such a way, as might promote the ends, for which the royal society was instituted, by building a handsom fabric for their use, with a library, repository, laboratory, and other conveniences for making experiments; and by founding and indowing a physico-mechanic lecture, like that of Sir John Cutler.' There were also the Boyle lectures established under the terms of the will of

Robert Boyle. These were given in church and delivered by a series of divines from 1691 to 1732. See Letsome and Nicholl (1739) and Hunter (1989). Jacob (1974) p. 142 examined the significance of the Boyle lectures taking as the starting-point 'Most of what contemporaries knew about Newton and his natural philosophy they learned not from the Principia or even from the queries to the Opticks, but from his followers who gave the Boyle lectures.' Once the lecturers on natural philosophy were well established, say by 1715, it seems likely that there were more people attending these lectures than those given under the terms of Boyle's will.

[28] See Taylor (1954).

[29] See Goodison (1977) especially Chapters 2 and 3. Addison composed a poem in Latin on the barometer. Samuel Johnson commented that it was a poem 'he would not have ventured to have written in his own language'. Johnson (1925) Vol. 1, p. 329.

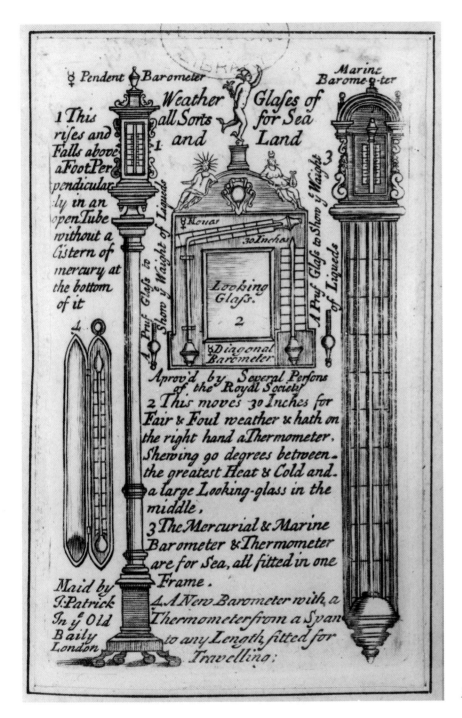

Patrick's trade card, c. 1710.

The Weather-Glass being of late grown into common Use, and in most Houses of Figure and Distinction, hung up as a Philosophical, or Ornamental Branch of Furniture. . .[30]

In turn the wider consumption of these items, ornamental or practical, created a demand for instruction in how to make them or how to use them, the need that Saul responded to with his work, *An Historical and Philosophical Account of the Barometer.*[31]

The growing demand for clocks provides another example of the changes that took place. Around 1700 they became commoner in domestic settings judging by

[30] Saul (1735) p. 12. Saul had been a student of Keill's at Oxford and had become a private tutor. One maker, Patrick, had specialized in the making of barometers. See Taylor (1954).

[31] See Smith (1688) for one example.

evidence from inventories, confirming that more people were buying and using them.[32] A book on clockmaking by William Derham shows the different audiences interested in such a work. Derham had written his book for gentlemen who wanted to understand the principles of clocks but he made his work short—and therefore cheap—so that workmen would also have access to it.[33] The readership for these practical works was growing and heterogeneous: it included both makers and users, people engaged in a trade involving the use of instruments such as surveyors as well as those with only a passing interest. These different groups were variously clients for makers of instruments, customers for pamphlets, and a potential audience for the lectures on natural philosophy. But it was not just the audience that overlapped in this way. A few instrument-makers became particularly involved with lecturing. In some cases they supplied the lecturers with apparatus, an effective form of advertising, or they themselves gave lectures and gained another source of income.

In the early days of collaboration the instrument-makers occasionally provided a room to accommodate the lectures. Of course, other places could be used instead, such as rooms in taverns or coffee houses, but lecturing at the instrument-makers' premises was more practical since the instruments needed for the lectures were close at hand.[34] The case of James Hodgson, whose course has already been mentioned, illustrates how an instrument-maker could become involved in a lecture course.[35] The problems Hodgson faced as he started his new career were outlined by his acquaintance,

Abraham Sharp:

It is a great undertaking for one, especially so young a man, to teach all parts of experimental philosophy, and the apparatus or instruments in order thereto which he intimates is now making for him. It will in my opinion be too great a charge for a single purse.[36]

Sharp's comment suggests that though Hodgson had the major part in planning the course of lectures, the instrument-maker Hauksbee was involved as well, if only as the maker of the air pump and other items of Hodgson's apparatus.[37] However, it is quite likely that Hauksbee carried out the demonstrations at the lectures for he had been doing just that at the Royal Society where he was curator of experiments.[38] A further possibility is that Hodgson did not purchase the apparatus from Hauksbee and other makers as Sharp had thought; perhaps the arrangement was more of a collaboration between Hodgson and the instrument-makers whereby they provided the instruments in return for a share of the proceeds. For Hodgson such an arrangement would mean that he avoided the expense that Sharp complained of whilst spreading the risk of the venture.

Evidence that these lectures were such a joint venture comes from advertisements in the newspapers. The first advertisements mentioned that syllabuses were available from Hauksbee, Rowley, and Senex, but within a year the arrangements had changed. In January 1706 Hauksbee's name now appeared alongside Hodgson's in advertisements, astronomy was no longer included in

[32] See Weatherill (1988) p. 28 who records that clocks were mentioned three times more frequently in probate records in 1715 than in 1685.

[33] See Derham (1714).

[34] In 1705 Hodgson had begun at Mr More's, formerly Ayre's. Then he had given courses at the Queen's Head Tavern before using Hauksbee's new premises in Wine-Office Court from 1707, a sequence of moves which suggests that Hauksbee's involvement with the courses grew greater as time went on.

[35] Before he began to lecture on natural philosophy Hodgson had been an assistant to Flamsteed, the Astronomer Royal. For Hodgson see DNB, Taylor (1954), Rowbottom (1968). Hodgson (1706) Preface mentions that it was at the observatory at Greenwich 'where I had the Happiness to have my Education'.

[36] The letter is from Sharp to Flamsteed and dated 25 October 1704. Sharp was Flamsteed's assistant before Hodgson. See Cudworth (1889) pp. 81–2 and Rowbottom (1968) p. 199.

[37] Hutton (1795) Vol. 2, pp. 445–8 describes Sharp as 'one of the most accurate and indefatigable computers that ever existed'.

See the advertisement quoted above. Cf. the account in Taylor (1954). Hauksbee had given demonstrations before the Royal Society by this date, so clearly it would have been possible for him to be involved in giving this course with Hodgson.

[38] See Hauksbee (1719). Hauksbee also carried out experiments for Newton, see Westfall (1980) pp. 684–6. Hauksbee was asked to bring his air pump to show his experiments to 'Philosophical Persons' at Newton's house receiving 2 guineas for his pains, Newton Correspondence Vol. 4, p. 446, Newton to Sloane 14 September 1705. Though a Fellow of the Royal Society from that year, his receiving payment from Newton (and on other occasions from the Royal Society, see Heilbron (1979)) indicates that he was not at the same social level as the other Fellows.

the course and Rowley, a well-known globe and instrument-maker, no longer distributed syllabuses.[39] It looks as if Rowley had dropped out of the venture—perhaps this was why astronomy was omitted—while Hauksbee became more involved, another indication of the nature of the arrangement—a joint collaboration rather than the initiative of one individual.[40]

Advertising

Within a few years Hodgson gave up lecturing in public when he became Mathematics master at Christ's Hospital in 1709.[41] Hauksbee then continued with his own Courses of Pneumatical and Hydrostatical Experiments at his premises in Wine-Office Court, Fleet Street, until his death about three years later. Nevertheless, in the winter of 1711–12 matters became difficult for Hauksbee, for now he faced direct competition from another lecturer, his nephew and namesake, Francis Hauksbee. In October Hauksbee senior advertised that his course would begin '. . . so soon as a convenient number have subscribed' but as subscribers were slow to

sign up he did not announce a starting date for several months.[42] But when Hauksbee junior and Humphrey Ditton advertised their new course to start on Monday 14 January,[43] Hauksbee senior was galvanized into action and that day he advertised that his own course would start a week later.[44]

However, Hauksbee senior died within a year, in early 1713, and his place as a lecturer was taken by J. T. Desaguliers who had returned to London from Oxford.[45] In May that year Desaguliers lectured at the premises of the late Hauksbee senior having come to an agreement with Mrs Hauksbee.[46] But by that time another lecturer, William Whiston, had also come to London, having been forced out of the Lucasian chair at the University of Cambridge.[47] Whiston joined forces with Hauksbee junior and gave lectures in collaboration with him from March.[48] Consequently the Hauksbee family wrangle continued as Whiston and the younger Hauksbee competed with Desaguliers and Mrs Hauksbee, with their advertisements in the *Daily Courant* bringing the feud to the attention of the public.

The rivalry between the Hauksbees was not just confined to lectures; it extended to the apparatus that

---

[39] See *Daily Courant* Friday 4 January 1706 No. 1162 p. 2. The lectures on experimental philosophy at the Queen's Head Tavern were on Mondays and Thursdays at 5 in the evening and the course cost 2 guineas.

[40] For Rowley see Taylor (1954) pp. 294–5. It is quite possible that Rowley may have supplied some of the apparatus that Hodgson used in his lectures on astronomy. Rowley made some of the first orreries a few years later. See King and Millburn (1978) pp. 154–7. At the time there was one other person advertising lectures on similar topics. This was John Harris, the author of the *Lexicon Technicum*, engaged by Charles Cox to give the series of public lectures mentioned above. For Harris see DNB. The day after Hodgson began his first course in 1705, Harris started to lecture at the Marine Coffee House in Birchin Lane on the 'Use of globes' on Tuesdays; on Fridays he lectured on 'Algebra in the former method'. See *Daily Courant* Thursday 18 January 1705 No. 861 p. 2. (cf. *Daily Courant* Friday 25 January 1706 for books for the course sold by Midwinter.) By October 1707, however, the arrangements had changed. Hodgson now took Harris's place and gave the 'Publick Mathematical Lecture' at the Marine Coffee House. See Taylor (1954) p. 141. For Hodgson's advertisements, see *Daily Courant* Tuesday–Friday 21–24 October 1707. That course began on the Friday. Hodgson also gave these lectures in 1708 *Daily Courant* 13 May No. 1946 and 20 September No. 2055, and Hauksbee advertised his own 'Course of Pneumatical and Hydrostatical Experiments' to start at his premises in Wine-Office Court, Fleet Street. It was advertised to begin on the

21st. However, the audience for Hauksbee's course took a week or two to respond for he advertised that the course would 'begin without fail' on Monday 10 November in the *Daily Courant* Tuesday 4 November No. 1785 p. 2, with lectures every Monday and Thursday at 6. As Hauksbee's advertisement does not mention Hodgson it seems unlikely that Hodgson was involved with that course, though they did collaborate for the following two years. See also Uffenbach (1934) p. 77 for an account of his visit to Hauksbee in 1710.

[41] See DNB and Taylor (1957) p. 118 No. 479.

[42] See *Daily Courant* Saturday 27 October 1711 No. 313–p. 2. This is in marked contrast with what happened in 1710 when Hauksbee advertised on Monday 6 November a course to begin on the 13th and repeated the advertisement from the Tuesday to the Friday and the following Monday.

[43] See *Daily Courant* Tuesday 8 January No. 3194 p. 2.

[44] This skirmish spilled over into *The Spectator*, a new periodical of the time. See the issue for Friday 12 January 1712 No. 272 etc.

[45] See Rowbottom (1968).

[46] See *The Guardian* Tuesday 5 May 1713 No. 47 p. 2. The course was to begin on 14 Thursday May.

[47] The *Daily Courant* for Wednesday 4 February 1713 No. 3530 carries an advertisement for Whiston's 'A scheme of the Solar System. . .' which was available from the author in Cross-Street, Hatton Garden. Whiston must have been in London. See also Farrell (1981), Force (1985), and Gowing (1983).

[48] *Daily Courant* 23 March 1713.

they both sold, illustrating again that lecturing was linked to the making and selling of instruments. First Hauksbee senior had claimed that his air pumps were superior, 'He being the only Person to whom the late Improvements are owing. . .'.[49] The following season the rivalry spilled over into Richard Steele's *Englishman* when Whiston and Hauksbee junior advertised their course of philosophical experiments on Wednesday 6 January 1714 (NS).[50] The following day Mrs Hauksbee advertised air pumps made by Richard Bridger, her late husband's apprentice. These pumps had an important advantage over others, she noted:

> *N.B. Whereas other Air Pumps exhaust the Receivers by the Airs lifting up Valves, these Pumps, not having that Inconveniency, are more valuable than any yet made, upon the Account of a Contrivance in the Pistons that gives the Air a free Passage at each Exhaustion. This new Improvement, which the late Mr. Hauksbee made a little before he died, was communicated to no other Workman besides Richard Bridger aforesaid, who now makes all the Pumps with it.*[51]

A few days later Mrs Hauksbee announced the start of another course of lectures on mechanical and experimental philosophy given by Desaguliers at her premises.[52] These developments annoyed the younger Francis Hauksbee. He objected to the claims his aunt had made for Bridger's pumps and issued a challenge, in the form of an advertisement, offering to test his own pumps against any other.

> *N.B. Whereas in the Courant of January the 9[t]h 1713 14, there is a Remark at the Close of an Advertisement concerning Air Pumps of a particular Contrivance in the Pistons, by which the Operation of exhausting the Air is more accurately performed than by any other Air Pumps heretofore made without that Contrivance: And it is also there asserted, that this Contrivance was communicated to no other Workman but Richard Bridger. Now, by their*

saying this, I suppose they would endeavour to insinuate, that no other Workman knows it, but I affirm that I also know that mighty Secret of theirs, together with its Imperfections and Deficienc[i]es; and that I also know how to remedy all this by the Substitution of a much better Contrivance: But in the mean time I will undertake to produce before competent and impartial Judges, an Air Pump with Valves of my making, which shall exhaust the Air from a Receiver to as great an Exactness as they can by any of their pretended Nostrums.[53]

The Hauksbee family quarrel illustrates how closely the retailing of instruments was connected with the business of lecturing. A lecturer who demonstrated apparatus was, in effect, advertising it, particularly if the advantages of machines made by one particular maker were highlighted by the demonstration or stressed by the lecturer. As both Mrs Hauksbee and her nephew realized, the differences between the rival machines then provided a basis for advertising copy. Through the advertisements, details of the family quarrel also became public knowledge.

The Connection between Lecturing and Instrument-making

The Hauksbee clan had become involved in courses of lectures partly because of the lucrative connection between lecturing and instrument-making. But they were not the only ones to benefit. As well as contributing to the growth and prosperity of the instrument-making trade, the lecturers also contributed to the well-being of the Royal Society. Hodgson had become a Fellow shortly before he began to lecture but, more importantly, Hauksbee senior, with his background as an instrument-maker, also became a Fellow and acted as

---

[49]   See *The Spectator*, Friday 1 February 1712 No. 290 p. 2. Hauksbee junior probably learnt the business from his uncle.

[50]   See *Daily Courant* Wednesday 6 January 1714 No. 3807 p. 2. See also Blanchard (1955).

[51]   *The Englishman* 7–9 January 1714 No. XLII p. 2 and also the *Daily Courant* for 9 January 1714 No. 3810.

[52]   *The Englishman* 14–16 January 1714 No. XLV p. 2 and *Daily*

Courant Monday 18 January 18 No. 3817 p. 2. 'wherein the Principles of Mechanicks, Hydrostaticks, Pneumaticks, and Opticks, are rationally demonstrated, and prov'd by more than 300 Experiments'.

[53]   *The Englishman* 21–23 January 1714 No. XLVIII p. 2. See also *Daily Courant* Tuesday 19 January 1714 No. 3818 p. 2.

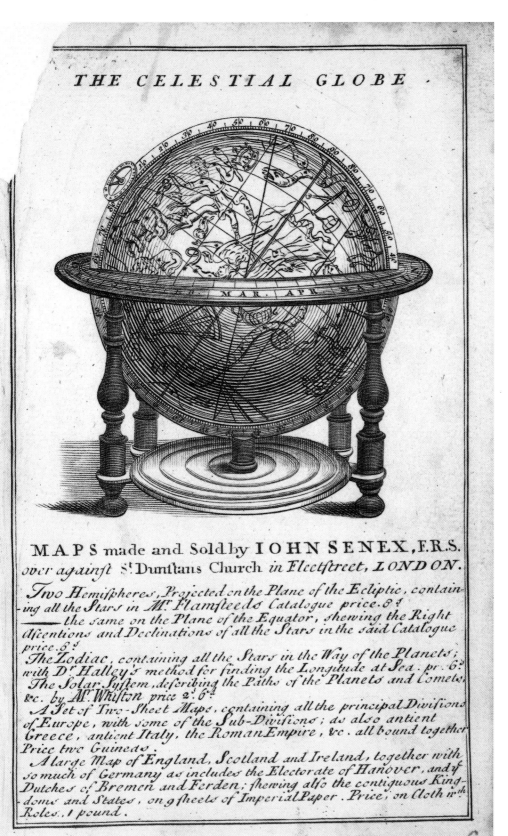

Senex trade card,
c. 1750.

curator of experiments for the Society.[54] After Hauksbee's death in 1713, Desaguliers succeeded him as curator of experiments and a Fellow. Between them Hauksbee and Desaguliers provided many of the demonstrations shown to the Royal Society.[55] The relationship between the lecturers and the Society was symbiotic; the demonstrations shown to the Society one week could be shown to the public the next.[56]

Within ten years of Hodgson and Hauksbee having given the first lectures on natural philosophy in London, Desaguliers, Hauksbee the younger, Ditton, and Whiston were all giving regular courses of lectures. These individuals had set the pattern of natural philosophy lectures by 1715 or so—a pattern of lecturing which, as already pointed out, involved the collaboration of a lecturer, often someone who had attended university at Oxford or Cambridge, with an instrument-maker who provided the apparatus and might also demonstrate it. In 1714, for example, there was a course of lectures 'To be perform'd by Francis Hauksbee, and the Explanatory Lectures Read by William Whiston MA' as their advertisements explained,[57] and in 1718 when Benjamin Worster and Thomas Watts offered a course of 'Mechanical, Hydrostatical, Pneumatical and Optical Experiments', they followed the same formula. The lectures were given by Worster (and later James Stirling) and the instrument-maker William Vream was involved.[58] Later this division of labour between lecturer and instrument-maker changed and by the time of the next generation, which included Benjamin Martin and James Ferguson, these roles were combined so that the lecturer and maker tended to be one and the same

J. T. Desaguliers Legum Doctor, Regiæ Societatis Londinensis Socius, Honoratißimo Duci de Chandos à Sacris, Philosophiæ Naturalis Experimentorum ope Illuftrator.

J. T. Desaguliers in 1725.

person.[59] The pattern of lectures established by the individuals mentioned above, consisting of demonstrations on mechanics, hydrostatics, pneumatics, and

54 Hauksbee does not seem to have had the title curator of experiments. See Record (1912) p. 32. Hauksbee junior was not a Fellow but had close connections with the Royal Society none the less, becoming its clerk. In Whiston's case, though Newton had supported him to be his successor in the Lucasian chair in Cambridge, he would not have him in the Royal Society because of the forthright and public way in which Whiston had expressed his religious views.

55 Cf. Abraham Sharp's remark quoted above. The editors of the Newton Correspondence, Vol. 5 p.xl comment '. . . but for the energy of Hauksbee, Papin and Desaguliers the meetings would have been sterile . . .'. See Heilbron (1983) pp. 20–1 etc.

56 One other body with an explicit interest in natural philosophy at that time was the Spitalfields Mathematical Society, see Cassels (1979).

57 *The Englishman* Saturday 6 February to Tuesday 9 February 1714 No. 55 p. 2.

58 The room that they used for the lectures was first called the 'Accomptant's Office for qualifying young Gentlemen for Business' in Little Tower Street but was later renamed with the grander-sounding title, the 'Academy in Little Tower Street'. Courses of lectures on natural philosophy were given there for many years. See *Daily Courant* Saturday 8 November 1718 No. 5320 p. 2. Stirling was a supporter of the House of Stuart, who had found Oxford University uncongenial after he had spent some time in jail for his part in a riot on the birthday of the new King, George I. See Tweedie (1922) and Rigaud (1831).

59 See Chapter III pp. 70–5. Alternatively, a lecturer would commission equipment from instrument-makers.

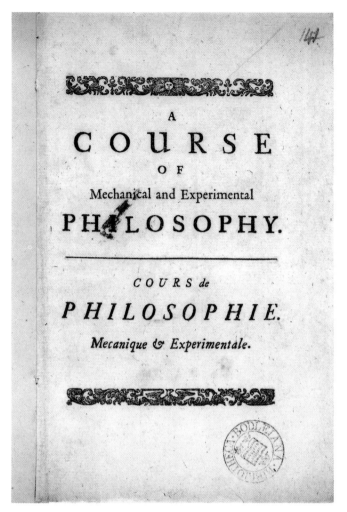

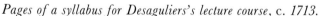

Pages of a syllabus for Desaguliers's lecture course, c. *1713.*

optics, was used for well over a century with only slight variations.[60]

These courses of lectures on natural philosophy were one of the main ways in which knowledge about

scientific matters was both formulated for an audience and disseminated to it throughout the eighteenth century.[61] Though the London lecturers were not the first nor the only ones giving courses in this period, they

[60] See Heilbron (1979) and Heilbron in Rousseau and Porter (1980). Another feature of early lecture courses was that they covered just mechanics, hydrostatics, pneumatics, and optics. See Cantor (1983) for his interesting argument about the role of the lecturers on Optics. Astronomy, for example, though it had been offered by Hodgson in his first season and later by Whiston and Hauksbee, was covered only occasionally. Lectures on the solar system eventually became a staple of the courses after the orrery and other devices for showing the motion of the planets became known and available. See Whiston and Hauksbee 'An experimental course of Astronomy. . .' London 1730. See also advertisements for such a course *Daily Courant* Monday 9 November 1719 No. 5632 p. 2. A similar course may have been offered in 1713. See *Daily*

Courant Tuesday 17 March 1713 No. 3564.

[61] See Hans (1951), Inkster (1980), and Schaffer (1983). For a general discussion of the coming of print and its effect on technical literature for an earlier period see Eisenstein (1979) Vol. 2, pp. 520 *et seq.* Many of her concerns seem relevant to a study of natural philosophy in the mid-eighteenth century. Printing had many effects on natural philosophy. It made possible syllabuses and books, some published by the lecturers under discussion, but there were a host of others, the texts of lectures by 's Gravesande, and Nollet, for example. These were read and assimilated or plagiarized, according to taste, but they did help to establish patterns for lecture courses in experimental philosophy.

were very influential, with their example being copied by Jurin in Newcastle, Triewald in Sweden, and 's Gravesande in Leiden, amongst others.

Desaguliers

Desaguliers can be seen as one of the key figures in the changes that took place in the business of lecturing on natural philosophy. From a Huguenot family, Desaguliers went to Oxford University. He went through a course of Experimental Philosophy with John Keill in 1708 and in the next few years Desaguliers gave that course for Keill who was visiting America.[62]

When Keill returned, Desaguliers moved to London, giving his first course there on Mechanical and Experimental Philosophy in January 1713 (NS) and linking up with Mrs Hauksbee a few months afterwards.[63] Thirty years on, towards the end of his life, Desaguliers reflected on the similarities and differences between the early courses of Keill and Hauksbee. He wrote:

Dr. John Keill was the first who publickly taught Natural Philosophy by Experiments in a mathematical Manner. . . He began these courses in Oxford, about the Year 1704 or 1705, and that Way introduc'd the Love of the Newtonian Philosophy. There were indeed, about the same time, Experiments shewn at London by the late Mr. Hauksbee, which were electrical, hydrostatical, and pnematical: But as they were only shewn and explain'd as so many curious Phaenomena, and not made Use of as Mediums to prove a Series of philosophical Propositions in a mathematical Order, they laid no such Foundation for true Philosophy as Dr Keill's Experiments; tho' perhaps perform'd more dexterously and with a finer Apparatus: They were Courses of Experiments, and his a Course of Experimental Philosophy.

His work for the Royal Society, his translations, his

textbooks, his trials of steam engines and his interest in water supply all illustrate the range of the involvement he had with contemporary scientific and technological developments. He was also very active as a Freemason but the effect of this on natural philosophy is unclear.[64]

The influence of Desaguliers as a public lecturer was immense; he was one of the first to travel both in Britain and abroad to give lectures, and his example inspired many to emulate him. Late in life he claimed that of the eleven or twelve individuals who then gave courses of lectures in Europe, eight were his scholars.[65] Another who could be added to these totals is Stephen Demainbray, who began lecturing after Desaguliers's death, a person whose career and apparatus is described in later chapters.

All these lecturers appealed to an audience with a polite interest in the subject matter and who were both able and willing to pay a fee for the privilege. Attending lectures on natural philosophy had much in common with other forms of polite culture. In London and elsewhere such attendance had become a form of consumption; lecturing was now a commercial enterprise, with the audience as the consumers.

While there were forms of consumption such as the purchase of newspapers and periodicals or instruments that directly or indirectly benefited the lecturers on natural philosophy, other forms of consumption had an ambiguous effect. The lecturers on other subjects, such as chemistry or anatomy, for example, were allies on occasion and rivals on others.[66]

Lectures on Chemistry and Anatomy

Public lecturing on both chemistry and anatomy started in London around the time it began on natural philosophy which suggests that there were common factors involved. Among the first to lecture on anatomy

[62] See Desaguliers (1744) Vol. 2, p. 404. See also Rowbottom (1968) and Hurst (1928). Hutton (1795) Vol. 2, pp. 1–2 says that Keill was Decipherer to Queen Anne and King George I.
[63] See *Post Boy* Tuesday 6 January–Thursday 8 January 1712–13 No. 1756 p. 2. It was held at Mr Brown's and consisted of mechanics, hydrostatics, pneumatics and optics.
[64] Desaguliers was an active Freemason and while this was an important activity for him, at least in social terms, the

possibility that it had a much wider significance, in spreading practical concerns to an aristocratic audience, is at best unproven. See Rowbottom (1968), Hans (1951) (who argues the strongest case, associating Desaguliers with a new type of Freemasonry), and Hamill (1986).
[65] See Desaguliers (1744) Vol. 1, p. x.
[66] The discussions about determining longitude that occurred in this period would be another topic to consider.

was George Rolfe in 1701, but when he left for Cambridge a few years later there were no private lectures on anatomy in London until around 1711 when William Cheselden began to teach.[67] Similarly, chemistry courses were first given in 1696 and such courses were regularly offered by both James Robertson and George Wilson by 1705.[68] To judge by their newspaper advertisements, these courses were aimed at physicians and medical students who wanted to learn pharmacy.[69]

To an extent there were similar impulses behind all these different courses of lectures on anatomy, chemistry, or natural philosophy—each lecture included demonstrations as an aid to understanding; knowledge was presumed to come from the experience of seeing some operation carried out.[70] The newspaper advertisements for the courses often emphasized the number of demonstrations. Wilson had 100 'operations' in 1706 compared with Bright's near 200 in his 'Compleat Course of Chymistry' six years later, and Desaguliers crammed 300 experiments into his course on Experimental Philosophy in 1715.[71] However, despite the similarities, the subsequent development of lectures on natural philosophy was quite different from that of anatomy and chemistry.

The courses differed in one important respect; unlike the courses on natural philosophy, the courses on chemistry and anatomy had a strong vocational element, and the people who attended them usually joined a trade or profession where the technical knowledge gained from the course would be useful. However, because trades and professions such as apothecary, surgeon, or physician were still fluid in Augustan England, the question of what constituted the proper training for a professional was not an easy one to resolve. In the event, the lecturers on chemistry and anatomy were able to take advantage of this situation and the vocational aspects of these courses became greater as time went on. The lecturers on natural philosophy were not in a position to emulate this example. This important difference between the purposes of the courses on natural philosophy and those on chemistry or anatomy became accentuated as the century went on. For example, when the arrangements for medical education became institutionalized, the lectures on anatomy became an important part of medical training. By contrast the public lectures on natural philosophy remained educational at an informal and general level without developing towards a vocational training, at least in this period.[72]

While the trajectory of the educational role of natural philosophy was different from that of anatomy and chemistry, the boundaries between them were indistinct on occasion. For example, the infrastructure that supported the different courses had many common features. All the lectures were advertised in the same newspapers. Some items of equipment were common to the different courses of lectures such as the air pumps sold by both Hauksbees that could be used for demonstrations in pneumatics or for blood letting.[73] There was also an overlap in respect of the audience who attended both anatomy and natural philosophy lectures, and of the teaching techniques. Alexander Monro *primus*, for example, who played an important role in establishing the Edinburgh medical school in the 1720s, attended Hauksbee's and Whiston's lectures on natural

---

[67] See Peachey (1924) p. 12 for Rolfe. Peachey also mentions that the first private anatomy lectures in London are attributed to Buissière, a Huguenot, *c.* 1700. See Holmes (1982) p. 198. Cope (1953) p. 4 says that Cheselden was '. . . the first regular and established teacher of anatomy in London.' Rolfe went to Cambridge to take up the first chair of anatomy in Britain, illustrating the close connection between the developments in public lecturing and what happened in universities.

[68] See Taylor (1970) p. 138, *Daily Courant* Thursday 9 November No. 802 p. 2. See *Daily Courant* Thursday 11 January No. 855 p. 2. See Gibbs (1952) p. 125 who states that Wilson was active in chemistry from 1660 to 1711 when he died.

[69] In this context, Golinski advances an interesting argument that if physicians had a knowledge of chemistry they might be better

able to direct apothecaries and that Peter Shaw, for one, appreciated this point. See Golinski (1983) p. 20.

[70] This was one outcome of the dispute between the Ancients and Moderns. Cf. Swift's 'Battle of the Books'. His *Tale of a Tub* was offered for sale alongside Newton's *Opticks* in 1704.

[71] Wilson *Daily Courant* Tuesday 15 January 1706 No. 1171 p. 2. Bright *Daily Courant* Monday 3 November 1712 No. 3451 p. 2. Desaguliers *Daily Courant* Saturday 1 January 1715 No. 4115 p. 2.

[72] See Chapter III below. An exception is the Woolwich Academy set up in 1741. See Guggisberg (1900) and Hutton (1795) Vol. 1, pp. 16–17.

[73] See advertisements by Hauksbee in *The Spectator* such as Friday 11 January 1712 No. 272 p. 2.

philosophy as well as anatomy demonstrations by Cheselden.[74] The Fellows of the Royal Society saw demonstrations by both Cheselden and Desaguliers.[75]

Furthermore, a development late in 1720 shows that the lecturers on natural philosophy had more in common then with contemporary lecturers on anatomy than later history suggests, which indicates that the different courses were diverging and not converging.[76] In the autumn of that year Hauksbee and Cheselden combined forces to offer a course of anatomy for entertainment. Cheselden may have been motivated to embark on this course in order to reach a new audience. The same was probably true for Hauksbee, though he may also have wanted to compensate for the effect of competition from other lecturers on experimental philosophy including, ironically, his one-time collaborator, Whiston.[77]

In December 1720 Cheselden and Hauksbee placed an advertisement in the *Daily Courant*.

A Course of Anatomy designed for the Entertainment of Gentlemen, in which will be shewn all the known Mechanisms of the Human Body, together with the Comparative Anatomy of Birds, Beasts, and Fish, with the various Contrivances for their different Ways of Life; the whole to be illustrated by Variety of Mechanical Experiments, there being a new Set of Instruments made for this Purpose. This Course to be performed by Wm. Cheselden, Surgeon, FRS and Fra Hauksbee at his House in Crane Court, Fleet-street, ...[78]

The fact that the course of anatomy was offered until

1722 by Hauksbee and Cheselden shows that they had some success.[79] The course gave Cheselden valuable publicity, but its novelty came with the attempt to present anatomy as 'entertainment'. However, this novel approach soon faded away, leaving the more established lecture courses and textbooks to cover the subject for the interested audience.[80]

Publishing

The information that we have about the course of anatomy for entertainment by Cheselden and Hauksbee, though limited, does bring out issues related to the representation of the subject matter. In fact, the development of courses on anatomy and natural philosophy in the early eighteenth century raises general questions concerning the development of a technology for the lectures, both to represent and to define the subject matter in a way which most people saw as accessible and effective.[81] In the case of natural philosophy, the demonstrations together with the text of the lectures provided the basic elements for this 'technology of representation'. In the case of the demonstrations, much of the apparatus was specifically designed for lecturing and became standard as time went on, such as the orrery developed by Rowley and others.[82] Many of the designs for this apparatus were codified in the textbooks which often gave details that would help someone wanting to replicate the equipment.[83]

74 Cope (1953) p. 97.
75 Both were excused their dues as a result. See Heilbron (1983) p. 19, fn. 53.
76 See Hanna (1982) for the case of Polinière, who taught anatomy alongside natural philosophy. However, the boundaries between the various subjects were still fluid and, for example, the work of Hales in this respect is particularly relevant.
77 A few weeks before the start of this anatomy course, Hauksbee had joined up with Whiston to offer a course on mechanics, optics, and hydrostatics from the 16 November. However, Worster and Watts advertised a rival course to start on Friday 25 November at 6p.m. (*Daily Courant* Friday 18 November 1720 No. 5952) which Hauksbee and Whiston tried to counter by readvertising their course to begin at the same time on that date. Since they did not advertise on the day that their course was due to begin, it seems likely that it did not actually take place (*Daily Courant* Wednesday 23 November 1720 No. 5956). So for Hauksbee his course on anatomy with Cheselden was in part a substitute for a course on natural philosophy.
78 See *Daily Courant* Tuesday 6 December 1720 No. 5967 p. 2.
79 Peachey (1924) pp. 18–19 records similar advertisements on 21 March 1721, also in the following November and in January 1722. See Holmes (1982) p. 234.
80 Another who is relevant here, Spence, a lecturer who gave courses on both natural philosophy and medical subjects, *Daily Advertiser* Thursday 25 November 1742 No. 3698 p. 2. The advertisement by Dr Spence mentions courses in Experimental Philosophy, Astronomy, Chymistry, Anatomy, Physiology. Spens, Spence, or Spencer eventually emigrated to America where he sold his apparatus to Benjamin Franklin, see Franklin (1986) p. 171 and fn. and Heathcote (1955). See also Appleby (1990) p. 379.
81 Cf the 'literary technology' of Shapin and Schaffer (1985).
82 See King and Millburn (1978) for a history of the orrery.
83 See Adams (1761a) and 's Gravesande (1747), for example. Nollet is another good example.

At first the only printed material associated with the lectures were syllabuses, sometimes with plates illustrating the apparatus, material which could be used as advertising or as a form of *aide-mémoire* during, or after, the lectures. Later, textbooks were published based on the lecture courses; the works by Desaguliers and 's Gravesande, to give two influential examples, consisted of the text of their lectures together with plates representing the apparatus.[84] These textbooks helped to stabilize the pattern of other lecture courses and the apparatus used in them. They were also another form of representation of the subject matter.[85]

The case of the anatomy courses seems similar. Dissections of the body were the counterpart of the demonstrations in the natural philosophy lectures. Anatomy textbooks, too, were another way of representing the subject matter of the courses, just as textbooks on natural philosophy had become an important way of representing that subject. The one published by Cheselden, for example, was reprinted many times during the rest of the century. The coincidence of the growth of lecturing and the appearance of textbooks for both natural philosophy and anatomy is very striking: the subject matter and the approaches to it were being codified at the same time.

In contrast with the successful use of instruments by the lecturers on natural philosophy in representing their subject as rational entertainment to their audience, an attempt to use similar instruments to present anatomy as entertainment did not prove popular with the intended audience. Though Cheselden and Hauksbee had devised and used a new 'Set of instruments' to illustrate anatomy and the mechanisms of the human body, rather than rely on dissection which might offend the sensibilities of their audience, that way of 'presenting' anatomy did not become accepted as polite entertainment. The 'technology of representation' that Cheselden and Hauksbee developed for anatomy never became an alternative to dissecting corpses in the existing anatomy courses for medical students, though it could be used to supplement dissections.[86]

Another example which illustrates the attempt to find entertainment value in representations of anatomy in this period concerns the surgeon Abraham Chovet, who gave courses on anatomy. Chovet made a figure of anatomy, a woman 'chain'd down upon a table', which showed the circulation of the blood.[87] In the 1730s Chovet put the figure on view to an audience with a serious and high-minded interest in anatomy. (This model with its glass veins must have been similar to some of the 'technology' used by Cheselden and Hauksbee.) However, by 1747 this model had been acquired by Benjamin Rackstrow, an individual more interested in public display, selling statues, and offering beatification by electrical means than the serious study of anatomy.[88] Chovet's figure of anatomy remained the same, but the context in which it was displayed had changed, and in this process its entertainment value increased as its educational value went down.[89]

Perhaps the activities of groups such as the Barber

---

84 The point particularly applies to Desaguliers because the first version of his lectures was one prepared by Dawson, a protégé of Steele's. Desaguliers was reconciled once errors were corrected and the publishers had given him satisfaction. See Dawson (1719). 's Gravesande was a lawyer who came to London from Holland for the coronation of King George I. During his stay he met Newton and attended lectures by Desaguliers—see Desaguliers (1744) Vol. II, p. 484. On his return he became professor at Leiden and wrote a textbook 'Mathematical Elements of Natural Philosophy', the English translation being done by Desaguliers. This illustrates how aspects of the work of the London lecturers on natural philosophy spread abroad, a topic which deserves detailed study.

85 There were other possibilities, e.g. the Boyle lectures. Interestingly, one edition of 's Gravesande omitted details of the demonstrations in order to make the book cheaper, but that pattern did not catch on.

86 A contemporary list of seven experiments about the lungs of an animal is mentioned in C. J. Lawrence (1988). There are also some interesting resonances with the work of Stephen Hales.

87 See *Daily Advertiser* Thursday 31 October 1734 No. 1172 p. 1.

88 See *Daily Advertiser* Tuesday 21 April 1747 No. 5079 p. 2. As a 'marginal' figure Rackstrow is of great interest. His pamphlet on electrical experiments sums up his social position: 'But, methinks, I hear one say, What is this Fellow at? Is this a Preface? No! 'tis an Advertisement! He is puffing about his Company, and tells us of his Business, instead of saying something of his Book. I owe that I am not used to write: read the Book; and if it has not a much better reception from the public than I expect, I will sincerely promise never to scribble again.' Rackstrow (1748) p.iii. In Dublin a few years later he was well treated: '. . . the Colledge have built two handsome rooms next the Anatomy House for the reception of his waxworks.' James Simon to Henry Baker 9 November 1754. Baker Papers, John Rylands Library, Manchester. Later still, Benjamin Martin took over Rackstrow's premises off Fleet Street.

89 Perhaps this change in use of Chovet's figure was also a consequence of the fact that corpses for dissection were cheaper in the 1740s. See Linebaugh (1977) pp. 76–7.

Benjamin Rackstrow's trade card, c. 1740.

Surgeons who were anxious to preserve the privileges of their vocation also prevented anatomy from becoming an established form of polite entertainment. Furthermore public views about what constituted 'polite' entertainment (and personal squeamishness) ensured that the lecturers on anatomy did not follow the route taken by the practitioners of natural philosophy. Correspondingly, the natural philosophers lost out because they were not able to make their lectures into part of a vocational training, as the lecturers on chemistry or anatomy were able to do.

Another difference between the development of natural philosophy lecturing and anatomy lecturing in this period concerns the relationship between the lecturers and the organizations with vested interests in the subjects. In the case of the Royal Society there was, as mentioned above, considerable overlap in the early part of the century, between its activities and those of the lecturers, in terms of both the personnel involved (Hauksbee or Desaguliers) and the subject matter. By mid-century this had changed; a much smaller proportion of the lecturers were Fellows and the subject matter overlapped less. The Royal Society eventually abolished the post of curator of experiments; this effectively occurred when Desaguliers relinquished the post in 1743 though the Society did not get round to changing their regulations formally until some years later.

There was a similar overlap in the early part of the century between the Company of Barber Surgeons and the lecturers on anatomy, in terms of both personnel and subject matter.[90] But the overlap between the lecturers and the surgeons resulted in a convergence, rather than a separation, of their activities. Later on, once the barbers and the surgeons had split and the surgeons were better able to control their profession and police their membership, lecturing on anatomy became a regular though informal part of medical training. By contrast, the Royal Society did not develop in the same way. It did not become a professional organization and did not concern itself with training.

Quacks

In the public arena in early eighteenth century London there were many problems about status, some of which were linked to issues of what constituted the proper training for a particular career. The careers of James Jurin and his good friend Francis Hauksbee junior illustrate another aspect of how the developing professional boundaries affected those who lectured on natural philosophy while simultaneously involving questions of training and perceptions of quackery.

Jurin, after attending the Universities of Cambridge and Leyden, became Master of the Free Grammar School in Newcastle around 1710. Over the next few years he used his knowledge of natural philosophy to give courses of Mechanicks, of use to '..Gentlemen concerned in Collieries and Lead-Mines. . .'. Despite charging '. . .but half the lowest Rate of Private Teaching of Mathematicks in London. . .' Jurin was able to save £1000 to pay for further studies at Cambridge, enabling him to qualify as a physician, and he became a Fellow of the College of Physicians in 1719.[91] Jurin had followed a conventional training, even though the way that he financed it was unorthodox.

By contrast, Francis Hauksbee had an informal training. He was involved, like Jurin, in lecturing on

mechanics and other subjects. He had worked with Cheselden and later with Shaw giving lectures on other subjects which meant, to use Hauksbee's words, though '. . . no Practitioner in any Branch of Medicine, I have been for many Years Employed in publickly teaching some of those Parts of Knowledge upon which the rational Practice of Physick is founded, as Mathematicks, Experimental Philosophy, Anatomy, and Chemistry. . .'.[92]

Despite his experience, accusations of quackery were made against him in the early 1740s when he developed an alternative medicine for cases of venereal disease. Hauksbee had published an advertisement inviting physicians or governors of hospitals to send cases to him so that his medicine could be tested, his plan being to follow proposals suggested earlier by Jurin to test inoculation for smallpox. As Hauksbee explained:

Matters of Fact, in regard to the good or bad Effects of any new Medicine, are very difficult to come by, and require a great deal of Time; especially to him, who being looked on as an Intruder in Physick, is destitute of the proper Assistance from the Faculty for that Purpose, and is left to proceed by himself. Nevertheless, a faithful and impartial Register of such Facts, when obtained, is highly necessary to be drawn up and published;. . .[93]

However, physicians had not provided patients in response to Hauksbee's appeal but instead had branded him a quack.

For altho' the above Advertisement was drawn up with the utmost Caution . . . it has nevertheless had the hard Fate to be treated as a Quack Advertisement, the Medicine as a Quack Medicine, and the Pamphlet, containing the first Thirty Cases, has been called Hauksbee's Quack Bill even by those who, I think, should know better, and who I am confident would blush was I to name them.[94]

Hauksbee had transgressed the professional boundaries by attempting to act like a physician without being one.

---

90 The Royal Society also had rights to use bodies for dissection. There was friction in the case of Cheselden and the Company of Barber Surgeons because the success of his lectures meant that he used most of the available supply of corpses which made it difficult for the Barber Surgeons to acquire those they needed for the lectures they arranged.

91 He also became President a few weeks before his death in 1750. For the details of Jurin's life, see DNB, Holmes (1982) pp. 209, 221 etc., Underwood (1977) p. 127. The details about Jurin's

course of Mechanicks comes from the proposals he published, *Newcastle Courant* 24/26 January 1712. Jurin and Hauksbee both had posts at the Royal Society.

92 Hauksbee (1743) p. xviii.

93 Hauksbee (1743) p. iii. Interestingly, Hauksbee's pamphlet carries the motto 'Nullius in Verba', the motto used by the Royal Society, on the title page.

94 Hauksbee (1743) p. xiv.

Mathematical playing cards, late seventeenth century.

He had not undergone the recognized training and appeals to rational experiment could not compensate for this lack. Both Hauksbee and Jurin had made use of natural philosophy to enter the world of the medical men but in different ways. Unlike Jurin, Hauksbee had had an informal training in medicine. While that kind of training was enough to allow him to lecture on natural philosophy, it was not enough to allow him to behave like a London physician.[95]

An unorthodox training was not the only reason for being branded a quack—guilt by association was another possibility. Because lecturers advertised in newspapers there was a risk that unfortunate conclusions would be drawn from this for 'quacks' advertised

---

[95] Hauksbee was not without insight into his situation. His pamphlet pointedly mentioned that the 'middling Expence' for a course of salivation offered by the physicians for venereal disease was 40 guineas whereas his pills cost much less. One of the patients treated by Hauksbee's pills was J...k K...h, who found corpses for surgeons to dissect. Even though he had access to many medical men, none had been able to cure him. Hauksbee's pills had worked well but the case had become more complicated when he became reunited with his wife. A close relative had died and the reconciliation between man and wife had led to the corpse being sold for 50s. Hauksbee went on 'I hear since, the Separation takes place again, and is like to continue till another Relation is hang'd.' Hauksbee (1743) p. 16.

extensively in the same columns.[96] An interesting example of the difficulties about the status of 'quacks' compared with that of lecturers arose in 1742 when Jo Douglas, a surgeon, was accused of being a 'quack' for advertising his cure for syphilis. An advertisement in his defence listed the eminent medical men who advertised their lectures and implied that these men were respectable members of society despite advertising their services.[97] Douglas ended his list of 'respectable' lecturers with Dr D[esagulier]s and Mr W[histo]n who advertised courses on natural philosophy. However, while Desaguliers may have been considered respectable by many at the time, the status of others who advertised, including Whiston, was not so obvious then.[98]

Consumption

If the social status of the lecturers was open to question, then the standing of the lectures was too. In so far as lectures on natural philosophy or on other subjects were a form of consumption in the early eighteenth century, they were offered in the market-place alongside other commodities and their value would be compared with other related products. Within the home, items such as playing cards offered another form of consumption of natural philosophy, 'Mathematical cards: wherein all the Instruments are exactly delineated', or people might choose to buy barometers and clocks as an expression of interest in natural philosophy. There were books and pamphlets also.[99] The retailing of spectacles was another example of how philosophical argument was linked with pecuniary gain. One person who made and sold them, John Marshall, took the opportunity in his advertisements to damn spectacles made by the unlearned while mentioning that he had received the approbation of the Royal Society, implying that their endorsement was commercially valuable.

John Marshall, Maker of Optick Glasses to his Majesty. . . having by long Experience, observed that abundance of Peoples Sight is spoiled by using false Spectacles sold by Toymen and other ignorant Persons. . . .By the use of true Spectacles the Sight is much preserved, so that many Persons of 60 or 70 Years of Age, can on occasion read without them. Also Reading-Glasses and Spectacles of Rock Crystal, as well as Glass, Microscopes, Telescopes, and all other Optick Glasses, made to the exactest Perfection by him who is the only Person approved of by the Royal Society for the Art of Glass Grinding.[100]

Although appeals to optics or the authority of the Royal Society might help to sell spectacles, spectacle-makers and their customers experienced competition from others with an interest in sight in this period. While lecturers on natural philosophy explained the workings of the eye, so did anatomists; instrument-makers or toymen sold spectacles, practitioners couched cataracts, and pharmacists sold remedies. All these groups had their own special interests to do with sight. Clear boundaries had yet to be constructed between the interests of lecturers on natural philosophy and their rivals.

Polite Entertainment

As the phenomenon of public lecturing emerged in London in the early years of the eighteenth century the lecturers on each subject had to develop different strategies to secure an audience. Unlike the lecturers on other subjects, the lecturers on natural philosophy were unable to make their courses into part of a vocational training; instead their lectures remained polite entertainment. Rather than being sponsored by a single patron, such as Cox or Cutler, they were patronized by a wider public, the people who would become their subscribers.[101] The lectures on natural philosophy were

---

[96] See Porter (1989). The printers of the papers had a distribution network which could be used to sell small expensive items such as medicines. See Cranfield (1962) p. 249. John Newbery, for example, who had an interest in the *Reading Mercury* and published the early works of Benjamin Martin, also had a share in Dr James's Fever Powder.

[97] See *Daily Advertiser* Monday 1 November 1742 No. 3727 p. 1.

[98] See Scriblerus (1731), for example.

[99] See *Daily Courant* Friday 4 November 1715 p. 2. Carving cards showing how to deal with Fowl, Fish, and Flesh were also available.

[100] *Daily Courant* Saturday 12 November 1720 No. 5948.

[101] A similar process happened in the case of the novel.

also distinguished from other contemporary developments, such as the lectures provided under the terms of Robert Boyle's will, by a combination of subject matter, the approach taken, and the venue.

The value of lecturing on natural philosophy as a form of polite and rational entertainment was common to other forms of consumption that developed in the early eighteenth century, such as the buying of periodicals. For example, in his contributions to periodicals such as *Town Talk* and *The Englishman*, Richard Steele set out to promote polite behaviour while entertaining his readers. Steele's intention was later recognized by Samuel Johnson to have a parallel in the activities of the early Royal Society. Johnson wrote:

It has been suggested, that the Royal Society was instituted soon after the Restoration to divert attention of the people from public discontent. The Tatler and Spectator had the same tendency; they were published at a time when two parties, loud, restless, and violent, each with plausible declarations, and each perhaps without any distinct termination of its views, were agitating the nation. . . . The Tatler and Spectator adjusted . . . the unsettled practice of daily intercourse by propriety and politeness. . .[102]

The same moral and political values lay behind both the Royal Society's interests in natural philosophy and the new attitudes conveyed by these periodicals, which is hardly surprising because the audience for the periodicals overlapped with that for natural philosophy. This overlap is apparent in a number of ways. Through advertising or occasional editorial comment, periodicals such as Steele's *Englishman* brought the activities of natural philosophers to the attention of their readers.[103]

Since I have raised to my self so great an Audience, I shall spare no Pains to make their Instruction agreeable, and their Diversion useful. For which Reasons I shall endeavour to enliven Morality with Wit, and to temper Wit with Morality, that my Readers may, if possible, both Ways find their Account in the Speculation of the Day. And

to the End that their Virtue and Discretion may not be short transient intermitting Starts of Thought, I have resolved to refresh their Memories from Day to Day, till I have recovered them out of that desperate State of Vice and Folly into which the Age is fallen. The mind that lies fallow but a single Day, sprouts up in Follies that are only to be killed by a constant and assiduous Culture. It was said of Socrates, that he brought Philosophy down from Heaven, to inhabit among Men; and I shall be ambitious to have it said of me, that I have brought Philosophy out of Closets and Libraries, Schools and Colleges, to dwell in Clubs and Assemblies, at Tea-Tables, and in Coffee-Houses.[104]

Steele himself was directly involved with several of the lecturers on natural philosophy; for instance, he encouraged Whiston when he first came to London and enabled him to start lecturing at Button's Coffee House.[105] Steele was also involved with the lecturers in natural philosophy in another way. One of his main concerns in this period was to arrange 'improving' entertainments in London, partly because he disliked the current fashion for Italian opera.[106] To promote the kind of entertainment he favoured, Steele had a project for a 'Censorium' in York Buildings where concerts that met with his approval were given in a room decorated with pictures of 'Truth' and 'Eloquence'. The first performance took place in May 1715. According to Steele in an article in *Town Talk*, supposedly in a letter written to a lady in the country:

The Disposition, which you call a Project, has nothing in it more Chimerical, than to suppose, that there are Two hundred Persons in this Town, who will be glad to meet, when they are summon'd to be entertain'd for Two Hours and a Half, at a lower Expence than seeing an Opera, with all the Pleasures which the Liberal and Mechanick Arts, in Conjunction and in their Turn, can produce—Musick, Eloquence and Poetry, are the Powers which do most strongly affect the Imagination, and influence the Passions of Men. The greatest Masters in these Sciences will find their Account, in turning their thoughts towards the

[102] Johnson (1925) Vol. 1, pp. 324–5.
[103] For example, Whiston and Ditton had letters published about their proposals for solving the longitude problem. Meanwhile in Parliament Steele helped to set up the Board of Longitude.
[104] *The Spectator* No. 10, Monday 12 March 1711. Cf. Sprat's comment quoted at the beginning of this chapter which talks of '. . .Philosophy being admitted into our Exchange, our

Church, our Palaces, and our Court. . .'.
[105] See Loftis (1950–1), Force (1985), and Farrell (1981).
[106] In Steele's view, the purpose of language was to convey meaning and so Italian opera had the double disadvantage of being in a foreign language and having music that drowned out the words. See Wilson (1984) for other aspects of this debate.

Entertainment of this select Assembly, which is to consist of a Hundred Gentlemen, and as many Ladies, of leading Taste in Politeness, Wit and Learning.[107]

During March 1716 one of those who entertained the assembly was Whiston, who lectured on the 'surprising appearances' in the sky earlier that month.[108] By 1719 the arrangements were more elaborate. In December that year Desaguliers and Thomas Watts delivered a course on natural philosophy at 12 noon in French and, in the evening at 6 p.m., Worster and Watts gave the same course in English.[109]

Steele also used other opportunities of making known his views on natural philosophy. In Steele's view the sciences should be pursued for the practical application they had. As he explained:

The Arts and Sciences (in which I pretend to no accurate Skill) should always be employ'd in Enquiries that may tend to the general Advantage, and they must lose the Name of Liberal, when the Professors of them seclude themselves from Society, or live in it without applying their Abilities to the Service of it. For it is by the joint Force of Men of different Talents, that useful Purposes are best accomplish'd; and a certain Felicity of Invention in one, join'd to the Experience and practical Skill of another, may bring Works to Perfection, which would be so far from Growth, that they would not so much as have had Birth, but from the good Intelligence between Persons of unlike Abilities, whose Good Will towards each other united their Endeavours.[110]

Steele believed that the benefits of the sciences were maximized if the knowledge was public and available to 'men of different talents'. Lecturing fitted well with this conception of the science as public knowledge. It is not surprising that Steele advanced these views for he was drawing on his own experience when he wrote. About 1713, after he had seen experiments on an air pump, he conceived '. . .a Design to form a Vessel which should preserve dead Fish from Corruption a longer Time than usual. . .'.[111] To this end he and a mathematician, Joseph Gillmore, carried out several experiments using model boats with promising results. Though full-scale ships were built, the scheme was not successful in the long run.

Steele's aspirations were shared by others and around 1720, the time of the South Sea Bubble, there were many other speculative ventures.[112] One scheme attempted to raise a capital of £1 million for a wheel for perpetual motion. Presumably, the lecturers on natural philosophy felt qualified to pronounce on such a scheme—though it is not obvious what their advice would have been.[113] One consequence of the bursting of the Bubble and the collapse of these schemes might have been to make people more interested in exploring the technical aspects by going to lectures.[114]

A Changing Audience

The audience for the lectures on natural philosophy seems elusive today, and surviving accounts of what a member of the audience made of a lecture are scarce and brief.[115] However, because that audience read newspapers and periodicals which carried the advertisements for the lectures, some characteristics of the audience are discernible from the newspapers—particularly the lecture advertisements.[116]

In order to improve their situation—or at least

---

[107] *Town Talk* No 4. Loftis (1950–1) p. 57 quotes Steele's draft for this article as 'All Works of Invention, All the Sciences, as well as the mechanick Arts will have their turn.' Interestingly, it looks as if Steele used 'Masters of these Sciences' instead of the 'Sciences' of his draft—a recognition, perhaps, of the role of the lecturer.

[108] See *Daily Courant* Thursday 15 March 1716 No. 4493. Steele was a supporter of the House of Hanover and had received a knighthood from King George I. One reason for his having Whiston lecture was that opponents of the King interpreted the 'surprising appearances' as a bad omen for the King. George I had refused to pardon some Jacobite lords and an aurora borealis had appeared on the eve of their execution. See Rogers (1978) p. 95. See Whiston (1716).

[109] See Loftis (1950–1) pp. 61–2.

[110] Steele and Gillmore (1718) p. iv.

[111] See Steele and Gillmore (1718) p. 3.

[112] See Brewer (1989) Chapter 4, Stewart (1986a), Dickson (1967). A list of Bubble schemes is given in Mackay (1956) p. 61.

[113] Rowley and 's Gravesande were both enthusiastic about a scheme for perpetual motion. See Desaguliers (1744) Vol. 2, p. 183.

[114] See Jurin quoted above and the discussion in Chapter III below.

[115] See, for example, the diaries of Richard Kay (1968), and John Byrom.

[116] See Chapter III for a more detailed examination of these matters.

prevent it from deteriorating—the natural philosophy lecturers tried to attract greater numbers to their lectures. Judging from the advertisements and their new references to ladies, one way that they did this was to encourage women to attend. In 1741 Desaguliers pointed out 'Note, Ladies attend the Lectures as well as Gentlemen'.[117] Though the idea that women might take an interest in natural philosophy was not a new one with, for example, John Harris publishing in 1719 his *Astronomical Dialogues between a gentleman and a lady* aimed at women readers, the formula was revived by Algarotti for his *Newtonismo per Dame* which appeared in English translation in 1739.[118] The publication of this work seems to have encouraged women's interest in natural philosophy which the lecturers were keen to exploit by attracting them to lectures. The contrast with Hodgson's appeal in his first advertisements in 1705 to 'inquisitive gentlemen' shows how much the audience for the lectures had changed.

Another changing feature of the audience—and a measure of the lecturers' success—was that the audience's knowledge of the terms used in natural philosophy had increased, a point made in the 1720s by several authors. In *A view of Sir Isaac Newton's Philosophy* published in 1728, Henry Pemberton mentioned that 'terms of art' were becoming familiar '. . . from the great number of books wrote in it upon philosophical subjects, and the courses of experiments, that have of later years been given by several ingenious men'.[119]

Conclusion

In the early part of the eighteenth century the lectures on natural philosophy became a regular feature of life in London and other large towns. In London the activity of lecturing brought together resources of different kinds: the skills and entrepreneurial talents of instru-ment-makers together with the status of the virtuosi of the Royal Society. However, the pattern was not unique to the capital. Against a background of rising consumption, both of goods and services, and growing interest in technical matters, a group of lecturers in natural philosophy emerged who established a pattern for these lecture courses.[120] The courses were non-vocational, that is they did not lead to a particular qualification or occupation. They were a form of rational entertainment for a general audience. But, as rational entertainment, they were only one of several diversions in competition with each other and the lecturers had to work hard for their audiences.

Some evidence of the competition between the lecturers and other purveyors of entertainment comes from the advertisments in the newspapers. Sometimes the competition was too close. On one occasion Desaguliers reported the activities of a strongman that he saw off by exposing his tricks: 'I don't hear that any of these Sampsons have attempted to impose upon People in the same manner in or near London'.[121]

Another example of the diversity of competition comes from Gerard Winstanley.[122] His house at Little-berry in Essex was open to the curious:

> *It is known by a Lanthorn on the Top of it, and a Model of Eddystone Light-House in the Gardens; and a Dial at the Door, shewing many Curious Motions, viz. The Age and Bigness of the Moon, the Elevation of the Planets and several Stars: And for the Benefit of Travellers, it shows how many Hours it is to Sun-set all the Year. This is seen from the Inns where the Stage-Coaches stop at Dining-time.*

Winstanley had also run a water theatre off Picaddilly with many curious displays. Winstanley's interests and the activities of the lecturers on natural philosophy had a lot in common. They were all interested in the phases of the moon and in water pumps. Desaguliers, for example, as well as lecturing on astronomy, was

---

117   See *Daily Advertiser* Friday 20 November 1741 No. 3381 p. 1. On Monday 18 January 1742 No. 3431 Desaguliers now said 'Note, Ladies are admitted to these Lectures as well as Gentlemen'.

118   See Harris (1719). Harris was following the still earlier example of Fontenelle. See Meyer (1955) p. 25 and Algarotti (1739).

119   Pemberton (1728) Preface. Pemberton himself gave courses on chemistry. See also Shaw's preface to his edition of Boyle's works, '. . . as Mr Boyle was the introducer, or, at least, the great restorer or mechanical philosophy amongst us; so, by

endeavouring to deliver himself in the most full and circum-stantial manner about it, he has spun out his works to what, now that philosophy is more generally known, appears an immoderate length.' Shaw (1725) Vol. 1, p. i.

120   See Brewer (1989) Chapter 8 who discusses a new view of the world.

121   Desaguliers (1744) Vol. 1, p. 265.

122   See DNB.

consulted about the Edinburgh water supply, designed water works for his patron, the Duke of Chandos among many practical projects.[123] Winstanley had worked on the building of the Eddystone Lighthouse, actually meeting his death there in the great storm of 1703. Desaguliers was himself involved in many practical projects. However, the combination of entertainer or lecturer with civil engineer in the career of one individual did not endure. Years later when the Eddystone Lighthouse came to be rebuilt and the capacity of the Edinburgh water supply was increased, it was not a Winstanley or a Desaguliers who carried out the work. In both cases it was John Smeaton, one of the first of the new breed of specialized civil engineers.

During the first quarter or so of the eighteenth century a pattern of lecturing on natural philosophy became fixed. However, by 1744, the year of the death of Desaguliers, the pattern of lecture-demonstrations established by these lecturers had begun to change. The next generation had to adapt to the new circumstances it faced. Natural philosophy lectures remained a form of polite entertainment, whilst other subjects became parts of vocational training taught within institutions allied to the professions.

[123] See Rowbottom (1968). For an interesting discussion illustrating the many connections between Desaguliers and various projects of his time, see Stewart (1986a).

III

A NEW GENERATION OF LONDON LECTURERS ON NATURAL PHILOSOPHY

A study of newspaper advertisements 1745-70

Besides the . . . Kinds of Learning for the Instruction of Youth, and forming their Minds for the Service of their God and Country, as well as an universal Benevolence to Mankind in general; there are divers philosophical Lectures read in the several Parts of the City and Suburbs, by Men of great Learning, Knowledge and Experience; who, at a small Charge, explain and demonstrate to their Auditors Doctrines in Experimental Philosophy, to their very great Advantage and Improvement.[1]

Introduction

$\mathcal{S}$O FAR THE WORK OF THE FIRST LECTURERS on natural philosophy in London has been considered: individuals who, in the early years of the eighteenth century, established a pattern for this kind of lecturing where lectures were accompanied by demonstrations and arranged in courses. It was a form of polite instruction mixed with entertainment. Though there were serious educational overtones, the courses were not allied to any vocational training nor were they associ-

ated with any particular institution. There was, however, a close connection between the work of lecturers and the business of retailing instruments in the capital.

During the 1740s the first generation of lecturers gave way to a younger generation who had to adjust to changing circumstances. In several respects the portents around 1750 were auspicious, such as a significant increase in instrument-making that might contribute to an expansion of lecturing.[2] But there were also events that worked to hinder the lecturers. The Jacobite Rebellion in 1745 caused the number of lectures of all

---

[1] See Maitland (1772) Vol. 2, Book v, p. 1278.
[2] See Bryden (1972a) p. 26, Fig. 9 for the number of instrument-makers in Edinburgh and Glasgow. Fairclough (1975) p. 305 gives figures for the number of instrument-makers in Liverpool

from 1660–1800. The number rises from four in 1750–60 to twelve in 1760–70. See also Macleod (1988) p. 119 who notes that the number of patents granted to people in London doubled between the 1750s and the 1760s.

A Lecture on Geography and Astronomy. Plate from the Universal Magazine, *1748.*

types to fall temporarily while both the audience and the lecturers had more pressing concerns. In the aftermath of the '45 a number of lecturers considered moving from the provinces to London. There the number of lecturers on natural philosophy rose, to judge by newspaper advertisements in the 1750s, as their opportunities improved. But their success was short-lived; the interests of their audience changed with the consequence that attendance at the lectures dropped until the numbers attending were not large enough to provide a livelihood for all the lecturers and their numbers dwindled.[3]

This decline in the numbers lecturing on natural philosophy in London became marked in the 1760s—indeed, some of the lecturers themselves commented on

it. Recently this decline has been explained predominantly as the result of competition between the individuals concerned, combined perhaps with new and attractive opportunities to lecture in the provinces.[4] However, the results of a detailed study of the newspapers of the period (presented below) reveal the significance of other rivals for the audience. For example, there were numerous advertisements for a variety of activities which competed with the lecturers on natural philosophy, ranging from the extravagant advertisements for wonders of the age, such as hermaphrodites and intelligent dogs, to the more sober advertisements for the meetings of the Society for the Encouragement of Arts, Manufactures, and Commerce founded in 1754, or for lectures on medical subjects. All

---

[3] See Rudé (1971), especially pp. xi and 69, and Paulson (1979).

[4] See Heilbron (1979) p. 162 and Cantor (1983) p. 51.

these may have taken a significant share of the audience of lecturers such as Stephen Demainbray, one of those who gave up lecturing.[5]

Whiston, the last of the early generation of lecturers, died in 1752.[6] Consequently, when Demainbray settled in London late in 1754 he encountered a new generation vying for the places left vacant by Desaguliers and Whiston. This younger generation included Erasmus King, Benjamin Martin, and James Ferguson, and they, together with Demainbray, were the four most active lecturers on natural and experimental philosophy in the capital in the 1750s and 1760s. However, in the 1750s Martin and Ferguson published profusely and, because of recent work by Millburn, much is known about their activities.[7] Less is known about the work of Demainbray and King, and other lecturers are even more obscure and transient.[8] In many cases we have to rely on a study of newspaper advertisements to provide the limited information we have about them: details about the audiences they attracted are virtually non-existent and in some cases the content of their lectures is unknown.

The intention in this chapter is to examine the evidence from advertisements in the *Daily Advertiser* and other sources to see how this can add to our knowledge about the lecturers working in London and their audience from 1745 to 1770.[9] By looking at the changes in the advertisements from year to year, a picture can be built up of the fluctuating fortunes of the lecturers— their growth or decline in popularity. Advertisements, too, provide clues to the social milieu of the lecturers and some indication of how the audience changed over the years. They sometimes also suggest possible reasons for the success or failure of the lecturing enterprise.

Newspapers

The newspapers that proliferated during the eighteenth century provided a means for the lecturers to announce their courses. Furthermore, by carrying items about unusual natural occurrences and, in peacetime when news was short, occasional articles on natural philosophy, newspapers also helped to create and sustain an interest in these matters.[10] Attracting an audience to a lecture on events or topics covered by the newspapers became easier and the lecturers benefited. To a large extent the spread of lectures was dependent on the spread of newspapers.

In London the growth of newspapers and periodicals which began at the turn of the century continued in the middle years of the eighteenth century. While there were some 25 periodicals in the capital around 1700, by 1760 there were about 100.[11] One feature of the expansion of the London press was the appearance of papers like the *Daily Advertiser*, founded in 1731. It, along with its rival the *Public Advertiser*, as their titles imply, carried more advertising than news; advertisements taking up about three of the four pages in each issue. The *Daily Advertiser* was very successful, having a circulation in 1746 of around 2500 copies, sold mostly to taverns and coffee houses rather than to private families.[12] It carried more advertisements for lectures in the 1750s than any other London newspaper. As the patrons of the coffee houses both read the newspaper advertisements and formed part of the lecture audiences, the newspapers, the lectures, and the coffee houses, together formed a niche in the cultural life of that period.[13]

---

5   His career is covered in Chapter IV.

6   Hodgson died in 1755 but for many years he had been teaching at Christ's Hospital and had not given public lectures.

7   See Millburn (1976, 1983, 1985, 1986, 1988) and Millburn and King (1988).

8   These individuals seem to have had no significant dealings with institutions connected with science. Consequently neither the records of an institution nor the ready-made cohort of members exist which would allow, for example, immediate use of the approaches demonstrated by Shapin and Thackray (1974) and Thackray (1974). But because these individuals were working before many of the institutions that later existed to sustain natural philosophy, in this respect the lecturers have a particular interest. For later developments see Inkster

(1977), the articles by both Hays and Inkster in Inkster and Morrell (1983), and also Porter in Bynum and Porter (1985).

9   If the outline of these changes can be established, then it may be possible to make a more detailed study of them by looking at the individuals who were members of the Society of Arts or those who took a medical training.

10   See Cranfield (1962) p. 216.

11   See Rudé (1971) p. 78. Similarly, in the rest of England there were no newspapers in 1700 and about 130 by 1760. See Cranfield (1962) p. v.

12   See Harris (1987) p. 190 and Werkmeister (1967) p. 30.

13   See Brewer (1976) Chapter 8 for a discussion of the political significance of the press in that decade.

Advertisements for Lectures on Natural Philosophy in London 1745–70

Of the lecturers active in the 1750s, Erasmus King was the first to advertise in the *Daily Advertiser*. King lectured in Westminster at his house in Duke's Court near the Mews (today Trafalgar Square) from around 1741, though he had lectured elsewhere in the 1730s.[14] By 1748 another regular advertiser of lectures was James Ferguson. Like both Demainbray and another lecturer, Richard Jack, Ferguson had spent some time in Edinburgh.[15] He moved to London in 1743 and around 1747–8 he began to lecture and publish on astronomy and other topics.[16] The next to arrive and advertise was Stephen Demainbray who came from Paris late in 1754. By that time Demainbray had been lecturing both in the British Isles and France for about six years, although he had had some experience in his youth of the business of lecturing in London because he had been brought up in the household of Desaguliers.

Benjamin Martin also advertised, announcing a course of lectures at his house in 1756, a few months after he had arrived in London to set up a shop in Fleet Street where he sold scientific instruments.[17] Martin had lectured from around 1740 in towns such as Chichester and Reading. He tried his luck in London in the winter of 1744–5, shortly after the death of Desaguliers, but it seems he was unsuccessful and he went back to the provinces where he toured for more than 10 years.[18]

Between 1747 and 1753 most of the London courses of lectures on natural philosophy were given by King. In

addition Ferguson, Sowerby, and Jack, the only others then lecturing on these subjects, gave a few courses. But within two or three years competition had increased. Both Demainbray and Martin had arrived in London and they, together with others such as Griffiss and Newberry,[19] all advertised courses rivalling those of King and Ferguson. The result was that in the seasons 1757–8 and 1758–9 there were more individuals lecturing on experimental philosophy than in either the years before or after. The rivalry which existed between these different lecturers can be traced from the advertisements they inserted in the *Daily Advertiser*.[20] A summary of the advertisements for lectures in November and December each year from 1745 to 1770 is shown in Table I.

Table I. Those advertising lectures in the *Daily Advertiser* in November or December in selected years between 1745 and 1770

Croker	1760
Demainbray	1755–8, 1761
Ferguson	1748–1767, except 1765
Griffiss	1757–60
Jack	1751, 1754
King	to early 1756
Martin	1756 on
Newberry	1754
Oliver	1750
Robertson	1747
Sowerby	1747–9
Spence	1741–2

---

[14] See Appleby (1990) p. 379. See Millburn (1986) p. 10. As well as lecturing, King had been involved in trials of a sea-gauge designed by Hales in the Baltic. See King (1750). He had also carried out some experiments on electricity after Demainbray had reported the results of his own experiments on the electrification of myrtle bushes performed while he was in Edinburgh. See pp. 90–2.

[15] For Jack see DNB.

[16] Ferguson was born in 1710, the same year as Demainbray. As well as lecturing in London he toured the provinces. See Millburn (1985) and Millburn and King (1988). He was the only one of these lecturers to be a Fellow of the Royal Society. He was elected in 1763 at his own suggestion made in a letter which delicately mentioned that he was able to pay his subscription. Ferguson died in 1776.

[17] He continued to lecture until 1773 when he retired, afflicted by

gout, and died in 1782. From 1777 Benjamin Donn gave lectures using Martin's premises. See Millburn (1986) p. 38.

[18] See Millburn (1976) p. 35.

[19] Newberry is a shadowy figure. It is unlikely that he was a close relation of John Newbery, the publisher, since none of the standard works dealing with the publisher mentions this Newberry. The *Daily Advertiser* for Tuesday 23 November 1762 No. 9946 contains an advertisement from Newberry mentioning that he was Professor of Mathematics and Navigation, Corps of Noble Cadets at St Petersburg, a post he may have held from around the time of an earlier reference to him in an advertisement by Richardson, *Daily Advertiser* Wednesday 1 October 1755.

[20] For another example of rivalry between lecturers recorded in newspapers see Robinson (1970).

Map of the area around Fleet Street by Roque, 1746.

The pattern of advertising adopted by the lecturers in the first half of the century had been to insert two or three advertisements to announce the beginning of each course, though someone such as King, who often ran two concurrent courses in different places, would advertise more frequently, perhaps twice a week. But during the period of greatest competition between the lecturers, 1755–9, this pattern of advertising changed as minor skirmishes between individuals gave way to a full-scale advertising war involving all the lecturers. Then almost every lecture was brought to the attention of the readers of the *Daily Advertiser*. As each individual lectured about three times a week there were often three advertisements a day for lectures.

It is clear from the advertisements that each lecturer, driven by the intense competition, took a different approach to his audience.[21] For example, the length of courses ranged from the six lectures offered by Ferguson to 34 by Demainbray. Though King had given a course of 24 lectures for several years he advertised a course of 32 for the first time late in 1755 in response to Demainbray's offering 34 a few months earlier.

Along with length of course, the cost varied from one and a half guineas for Demainbray's 34 lectures to 12 shillings charged for 12 lectures by Martin.[22] It was usually possible to attend a single lecture for a shilling but this concession was an innovation—subscribers to the earlier courses had been asked to pay the whole fee in one or two instalments. As with the earlier courses, the high cost implies that the audience came from the nobility, gentry, and better-off tradesmen, since only they could afford these fees.[23] In 1755 both Demainbray and Ferguson charged one shilling and sixpence a lecture when it seems that they were confident they could attract an audience. However, before long they reduced their prices. Perhaps this was an early indication of the problems they would soon suffer more acutely.

Following the strategy of his mentor Desaguliers, Demainbray offered long and expensive courses aiming for the top of the market. To begin with he was successful, for in May 1755, a few months after he started to lecture in London, he was employed to give a course of lectures to the Prince of Wales (the future George III) and Prince Edward and, once established, the royal connection later featured in many of his advertisements. But Demainbray's success was short-lived. Still aiming for the top of the market, in 1758 he tried to restrict his course to just those subscribers who were willing to pay for the whole course. This strategy miscarried when he failed to attract a large enough audience and these lectures were among the last he advertised.[24]

Competition between lecturers extended beyond the price of lectures and the length of the courses. Early in 1755 Demainbray advertised a concession: every gentleman who subscribed would receive a free ticket for a lady.[25] King offered daytime courses for Ladies (describing one as 'Mr King's Philosophical Entertainment for the Ladies'[26] and so did Griffiss. The timing of lectures was an important factor too. Demainbray began by offering daytime courses but later switched to evenings, the time of day chosen by the other lecturers, including King, for their courses. Usually, one or two lectured on any one day at different hours. However, in April 1758 the competition became particularly intense with Demainbray, Martin, and Griffiss all lecturing at 7p.m. on Wednesday and Friday evenings.[27]

The venue was another selling point and an indicator of status. King lectured at his 'Experiment-Room' in his

---

21 See McKendrick in McKendrick *et al.* (1982) for another example of how advertising campaigns were conducted in the eighteenth century.

22 By comparison, a woman silkwinder in Spitalfields was hired for 3 shillings a week, labourers and journeyman in the lower-paid trades received 9–15 shillings, and a jeweller, instrument-maker, or chair carver, £3–4 a week. Rudé (1971) p. 88. Rudé also says that in the 1730s tickets for the theatre were between 1 shilling (gallery) and 3 shillings (in a box) and that admission to Ranelagh was two shillings and sixpence in 1742.

23 One of the few explicit references to the audience in the advertisements in the *Daily Advertiser* is in the issue for Saturday, 29 December 1750 No. 6232 p. 2 where King informs the

'Nobility and Gentry' that he has finished the most complete Astronomical Apparatus in Europe.

24 Demainbray's course was advertised to begin on Monday 20 November 1758. There were no advertisements for it after Friday 24 November which suggests that he stopped lecturing around then. He seems to have made one final attempt to lecture in November 1761 but again there were only a few advertisements which confirms that his course was not successful. See pp. 111–13.

25 *Daily Advertiser* Saturday 8 February 1755 No. 7492 p. 2.

26 *Daily Advertiser* Friday 1 March 1754 No. 7215 p. 2.

27 *Daily Advertiser* Wednesday 5 April 1758 No. 8497 p. 2.

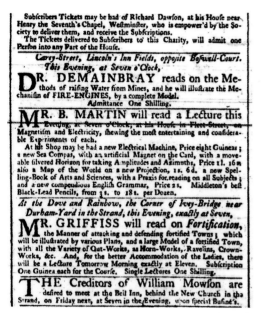

Advertisements for competing lectures. Daily Advertiser, *Wednesday 5 April 1758.*

one out. During one week in November 1755 he lectured at the Bear Tavern in Westminster on Monday, the Crown and Magpie in Whitechapel on Tuesday, and North's Coffee House on Friday, as well as giving several lectures at his own house.

The variety of the apparatus used in the lectures, particularly novel or unusual items, was often stressed in advertisements to catch the eye of the audience. Ferguson mentioned that he had a new orrery, recently constructed.[32] Demainbray had a model of the water-mill at the Bazacle in Toulouse, '. . .the only model to be brought to England'[33] and Griffiss countered 10 days later with a 'Model of a curious Mill, not in the Possession of any Person but himself'.[34] King regularly advertised his apparatus as being better than 'any other apparatus in Europe'[35] and Griffiss explained that he had '. . . been above thirty Years travelling thro' most known Parts, and collecting Machines in the several Branches of Experimental Philosophy, Chymistry, Fortification, &c'.[36]

Of all the lecturers, Ferguson drew most attention to his teaching skills. To attract individuals who knew little about the subject, Ferguson claimed that he made astronomy, the 'sublimest of Sciences familiar and easy to those who will only look and be attentive'.[37] He encouraged his potential audience by explaining, 'No hard Terms of Art will be used, nor any Thing but what generally happens in common Discourse; by which Means the whole Subject will be intelligible'.[38]

The subject matter of the courses offered by each lecturer was similar but was not standardized.[39] Astronomy and Globes were common topics, and indeed Neale and Richardson lectured only on these. Other subjects regularly offered by lecturers were Optics, Hydrostatics, Pneumatics, and Laws of Motion. King

house at Duke's Court near the Mews, and later his rival, Demainbray, close at hand in Panton Street, advertised his own premises in the same way.[28] Newberry had his 'Academy'[29] and though both Martin and Ferguson lectured in their own houses, they chose not to dignify them with a grand title. Other venues used by Ferguson and King were coffee houses, the setting for many courses earlier that century.[30] Griffiss catered for popular tastes, lecturing in a tavern, the Dove and Rainbow near Durham Yard in the Strand. During cold weather Demainbray and the others took pains to point out that 'The room is made very warm'.[31]

While all the lecturers advertised to attract an audience, King was more active than others in seeking

28 Desaguliers had advertised his premises as an Experimental Room. Perhaps it was unfinished!

29 *Daily Advertiser* Friday 8 November 1754 No. 7413 p. 2.

30 See Stewart (1986a).

31 *Daily Advertiser* Wednesday 19 November 1755 No. 7753 p. 2.

32 *Daily Advertiser* Thursday 2 November 1758 No. 8678 p. 2. Ferguson had sold an earlier one by subscription. See *Daily Advertiser* Friday 8 March 1754 No. 7221 p. 2. Another novel piece of apparatus was the Glass Sphere invented by Dr Long, Cambridge. See *Daily Advertiser* Monday 24 March 1755 No. 7547 p. 2.

33 *Daily Advertiser* Monday 20 March 1758 No. 8483 p. 2. This model is inventory No. 1927-1641, catalogue No. D17.

34 *Daily Advertiser* Thursday 30 March No. 8492 p. 2.

35 *Daily Advertiser* Tuesday 18 November 1755 No. 7752 p. 2.

36 *Daily Advertiser* Tuesday 8 November 1757 No. 8370 p. 2.

37 *Daily Advertiser* Saturday 22 December 1753 No. 7156 p. 2.

38 *Daily Advertiser* Saturday 15 December 1759 No. 9018 p. 2.

39 In 1755 Newberry offered a course on geography and astronomy. *Daily Advertiser* Wednesday 12 March. Ferguson had begun by lecturing just on astronomy but after he bought further apparatus from King and others he included other topics. Millburn has suggested that some of apparatus—which may have originally belonged to Desaguliers—was sold by King to Ferguson in April 1758. Millburn (1985) p. 403, Millburn and King (1988) p. 90.

Model of the mill at the Bazacle at Toulouse, catalogue number D114.

offered a short-lived course to instruct his listeners in '. . . the Elements of Euclid, and shew their Use in the various Purposes of Life'.[40] But some of the differences between courses are significant for they show how the lecturers responded to changing circumstances; for example, the tensions leading to the outbreak of the Seven Years War encouraged King to lecture on naval and military architecture, a subject, he said, that 'every Person should be acquainted with that proposes to distinguish himself in the Service of his King and Country'.[41] Similarly, responses to the Lisbon earthquake of 1756 were significant with both Demainbray and King giving many lectures in the month following the 'late melancholy Catastrophe at Lisbon'.[42]

Sometimes the founding of institutions presented the lecturers with new opportunities. For his part Demainbray anticipated the opening of the British Museum by offering a novel course late in 1756. It was on natural history so that his '. . . Auditors may view, with an intelligent Eye, the British Museum'.[43] While it may be that Demainbray was taking advantage of an increasing interest in natural history amongst the lecture-going public, it also seems he offered this course because of poor attendance at his lectures on natural philosophy— already he was suffering the effects of competition. A few months earlier, in August 1756, he alluded to these problems in a letter he wrote to John Ellis, the naturalist.

I shall shortly read a Course to the Publick of Twelve Lectures on mixt Subjects of Natural History, in which I hope to be the Echo of your Writings and of all that is good on that important Branch of Science; I apprehend that it will be the first Attempt of that nature made publick in

London, as I mean well, I am quite indifferent as to the Reception they may meet with from my Countrymen, whose Inattention to Scientific Pursuits, I am but too well acquainted with.[44]

Demainbray offered his course on natural history for two seasons, 1756–7 and 1757–8. It seems likely that his example persuaded both Griffiss and Martin to offer lectures on natural history in 1757.[45]

The Outcome

When the fury of the advertising war had abated, the outcome was clear. The audience was not large enough to 'employ' all those who lectured on natural philosophy.[46] The first casualty, King, gave up lecturing sometime in 1756 and died four years later.[47]

Around the end of 1757 James Ferguson also considered giving up lecturing to return to Scotland. He described his situation when he wrote to Alexander Irvine in Elgin the following January.

And as to Astronomy, there are at present more than double the number which might serve the place—people's taste lying but very little in that way; so that unless something unforeseen happens, I believe my wisest course will be to leave London soon—everything being so excessively dear, and the taxes so oppressive, that there is no living.[48]

However, Ferguson eventually decided to stay in London and continue to lecture, high taxes or no.[49]

Later that year, 1758, Demainbray gave up lecturing.[50] The significance of his action was not lost on one contemporary, Benjamin Franklin. In a letter written the following July, Franklin pointed to the

---

[40] *Daily Advertiser* Friday 27 September 1754 No. 7395 p. 2.
[41] *Daily Advertiser* Wednesday 5 March 1755 No. 7531 p. 2. Both Martin and Oliver also offered lectures on fortification.
[42] See *Daily Advertiser* Thursday 22 January 1756 No. 7808 p. 2 for advertisements for both King and Demainbray's lectures on earthquakes.
[43] *Daily Advertiser* Tuesday 30 November 1756 No. 8076 p. 2.
[44] Demainbray to Ellis 20 August 1756. Ellis Correspondence Linnean Society. I am grateful to the Linnean Society of London for permission to quote from this correspondence.
[45] *Daily Advertiser* Saturday 19 November 1757 No. 8380 p. 2.
[46] One way in which the press exacerbated the problems of those involved in the trade of natural philosophy was by charging more for advertising in the newspapers. The charge for one advertisement of 'reasonable length' in the *Daily Advertiser* was

2 shillings at the beginning of the 1750s and 3 shillings by the end, a result of the Stamp Act of 1757 which raised the tax on an advertisement by 1 shilling. As this cost was equivalent to the income from two, and later three, people at the lecture each evening at a time when audiences were dwindling, it is hardly surprising that the long and frequent advertisements found at the start of this period gave way to fewer and terser ones by the end.
[47] See London Magazine **29** (1760) 107, the issue for 27 January 1760 which records King's death.
[48] Quoted in Henderson (1870) p. 225 and Millburn and King (1988) p. 93.
[49] See Millburn and King (1988) p. 94.
[50] See pp. 111–13 for the details.

decline in the audience for natural philosophy as a reason for Demainbray's decision. In the letter, Franklin cautioned Ebenezer Kinnersley about the insecurity of lecturing.

> *I once thought of advising you to make Trial of Your Lectures here, and perhaps in the more early Times of Electricity it might have answer'd; but now I much doubt it, so great is the general Negligence of every thing in the Way of Science that has not novelty to recommend it. Courses of Experimental Philosophy, formerly so much in Vogue, are now disregarded; so that Mr Demainbray, who is reputed an excellent Lecturer, and has an Apparatus that cost nearly £2,000, the finest perhaps in the World, can hardly make up an audience in this great City to attend one Course in a Winter.*[51]

Demainbray and Ferguson were not the only lecturers affected by the declining audience. At the end of 1758 Griffiss advertised that he would give only '. . . one Course this Winter, being engaged elsewhere' though he may have continued to lecture early the following year in London.[52] By 1766 Griffiss had sold his apparatus to Adam Walker.[53]

On this evidence, in the late 1750s, two of the most consistent lecturers on natural philosophy in London, King and Demainbray, gave up lecturing in London and a third, Ferguson, gave serious consideration to that course of action. The dwindling audience was left to Martin (and to Ferguson once he had decided to stay).

Other Attractions

While direct competition between the lecturers on natural philosophy must have influenced the decisions by some to give up, there were other considerations as well. Ironically, newspaper advertising, which had provided the means for the earlier lecturers to attract their audiences, probably hastened the demise of some of the later lecture courses by publicizing other kinds of entertainment. For alongside advertisements in the newspapers for lectures on natural philosophy were those for rival attractions such as intelligent dogs, automata, or, in one case, '. . . marvellous pictures in wood attested by Demainbray'.[54]

It was not just the audience with only a fashionable and passing interest in natural philosophy who were attracted to these novelties. An advertisement for the 'Porcupine Man and Son' referred to articles about them in both *Philosophical Transactions* and *Gentleman's Magazine*, showing that some members of more élite groups also took an interest in this attraction.[55] However, by the 1750s there was a boundary, now better defined than earlier in the century, between the lecturers and the showmen, and sometimes this was defended vigorously. When a mermaid on exhibition was exposed as a fake by Dr James Parsons, he 'caused the show-man to be turned out of town'.[56]

But for the most part, competition for the more serious members of the audience for natural philosophy came from other quarters. Continuing the trend of the earlier century, there were lectures on other subjects, midwifery, anatomy, or chemistry. No doubt aware of this competition, King offered lectures on chemical processes '. . . all carried on in the same Order as larger Operations'[57] and later Griffiss talked in a lecture on chemistry about the 'frauds of several trades considered'.[58] These lectures by King and Griffiss were in competition with those given by Lucas[59] and M. Morris MD[60] who regularly offered courses on chemistry[61] and on pharmacy. However, the number of lecture courses on chemistry declined in the 1760s

---

[51] See Labaree (1965) p. 416.

[52] *Daily Advertiser* Tuesday 14 November 1758 No. 8688 p. 2. See the anonymous advertisements e.g. *Daily Advertiser* Friday 2 March No. 8781 p. 2 for a lecture at the Dove and Rainbow, the venue used by Griffiss for lectures on similar subjects.

[53] See Fawcett (1972) p. 591, who quotes Griffiss as advertising in Norwich in 1764 that his '..Instruments to be disposed of, or a Partner will be accepted'. See Musson and Robinson (1969) p. 105 and Hans (1951) pp. 146–8.

[54] See the advertisement by Snead, *Daily Advertiser* Friday 25 November 1757 No. 8385 p. 2. See also Altick (1978),

especially Chapter 2.

[55] *Daily Advertiser* Monday 12 December 1757 No. 8399 p. 3.

[56] Nichols (1966) Vol. 5, p. 487.

[57] *Daily Advertiser* 5 January 1751 No. 5925 p. 1.

[58] *Daily Advertiser* Monday 10 April 1758 No. 8501 p. 2.

[59] *Daily Advertiser* 14 January 1755 No. 7470 p. 2 'Physico-Chemical Lectures'.

[60] *Daily Advertiser* Monday 7 March 1757 No. 8159 p. 2 'Chemistry and Pharmacy'.

[61] See Golinski (1988).

suggesting that the lecturers on chemistry had similar problems to their competitors who lectured on natural philosophy.

Medical Lectures

Most of the other lectures rivalling those on natural philosophy were on medical subjects.[62] For example, in one week in September 1759, admittedly untypical, at a time when only Ferguson and later Martin advertised

Table II. Lecturers advertising in the *Daily Advertiser* in November and December, 1745–70.

Year	Natural Exp. Phil.	Medical	Chemistry	Other
1745	1	1	0	3
1747	3	4	1	1
1748	3	3	0	2
1750	1	1	0	1
1751	1	4	0	0
1752	1	2	1	0
1753	2	3	2	2
1754	3	7	0	1
1755	2	2	1	0
1756	3	6	1	1
1757	4	9	0	1
1758	4	5	1	0
1759	2	4	1	0
1760	4	7	2	0
1761	3	2	2	2
1762	3	6	0	2
1763	2	5	0	0
1764	2	5	0	0
1765	1	6	0	0
1770	1	6	1	2

courses, four people were lecturing on anatomy, two on midwifery, and one on human physiology, implying that there was a greater demand for lectures on these subjects than for lectures on experimental philosophy.

From Table II it is clear that throughout the period 1745–70 the numbers lecturing on medical subjects were usually greater than the numbers lecturing on natural philosophy. But, unlike the case of the natural philosophy lecturers, the numbers of medical lecturers did not decline in the 1760s. This was a crucial difference and the contrasting fortunes of the two groups of lecturers reveal something of the changes that were taking place in these years.

Some medical lecturers like William Hunter were extremely successful—though most of his income came from consultations and not from lecturing.[63] Nevertheless, Hunter's example must have encouraged others with similar provincial backgrounds to make a living by gathering an audience in the capital.[64] The natural philosophy lecturers aspired to capture the audience attending medical lectures. Demainbray made a bid, probably in vain, to attract the medical audience in 1755 when he advertised 'Students in Physick will be free to all such Lectures as may be publickly read by the Doctor during the ensuing Winter Season, on paying Two Guineas and a Half.'[65]

In the 1750s lectures on medical subjects continued to attract a greater audience for several reasons. For one thing, demand for medical training increased on the outbreak of the Seven Years War in 1756. In December that year Jenty, a lecturer on anatomy, advertised his course saying, '... especially at this critical Juncture there is Occasion for a great many surgeons to go abroad'.[66] But of greater significance for the medical lecturers were the new opportunites that came with the opening of several hospitals in London during the second quarter of the eighteenth century when both the number of medical posts and the need for teaching increased. The links between medical courses and medical careers became greater as time went on, further

---

[62] For example, one London lecturer in the 1740s offered courses in both medical topics and experimental philosophy. See the references to Spence on p. 56.

[63] Brock in Bynum and Porter (1985) p. 38 suggests that Hunter had an income of a few hundred pounds a year from lecturing, after expenses.

[64] Such as Ferguson or Martin, for example.

[65] *Daily Advertiser* Monday 1 September 1755 No. 7685 p. 2.

[66] *Daily Advertiser* 16 December 1756 No. 8090 p. 2. See also a further advertisement by Jenty and Ingram, Saturday 18 November 1758 No. 8692 p. 2 'Gentlemen continue to be speedily qualified either for the Navy or the Army, &c.' Several of the natural philosophy lecturers took the opportunity to advertise lectures on fortification.

accentuating an important difference between the medical and natural philosophy lecturers.[67]

Until the early nineteenth century the medical courses advertised in the press were an important, though informal, part of medical training in Britain.[68] To become licensed by either the Society of Apothecaries or the Company of Surgeons the candidate had to follow an apprenticeship and then take an oral examination. The medical lectures often formed part of this training.[69] Even if someone wanted to practise without being licensed, which commonly happened outside London, attendance at these courses provided credentials to impress new patients.[70] The arrangements for medical lecturing became more formal during the eighteenth century: groups of lecturers collaborated to provide courses and lecturers who had hospital appointments were able to use hospital premises for their lectures, and eventually medical schools were set up.[71] The development of medical lecturing was an integral part of the growth of the medical professions.

In marked contrast to the medical situation, lecturing in experimental philosophy was not a well-established occupation in the 1750s. Apart from a handful of exceptions, such as some posts connected with the Royal Society, there were hardly any employment opportunities in London for which a background in natural philosophy was required.[72] As if to demonstrate this lack of opportunity for lecturers on natural philosophy, the Royal Academy of Arts, founded in 1768, employed William Hunter to lecture on anatomy as one of their four professors and the Society of Artists had a professor

of chemistry, but neither organization had anyone to lecture on natural philosophy.[73]

More importantly, attendance at lectures on natural philosophy did not qualify anybody, formally or informally, for a career. Natural philosophy courses in this period did not develop into vocational training as did those set up for medicine and for pharmacy. An illustration of this comes from Emerson in his book 'The Principles of Mechanics . . .' published in 1754. There he explained that knowledge of subjects like mechanics would be useful to architects, engineers, shipwrights, millwrights, and watchmakers.[74] Though practitioners of all these trades would find a lecture course on natural philosophy relevant, such a course could not be a part of, or a substitute for, an apprenticeship. In the case of the natural philosophy lecturers there was no well-established link with particular trades or professions like the symbiosis between the medical lecturers and the medical professions.

Boundary Problems

Though the activities of the medical lecturers and the settings for their lectures became more distinct from those of the lecturers on experimental philosophy as time passed, on occasion the boundaries between the two groups were not so clear. Sometimes the demonstrations or the subjects shown in the different lecture courses overlapped. King dissected an eye and later Griffiss and Martin did the same, and though these

[67] Rudé (1971) p. 84 who mentions the following hospitals, Westminster 1720–4, Guy's 1725, St George's 1734, London 1740, Middlesex 1745. See also Bynum in Bynum and Porter (1985).

[68] Authors who deal with these issues are Gelfand in Bynum and Porter (1985) and S. C. Lawrence (1988).

[69] The Royal College of Physicians had a different arrangement—an MD was required, that is being a student at Oxford or Cambridge.

[70] See the advertisement Daily Advertiser Monday 24 December 1750 No. 6227 p. 2 by Theophilus Lobb who offered courses for those for 'Sea or Country service' and for 'Surgeons, Apothecaries who act as Physicians'.

[71] St Bartholomew's Hospital ran courses on anatomy from 1738 and on surgery from 1765. See Rudé (1971) p. 84.

[72] Examples are James Jurin (1684–1750), who lectured in Newcastle and became a Secretary of the Royal Society, Francis Hauksbee jr and J. Robertson, lecturers who were the

Librarian and Clerk at the Royal Society in succession. Another lecturer who found other employment was Benjamin Robins, a candidate for professor at the Military Academy at Woolwich, who worked for the East India Company and died at Madras in 1751. See Gunther (1984) and DNB. In discussing whether there was a group identity for those who lectured on natural philosophy it must be significant that on the rare occasions that they described their trades both Jack and Ferguson called themselves Teachers of Mathematicks. Later, when natural philosophy was found a setting in institutions towards the end of the century, new institutions had to be created. See Inkster (1977) and Hays in Inkster and Morrell (1983).

[73] See Ward (1989) p. 77 and Daily Advertiser Thursday 20 December 1770 No. 12476 p. 1 for an advertisement by Awsiter, professor of chemistry for the Society of Artists.

[74] See Emerson (1758) title page.

dissections occurred during a lecture on Opticks, they were close to what a lecturer on anatomy might do.[75] Interestingly, a few years before when Sowerby lectured on optics, he had not dissected an eye himself; it was done for him by Hamilton, a surgeon at the Middlesex Hospital.[76] Another who dissected eyes was Taylor, the son of Chevalier Taylor. Both he and his father were 'quacks' who specialized in the treatment of eyes, much to the concern of the medical establishment.[77] As natural philosophers, anatomists, and 'quacks' all felt equipped to perform the dissection, the professional boundaries were still slightly indistinct.[78]

A further demonstration which straddled the ill-defined boundary beween natural philosophy and anatomy was repeated by Newberry, who placed a rat on an air pump.[79] This had a counterpart in anatomy courses.[80] The context in which an air pump or other kinds of equipment was used was significant: no one group had established a monopoly on subject matter or apparatus.[81] Who could successfully lay claim to particular combinations of apparatus and subject matter to make up a course of lectures on anatomy or experimental philosophy was an important issue. In the long run the lecturers on medical subjects were able to consolidate their hold on the subject-matter; the commercial lecturers on natural philosophy could not.

In the 1750s the relationship between the lecturers on natural philosophy and the lecturers on medical subjects was still in a state of flux. On occasion natural philosophy lecturers tried to accommodate or exploit the competition from the medical lectures by making natural philosophy a practical and useful supplement to them. Demainbray, for instance, gave a lecture in which he discussed the wind gun, the action of the lungs, and a cock for anatomical injections.[82] But it was not just the people who lectured who were involved in setting subject boundaries. For example, Benjamin Rackstrow, a dealer in statues and other items, offered to treat rheumatic disorders with electrical machines—a case where the apparatus of natural philosophy was used in the treatment of patients.[83] The drawing and redrawing of the boundaries about who had a 'professional' interest in these matters in the mid-eighteenth century was a lengthy business.

New Institutions

The success of medical lecturers was not the only factor which made life difficult for the lecturers on natural philosophy in the 1750s. Their fate was also bound up with wider developments, such as the changing patterns

[75] King, e.g. *Daily Advertiser* Saturday 12 October 1754 No. 7390 p. 2, Griffiss 8 December 1757 No. 8396 p. 2, and Martin Friday 5 December 1760 No. 9333 p. 2. Interestingly, King also had another lecture in which an 'Air Pump and Condensing Engine will be dissected' *Daily Advertiser* Saturday 9 December 1751. This suggests that he thought this was the correct procedure to adopt. It seems likely that a real eye was used in these lectures rather than a model though the available evidence is not conclusive. Adams (1794) Vol. 2, p. 292 mentions dissecting an ox eye. See also King (1750) and Algarotti (1769) pp. 2–3.

[76] *Daily Advertiser* Wednesday 23 November 1748 No. 5575 p. 2.

[77] *Daily Advertiser* Monday 5 December 1748 p. 2 No. 5589 and Saturday 12 October 1754 No. 7390 p. 2.

[78] Eye problems were a source of income for those who offered treatment as well as those who offered a description of the workings of the eye. In addition to the surgeons and 'quacks' who offered surgical solutions there were also those who offered chemical treatment. See the *Daily Courant* Monday 13 November 1727 No. 8141 p. 2 where there is an advertisement for a 'chymical liquor for incipient cataracts, soreness of the eye and which strengthens the Optick Nerves'.

[79] *Daily Advertiser* Wednesday 11 December 1754 No. 7441 p. 2.

[80] Alexander Monro *primus*, who had attended a course on experimental philosophy given by Whiston and Hauksbee in 1717, later offered experiments on lungs using an air pump in his course of anatomy in Edinburgh. See Erlam (1954) and C. J. Lawrence (1988). The institution of Edinburgh University with its Medical School and Professorship of Natural Philosophy are an interesting contrast with the London situation.

[81] An interesting example connected the dissection of bodies is that after the Barbers and Surgeons had gone their separate ways, the Company of Barbers Theatre for Anatomical Lectures had to be pulled down. See *Daily Advertiser* Monday 23 September 1754 No. 7391 p. 2.

[82] *Daily Advertiser* Thursday 17 March 1757 No. 8168 p. 1. Demainbray's Wind Gun is inventory No. 1927-1472, catalogue No. D45.

[83] *Daily Advertiser* Monday 7 March 1757 No. 8159 p. 2. Earlier, Rackstrow had disturbed other boundaries by offering beatification by electrical means. He had also shown a 'Figure of Anatomy by Mr Abraham Chovet, surgeon' *Daily Advertiser* Tuesday 21 April 1747 No. 5079 p. 2. See p. 57 for another mention of this model.

of leisure and the growth of new institutions.[84] One change in leisure apparent from newspaper advertisements was the proliferation of magazines during the 1750s. These magazines took up both the time and the interest of readers which otherwise might have been devoted to lectures.[85] However, the setting up of institutions such as the Society for the Encouragement of Arts, Manufactures, and Commerce (today the Royal Society of Arts) in 1754 and the British Museum (open to the public in 1759) was even more significant. As already mentioned, Demainbray commended his new course on natural history so that his 'auditors' may view the British Museum with 'an intelligent Eye',[86] and in 1755 King responded to interest in manufactures with a course of 'Mr King's Experiments, which shew the Use of Natural Philosophy in several Trades . . .'.[87] By stimulating public interest in practical matters the Society of Arts helped to create the situation which led to a growth of the numbers lecturing on natural philosophy in the mid-1750s. However, by 1760, the Society, rather than the lecturers, had succeeded in capturing the new audience; the Society was then a rival to the lecturers and contributed to their decline.

The actions of Demainbray and King exemplify the attitudes of some of the lecturers to the Society of Arts. They joined it in the autumn of 1755, which made them among the first hundred members or so, but let their membership lapse the following year.[88] Perhaps they recognized that while the Society provided a reservoir of subjects and members for lecturers to draw on, it was also winning the competition for that audience.

The interests and activities of the people involved in the Society of Arts in its early stages were similar to those of the lecturers. Stephen Hales and Henry Baker are two prominent examples.[89] Another is Dr James Parsons, who illustrates the transition from lectures involving a small audience to a society catering for a larger public.[90] In the early 1750s he was '. . . giving weekly an elegant dinner to a large but select party' of his friends, who included Hales and Baker, but as well as that,

Weekly meetings were formed, where the earliest intelligence was received and communicated of any discovery both here and abroad; and new trials were made to bring to the test of experience the reality or usefulness of these discoveries.[91]

In both subject matter and style these gatherings were similar to ones held by the lecturers. But the context in which Parsons and his circle pursued their interests altered around 1754: from a small informal group of individual enthusiasts they became members of a society with much greater influence.[92] Indeed, the success of the Society of Arts in its early years was dramatic: the membership grew from 17 members in 1754 to 708 by

---

[84] One facet of the changing audience for natural philosophy was that children began to be seen as a potential audience for the lecture courses. This was reflected in new literature written especially for children such as the first volume of the lectures of Tom Telescope published by John Newbery in 1761. See Secord (1985). Martin's *General Magazine* had a section aimed at Young Gentlemen and Ladies, a slightly older age group. See Millburn (1976) pp. 71–5.

[85] In 1760 the issues of the *Daily Advertiser* regularly carried notices for more than a dozen different magazines—*The Gentleman's Magazine, London Magazine, Universal Magazine, General Magazine, British Magazine, Imperial Magazine, Royal Female Magazine, Lady's Museum, Royal Magazine, Christians Magazine, Musical Magazine,* and *Monthly Melody.* There was even a *Magazine of Magazines.* One lecturer, Martin, had himself been involved in publishing magazines in this period. In 1755 his 'The General Magazine of Arts and Sciences, Philosophical, Philological, Mathematical, and Mechanical. . .' appeared which he continued until early 1764. See Millburn (1976) pp. 69–83.

[86] Demainbray was drawing an interesting distinction between the senses of hearing used for lectures and sight used in the

museum. One irony of history is that while Demainbray's attempt to capitalize on the collections of the British Museum was unsuccessful, his own apparatus has finished up in the Science Museum.

[87] *Daily Advertiser* Saturday 27 September 1755 No. 7708 p. 2.

[88] Ferguson had his own brush with the Society of Arts. See Millburn and King (1988) p. 93.

[89] See Wood (1913) and Allan (1979). In the 1750s Hales was still designing ventilators for ships and prisons, a subject dear to the heart of Desaguliers. Baker wrote on microscopes and Short was a noted maker of telescopes. See Turner (1967b, 1974a). They were also all Fellows of the Royal Society.

[90] Hales, Baker, and Parsons were all Fellows of the Royal Society. In order to see how generally the points raised here can be applied to London society it will be essential to examine the activities of members of both the Royal Society and the Society of Antiquaries. I am indebted to Dr Allan of the Royal Society of Arts who brought Parsons to my attention.

[91] See Nichols (1966) Vol. V, p. 479 quoting Maty. See also pp. 477–81.

[92] Parsons himself was the Chairman of the Agriculture Committee of the Society of Arts.

1758 when the meetings overflowed.[93] With its member-ship the Society had at least one major advantage over the lecturers—an established audience interested in the results of its labours. One of the active early members of the Society, Robert Dossie, later emphasized this point.

But vain are all these labours to cultivate experimental research, or collect information, if a proper channel be wanting to transmit the produce of them to the public;. . .[94]

The Society was taking over one of the roles that lecturers such as Demainbray had played, as transmit-ters of knowledge about machines and other devices and of improvements to them.

The Society of Arts was able to capitalize on the growing interest in practical matters in the mid-1750s in a way that the lecturers on natural philosophy could not.[95] The meetings of the Society provided oppor-tunities for a member to be actively involved in its work and these opportunities may have appealed more than passive attendance at lectures. Perhaps individuals who had previously been part of Demainbray's audience began to enrol with the Society. After all, one year's subscription to the Society was two guineas, a little less than the two and a half guineas required for a season ticket for his lectures.

However, there was much more at stake than the success of a society or the income and prospects of a handful of lecturers. As Brewer has emphasized, the clubs and societies that were such a feature of eighteenth century England had a range of purposes; while their main activities were usually to hold convivial gather-ings, they often encouraged business or the spread of political ideas at the same time.[96] The Society of Arts fulfilled similar functions for its membership, though in its case the members were exposed to ideas about the

development of technology and to views of progress. While attracting an audience who might otherwise have attended lectures on natural philosophy, the Society of Arts was doing more than providing an alternative form of 'rational' amusement. The members of the Society dealt with a wider set of topics, namely the arts, commerce, and manufactures of its title, and approached these in new ways, for example, by awarding premiums. Through these activities the Society helped to define these subjects and to publicize novel ways of pursuing them.[97]

Brewer has pointed out that many of the societies active in promoting bourgeois social and economic aspirations later turned to radical politics during the upheavals connected with John Wilkes and his attempts to take his seat as a Member of Parliament in the later 1760s.[98] The Society of Arts with its membership amongst the aristocracy, however, does not fit this pattern. The views of its members on promoting technological change seem to have been connected to an aversion to radical politics.[99]

A case-study of these tensions is provided by the first sawmill in London built in the 1760s. Such mills had been used on the Continent for many years and possibly Demainbray described one in the lectures he gave in the 1750s. In the following decade the Society of Arts awarded several premiums to one of its members, Charles Dingley, who had a sawmill erected at Lime-house. Dingley, as well as sharing with Demainbray an interest in sawmills, also had similar political views for he, too, was a supporter of the Court and an anti-Radical. In fact, Dingley stood against Wilkes in one election. (Demainbray was much embarrassed by his brother-in-law, John Horne Tooke, who was for a while a staunch ally of Wilkes.)[100] However, these views on

---

93   Rousseau (1982) p. 31 quoting Smollett.
94   Dossie (1768) Vol. 1, p. xi. See Gibbs (1952*a*) for Dossie.
95   In 1752 some physicians set up a medical society for collecting and publishing information which they thought useful. See Corner and Booth (1971) p. 281, fn. 2. Artists had a different experience, for the example of the Society of Arts seems to have galvanized the artists into setting up their own organizations, eventually including the Royal Academy. See Ward (1989) p. 76 and Paulson (1979) p. 37.
96   See Brewer in McKendrick *et al.* (1982) pp. 197–262.
97   Through these processes the Society helped to reinforce and promulgate some of the distinctions between science and technology that later became so marked. In this context it is interesting that the word 'Sciences' was quickly dropped from

the title of the Society and that agriculture became one of the main interests, presumably a consequence of the number of peers who became members. See Shapin (1974) for the related example of the peers and science in Edinburgh.
98   Brewer in McKendrick *et al.* (1982) pp. 201, 233.
99   If the activities of the Society of Arts did undermine the work of the lecturers and their programme for natural philosophy and if the members of the Society were opposed to radical ideas, these factors may have contributed to the connection between natural philosophy and radical ideas in a later period to which Schaffer (1983) has drawn attention.
100   See p. 106. See DNB for John Horne Tooke and also Rudé (1983).

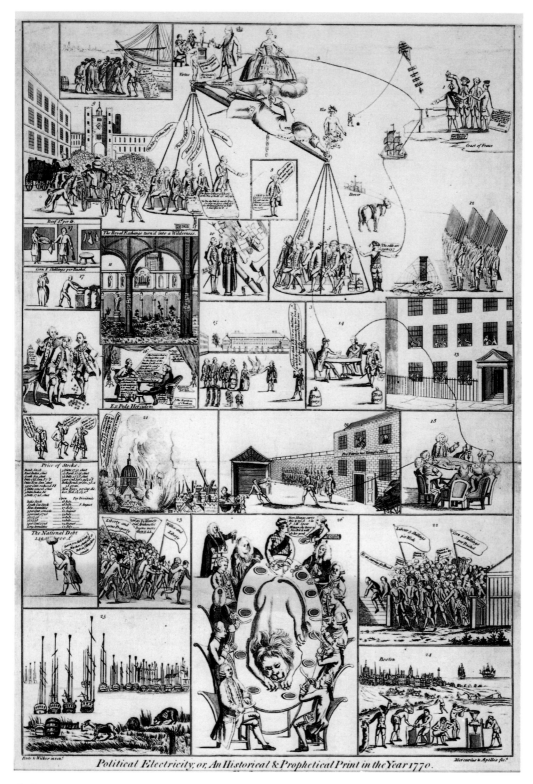

Political Electricity. This satirical print portrays Lord Bute (top right-hand corner) as an electrical machine linked to John Wilkes, looking out of prison (No. 19), 1770.

both politics and technology were not shared by the crowd who damaged Dingley's sawmill in 1768.[101]

While the Society was interested in ways of improving the nation through arts, commerce, and manufactures, these policies did not include the kind of changes wanted by the supporters of Wilkes, such as an end to aristocratic patronage and the place-seeking that went with it.

While the Society of Arts was the most significant of the organizations in the early 1750s with interests in encouraging new developments in the arts and manufactures, there were others with related aims, such as the Honourary Purple Society for promoting the Liberal Sciences[102] and the Antigallican Society. The Antigallican Society, for example, anticipated the system of premiums employed by the Society of Arts, awarding them to encourage English production of goods such as lace to replace French imports. But in other respects the Society of Arts did not emulate the Antigallicans. Later on the Antigallicans furthered their aims by chartering a privateer to provide more forceful competition for the French.[103] Perhaps it was this desire to take matters into their own hands that made the Antigallicans one group to support Wilkes, contributing £3000 to funds for an election campaign.[104]

Old Institutions

The new institutions that had sprung up to challenge the lecturers for their audience were not the only institutional threat. In contrast to the Society of Arts, the Royal Society was in a reduced state in the 1750s. It continued to change, though some of the changes illustrate the declining status of the lecturers. The post of curator of experiments lapsed when Desaguliers

vacated it in 1743 though it was not formally abolished until new statutes were published in 1776.[105] One reason given for abolishing the post was that the skills required for the work were now more common than they had been:

> . . . the experimental part of Philosophy having begun, even in the early years of the Society, to be cultivated, several gentlemen procured apparatuses to satisfy themselves in private; so that many years had not passed, before a considerable number of the Fellows were so well acquainted with the mode of making experiments, that such accomplished Curators have not been found necessary; and when experimental philosophy arrived at such high popular regard as to be frequently explained in public lectures, that kind of knowledge became more generally diffused. . .[106]

In this instance the lecturers had become victims of their own success.[107]

New Possibilities

It was not just that the early lecturers had done the work of teaching their audiences too well. They were being supplanted in another of their roles, namely the collecting and transmitting of scientific information. Increasingly, the Fellows of the Royal Society had their own collections of instruments because the instrument-making trades could supply demonstration equipment off the shelf or more expensive items to order if required. There were books to accompany these items which showed how to carry out the demonstrations and the results to be expected. Around 1750 Adams was advertising sets of demonstration equipment for sale and in 1755 he advertised new globes with a book to go with them—another example of 'do it yourself' lecturing with an instrument-maker providing the where-

[101] See Rudé (1971) p. 194. A further insight into the social bias of the view of technology espoused by the Society of Arts comes from a letter to Henry Baker about a hand-mill for grinding corn, a device for which the Society offered a premium: 'I know a Gentleman in this County who procured a Hand Mill made after the Best Manner, wch preformed (sic) tollerably well, he got it set up, but his Servants told him, if he expected they should grind with it, they would leave him to a Man, they were not hired to be slaves.' Arderon to Baker 29 December 1758. Baker Correspondence Vol. VII, f76.

[102] *Daily Advertiser* Wednesday 23 November 1748 No. 5575 p. 2.

[103] See Allan (1989). In fact both the Antigallicans and the Society of Arts had copied the system of premiums from the Dublin Society.

[104] Brewer in McKendrick *et al.* (1982) p. 234

[105] Miller (1989) and Weld (1848) Vol. 2, pp. 85–8.

[106] Weld (1848) pp. 87–8.

[107] A different example is provided by the members of the Spitalfields Mathematical Society who in 1746 resolved to buy their own electrical apparatus and towards the end of the century had courses of lectures. See Cassels (1979) pp. 242, 245.

withal.[108] Henry Baker, one of the early members of the Society of Arts, wrote on microscopes and recommended that his correspondents buy them from Cuff. But George Adams soon published his own work on microscopes, much to Baker's annoyance. While some of these activities might encourage individuals to attend lectures, it could be an alternative to what the lecturers offered, and furthermore an alternative that allowed others to reap the economic benefits.

Faced in the late 1750s with the situation of being squeezed by the activities of the Society of Arts and the instrument-makers, lecturers on experimental philosophy found it difficult to make a living just from lecturing and they themselves followed some of these alternatives to lecturing to supplement their income. For example, Martin and Ferguson published extensively. King proposed publishing his lectures by subscription but nothing seems to have come of it[109] and neither he, Demainbray, nor Griffiss published more than syllabuses. Their contemporary William Hunter, who lectured on anatomy, was wary of publishing because it would 'cheapen the goods' and make people less inclined to pay to attend his lectures.[110] While this may have been the attitude of some lecturers on natural philosophy, it was clearly not shared by Ferguson and Martin. Their enthusiastic attitude towards publishing perhaps reflects both their aptitudes and the sparse rewards from lecturing.

In fact, the two who continued to lecture in Westminster on natural philosophy after 1759, Martin and Ferguson, always had other ways of earning a living and so were never solely dependent on their income

from lecturing. Neither Demainbray nor King, two who gave up, seem to have had occupations concurrent with their lecturing.[111] When Demainbray stopped lecturing he eventually found the patronage he sought, a reminder of the important role that patronage played in eighteenth century life.[112] He became an official of the Excise, an employment he found much more lucrative than giving lectures on natural philosophy. Apparently he abandoned physics and would not even show his instruments to Franklin, though later he became superintendent of the observatory built for King George III.

Other Activities

While Martin lived in London his main business was retailing instruments and his numerous published works were sold alongside instruments at his shop in Fleet Street. His publications, instruments, and lectures were advertised together in the newspapers in an economical arrangement.[113] While Martin was the only one to combine the roles of lecturer and instrument retailer on a large scale, the other lecturers did have connections, probably financial, with other instrument-makers. Jack's design of a sea quadrant was made and sold by George Adams.[114] Demainbray lectured on a pump invented by Jeremiah Sisson which he 'puffed' in the *Daily Advertiser*,[115] and Newberry had subscriptions taken by Watkins.[116] The lectures acted as a form of advertising for the instrument-makers.[117]

108 *Daily Advertiser* Wednesday 29 October 1755 No. 7735 p. 3. See pp. 22–6 for more details of Adams's book on Globes. See Turner (1974a) for a discussion of the work of Adams and Baker on the microscope.
109 *Daily Advertiser* Tuesday 2 January 1750 No. 5922 p. 2.
110 See Porter in Bynum and Porter (1985) p. 25 who mentions Joseph Black as another example.
111 King's wife ran a lace shop which may or may not have a bearing on this argument.
112 Ferguson was another who enjoyed patronage. He received a pension of £50 a year in 1761 from King George III. See Millburn and King (1988) p. 140.
113 An air pump 'fit to adorn a Museum' was to be sold by subscription in 1758. See *Daily Advertiser* Friday 24 March 1758 No. 8487 p. 2.
114 *Daily Advertiser* Monday 23 December 1751 No. 6538 p. 3. George Adams, for his part, borrowed from others in his

published works but did not himself lecture. Perhaps he kept to instrument-making because he was busy with his work for the Board of Ordnance. See Brown (1979) and Millburn (1988).
115 *Daily Advertiser* Wednesday 8 December 1756 No. 8083 and *Gentleman's Magazine* 28 (1758) 80. This pump and another by Sisson are inventory Nos 1927-1251 and -1252, catalogue Nos. D37, D38.
116 See *Daily Advertiser* 13 November 1754 No. 7417 p. 2.
117 One important aspect of London in the mid-eighteenth century was the development of occupational groups. It is significant that the 'Optical and Mathematical Instrument-Makers' had meetings during this period *Daily Advertiser* Saturday 30 October 1756 No. 8050 p. 2, Saturday 8 January No. [8110] p. 2. Perhaps this indicates that they now had common interests that fell outside the scope of the different city livery companies of which the instrument-makers were members. See Brown (1979) and Brewer (1982).

Benjamin Martin's trade card, c. 1760.

Instrument-making

The interface between lecturing and instrument-making was still important though it had shifted since the early part of the century. The lectures were no longer a collaboration between maker and a lecturer as was the case earlier. By mid-century when the apparatus had to be manipulated it would be done by the lecturer or a servant who may have had the skill of an instrument-maker, but the services of someone on a par with the lecturer were no longer required. The tacit knowledge that the instrument-maker possessed about the operation of the equipment was now more widely known. A similar point was made about the instruments required by the Royal Society.

. . .at the institution of the Society, the fabrication of instruments for experiments was not commonly known to workmen; and therefore some Statutes were made concerning Operators. . .but the Experimental Philosophers having long since taught and familiarized such general knowledge, in working on things formerly unknown, particular Operators to the Society have been for many years found useless; as the different artificers employed by them can now readily furnish whatsoever is wanted.[118]

Indeed, those who lectured in mid-century had a variety of relationships with makers. Though Demainbray had a hotchpotch of instruments collected on his travels from different makers, he dispensed a form of patronage by recommending instruments made by particular makers to his audience or correspondents.

[118] Weld (1848) p. 88.

Martin used the instruments sold in his shop, though they may have been made by others. Ferguson had items he made himself, 'Although his apparatus be simple, an[d] no Ways showy..' and he designed various astronomical devices which he sold.[119] For a brief period, from 1755 to mid-1757, Ferguson retailed globes having bought the plates to make them from John Senex's widow, Mary. In turn, Martin bought these plates from Ferguson during 1757.[120] However, Ferguson's main occupation during these years, as he regularly reminded the readers of the *Daily Advertiser*, was that he made portraits in ink for 15 shillings (frame included) though this work, too, had its snags; one advertisement in November 1753 stated 'He draws no Pictures of Children under seven Years of Age', presumably because he had come off the worse in a wrangle with a child of six.[121]

The career of John Cuff offers some insights into the trade in instruments and the relationships between the lecturers, makers, and writers. In the early 1750s Cuff, a well-known maker of microscopes, was bankrupt though he managed to continue to trade. Some of his difficulties stemmed from the fact that he made expensive bespoke instruments and he lost business to cheaper retailers even though his instruments were recommended by Henry Baker, who wrote on microscopy. Cuff's problems became particularly acute in the mid-1750s when Martin set up shop next door to him in Fleet Street. Soon Cuff moved shops and Baker received a letter from his correspondent Arderon about Cuff's situation.[122]

I know Martin very well, and believe he never can do Mr Cuff any damage with Men of Judgment and Tast[e], as his Instruments bear no Comparison either in neatness or Goodness, yet he certainly hath done well to remove out of his Neighbourhood, for there are many, if they can but get things cheap, they pay no regard to their correctness or Beauty.[123]

However, those with 'judgment and taste' were not numerous enough to sustain Cuff and eventually he gave up his shop in London. Baker continued to try to help him and in 1761 Arderon reported on Cuff's prospects were he to set up shop in Norwich where Arderon lived. Arderon's opinion of such a plan was not encouraging.

I fear Mr Cuff would find but poor Success at Norwich, as there are now Several persons who sell Optical Instruments of an inferior Sort and of late, Riders are imployed (sic) as in many Other Trades, who comes (sic) to seek out Customers & get orders.[124]

In any case discerning customers were already served by Cuff whom Arderon had recommended to them. Clearly, the distribution networks for instruments had improved greatly by that time, one result of the growth of turnpikes.[125] Another indication of improvements to the means of distributing these goods was that around 1760 Martin's advertisements began to mention that he sold instruments wholesale.[126] While this may have been a response to a declining audience at his lectures and falling sales of instruments to retail customers, it also reflects the extension of the market for instruments throughout Britain so that a supplier like Martin now

---

[119] *Daily Advertiser* Monday 3 December 1759 No. 900[7] p. 2.

[120] Millburn and King (1988) p. 90 suggest that there was a complicated arrangement between Ferguson and Martin—that in return for the plates for the globes, Ferguson received instruction from Martin on optics and optical instruments.

[121] *Daily Advertiser* Saturday 10 November 1753 No. 7120 p. 3.

[122] See Millburn (1976) pp. 107–8. Henry Baker wrote '..poor Cuff has still the Pride to suffer Nothing to go from him, but what is exececuted in the best Manner; and this has been his ruin; for would he have made bad Instruments at low Prices, as his Neighbours have done, he would have sold to the Multitude, and might still have flourished.' Baker Correspon-dence Vol. VII, f220 draft, May? 1760. Cuff's other neighbour at the time that Martin moved in was Benjamin Rackstrow, who dabbled in electrical displays and also sold statuary. See p. 79.

[123] Arderon to Baker, Norwich 25 July 1757 Baker Correspon-dence Vol. VI, f344.

[124] Arderon to Baker, Norwich 11 July 1761 Baker Correspon-dence Vol. VII, f310.

[125] See Albert (1972) for a discussion of turnpike mania in this period.

[126] See for example Martin in *Daily Advertiser* Saturday 8 November 1760 No. 9320.

sold instruments to retailers in other parts of the country or abroad without great difficulty.[127]

As the market for instruments grew, the range of items sold became wider.[128] Some items were large and expensive, and rich individuals or institutions would have been the purchasers. This was the case with large orreries and various astronomical devices.[129] Bernouilli, who travelled round European observatories in 1770, provides details of the many expensive instruments that had been installed, many from the workshops of the London makers, demonstrating how they dominated the market for these items. However, in the longer term, the fact that institutions could assemble collections of apparatus beyond the means of an individual lecturer was another factor in settling the fate of the lecturers.[130]

Conclusion

The evidence presented in this chapter shows a rise followed by a fall in the number of people lecturing on natural philosophy in London between 1750 and 1765. For those who lectured it was a case of having to adjust to changing circumstances, and the activities of Demainbray, Ferguson, King, Martin, and the others who lectured can be seen as attempts to take advantage of whatever opportunities presented themselves. Perhaps in this respect they are like William Hunter whom Porter describes as having an entrepreneurial attitude to his circumstances.[131] But, unlike Hunter, Demainbray and the others were not financially successful. However, their experience may only underline the significance of entrepreneurial attitudes for the lec-

turers, for if there are successful entrepreneurs, there also have to be ones who fail.

Among the factors which influenced the rise and fall in the number of lecturers, two have been singled out for comment: the continuing attraction of medical lectures in the period and the founding of the Society of Arts in 1754. Medical lecturing was not subject to the same rise and fall; this emphasizes that the market for medical lectures was significantly different from that for natural philosophy lectures. The Society of Arts provided another kind of competition. By offering a different approach to technical subjects, the Society seems to have increased the size of the public interested in these matters but at the same time managed to keep this growing audience to itself.

The market for natural philosophy was changing. Though the 'useful arts' were no longer the property just of their practitioners (instrument-maker, engine builder, or civil engineer), the attempt by enterprising individuals to make a living in London by lecturing on them had succeeded only for a time. Other arrangements—the printed text, the British Museum, or the Society of Arts—were among several rivals that laid claim to the audience for lectures on natural philosophy. If the 'Inattention' of his countrymen Demainbray mentioned in a letter in 1756 had made his career as a lecturer in London a short one, then the attention of that audience was given to some new institutions. The individual entrepreneur with his own instruments, renting or using his own premises and reaching his audience by advertising in the newpapers, was displaced by institutions with extensive facilities such as their own premises and collections of material together with a membership who were more or less actively promoting the work of these bodies.

---

[130] Harvard was one example of an institution buying large numbers of instruments. Martin and other makers had large orders from Harvard College to replace apparatus lost in a fire in the 1760s. See Millburn (1986) and Wheatland (1968). The Scottish universities are interesting in this respect because they moved from a situation where the professor owned the apparatus and collected the class fees to one where the institution owned the apparatus. The Spitalfields Mathematical Society resolved to buy electrical apparatus in 1746. See Cassels (1979) p. 242.

[131] See Porter pp. 31–2 in Bynum and Porter (1985).

[127] Henry Baker passed on orders to people like Martin from his correspondents in Italy and Poland etc. There may also have been a falling off of sales of instruments to customers in London at the time which would encourage Martin to seek other outlets for his instruments. Cuff, for example, mentioned the 'present Decay of Business' in an advertisement in the *Daily Advertiser* Monday 21 December 1767 No. 11537 p. 2 when he sought election as a Ward beadle.

[128] See Bryden (1972a) for a discussion of the different markets for instruments.

[129] Perhaps the origin of 'Big Science'!

IV

THE CAREER OF S. C. T. DEMAINBRAY (1710-82)

I could also wish, for the honour of the nation, that there was a complete apparatus for a course of mathematics, mechanics and experimental philosophy; and a good salary upon an able professor, who should give regular lectures on these subjects.[1]

THE ABOVE SUGGESTION came from Matthew Bramble, one of the main characters in the novel *The Expedition of Humphry Clinker* by Tobias Smollett, published in 1771 shortly before the author's death. The remark was made after Bramble's visit to the British Museum and expressed his opinion that the exhibitions could be usefully supplemented by a professor or lecturer giving demonstrations using special apparatus. This view illustrates the contemporary interest in how to present scientific subjects to an audience. By coincidence, two years before the publication of Smollett's novel and a few miles from the British Museum, a lecturer had been installed in a new building complete with a large collection of apparatus. This was Dr S. C. T. Demainbray who occupied the observatory built in 1769 at Richmond for King George III. Whether Smollett was aware of Demainbray's situation when he was writing his novel is not known to us, but even if he was not, Demainbray's life and work still provides an example to contrast with the views of contemporaries like Smollett who welcomed the idea of this kind of lecturing. The career of Stephen Demainbray illumi-

nates aspects of the development of natural philosophy in the mid-eighteenth century, in particular the opportunities, or sometimes the lack of them, for such development in London society of the time. In particular it demonstrates the power of patronage in his social world. The career of Demainbray is also important to us because, as described earlier in this work, it is his instruments which form part of the King George III Collection.

The Early Career of S. C. T. Demainbray

Stephen Demainbray was born in St Martin's, London, in 1710, to Huguenot parents who had come to Britain with King William III.[2] His parents died when he was a small child and Demainbray was brought up by his uncle, an officer in the English Service and page of honour to Queen Mary.[3] In these years Demainbray lodged with another Huguenot family. As this was the family of J. T. Desaguliers, who was to become the most

---

[1] Smollett (1771) Matthew Bramble to Dr Lewis pp. 133–4.
[2] The only extensive account of Demainbray's life is Rigaud (1882) from which most of the information about his early years

has been taken. The article about Demainbray in DNB covers much the same ground.
[3] Lysons (1811) Vol. 2, Part 2, pp. 589–90.

influential lecturer in natural philosophy in England, Demainbray must have been familiar with the business of lecturing in his youth.[4] He was educated both by Desaguliers, who instructed him in mathematics and natural philosophy, and at Westminster School. In 1727, aged 17, Demainbray left London for the Continent, perhaps as the consequence of an unwise marriage. He may have gone to Leiden as the biographical article by his great-grandson suggests, though there is little evidence to support that assertion. Demainbray himself mentioned in a later advertisement that he had stayed in Paris.[5] However, around 1731 he became a Freemason, a member of a French Lodge in London, which suggests that for at least some of this period he was in London.[6] But apart from these clues nothing else is known about his activities in these years.

Many years later, in the early 1740s, Demainbray appeared in Edinburgh running a boarding school where young ladies were taught French and English. He advertised in the *Caledonian Mercury* in July 1744:

Mr Demainbray, French and English Teacher, just arrived from London, takes this Method of acquainting all Lovers and Encouragers of Learning, That he teacheth French and English in their utmost Purity and Perfection, in an easy and expeditious Manner, as several Gentlemen and Ladies will testify, if required, having taught with success upwards of 17 Years, both in London and Paris.— The Methods, whereby he teaches, being so very plain and easy, that Persons of the meanest Capacities, tho' unacquainted with the first Rudiments of grammar, need not doubt of speedy Improvement and Success.[7]

Demainbray's teaching career was temporarily interrupted when he fought in the battle of Prestonpans as a Government volunteer.[8] But this opportunity of showing his loyalty to the House of Hanover would stand him in good stead in later life.

Some months later, back in Edinburgh, Demainbray carried out experiments on the effect of electricity on the growth of plants. He published accounts of these early in 1747 in letters sent to the *Caledonian Mercury* and the *Gentleman's Magazine*. Demainbray wrote:

On the 20th of December last, I had a MYRTLE from Mr Boutcher's green-house, which since that time I have electrified seventeen times, and allowed the shrub half an English pint of water each fourth day, which you'll please to observe was kept in the room most frequented of my house, and consequently the most exposed to the injuries of the air, by the doors and windows being oftenest opened.

This myrtle hath since, by electrifation (sic), produced several shoots, the longest measuring full three inches; whereas numbers of the same kind and vigour left in the said green-house have not shewn the least degree of increase since that time.

Having now undertaken a further and more satisfactory experiment of the same nature, I am in hopes of communicating soon to the publick some proofs still more evident of the present hint, which I must leave to be improved by men of more extensive knowledge, and of talents superior to, & c.[9]

The editor of the *Gentleman's Magazine* commented that Demainbray's account was 'deficient' because he had not allowed the plants equal amounts of water. A few weeks later Demainbray took up this point when he wrote giving the results of his further and more satisfactory experiment.

On the seventeenth of January last, Mr Boutcher favoured me with two myrtles of the greatest equality of growth, vigour, &c. he could chuse; these I placed in the same room, and allowed them each an equal quantity of water. On electrifying one of them, it hath produced several shoots full three inches.

---

4 The *Newcastle Courant* for 11/18 March 1749 mentioned that Demainbray's '. . .Knowledge and extensive Apparatus are allowed to be little inferior, if not equal to those of his Master the late Doctor Desaguliers. . .'

5 Rigaud (1882) p. 279. There is no reference to Demainbray in the published Leiden records. I should like to thank Peter de Clerq for this information. DNB says that there is no record of Demainbray at Leiden.

6 See Williams (1928) p. 38.

7 *Caledonian Mercury* 12 July 1744 No. 3716. See also other advertisements No. 3737 30 August 1744. See also Law (1966).

8 *Caledonian Mercury* 9 October 1745 No. 3899 advertises the start of their classes on 14 October. The issue for 7 March 1746 No. 3963 announces the publication of *An attempt towards rendring the different branches of a lady's education more easy and effectual. Extracted from the best Authorities, and calculated for the use of Mr Demainbray's school.* See Rigaud (1882) p. 280.

9 Letter dated 19 January from Edinburgh reprinted from the *Caledonian Mercury* in the *Scots Magazine* **9** (1747) 40. A slightly different version dated January 19 appeared in the *Gentleman's Magazine* **17**(1747) 80–1.

Plate from Martin's General Magazine showing a scheme for electrifying plants, 1759.

The other shrub (which I did not electrify) hath not shewn any alteration since I first had it.
As the business of my school [a boarding-school for Ladies] does not allow me the necessary time of attending this now certain discovery, I submit it to those whose leisure will permit them to pursue a hint which may hereafter be highly beneficial to society.[10]

Although Demainbray did not continue with this work, at the time these letters were sufficient to bring him to the attention of savants such as Hales and Nollet. Stephen Hales knew of contrary experiments which challenged the views of the 'ingenious Scotchman' as Demainbray was misleadingly described in the *Gentleman's Magazine*. Dr Hales found

... his suspicion, that electricity will not promote vegatation, confirmed by several experiments made by Mr King, at his experiment room, near the king's Meuse, London, and by Mr Yeoman at Northampton.[11]

In their experiments King had electrified new laid eggs and a frog, and Yeoman a man, without noticing any effects. This difference of opinion between Demainbray and King foreshadows their rivalry in the mid-1750s when they were in competition for the audience for lectures which they both gave within a few hundred metres of each other.

However, Abbé Nollet seemed to agree with Demainbray about the effects of electricity on vegetation but in this case there arose a modest dispute about priority. In

---

[10] Reprinted from the *Caledonian Mercury* 20 February 1747 in the *Scots Magazine* **9** (1747) 93.

[11] *Gentleman's Magazine* **17** (1747) 200. Demainbray was referred to as an ingenious Scotchman on p. 102 where a letter dated 20

February is alluded to though not printed. It seems likely that this letter was similar to the second letter he published in the *Caledonian Mercury*. For details of King see pp. 70–5. For Yeoman see Robinson (1962).

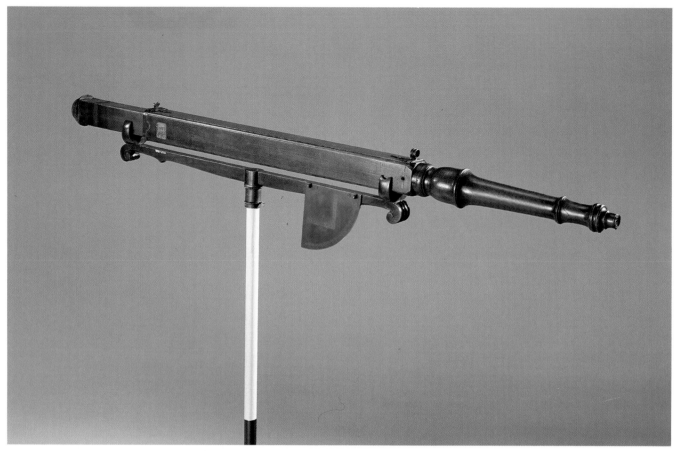

Telescope by William Robertson, an Edinburgh maker, used by Demainbray, catalogue number D68.

The History and Present State of Electricity, first published in 1767, Priestley described Demainbray's work and, defending British honour, suggested that this was the source of ideas for some experiments by Nollet.[12] Priestley wrote:

> *The electrification of growing vegetables was first begun in Britain. Mr Maimbray at Edinburgh electrified two myrtle trees, during the whole month of October 1746; when they put forth small branches and blossoms sooner than other shrubs of the same kind, which had not been electrified. Mr Nollet, hearing of this experiment, was incouraged [sic] to try it himself.*[13]

However, in 1749, much earlier, Nollet had published his own account in which he claimed that he had had similar ideas before he knew the details of Demainbray's experiments.[14]

Returning to Demainbray, for some reason he forsook this line of enquiry and abandoned research on electricity. Curiously, he hardly mentions the subject in his later lecture courses even though it was popular, particularly after the arrival of the Leyden jar in 1746.[15]

[12] Rigaud (1882) p. 280 states that in 1746 Demainbray '... resumed his lectures and published his discovery of the effects of electricity upon the growth of vegetables. This discovery was afterwards claimed by Abbé Nollet, but is very properly assigned to Dr Demainbray by Priestly (sic) in his history of Electricity'.

[13] Priestley (1767) p. 140, (1775) p. 172.

[14] 'J'étois occupé de cette pensée, lorsque j'appris qu'en Angleterre on avoit électrifé des plantes & des arbustes, qui s'en étoient ressenti de maniere à faire croire que la vertu électrique favourise ou hâte la végétation; mais comme il ne nous est venu aucun détail de ces expériences, (a) je n'ai pû en tirer d'autre avantage, que celui de m'enhardir dans le dessein ou j'étois me livrer à ces épreuves.' Footnote (a) is '(a) J'ai appris depuis, que cette expérience a été faite à Edimbourg par M. Mambray, que deux myrthes ayant été électrifés pendant tout le mois d'Octobre 1746, pousserent à la fin des petites branches & des boutons; ce que ne firent pas de pareils arbustes non électrifés.' Nollet (1749*a*) pp. 356–7.

[15] Heilbron (1979) pp. 309–323.

Lecturing on Natural Philosophy

During the winter of 1748–9 Demainbray took up a new career by giving a course of lectures in natural philosophy in Edinburgh. The reasons why he took up lecturing on natural philosophy are not clear from the surviving sources. French may have not have been a popular subject in Edinburgh after the failure of the Jacobite uprising in 1745, and perhaps this was a factor in making Demainbray look for other opportunities to use his teaching skills. Another factor was the death of Desaguliers in 1744 which may have given Demainbray, among others, the idea of trying to give lectures in his place.

However, the opportunities in Edinburgh for a lecturer on natural philosophy were probably very limited at that time. For though the Philosophical Society of Edinburgh had flourished for a few years in the late 1730s when it was formed out of the Edinburgh Medical Society, the '45 had put an end to its activities.[16] One of its most active members, Colin Maclaurin, Professor of Mathematics and Natural Philosophy at the University, died in 1746. Because he had helped to fortify the city against the rebels he had been forced to flee, becoming ill during his escape.[17] Other indications of the situation in Edinburgh come from Demainbray's contemporaries, James Ferguson and Richard Jack, who both left for England. Ferguson, a friend of Maclaurin's, gave some lectures in Edinburgh before moving to London in 1743, presumably because he thought that there would be better opportunities for him there. Jack was a teacher of mathematics who was in England in 1746 and unsure about when he would return to Scotland.[18]

By early 1749 Demainbray had also decided to leave Edinburgh. The *Caledonian Mercury* carried an advertisement:

On Tuesday March the 21st 1749 will begin a Rouping of the Household Furniture of Mr Demainbray, (who leaves off Business) . . . The Furniture bought new within these three Years, and warranted free from Buggs.[19]

Shortly thereafter he arrived in Newcastle, advance notice being given in the *Newcastle Courant*:

We hear for Edinburgh, that Mr DEMAINBRAY, who gave a Course of Natural and Experimental Philosophy consisting of 46 Lectures, during the Winter Session in that City, and whose Knowledge and extensive Apparatus are allowed to be little inferior, if not equal to those of his Master the late Doctor DESAGULIERS, is set out for Newcastle, where he proposes to lecture till the Scarborough Season, and will be here the latter End of this Week, his curious Collection of Mathematical Instruments, deemed the most considerable in any private Hands in Europe, being shipped for this Town.[20]

Demainbray's course began on the 6 April and he gave a lecture each day at 4p.m. for about an hour in the large dining room at the White Hart Inn. The cost of the course of 46 lectures was 25 shillings; a single lecture cost one shilling. The course lasted until around the end of May when Demainbray left for Durham.[21] From Durham Demainbray travelled to Sunderland.[22] By December he was in York where he advertised a course of 40 lectures to begin on 2 January, the course being described in a printed syllabus.[23] In York it seems that he met Henry Hindley, the instrument-maker, for soon Demainbray was using a pyrometer by Hindley in his

---

[16] McClellan (1985) pp. 105,145–6, Emerson (1979).

[17] See the account of Maclaurin's life in Maclaurin (1801). See also Hutton (1795) Vol. 2, pp. 60–1.

[18] Letter from A. Hume Campbell to Henry Baker, Stanwell 27 September 1746. Baker Correspondence Vol. 2, f260. In addition, Ebenezer M'Fait, taught natural philosophy and other subjects in Edinburgh from around 1745. See Bryden (1972*b*) p. 169. Perhaps there was too great competition for someone, like Demainbray, who was just starting out.

[19] See *Caledonian Mercury* 20 March 1749 No. 4437.

[20] *Newcastle Courant* 11/18 March 1749. Interestingly, the report next to the one about Demainbray gave the movements of Chevalier Taylor who was making a circuit similar to Demainbray's but in the opposite direction, travelling from York to Edinburgh via Newcastle.

[21] *Newcastle Gazette* 5 April 1749, *Newcastle Journal* 8 April 1749 p. 3, 13 May 1749 p. 2.

[22] Robinson 1972 Vol. 2. Demainbray is indexed as D187.

[23] *York Courant* Tuesday 19 December 1749 No. 1262 p. 3 and Tuesday 26 December 1749 No. 1263 p. 4.

lectures.[24] Demainbray gave his next course at the Assembly Rooms in Leeds starting in April 1750.[25]

Demainbray's reputation spread and he was awarded an Ll.D. by King's College, Aberdeen, in July 1750, an event reported in the *Edinburgh Evening Courant*:

They write from Aberdeen, that the Degree of Doctor of Laws was lately conferred on Mr. Stephen Demainbray. His Course of Natural and Experimental Philosophy hath met with the greatest Encouragement in England, and no wonder; for now he has so much improved and augmented his Apparatus, that it is reckoned the compleatest that ever was in Britain.[26]

This new career of lecturing on natural philosophy sustained Demainbray for most of the next decade until public interest in the subject waned and he had to find other sources of income.

After his travels in the north of England Demainbray went to Ireland where he was well established in Dublin by August 1751. A summary of his activities there is unremarkable; he gave several courses of lectures, ordered some new apparatus, and left bound for France. However, for the first time in Demainbray's case it becomes possible to glimpse some of the circumstances surrounding his lectures because they feature in some correspondence between James Simon, a Dublin wine-merchant and antiquarian, and Henry Baker, a teacher of the deaf and writer on the microscope, in London.

Simon was a key figure in Demainbray's stay in Dublin and may have been responsible for his going there. Late in August 1751 Simon wrote to Baker:

My Good friend Dr. Demainbray goes on very successfully here in his lectures on Natural & Experimental Philosophy he is daily improving his Apparatus (the best & largest I ever saw & I believe now extant) he has lately improved a clepsydra & brought it to a Minute of time, which must

---

24 For the pyrometer see inventory no. 1927-1184 catalogue No. D26. Demainbray was also engaged in microscopical studies in the company of a York doctor for some months in the winter of 1749–50. For Hindley see Law (1970–2a,b), Setchell (1970a,b).

25 See *Leeds Mercury* Tuesday 17 April 1750 No. 1296 p. 4. The course was advertised as being on the 20th.

26 *Edinburgh Evening Courant* 28 August 1750. Demainbray's Ll.D. diploma dated 26 July 1750 is in Aberdeen University Archives (MSS 2045/1) See also Anderson (1893) p. 111 for an entry 'Doctor of Laws 1750 July 26 Dr Stephanus Demembray, Armiger'. I am indebted to Dr J. S. Reid for this last reference.

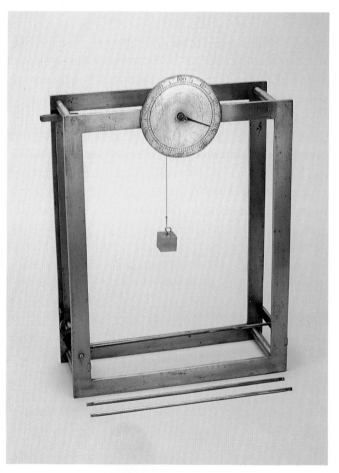

Mr. DEMAINBRAY,

Begins his Course of NATURAL and EXPERI-
MENTAL PHILOSOPHY,

On *Friday* next, the 20th Instant.

HIS Lectures (being 40 in all) will be held at the *ASSEMBLY-ROOM*, in LEEDES, on *Mondays*, *Wednesdays*, *Thursdays*, *Fridays* and *Saturdays*; to continue from Twelve to One o'Clock each Day.

Every Subscriber to pay 25 Shillings on the Delivery of his Ticket, which will be transferable to Ladies. Six-pence being the Servant Assistant's usual Perquisite.

SUPSCRIPTIONS will be taken in at the *Assembly-Room*, on *Monday* 16th, on *Wednesday* 18th. and on *Thursday* 19th Instant, from Twelve to One, where the *Syllabus* will be given *Gratis*, and the *Apparatus* may be Seen, and at Mr. *Demainbray*'s Lodgings at Mr. *Harrison*'s Chandler, in *Briggate*, *Leedes*.

Demainbray's advertisement in The Leeds Mercury, *1750.*

Pyrometer by Henry Hindley used by Demainbray, catalogue number D26.

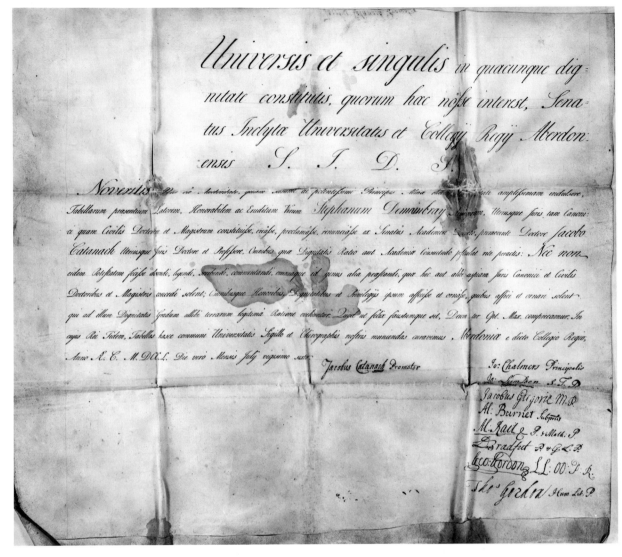

Demainbray's Ll.D. certificate from the University of Aberdeen, 1750.

certainly confirm the truth and certainty of the Antient Astromomical Observations: he certainly deserves all encouragement. I wish the Royal Society would accept of his 20 Guineas. I delivered him about a fortnight ago a very kind Invitation from the Royal Academy of Sciences at Bordeaux to go there this Winter; but I have engaged him two Courses during the next Session.[27]

Simon's statement that he had 'engaged him two Courses' suggests that he was involved in the arrangements for Demainbray to lecture in Dublin. However, it is not clear from the correspondence what these arrangements were. Simon could simply have acted as an agent for Demainbray, or it may be that Simon was acting on behalf of others for at the time he was a member of the short-lived Physico-Historical Society and Demainbray may have lectured under their auspices.[28]

Demainbray spent 18 months or so at Trinity College Dublin 'reading courses of lectures on natural and experimental

[27] James Simon to Baker, Dublin 27 August 1751. Baker Correspondence Vol. V, f72.

[28] See Berry (1915) p. 45. According to Rigaud (1882) p. 280,

Simon's remark that he passed on an invitation from the Bordeaux Academy is open to several interpretations. Simon, a wine-merchant, was very likely to have close connections with Bordeaux and a letter from there to Demainbray might simply have been sent via Simon's agents. However, two years before, Simon had sent a copy of his book on Irish coins to Montesquieu, then President of the Bordeaux Academy, so there may have been additional reasons for sending an invitation via Simon; he may have provided a link between the Bordeaux Academy and organizations in Dublin such as the Physico-Historical Society.

Societies

For Simon and for Baker the significance of the various societies they belonged to is apparent from their correspondence, but their letters also reveal the significance of these societies for others like Demainbray who only came in contact with them in passing. Simon had become a Fellow of the Royal Society of London with Baker's help in 1748. A few years earlier Baker had explained to another correspondent the qualifications needed to become a Fellow. These were:

> . . . some Knowledge in Mathematics & Natural Philosophy, an ardent Desire to search into & Discover the Properties of every Thing about us, & some Character in the World for such like Knowlege & Inclination. [29]

While such a description would apply to Demainbray he did not join, even though one Fellow, Simon, in the letter quoted above, was anxious that they should accept his twenty guineas. Perhaps Demainbray had his

own reasons for avoiding the Royal Society. A few weeks after Simon suggested the Royal Society should accept Demainbray's membership fee, Demainbray himself wrote to Baker:

> You are so very kind, Sir, as to offer to introduce any Paper I might send to the Royal Society, for which you'l allow of my most grateful Thanks; Conscious of my own Incapacity, I shall never presume on a Task so unequal, what my numerous warm Friends may call Improvements in the practical Way of Arts, might be deemed amusing when viewed in the Models or executed, but these are in no Manner worthy of being laid before so learned a Body of Gentlemen, hence you must allow me not to loose the favourable Opinion you seem to conceive of me, thro the undeserved kindness of my Friend.
>
> But as for the next Summer, I am invited to France with the Countenance of the Academy of Sciences at Bordeaux; if ought there has lately occurred in the learned Way, I should proudly embrace the Opportunity of acquainting you with it, as a private & not publick Correspondent, as likewise had you any Commands for those parts, none should more joyfully embrace them. . . .[30]

Demainbray saw himself as someone whose task it was to draw the attention of the audience at his lectures to new developments, developments he did not initiate himself. His concerns were with practical matters rather than theoretical, and with illustrating technology rather than with developing new machines. But if Demainbray did not belong to the Royal Society, what social position did he occupy? Judging by the Simon-Baker letters,[31] it seems his position was intermediate between individuals like instrument-makers and the Fellows of the Royal Society.

---

philosophy with his usual good fortune of public approbation and profit.' But there seems to be no record of Demainbray's connection with Trinity College. Demainbray may have had informal dealings with the Dublin Society, for when he was in contact with the Academy in Bordeaux from Ireland he was described as 'médicin anglais, membre de la Societé de Dublin'. Barrière (1951) p. 44, *see also* p. 68. However, the Minute Book of the Dublin Society '2nd Day of April 1750 to 13 December 1753', does not record Demainbray being present at any meetings. I am grateful to Dr Charles Mollan for providing me with this information. Berry (1915) p. 13 mentions that Huguenots such as Maturin, the Secretary from 1736, were active in the Society and perhaps Demainbray had informal contact with the Society through them. Some circumstantial

evidence is that the later three of Demainbray's four surviving syllabuses each mention an 'improved four wheeled Carriage offered to the Dublin Society'. John Cam, an English maker of ploughs and carts, was employed by the Society, Berry (1915) p. 50. Simon did not become a member of the Dublin Society until November 1752 which makes it unlikely that Demainbray's lectures in 1751 had any formal connection with that Society.

29 Henry Baker to Arderon 28 May 1745. Baker Correspondence Vol. II, f49.
30 Demainbray to Baker, Dublin 12 November 1751. Baker Correspondence Vol. V, f100.
31 Though of course there were a few notable instrument-makers who became Fellows of the Royal Society in this period.

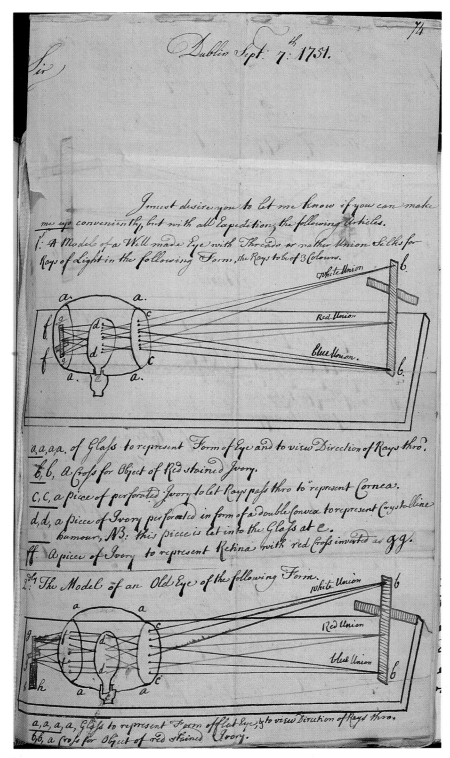

A copy of a letter from Demainbray ordering optical models from John Cuff, 1751.

John Cuff by Zoffany, 1772.

A JOURNAL of the WEATHER in *Dublin*, for the Year 1752.

Records of the weather in Dublin in 1752 compiled by James Simon and published in the Philosophical Transactions. There is a gap in September because the calendar or 'Style' changed then.

In his letter to Baker, which has already been quoted, Simon wrote,

Sr. James Lowther has also Invited him to Whitehaven to oversee his Works, he should not want Certificates (sic), Dr. Byram of Manchester, Mr. Walton of Newcastle and several Scotch Peers and Gent^n. are intimatly acquainted with him . . .[32]

Demainbray was someone open to offers of employment and had 'certificates' from people of good standing. But he was not considered to be on as low a level as a

[32] James Simon to Henry Baker, Dublin 27 August 1751. Baker Correspondence Vol. V, f72. Lowther was reputed to be the richest commoner in England and had interests in the coal and tobacco trades in West Cumberland. See Beckett (1981) and DNB. Dr Byram is probably Dr John Byrom see DNB.

Irish low-backed cart, catalogue number D20.

tradesman whom, in turn, he might employ. Some of these issues about status become clear when Demainbray wanted to augment the apparatus he was using in Dublin. He wrote to James Cuff, the well-known London microscope maker, to order three models showing the working of the eye, a normal eye, an 'Old Eye', and a 'bulging' or 'Myops'.[33] This commission became a source of friction between Demainbray and Cuff; as Cuff was slow to reply, Demainbray eventually took his business elsewhere. Cuff complained about his treatment to Baker who passed on his complaint to

Simon. Demainbray then replied to Baker in the following terms:

My good Friend Mr. Simon shewed me a Letter of yours in which you are so obliging as to make mention of me, and I hope you will not think it too great a Freedom, if I first endeavour to clear myself from what is laid to my Charge by Mr. Cuff, as I find he has not fully stated the Circumstances, as I should be very unhappy were you to harbour the least Doubt of my behaving with Severity to any Tradesman; hence, Sir, you must pardon an overflowing heart, which omits all Forms.

---

[33] Demainbray also enquired about a micrometer for a microscope that he heard Cuff had invented and a Fahrenheit thermometer. A copy of a part of Demainbray's letter to Cuff is in the Baker Correspondence Vol. V, f74. It is likely that this copy was sent to Baker a few weeks later during the dispute between Cuff and Demainbray.

Map showing Demainbray's travels 1749–55.

On the 7th. of last Sept. I wrote to Mr. Cuff and sent him the Commission mentioned in Yours; in mine I dealt, as I hope I ever shall, plainly, and told Mr. Cuff that tho' a Stranger to him, I was nevertheless acquainted with his Excellence as a Workman, and as such I had even but lately recommended him to one of my Auditors Mr. Weldone for a Job of between 9 and 10 Pounds but that I was likewise as well acquainted with his Remissness as to Country orders,

particularly I mentioned an Instance of a most inexcusable Delay in some Things sent for by a Physician at York, I did not enter into Particulars as not to my Purpose, but It is sufficient I believe to say that he kept us in Suspence about a nest of Tubes for Microscopes and some Lenses of particular Foci for a whole Winter, which hindered the Doctor and Myself from pursuing our Tryals, as we expected he would send them at every Return of the Stage

Coach for some Months in 1749/50 and therefore we would not write to any other Workman.

Nevertheless on all occasional Recommendations I did Justice, as I shall ever do, to his Merit and advised any Friends to such Things as might be got ready made in his Shop.

I shall frankly confess that my former Optical Workman having greatly blundered in my last Directions, and that having heard of Mr. Cuff's Misfortunes I ventured to write to him imagining they might have excited his Diligence and since he has made his Complaint, must insist on his producing to you the Letter I wrote to him on the 7th. of September; in which you will find that I bluntly let him know, that I am no Stranger to his Neglect of Country Commissions, and that I can admit of no Delays; that I shall not depend on his taking in hand my Concern, if I have not his immediate Answer by the Return of the Post, as I would not be dependent on the Caprice of any Workman, and therefore shall take his Silence as an absolute Refusal: In this Letter I likewise give an Order to pay on Sight whatever may be the Charge, and that he may depend on direct payment as I always deal for ready Money; and least he should think it hard to pay Postage for a Letter whose Contents he did not purpose to comply with, I paid the Carriage to London, as you'l find on the Superscription.

In mine you will find, Sir, more than once repeated, that his Silence should be construed into a Refusal at the Return of the Post; and as I had wrote to London by the same Post on some other Concerns and received an Answer dated the 14th. of September, I waited Two more Mails and then gave up all Thoughts of Mr Cuff, and accordingly set another Person to Work, and It is long since I got my Models delivered and executed to Satisfaction and according to the same Directions Mr. Cuff had received, which cost me 32 Shillings as per Receipt which I shewed your Friend Mr. Simon.[34]

While Demainbray's contact with Cuff had its difficult side, Demainbray was careful to bring out the positive role that he had in recommending people like Cuff to the gentlemen he met who might want to

purchase instruments. Indeed, a further example comes from the covering letter Simon sent with Demainbray's reply; in it Simon asked Baker for a telescope by Short, a maker recommended to him by Demainbray.[35] Baker also recommended makers to potential purchasers since he was in a position to do so because of his knowledge of microscopes.

The letters between Simon and Baker illustrate the networks that had grown up and which linked together individuals with different skills, such as instrument-maker or lecturer, with their actual or potential audiences—those with a polite interest in the subject matter. Some of these networks were based on societies: the Royal Society of London has already been mentioned in the case of Baker and Simon but they also extended far beyond these societies, geographically and socially.[36] But as these networks grew and developed, in turn they caused the societies to change too. One example of this process comes from the following year, 1752, when Simon became a member of the Dublin Society. He wrote to Baker to explain how he promoted the Society's views:

. . . I am pursuing as much as I can by encouraging artists to come over from france having beyond my expectation succeeded in forming a Society here for the encouragement of Protestant french Refugees, whom we settle in [..] indifferent parts of the kingdom, advance them Money Tools & materials to carry on their respective Trades.[37]

Simon expected these views to be passed on to William Shipley who was gathering support at the time for what became the Society of Arts. However, Simon's comments show how both his contacts with France (and with Demainbray—a Protestant of French extraction) could be used for a variety of purposes.

Demainbray in France

Demainbray travelled from Dublin to France, arriving in Bordeaux by May 1753.[38] According to Rigaud, he

34   Demainbray to Baker, Dublin 12 November 1751. Baker Correspondence Vol. V, f99.

35   See Simon to Baker 12 November 1751. Baker correspondence Vol. V, f101. For Baker connection with Cuff see Turner (1974).

36   They were also both members of the Society of Antiquaries.

37   Simon to Baker, Dublin 14 July 1753 f294.

38   A letter from Paris dated 30 April 1753 (Aberdeen University Archives MSS 2045/2) mentions that it had been agreed the previous year that 'de Mambray's' Cabinet had been allowed to enter France duty free on condition that it would be sent back to Ireland within 15 months.

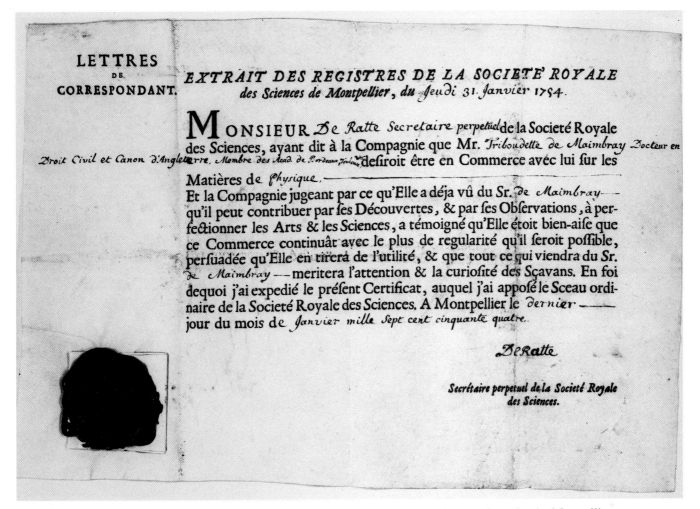

In 1753–4 Demainbray became a member of several academies in the French provinces such as that in Montpellier.

had been invited there by Barbot and Montesquieu, the President and ex-President of the local Académie Royale. He also had connections in the area through his mother's family.[39]

By the time of Demainbray's arrival the Bordeaux Academy had been interested in natural philosophy for some years. In 1741 they sponsored a course of lectures by Nollet and they also bought a cabinet of equipment from him. Nollet's lectures on electricity encouraged the Academy to hold a competition for an essay on that subject. The competition was won by Desaguliers in 1742.[40] Perhaps Demainbray, through his own connection with Desaguliers, was encouraged to visit Bordeaux, and because Desaguliers's links with Bordeaux were not restricted to natural philosophy but also owed something to masonic organization, this masonic connection probably operated for Demainbray as well.[41]

The provincial academies in Montpellier, Bordeaux, and Lyon were the most important in France after the Paris Academy. Demainbray made a circuit of these,

39 '. . . of whose station in life he was rather proud' Rigaud (1882) p. 280.
40 'Another vacuum, another conquest: only two others entered, and Desaguliers won.' Heilbron (1979) p. 293, *see also* pp. 280, 294, and Barrière (1951) p. 136, McClellan (1985) p. 135.
41 Demainbray was a member of a French Lodge in London around 1731. See Roche (1978) Vol. 1, pp. 257–8, Heilbron (1979) p. 294.

Demainbray's model of a boat-mill proposed for the Rhône by Dubost, catalogue number D39.

together with Toulouse, from around early 1753 to mid-1754. He delivered courses of lectures to the members of these academies and in recognition of his services he was made an associate member of each academy.[42] One reason for his success was the current interest in Newtonian philosophy amongst the members of these new and active academies, an interest which Demain-

bray was both able to promote and turn to his own advantage at the same time.[43]

There survive a few scattered details of Demainbray's experiences in France. His lectures in Bordeaux were attended by one Père Chabrol whose manuscript notes are now in the archives of the Bordeaux Academy.[44] During his stay in Bordeaux, Demainbray learnt

[42] For Bordeaux see Barrière (1951). On 13 May 1753 Demainbray was elected 'academicien associe' of 'l'academie des belles lettres, sciences et arts' at Bordeaux. (Extract of register, Aberdeen University MSS 2045/3). See Rigaud (1882) p. 280. Demainbray had had a 'lettre de correspondant' from 'La Societé Royale des Sciences de Montpellier' dated 31 January 1754. According to that extract 'Mr Triboudette de Maimbray' was also a member of the academies of Bordeaux and Toulouse by that time. (Extract Aberdeen University Archives 2045/4).

On 3 May 1754 'Mr Triboudet de Maimbray' became an associate academician of 'La Societe Royale de Lyon' (Aberdeen University MSS 2045/5). In Lyon 'il demontre avec applaudissement un Cours de Physique experimentale. . .' See Demainbray (1753, 1754a); Roche (1978) Vol. 1, pp. 257–8.

[43] 'La physique de Newton, discutée partout avant 1750, s'impose après; le retard sur Paris est de vingt à trente ans.' Roche (1978) Vol. 1, p. 380.

[44] These are in the Bibliothèque Municipale in Bordeaux.

Demainbray's model of the crane used at Ralph Allen's works near Bath, catalogue number D14.

something about ventilators and the economic advantages of their use in ships to transport slaves. Some years later he described his experience in a letter to Stephen Hales:

That in the Year 1753, Ventilators were put into the Vessels in the Slave-trade at Bourdeaux, and in other Ports of France; the happy Effect of which was, that instead of the Loss of one-fourth of those valuable Cargoes, in long Passages from Africa to the French Plantations, the Loss seldom exceeded a twentieth. And since my Return to England, I have been informed of a French Vessel, which by this self-evidently reasonable Precaution, saved 308 out of

312 Slaves, spite of most tedious Calms, and a long Passage.[45]

In Montpellier Demainbray's wife died and this event affected him deeply and may have hastened his return to England.[46] On this journey Demainbray went to Lyon where he took the opportunity to collect models of local machinery, including some copied from those in Marquis de 'Grollier de Servière's collection. He used these models later in his lecture courses in London.[47] In the autumn of 1754, on the eve of the outbreak of the Seven Years War and with relations between Britain and France at a low ebb, Demainbray left Lyon and returned to London, staying briefly in Paris on the way back.[48]

[45] Hales (1768) pp. 94.
[46] There Demainbray, according to Rigaud, refused an invitation to go to the Academy at Madrid and Earl Brooke proposed that he should return to England to '. . .become teacher to the Heir Apparent to the Throne in mathematics, philosophy, and natural history.' Rigaud (1882) p. 280.
[47] The models have inventory numbers 1927-1635, catalogue Nos

D5 *et seq.* Demainbray experienced high temperatures while in Toulouse. See the description of the thermometer, inventory No. 1927-1810, catalogue No. P21.
[48] Simon wrote to Baker 14 July 14 1753, 'My friend Demainbray is gone to Paris on an Invitation from his most Christian Majesty.' Baker Correspondence Vol. 5, f293.

Demainbray's Lectures in London

Whatever the reasons for his return to London, personal and professional, once there and established in rooms in Panton Street near the Haymarket (described by one contemporary as 'a good street inhabited by tradesmen'), in February 1755 Demainbray began his first course of lectures in the capital on natural and experimental philosophy.[49] He announced his course in the *Daily Advertiser* in the following terms:

The Concert Room in Panton-Street, near the Hay-Market, formerly Mr Hickford's, is completely fitted up for the Use of Dr. Triboudet Dimainbray, Fellow of the Royal Societies of Montpelier and Lyons, associated Member of the Royal Academies of Toulouse, Bourdeaux, & c. where he intends about the Middle of January next to read the Course of 34 Lectures on Natural and Experimental Philosophy, he read in the different French Academies he was called to, agreeable to the Syllabus he has publish'd, since his late Return to England.[50]

On the first of that month Demainbray remarried; his second wife, Sarah, was the daughter of Mr Horne, poulterer to his Majesty and the Royal Family, whose 'accomplishments were such as seldom were met with in a tradesman's daughter'.[51] Shortly afterwards Demainbray was employed to give a course of lectures to the Prince of Wales and Prince Edward.[52] One result of the royal connection was that Demainbray introduced his young brother-in-law John Horne (later John Horne Tooke) to the Prince of Wales, a connection that later

caused much embarrassment to Demainbray when Horne became a staunch supporter of Wilkes and an opponent of the King.[53]

The royal appointment was lucrative for Demainbray, at least in the short term. The Royal Account of the Privy Purse for 1755 records that on 14 May he was paid £210 for a 'Course of Natural & Experimental Philosophy for their Royal Highnesses.'[54] In addition, Demainbray had a pension of £100 a year awarded him in 1755 by King George II giving a total income of about £300.[55] However, Demainbray was to have great difficulty collecting this pension.

The following winter Demainbray's advertisement for his lecture course highlighted his royal connection. On 15 December 1755, it read '. . . Dr Demainbray, Teacher to their Royal Highnesses the Prince of Wales and Prince Edward. . .'.[56] But the terms of his employment as a teacher of the Princes cannot have satisfied Demainbray because he continued to seek additional posts. In January 1756 Brookes wrote to Lord Bute to ask if the Prince of Wales could grant further favours to Demainbray such as making him the Prince's Librarian.[57] At that time Bute had considerable influence over the Prince as he had just been given responsibility for George's education.[58] Brookes wrote in his inimitable way:

I have done myself the Honr to Call once or twice at your Lordship's Door to thank you (had I been so fortunate to have mett you at home) for the very obliging & handsome Manner you was so good as to reccomend Poor Doctr de Mainbray to the Prince's Charity & for the speedy releif that with so much Generosity has been given him. No one

---

[49] For the quotation about Panton Street (not quite the present-day Panton Street) by Strype see Phillips (1964) p. 88.

[50] *Daily Advertiser* December 1754. The course seems to have started on Tuesday 11 February, *Daily Advertiser* Saturday 8 February 1755 No. 7492 p. 2. The syllabus for this course of lectures is Demainbray (1754*b*).

[51] Rigaud (1882) p. 281. See DNB for John Horne Tooke for an explanation of how Horne senior became the royal poulterer. It was the outcome of a tussle between Horne and Frederick, Prince of Wales, whose servants made their way through Horne's property. Horne won a court case but continued to allow access and so was created royal poulterer. However, on Frederick's death Horne's bill was not paid. The marriage to Demainbray involved a financial transaction of £600.

[52] Rigaud (1882) p. 281.

[53] 'While a boy, he had been introduced at Leicester House, by

means of Dr de Mainbray, who was still caressed by the young monarch, and was accustomed to play with his present majesty, who was exactly two years younger than himself, once or twice a week.' Stephens (1813) Vol. 1, p. 36, see also pp. 8 and, especially, 56–7. Horrins (1835) p. 1 describes Demainbray as '. . . a gentleman of scientific abilities, but chiefly known at present by his marriage with the sister of John Horne Tooke having exposed him to some illiberal epithets from the popular demagogue, Wilkes, when they were pointedly opposed in politics.'

[54] Royal Archives RA 55692.

[55] The figure mentioned by Rigaud.

[56] *Daily Advertiser* 15 December 1755 No. 7775? p. 2.

[57] This Brookes may have been the Earl Brooke who had a hand in bringing Demainbray from France.

[58] See Brooke (1972) p. 43.

Demainbray's model of Vauloué's pile driver, catalogue number D16.

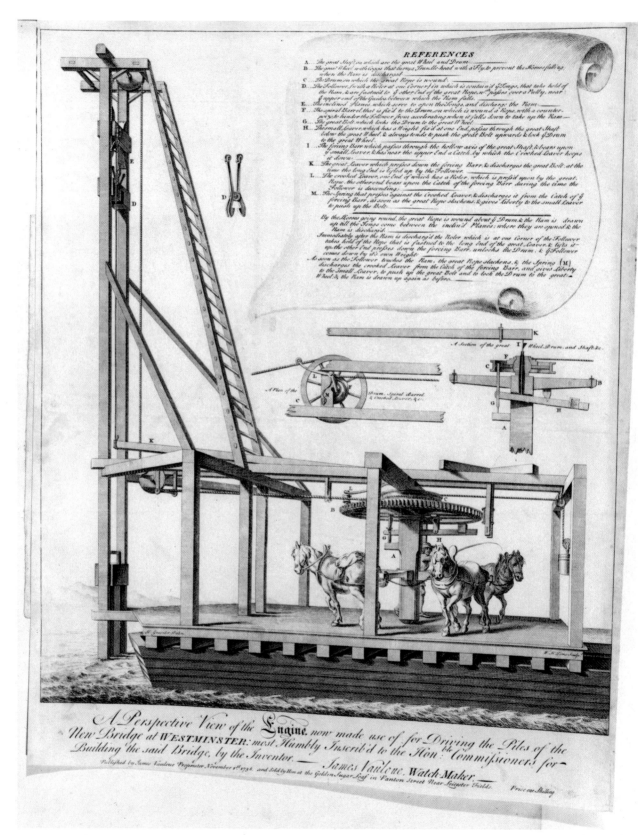

Print showing Vauloué's pile driver, c. 1740.

View of London by Samuel Scott, 1742, showing Vauloué's pile driver being used in the construction of Westminster Bridge.

feels more than I do whenever a request is speedily granted, I wish others like yourself had that Talent, which would do more Honr to themselves, & Afford happiness to those Concerned more than even the reward itself, as nothing is so dreadfull as the cruel suspence which Doctr. de Mainbray has undergone for these two last years & which As I was concerned for him I had my share in. . . . By rendering himself usefull in any degree, as his Correspondances are extensive amongst the Litterati & is greatly thereby enabled to give a proper Account of what things new & curious may appear, he would if he was allowed lay before yr Lordship from time to time such productions as perhaps might not be unworthy or unentertaining to his Royl Hss. and further if Without any View of Salary or incumberance of this sort he might have the Title conferred on him of his Royl Hss

Librarian it would make him the happiest of all men & greatly raise his spirits. I beg leave to mention this freely to yr. Lordp. & indeed my view seeing the Deficiency of these Studys in Detaining him here was that I imagined as No one has a better apparatus than he has that such a Person ready to be called in for the service of the Royal Family now Growing up would be agreable & particular(l)y as he is certainly I may venture to say in every Branch as able to convey Knowledge as can now be found & add to this & on what you may Depend one whose moral character is remarkably conscientious & good, which had it not been so I should have Never have spoke for him. He tells me he has a House in View just over right Kew . . .[59]

However, these suggestions seem to have fallen on deaf

---

[59]   Brookes to Bute. Bute Papers Grosvenor Square 18 January
1756.

Silhouette of Demainbray, inventory number 1991–109.

ears—Demainbray was not given a new post at that time. But more awkwardly the pension of £100 a year. awarded him by King George II did not materialize; years later Demainbray was still pursuing the promised but unpaid pension.[60]

Natural History

To compensate for the missing pension Demainbray had to rely on his income from lecturing.[61] As discussed in Chapter III, the later 1750s were not good times for the lecturers on natural philosophy and Demainbray had to resort to a variety of tactics to entice an audience to his lectures. One way was to flatter them, which is what he attempted to do in April 1757 when he expressed his gratitude to those who attended.

The Doctor thanks his Auditors for their Sanction during the Winter Season, and humbly hopes he has fulfilled his Engagement of being the faithful Echo of what has been produced in Sciences both at home and abroad, without having neglected the Discoveries of the Antients.[62]

Another tactic was to offer courses on new subjects. Earlier that season Demainbray had given lectures on natural history for the first time and he repeated these the following year, 1757–8. Though the courses he gave on natural history were short-lived, it seems that Demainbray's interest in the subject was more endur-

ing. Demainbray had contact with several individuals well known for their interest in natural history, including John Ellis whom Demainbray solicited for material to use in his lectures.[63]

I have not left a Shop, that I know of, unsearched for Pictures made by Ulva Marina and Corallines; One sheet would have been sufficient for a general Hint in my Lecture on that Head, as I am provided with your Essay, and had framed and glassed in my Lecture Room the Prints you complimented me with about a Year since.

I hope good Sir, youl (sic) pardon my Request for one of Yours, and that you would let Mr Cuff make some Sliders for the Microscope that I must beg you at your Leizure fill. . .[64]

Other contacts were with William Hunter, John Fothergill, and Lord Bute.[65] In August 1761 Demainbray offered Bute 'Chests of Corals, Keratophyta and other Marine Substances, being the first specimens of the natural history of Guadeloupe yet come to England.'[66]

Patronage

During 1758 'Dr de Mainbray' moved from Panton Street to an 'experiment-room' in Carey Street, but he cannot have been happy with his new establishment because he gave it up early in 1759 when he found it

[60] In May 1763 Demainbray wrote to Lord Liverpool, 'In the Month of January 1757 I received a Message from Lord Gage's Office to come & receive the payment of a Pension of £100 a Year granted me by his late Majesty, as having been at that Time the Teacher of his present Majesty for two Years (and permit me to add, continued as such till the Kings Accession to the Throne.) It is necessary to also observe that My lord Gage told the Earl of Warwick that the Monies were paid into his office for me, from the Treasury, & ready for my Receipt. But I was disappointed of the King's Bounty, since on my Application, I was told that the Warrant had been mislaid by the Officers concerned, & to this Hour not restored to me.' British Library Add MSS 38200, f334. See also f298, Add MSS 38204, f95, f96, Add MSS 38304, f65b, Add MSS 38397, f76, Add MSS 38469, f134. See also Demainbray to Bute 30 June 1762 from Brentford. Bute Papers.

[61] In a letter to Bute in December 1758, Brookes mentioned that Demainbray had, however, received other marks of royal bounty, though not presumably the promised pension. He

reported Demainbray's reaction, '. . . the poor Man came to me with Tears in his Eyes quite overcome by this great Generosity, & not less penetrated by the Condescending manner your Lordp was pleased on this & all occassions to receive him'. Brookes to Lord Bute Hill Street 11 December 1758. Bute Papers.

[62] *Daily Advertiser* Thursday 14 April 1757 No. 8192 p. 2.

[63] See p. 75.

[64] Demainbray to John Ellis 20 August 1756. Ellis Collection Linnean Society of London.

[65] In May 1768 Fothergill wrote to Hunter, 'It may perhaps be proper to acquaint Dr. Demainbray that the corals are just as they came out of the sea. They may be easily cleaned by putting them in warm water just acidulated with spirit of sea salt, and then again washing them in fresh warm water.' Corner and Booth (1971) pp. 281–2.

[66] British Library Add MS 5726 C, f178. Guadeloupe was one of the prizes in the Seven Years War so these specimens of corals etc. were one of the spoils of war.

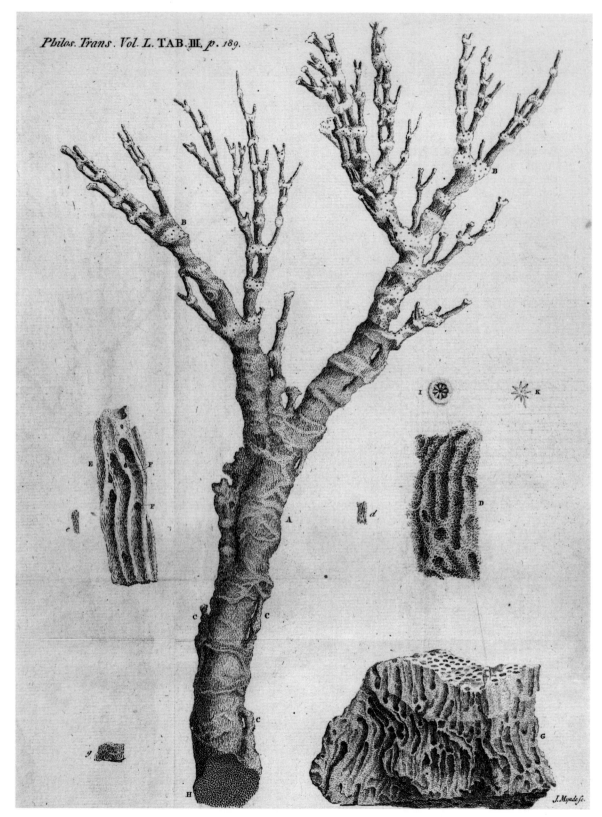

Philos. Trans. Vol. L. TAB. III. p. 189.

A contemporary illustration of coral from the Philosophical Transactions, *1757.*

difficult to attract an audience.[67] However, in 1760 when King George III, his old pupil, came to the throne and Bute was in a strong position to influence the King, Demainbray's hopes of royal preferment must have risen. In a letter to Lord Bute in November that year, two weeks after the death of King George II, Demainbray explained the vagaries of his employment and once again sought to have Bute help his case with the new king.

The Death of her Royal Highness Princess Elizabeth deprived me (the Autumn of last Year) from attending on his Majesty, when ordered to read Astronomy to him; and this Year Mr. Le Grand did not chuse I should be employed in reading to their Royal Hignesses his Pupils. I leave all Inferences from thence to Your Lordship's Consideration.

His Majesty's Collection of philosophical Instruments increases daily, and as it is finely chosen, requires one conversant to keep it in Order; would it be adviseable, Mylord, to apply, on the Grounds of the abovementioned Considerations, to be nominated as Keeper of the same, for to spend the Wane of Life, and to approach his Majesty's sacred Person, tho' as the lowest of his Servants, is my whole Ambition. . .[68]

Demainbray kept up the pressure in his search for employment. A few months later in July 1761 he wrote once more to Lord Bute though now to suggest that he be given the job of teaching English to Charlotte whom George III proposed to marry.[69] For Demainbray this would mean using the skills he had developed in Edinburgh where he had taught French and English.

In 1761 he made a last attempt to give a 'Popular and Practical Course of Experimental Philosophy' but it was not successful.[70] A few years later in 1763, Lalande, the French astronomer, while on a visit to Britain heard more of Demainbray's unsuccessful story; apparently Demainbray had abandoned physics and would not even take the time to show his instruments to Franklin.[71]

By that time Demainbray had been an official of the customs for a few years.[72] As Rigaud explained:

When Dr. Demainbray's salary of £300 a year ceased, he was recommended for employment in the Revenue; and, on appointment by the Ministry of the day, undertook the duties of three offices in the Custom house, London. In two of these offices the whole of the Crown Revenue arising from the East-India Company, and amounting to about one million a year, was under his care as principal of that department. His third office was that of one of the four examiners of the accounts of the receipts of monies of all the seaports of England, except London, which was supervised by His Grace of Manchester. The emoluments of these offices were sufficiently good for the Doctor to remark in one of his letters that under his present position he would be enabled (if he lived) to educate his family, and leave them, not destitute, as he once was, provided he then lived some few years.[73]

Stephen Demainbray was still not satisfied and continued to press Bute for further appointments, in part to make up for the promised but still unpaid pension. From Brentford in June 1762 Demainbray wrote to him:

Some Days before the Duke of Newcastle left the Treasury (between five & six Years since) Mylord Gage sent me

---

67 Westminster Local History Library Rate Books. pp. 75–6 above gives the details of the end of Demainbray's public lectures.

68 Demainbray to Bute Brentford 5 November 1760. Bute Papers.

69 'I most humbly hope for the Pardon of my Presumption in troubling Your Lordship with my Petition to be employed as a Teacher of English to the Princess, whom his Majesty has declared his Intentions of espousing. If your Lordship thinks me deserving of so great an Honour, I shall use every possible Endeavour to answer Your Lordship's Recommendation.' Demainbray to Bute 13 July 1761 British Library Add MS 5726 C, f177.

70 *Daily Advertiser* Monday 2 November 1761 No. 9617 p. 2. The course was due to start that day. There were no further advertisements after 3 November suggesting that he did not attract enough subscribers.

71 'M.Mainbrai est riche; il a un bon emploi à l'accise. Il est dégoûté d'en faire davantage. Il abandonne la physique. Il n'a pas meme voulu fair voir ses instruments à M Franklin.' Lalande (1980) pp. 44–5.

72 He had several appointments one of which was inspector of unrated East India goods in the Port of London from July 1757. See *London Magazine* **26** (1757) 365.

73 Rigaud (1882) p. 282. Demainbray's connection with the 'Commissioners' continued and in 1762 he received £30 for 'his extraordinary care and trouble in attending the Survey and preparing the Abstract of the prohibited East India Goods for the last half year. . .' and he was invited to apply for a like reward at the end of each half year. Copy of the Commissioners minute signed by Jos. Martin, dated 11 December 1762 Aberdeen University Archives 2045/6.

Orrery by Charles Butcher of Bedford, made in 1731–3, catalogue number D28.

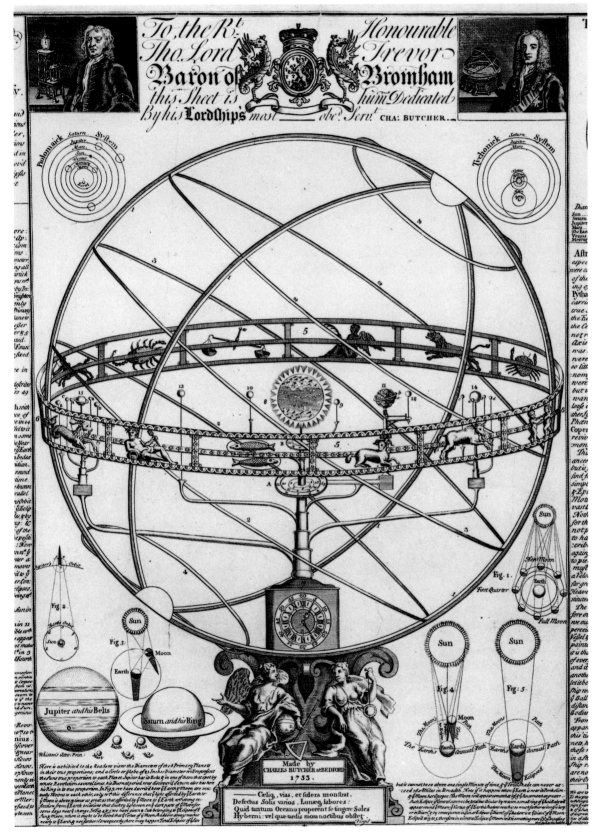

Print showing the orrery by Charles Butcher, 1733.

Word, by Mylord Warwick, to come and receive half a Year's Payment of a Pension of one Hundred Pounds, paid in his Office for me; Unfortunately, the Warrant was mislaid, and for Want of Interest at the Board of Treasury, not yet recovered; Whereby, I have been frustrated from the intended Royal Bounty: Wherefore I most humbly beseech that Your Lordship, in your great Goodness, would condescend to order Enquiries to be made, and You will find the Truth of my afflicting Case, since struggling with the Hardships of scarcely finding Necessaries for a very numerous, & increasing Family of helpless young Children, whose whole Subsistence depends on my Place of Inspector of East India Goods, in the Execution of which, by an upright & faithful Conduct, I have obtained such Vouchers (by the Board of Commissioners honourable, & recommendatory Minutes) as prove my real Services done to the Revenue, & my Sacrifice of private Interest to publick Utility. Since I have been chief Officer of the East India Trade, I have furnished the Board of Customs with Means of obviating gross Abuses, & Frauds, and were Your Lordship at Leizure, but for one Quarter of an Hour I could propose effective Means to prevent Frauds on India Goods, not to be stopt, as the Laws now stand: Believe me, Mylord, I should neither be absurd, or bold enough to request so high an Honour, were I not certain that what I have to disclose merits Attention, having closely studied the Branch of East India Customs, ever since I have been concerned with that Business. . . .[74]

Kew Observatory

Demainbray eventually found a sinecure, not with the customs as he had hoped but connected with his earlier career in natural philosophy. Rigaud records:

In the year 1769 the Transit of Venus was anticipated to be seen in England, and His Majesty King George III, took so much interest in the matter that in 1768 he commenced building an Observatory at Kew, and appointed Dr. Demainbray the Astronomer of it, who, when it was finished, adjusted the instruments there in time to observe the Transit of Venus, which happened in the year 1769.[75]

In the years after the transit, Demainbray spent his time recording the weather, perhaps giving a few lectures and checking the clocks that provided time for Parliament.[76] Though Demainbray finally found congenial employment at Kew, it was an arrangement that benefited himself, his family, and perhaps even members of the royal family, but hardly anyone else. While Demainbray had some knowledge, most of the instruments he could possibly want, and acquaintances who could assist him, he chose to live quietly in his last years, enjoying his sinecure rather than carrying out scientific endeavours of any significance. Demainbray died in February 1782 and was buried at Northolt, leaving his succession to be decided.[77] His son, who had been promised the post of superintendent at Kew by the

---

[74] Stephen Demainbray to Bute 30 June 1762 Brentford. Bute Papers. A year later, in May 1763, Demainbray pursued the matters of pensions and employment with Lord Liverpool: 'If I am unhappily never to recover the Pension, Lord Warwick tells me, I may hope for some Additional Place in the Customs such as a King's Waiters' in the Port of London, or a Comptroller's Patent in some of the out-ports, wither of which being executed by Deputies, would be no Impediment to my present Duties as Inspector.' British Library Add MSS 38200, f334.

[75] Rigaud (1882) p. 282.

[76] Rigaud (1882) p. 279. Shortly before the transit of Venus he was visited by Jean Bernoulli who was making the tour of European observatories he reported in his 'Lettres Astronomiques'. Bernoulli's enigmatic comment on Demainbray was '..un savant qui est bien vu de leurs Majesté..' Bernoulli (1771) p. 114. Demainbray received a diploma from the Royal Society of the Sciences at Gottingen dated 27 March 1771 (Aberdeen University Archives MSS 2045/7) which may have been Bernoulli's doing.

[77] Lysons (1811) Vol. 2, Part 2, p. 589. Grant of probate was made

to Demainbray's executors in 1782 (See Public Record Office PROB 8/175 PFF 2960). His goods and chattels were to be listed (and an inventory provided for the Court on or before 1 September 1782) and then sold. The proceeds were to be divided between his four daughters and one son after his wife, Sarah, had had a payment as compensation for a 'Bond or Obligation' executed on 30 January 1755 and then in the hands of the executors of Benjamin Horne. Such an inventory is not held by the Public Record Office today. At the sale of Queen Charlotte's Library in 1819 one of the lots consisted of some of Demainbray's manuscripts. Lot 1330 'Cours de Physique Experimentale, par Demainbray, 11 vol. 4to. MS. A System of Natural History, by Demainbray, 9 vol. 4to. MS. A System of Experimental Philosophy, by Demainbray, 8 vol. 4to. MS. Elements of Geometry, by Demainbray, 6 vol. 4to. MS. A Dissertation on Fire, by Demainbray, 4to. MS. Demainbray on Air and Gasses, 4to.MS. Demainbray, on Botany, 4to.MS. Demainbray's Lectures, 4to. MS.' These were all sold to someone called Osborne for £4 4s.

King, eventually got it. This Demainbray, also called Stephen, was assisted by his nephew S. P. Rigaud, who later became Professor of Astronomy at Oxford and developed an interest in the history of science.[78]

In contrast with Demainbray's work and interests, a few of his contemporaries did do research. Often these individuals taught in an institution, or were Fellows of the Royal Society, or had contact with other luminaries. These factors were probably necessary for research to be done (but not sufficient to guarantee it). However, Demainbray's career illustrates that patronage and place-seeking were important aspects of eighteenth century life, patterns that survived well into the nineteenth century.[79]

On one occasion Priestley drew attention to the role of patronage in the pursuit of natural philosophy.

Nature will not be put out of her way, and suffer her materials to be put into all that variety of situations which philosophy requires, in order to discover her wonderful powers, without trouble and expence. Hence the patronage of the great is essential to the flourishing state of this science.[80]

While Priestley had justification for his view that patronage was essential for the study of natural philosophy (even if his view was motivated by self-interest), this 'patronage of the great' was only one of many requirements that had to be met before science would flourish. Patronage had many effects and, as the example of Demainbray at Kew shows, not all of them were conducive to the development of natural philosophy. Other additional factors were required for the cultivation of natural philosophy and all of these, together with the networks of patronage, went to make up the complexity and variety of life in eighteenth century Britain.

One supporter of Lord Bute (for a time at least) was Tobias Smollett, mentioned at the beginning of this chapter as one who was in favour of the idea of an able professor working with a set of equipment for the honour of the nation. His own comment on that proposal was:

But this is all idle speculation, which will never be reduced

to practice—Considering the temper of the times, it is a wonder to see any institution whatsoever established for the benefit of the public. The spirit of party is risen to a kind of phrenzy, unknown to former ages, or rather degenerated to a total extinction of honesty and candour...[81]

Smollett was perhaps more right than he knew, at least in relation to Demainbray, whose plans to be a professor working for the honour of the nation were not 'reduced to practice' of any very useful, or enduring kind.

Conclusion—one end of natural philosophy

The establishment of the King's observatory at Richmond did little, directly or indirectly, to advance scientific knowledge, an unproductive situation that continued until the mid-nineteenth century when the building was taken over by the Royal Society and the newly founded British Association for the Advancement of Science to be used for more serious and lasting scientific work. In some respects this limited use of the new observatory and its resources was unremarkable, for Demainbray, its superintendent, had been looking for a comfortable situation to pass what he termed, 'the wane of life', whilst George III eventually became incapable of pursuing scientific interests because of the onset of madness. However, these biographical details, though important for the individuals concerned and influencing the course of events, only illustrate rather than explain the changes taking place in natural philosophy—even allowing for the fact that one of the protagonists was the King.

It was argued above, in Chapter III, that the years around the mid-century marked the high point of interest in natural philosophy by some of the middle and upper ranks of London society. This interest was expressed through attending lectures or purchasing the books and equipment used to carry out the demonstrations which illustrated the lectures. However, this interest found new channels later in the century. In part this was associated with changes within natural philoso-

[78] Interestingly, Rigaud chose to research the work of Bradley who had fallen out with Demainbray.
[79] See Turner (1976).
[80] Priestley (1767) p. xviii.
[81] Smollett (1771), Matthew Bramble to Dr Lewis pp. 133–4.

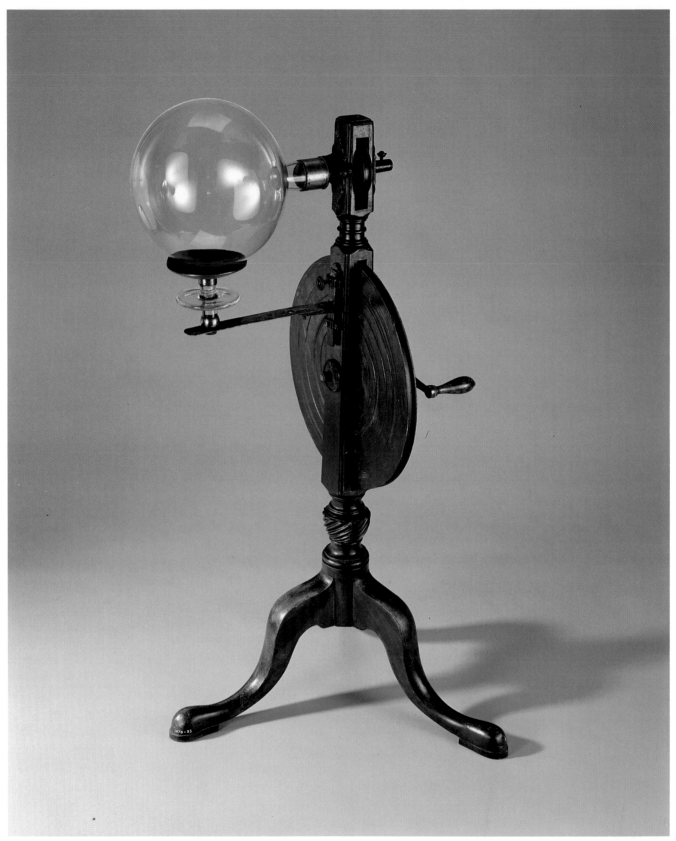

Priestley's electrical machine, inventory number 1970–23.

phy; the elementary aspects of natural philosophy began to be seen as more suitable for children, whilst more advanced areas of astronomy, for example, became less accessible to someone with a general interest once more advanced mathematical techniques came to be used. The attention of individuals was also diverted by other subjects which now had a broader appeal, such as chemistry and natural history.[82] The study of natural history had many attractions with opportunities for both local activities, such as field trips to collect specimens, and for participating in new national, or even international, organizations, such as the Linnean Society founded in the latter part of the century (and followed by many other specialized scientific societies). Natural philosophy could not offer such a range of activity and interest.

Another attraction of natural philosophy earlier in the century had been its possible relevance to an understanding of new machines and industrial processes. However, with the advent of organizations like the Society of Arts and a generation of engineers with experience of building canals, mills, and steam engines—Brindley, Smeaton, and Watt—improvements in technology seemed less likely to result from following a course on natural philosophy and more from hiring an engineer or joining the Society of Arts.

With the decline in interest in natural philosophy amongst people of the upper or middling sort, another effect came to the fore: interest amongst those with radical views in the ideas of natural philosophy. One prominent example was Tom Paine who, while he was an apprentice stay-maker, had attended Ferguson's lectures.[83] Later, his books, such as the *Rights of Man*, are suffused with the conviction that scientific reasoning could be applied to understanding the problems of society. As a consequence, institutions such as royalty, regarded as relics of less enlightened periods of history which had outlived their usefulness, would be swept away. This was hardly a reading of natural philosophy that would have found favour with George III or even Demainbray some years before.[84]

As Priestley deftly expressed the point:

This rapid process of knowledge, which, like the progress of a wave of the sea, or sound, or of light from the sun, extends itself not this way or that way only, but in all directions, will, I doubt not, be the means, under God, of extirpating all error and prejudice, and of putting an end to all undue and usurped authority in the business of religion, as well as of science; and all the efforts of the interested friends of corrupt establishments of all kinds, will be ineffectual for their support in this enlightened age; though, by retarding their downfal, they may make the final ruin of them more complete and glorious. It was ill policy in Leo X. to patronize polite literature. He was cherishing an enemy in disguise. And the English hierarchy (if there be any thing unsound in its constitution) has equal reason to tremble even at an air pump, or an electrical machine.[85]

Both George III and Joseph Priestley had air pumps to Smeaton's design, yet they had wildly different views on constitutional issues. Their understanding of the connection between natural philosophy and the constitution was also different, differences that became greatly exacerbated during the French Revolution. In Britain as elsewhere, a few had now grasped the point that while air pumps and electrical machines might help in understanding the world, the more important task to accomplish, a task which required other skills, was to know how the world could be changed.

---

[82] Priestley, to give one example, pointed out the great advances made in chemistry since Newton's time. See Priestley (1775–7) p. vi. For the study of natural history and chemistry at Cambridge in this period see the remarks in Gascoigne (1989) pp. 283–8.

[83] See Paine (1794) p. 37.

[84] See Schaffer (1983, 1986).

[85] Priestley (1775–7) p. xxiii.

Catalogue

The collection has been divided into five main groups for the purposes of this catalogue.

1   The apparatus of S. C. T. Demainbray

2   The material supplied for King George III by George Adams in 1761–2
    (a) pneumatics
    (b) mechanics

3   Other seventeenth and eighteenth century material

4   Nineteenth century material

5   Less significant items

PART 1
The apparatus of S.C.T. Demainbray

The objects in the King George III collection which belonged to Demainbray have been identified by a variety of means. A few items have Demainbray's name or initials on them, but in most cases the items have been identified from documents. These documents include four different syllabuses of his lecture courses dating from 1750 to 1754, notes and sketches made by one Père Chabrol at Demainbray's lectures in Bordeaux in 1753, and the 'Queen's Catalogue', an inventory of the items deposited by Queen Charlotte at Richmond Observatory. This last catalogue contains all the instruments that can be attributed to Demainbray using the other sources. According to Whipple in 1926 the contents of the Queen's Catalogue also featured in a larger catalogue entitled *A Catalogue of the Philosophical Instruments collected by S.D. in His Majesty's Observatory at Richmond*. This suggests that the Queen's Catalogue represents at least part, and possibly all, of the material used by Demainbray. It does not list the items made for George III by Adams in 1761 and 1762.

Having identified some of Demainbray's instruments using these documents, other related objects can then be

associated with him. Though Demainbray's objects are not as similar to each other as the objects by Adams, they do share a few characteristic features. Certain materials are common: green/blue painted tinplate and a dull brass with a high tin content, for example. In contrast with Adams, who used mahogany almost exclusively, the type of wood used for Demainbray's items varies and the workmanship is rough. Interestingly, apart from instruments by Jeremiah Sisson, only a small number are signed by a maker, and none of the unsigned instruments is of the highest quality. This is hardly surprising given that it was a collection to be used, packed up, and taken to the next town where Demainbray was to lecture.

Demainbray's collection includes items acquired on his travels. He bought several in Edinburgh, and collected a number in France including two instruments by 'Costa of Bordeaux'. We also know that Demainbray had instruments by Pease, and magnets by Knight, Morgan, and Hindley. Henry Hindley of York also made a pyrometer for Demainbray, possibly to the latter's design. In short, Demainbray's collection was the work of many people, some of whom were not highly skilled.

The scope of the original collection can be determined from the Queen's Catalogue and the syllabuses. The former contains a main inventory of 307 items; over half of which can otherwise be identified as items which belonged to Demainbray, although not all have survived. Some others cannot be directly associated with Demainbray's courses because they are not distinctive enough. For example, a funnel for pouring mercury or a lead weight are unlikely to feature in a syllabus or lecture notes. Leaving these and similar accessories to one side, there are about 50 items listed in the Queen's Catalogue which cannot be linked to Demainbray through the syllabuses or lecture notes. It is a reasonable assumption that the majority of these were acquired by Demainbray after 1754, the date of his last surviving syllabus. Certainly the 'Map of Venus's Transit 1769' and the Dollond prisms come into this category.[1] With a few exceptions we have assumed that the main inventory in the Queen's Catalogue represents the Demainbray collection as it was around 1770, though it is obviously possible that some items came from other sources.

Demainbray's syllabuses show how his collection changed during the early 1750s. A cursory glance shows the similarity of his courses to those given earlier in the century by Hauksbee and Whiston, Keill, Desaguliers, and 's Gravesande. In these courses it was usual to start with the properties of bodies, followed by mechanics, i.e. statics and 'motion'. The order after that varied, but usually the course continued through hydrostatics and pneumatics, optics, and then astronomy. By the mid-century magnetism was sometimes included as were electrical experiments. Within each topic, all lecturers covered some of the same ground. Attraction and adhesion were illustrated with capillarity tubes and plates, centre of gravity by the rolling double cone and Leaning Tower of Pisa models. Model carts on inclined planes demonstrated the superiority of large wheel diameters and fountains demonstrated air pressure. While a few lecturers were genuinely innovative, many, such as Demainbray, struggled for novelty.

The dating of Demainbray's syllabuses has been done as follows. He was awarded an Ll.D. in 1750 which is mentioned in each syllabus so they cannot date from before then.[2] He became a member of the Bordeaux Académie in May 1753, of Montpellier in January 1754, and of Lyons in May 1754.[3] The syllabus in English which does not mention any of these memberships must be the earliest and dates from between 1750 and early 1753. One syllabus in French lists membership of Bordeaux and Toulouse and so dates from between May 1753 and January 1754; a second in French lists memberships of all these academies and must date from between May and October 1754 when Demainbray returned to London. A fourth syllabus, in English, lists all the academies and is probably the syllabus referred to in the *Daily Advertiser* of 26 December 1754 as one 'he has published since his late Return to England'.

The earliest syllabus contains descriptions of items dropped from later versions. It is also the longest course with 40 lectures arranged as follows: eight mechanics, four motion, fourteen hydrostatics and pneumatics, one magnetism, eight optics, and five astronomy. The later syllabuses had 34 lectures: seven mechanics, three motion, four astronomy, twelve hydrostatics and pneumatics, one magnetism, and seven optics. Demainbray had reduced each section by one lecture and moved astronomy to follow mechanics. The link between these subjects was provided by a discussion of forces acting towards a centre. The rearrangement also allowed Demainbray to end his course on a triumphant note— Newton's theory of colour.

The first lecture of the earliest course contained a preliminary discussion on the properties of bodies. Although Demainbray had an electrical machine, a sketch of which is in Chabrol's notes, and an earlier interest in the subject, this is the only place in the course in which electricity is mentioned. The glass balls to float on water, lead balls to show attraction, and cohesion and capillarity tubes, all listed in the Queen's Catalogue, would have been used. In later courses this lecture was combined with the second. The next four lectures followed the well-trodden path of the simple machines: a balance, a false balance, a set of pulleys, a

---

[1] Unfortunately, these items are not part of the collection today.
[2] *Edinburgh Evening Courant* 28 August 1750.

[3] Aberdeen University MSS, 2045/3, 2045/4, 2045/5.

compound lever, an axis in peritrochio, and a wedge machine attributed to 's Gravesande. Most of the apparatus for the sixth lecture on compound engines has survived including the models of a rat's tail crane, a scorpion, and Vauloué's pile-driver. The last two lectures in the section on statics dealt with friction. Dr Desaguliers's friction engine and the wheel carriage on the inclined plane are featured here together with apparatus to show the friction of ropes and the strength of a man, the last called a Samson's Seat. The maximum power of a man in raising water was demonstrated—a central concern which arises throughout the syllabuses.

The first lecture on motion dealt with elasticity and Newton's first law of motion. Here the 'diagonal machine', a machine for illustrating two different motions, is the only apparatus mentioned. The second lecture had more apparatus including one to show the parabolic path of projectiles and another to show the trajectories of water from different heights in a tank. Both experiments were taken from 's Gravesande. Demainbray then departed from the standard programme to discuss a new scheme of forging cannons. The third lecture concerned pendulums and the shape of the Earth. Only a pyrometer by Hindley survives— pyrometers being important for the discussion of how to compensate pendulums for the effects of heat. The last lecture in this section dealt with central forces, but in later courses this topic was moved to the astronomy lectures. The main item was the whirling table.

The mechanics and motion sections are similar to those of Hauksbee and Whiston's lectures in the order of topics, the stress laid on friction, and some of the apparatus. The sections dealing with hydrostatics and pneumatics owe much to Desaguliers's *A Course of Experimental Philosophy* of 1744. This part was the longest and arguably the most important in the course in terms of the applications it covered. The first three lectures went through the basic principles of hydrostatics; Desaguliers was followed in detail, although not in his order. Though Desaguliers did not separate 'pneumatics' from hydrostatics since 'air is a fluid whose pressure and resistance acts like water . . .', Demainbray followed the more usual practice of dealing with the properties of air under the heading pneumatics. He tells us that his air pump was 'after Hauksbee's manner' and

a double-barrelled air pump is listed in the Queen's Catalogue.

The first lecture concerns the construction of barometers and includes standard experiments on the air pump such as the adhesion of marble planes. The second resembles Desaguliers's section on 'The Art of Living Under Water', even using models similar to those shown in his plates. The third lecture continues with air pump experiments such as the breaking of glass phials and the demonstration of artificial lungs, the wind gun, and the condensation fountain.

The last four lectures in this section cover a wide range of subjects of contemporary interest including 'smoak jacks', earthquakes, meteors such as 'flying dragons' 'ignes fatui', acoustical experiments, thermometers, hail and snow, and the rise of sap in plants. The acoustics material is well represented but most of the other items cannot be identified today. The most important topic, dealt with in the penultimate lecture, was 'fire engines' or what we would now call atmospheric engines. Demainbray had a Newcomen engine model on which to base his lecture.

Demainbray advertised the experiments in his lecture on magnetism as 'yet unpublished' although they were likely to have appeared in *Philosophical Transactions of the Royal Society*. In these he used a lodestone, and magnetic bars by both Knight and himself.

The optical apparatus, particularly the microscopes, was better in quality and quantity than the rest of the collection. The syllabus loosely followed that of Hauksbee and Whiston. The first lecture concerned reflection where the major items were the catoptrick cone and cylinder. Then came refraction which included the use of a zograscope, called a diagonal mirror and lens, and anamorphic pictures which Demainbray described in detail. The third lecture involved vision, and included the dissection of an eye and methods for couching a cataract. After this came the usual items: camera obscura, scioptic ball, magic 'lanthorn', and convex and concave mirrors.

Microscopy was a special interest of Demainbray's to judge by the enthusiasm with which he described the instruments in the two lectures he devoted to it. He used examples of Leeuwenhoek, Marshall, and Lieberkühn type microscopes among others. The Queen's Catalogue states that the first was by Leeuwenhoek himself, presumably bought at the sale of Leeuwenhoek instruments in 1747.[4] He discussed the methods by which the magnification could be measured, showed experiments for 'viewing in the solar way', and showed a portable microscope which could be contained in a snuff box. The uses of models of Buffon's speculum and of burning glasses were appended. The lecture on telescopes contained all the main types available, including the polemoscope. The last lecture in the optics section dealt with Newton's doctrine of light and colour and included a heliostat 'if the weather permit'. Later the heliostat was omitted.

The astronomical instruments comprised a terrestrial and celestial globe, an 'armillary sphere of an orrery', a planetarium, and a large orrery of four and a half feet diameter, made by Charles Butcher of Bedford, which could exhibit the motions of the planets 'in their true times'. The first lecture on astronomy consisted of an introduction on the globes. Then followed the various solar systems, the planets, the moon, the seasons, eclipses, and tides. The eclipses of Jupiter's satellites were applied to the problem of longitude. The five lectures were reduced to three in later courses.

The greatest changes were made between the first and second syllabuses; items were added and others omitted. Demainbray's collection now included items he designed himself. In the second and subsequent courses the first and second lectures on mechanics were combined. The electrical experiments were not mentioned but we know from a sketch in Chabrol's notes that they were still included. Helsham's property of the balance was added to the second lecture and a model of a water wheel at Newcastle was added to the third. In the sixth lecture Demainbray stressed the importance of friction but the only new item was a four-wheeled carriage model, a design presented to the Dublin Society. The sections on motion and astronomy remained unchanged in content throughout the syllabuses.

The first two lectures on hydrostatics were combined and the collection was embellished with models of water pipes based on Desaguliers's work on the Edinburgh water supply in the early 1720s. Plates in Desaguliers's *A Course of Experimental Philosophy* and sketches in Chabrol's notes show the apparatus. Two other models

were added to this section: a Dutch forcing pump to the third lecture and a model of the water mill at the Bazacle at Toulouse to the fourth. The first lecture on mechanics lists a greatly enhanced collection of barometers: portable, simple, wheel, and diagonal. The whole lecture was devoted to their construction and use, the air pump being mentioned only as an accessory. In the second lecture, 'The Art of Living Underwater', marble planes were now omitted, probably because they were unreliable, and their place was taken by Magdeburg hemispheres. Demainbray included an instrument of his own invention to take air underwater. In the fourth lecture Demainbray also used apparatus of his own design to show Hales's work on vegetation. The fifth lecture became even more of a mixed bag with hail and snow and acoustics being supplemented with 'a machine to fetch up sea water', or water bottle, and a sea gauge. A model of Savery's engine was added to the model Newcomen engine in the lecture on atmospheric engines.

Further evidence of Demainbray's additions comes from the lecture on magnetism and the section on optics. A 'magnetical magazine' was introduced into the lecture on magnetism following Gowin Knight's laminated magnets. In the optics section an innovation appeared in the fourth lecture when he mentioned 'machines' for showing the paths of light rays in the myops and presbyte or short- and long-sighted eyes. These 'machines' also appeared in the last lecture, now on telescopes and Newton's theory of colour.

It is likely that Demainbray added to his collection of microscopes at this time: the second syllabus mentions instruments by Cuff, Scarlett, Wilson, and Joblot and an aquatic microscope by Trembley in addition to those in the earlier syllabus.

The changes made in the third syllabus can be dated fairly accurately. They are largely models of French technology copied from the original. One source was the Marquis de Grollier de Servière's collection at Lyon, a collection of curios built up by the Marquis's grandfather in the second half of the seventeenth century. In his third lecture on mechanics Demainbray added a model of a Persian wheel, a model of an ordinary water wheel, and a model of a machine to raise water by means

---

[4] Engelsman (1983).

of ladles, copied from de Servière's collection.[5] Following a demonstration of the model of Vauloué's pile-driver in the next lecture, Demainbray added a model of de Servière's pile-driver—a comparison which could only favour the English model designed half a century later. The last lecture in this section contained a demonstration of two additional cart models.

There are fewer changes in the hydrostatics and pneumatics section. A model of the comet mill by Dubost for use on the Rhône was added incongruously to the last pneumatics lecture on thermometers and hygrometers. In the optics section Demainbray added a microscope by Clark of Edinburgh and re-introduced the heliostat. In the last lecture he mentioned a telescope with six lenses from Costa of Bordeaux.

Demainbray's course appears to have become fixed around this time as few changes were made later, and those that were, were in presentation rather than content. The selection of instruments used remained the same as far as can be ascertained.

The fate of the microscopes in the King George III Collection has been dealt with elsewhere.[6] Having been transferred to King's College with the rest of the collection in 1840, they were given to the collector Frank Crisp in 1887 in return for a modern microscope for the college. Although this was done on the understanding that the instruments would later be presented to the nation, the Crisp Collection was sold in 1925. Fortunately some were purchased by Thomas Court and subsequently presented to the Science Museum. As a result, several of Demainbray's microscopes can be found in the microscope collection at the Science Museum, including a Wilson screw-barrel, a Cuff aquatic, a Cuff solar, a pocket microscope by Sisson, and a microscope by Loft with 'Dr. Demainbray' engraved on the body. Another pocket microscope made by Sisson for Demainbray is at the Museum of the History of Science in Oxford.[7] It is possible that in future further instruments will be found to have connections with Demainbray.

Given the time span and the fluctuating circumstances of the collection it is fortunate that any instruments have survived at all. These instruments and the information we have about them provide a remarkable and unique insight into the practice of experimental philosophy in the mid-eighteenth century.

[5] See de Servière (1719).
[6] Whipple (1926) pp. 502–28.
[7] Chaldecott (1951) p. 83.

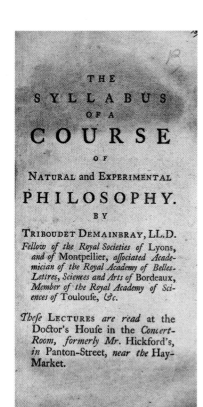

THE
SYLLABUS
OF A
COURSE
OF
NATURAL and EXPERIMENTAL
PHILOSOPHY.
BY
TRIBOUDET DEMAINBRAY, LL.D.
*Fellow of the Royal Societies of Lyons,
and of Montpellier, associated Acade-
mician of the Royal Academy of Belles-
Lettres, Sciences and Arts of Bordeaux,
Member of the Royal Academy of Sci-
ences of Toulouse, &c.*

These LECTURES *are read at the
Doctor's House in the Concert-
Room, formerly Mr. Hickford's,
in Panton-Street, near the Hay-
Market.*

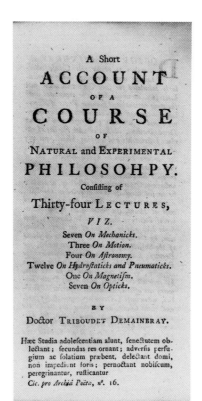

A Short
ACCOUNT
OF A
COURSE
OF
NATURAL and EXPERIMENTAL
PHILOSOHPY.
Consisting of
Thirty-four LECTURES,
VIZ.
Seven *On Mechanicks.*
Three *On Motion.*
Four *On Astronomy.*
Twelve *On Hydrostaticks and Pneumaticks.*
One *On Magnetism.*
Seven *On Opticks.*

BY
Doctor TRIBOUDET DEMAINBRAY.

Hæc Studia adolescentiam alunt, senectutem ob-
lectant; secundas res ornant; adversis perfu-
gium ac solatium præbent, delectant domi,
non impediunt foris; pernoctant nobiscum,
peregrinantur, rusticantur
Cic. pro Archiâ Poëtâ, n°. 16.

*The syllabus Demainbray published
on his return to London
in late 1754*

DOCTOR DEMAINBRAY *en-
deavours to render this Course entertaining and improving, being particularly calculated for such* GENTLEMEN *and*
LADIES, *as would chuse to be acquainted with
the more rational and sublimer Parts of Knowledge in the most expeditious and familiar Manner; and he attempts giving the most obvious
Demonstrations, illustrated with Variety of curious Instruments and working Models, not only
of such Engines in Mechanicks, Hydraulicks,
&c. as are known in Great-Britain, but of
such as are in Use abroad, and not in Practice
in these Kingdoms; which the Doctor has had
uncommon Opportunities of collecting, and which
he caused carefully to be modelled on the Spot,
according to regular Scales, having ever had
most at Heart, the becoming an useful Member of Society in this his native Country; some
slight Idea may be framed of his large Collection, by the Contents of the following Abstract
of the Subjects of each Lecture, in which, for
Brevity's sake, it is impossible to descend to the
Particulars of the Apparatus, which tho' so
extensive, is not only framed to fix the Eye,
but instruct the Mind, with what Truth Geometry can afford, without its apparent Difficulties, when these Lectures are regularly pursued,
and their Chain, as depending the one on the
other, followed without Interruption, the Ho-
nours these Lectures have received by the Ap-
probations of the different Royal Societies and
Royal Academies, before which they have been
read, may claim some Credit, since their Success has made the Doctor be received a Member of each to which he has been called; tho'
even in some Cases contrary to the usual foreign
academical Regulations.*

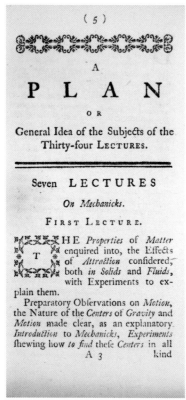

(5)

A
PLAN
OR
General Idea of the Subjects of the
Thirty-four LECTURES.

Seven LECTURES
On Mechanicks.
FIRST LECTURE.

THE *Properties of Matter*
enquired into, the Effects
of *Attraction* considered,
both *in Solids* and *Fluids,*
with Experiments to ex-
plain them.
Preparatory Observations on *Motion,*
the Nature of the *Centers* of *Gravity* and
Motion made clear, as an explanatory
Introduction to *Mechanicks, Experiments*
shewing how *to find* these *Centers* in all
A 3　　　　kind

(6)

kind of *Bodies*, with Illuſtrations concerning the *Line of Direction* in different *Poſitions of Bodies*.

II. Introductory Obſervations to the Knowledge of thoſe Engines called the *mechanical Powers*, with an *Engine wherein theſe Powers are combined*, ſo that a very conſiderable Weight may be ſupported by a very ſmall Power, to illuſtrate that univerſal Law in Mechanicks, whereby we find that we cannot alter Nature, ſince where we would gain Strength, we muſt loſe Time ; and where we would gain Time, we muſt employ more Strength : The *Balance* explained ; the *Frauds* that may be committed by *a falſe Balance* enquired into ; ſome Remarks on *Doctor Helſham*'s mechanical *Paradox*, illuſtrated by a *Balance of a new Conſtruction*: The different Kinds of *Levers* enquired into, and their *Combinations* ſet in a clear Light by *proper Inſtruments*.

III. *Pullies* illuſtrated with curious *Tackles*: The *Axis in Peritrochio*, *ſimple* and *compound*, explained by accurate *Models*, their Doctrine applied to the *Conſtruction* of a *Pocket Watch*, alſo applied to ſome ſimple but *conſiderable Improvements*

(7)

Improvements on *Water Wheels*, ſuch as for *draining Mines*, *raiſing large Loads* from Mines, &c. The *Perſian Wheel*, or a *Model* of the *Machine* with which Gardens, Meadows, &c. are watered in *Languedoc* and *Provence* ; and a *Machine of ſingular Conſtruction* copied from the Marquis *de Grollier*'s Collection at *Lyons*, to be built on the *Banks* of any *River*, in order to *raiſe Water* by means of the *Current* to the *Top* of any *Building* where the *Reſervoir* is placed, by means of *Ladles*, *Troughs*, *a Water-Wheel*, &c.

IV. Experiments to ſhew the different *Effects of the ſame Power*, *acting at the ſame Point of an Engine, in different Directions* ; on the *inclined Plane* ; the *Wedge* illuſtrated adequately to the Doctrine laid down, and the *Screw* variouſly exhibited, with *Models* of an *endleſs Screw*, and of *Archimedes*'s *Screw* : Conſiderations on the *Uſes* of a *Fly* applied to Machines.

V. Compound Engines examined ; common *Capſtanes* ; Rat's Tail *Cranes* ; the *Scorpion* of the Antients ; *Vauloüé*'s *Machine* for driving of Piles uſed at *Weſtminſter-Bridge*. alſo a *new Engine*
A 4 *for*

(8)

for driving of Piles, without the Aſſiſtance of either Men or Horſes, moved by the *Stream of a River*, which receives no Obſtruction from the Change of Tides, given to the Doctor (when at *Lyons*) by the Marquis *de Grollier*.

VI. Uſeful Obſervations and curious *Experiments* on the *Friction of Engines* ; Enquiries on the Reſiſtance of *Weights* and *Surfaces* by the late *Doctor Deſaguliers*'s *Machine*, the Reſiſtance of *Lines* and *Weights* over *ſquare* and *round Bodies* examined ; the Effects of *Friction* calculated in *one compound Machine* compoſed of all the *mechanical Powers*; curſory Hints on the Advantages of a *Cart* over a *Sledge*.

VII. The *Friction of Ropes* explained according to Monſieur *Amonton*'s Ideas ; *Wheel Carriages* examined, their Uſes and the Manner of applying them to the beſt Advantage proved with *Models* ; the Manner in which *Wheels* ſhould be *ſhod* with Iron, Monſieur la *Puyade of* Toulouſe's *Method* ; the Model of an *improved four wheeled Carriage* offered to the *Dublin Society*, the Reaſons why rejected ; the *Iriſh low back Carr* ; the *Strength of Horſes* conſidered,

(9)

ſidered, that of a *Man* illuſtrated by *a Machine* called *Sampſon*'s *Seat* ; with Obſervations and Experiments applied to different Uſes in Life, particularly a *Model* to ſhew the *Maximum of a Man's Power* in raiſing of Water.

_____

Three **LECTURES**
On Motion.

VIII. THE various Opinions (collected from the moſt eminent Philoſophers) on *Elaſticity*, laid down, with ſome Illuſtrations on its Effects : The *firſt Law of Motion* explained by Experiments and familiar Obſervations ; the *Law of compound Motion*, and the *diagonal Direction* of a Body ſo moved, rendered obvious by an *Inſtrument* ſelf-evident in its Contrivance.

IX. The *ſecond* and *third Laws of Motion* variouſly demonſtrated, in which the general Rules of the *Doctrine of Gunnery* will be explained, and the *Parabolick Motion of Projectiles* made evident to the Sight by S'*Graveſande*'s *Machine*, and by an *Engine to play Jets*, ſo as to deſcribe according to the *different Elevation*
A 5

(10)

Elevation of the Monture *different Randoms*; in this Lecture the *Doctrine of falling Bodies* will begin to be confidered; and *Action* and *Re-action* will be varioufly demonftrated; with a felf-evident *Experiment* laid down as an Objection to the new *Schemes of forged and other lighter Canons* propofed *for Shipping*, fhewing what Advantages, and Difadvantages may attend thefe modern Improvements.

X. The *Doctrine of Pendulums* examined, fome curious *Experiments* on the *Expanfion of Metals* by Heat, fhewn with a *Pyrometer* of a new Conftruction, fo accurate as to point out very confpicuoufly the abfolute Quantity of fuch an Expanfion in a moft exact and minute Manner; the Effects of the *Expanfion* and *Contraction of Pendulums* corrected by a *Model* as firft propofed by Mr. *Graham* the Watch-Maker, and by a *very exact Model* after that Manner, fince invented by Monf. *Julien le Roi*, the King of *France*'s Watch-Maker at *Paris*: Some Obfervations on the *firft Difcovery* by *Pendulums* of the *Earth*'s true *fpheroidical Figure*.

Four

(11)

Four LECTURES
On *Aftronomy*.

XI. CEntral Forces varioufly explained and illuftrated with inftructive *Experiments* on the *Whirling Tables*; Objections to *Defcartes*'s *Vortices*; an Explanation of the *Newtonian Syftem of Aftronomy*; an *Experiment* that will be an obvious *Conviction* of the *oblate fpheroidical Figure of the Earth*, agreeable to *Monfieur de Maupertuis*'s, &c. Obfervations at the northern Polar Circle, and to *Monfieur Bouguer* and the other *French Academician*'s *Meafure of a Meridian* at the *Equinoctial Line*.

XII. Introductory Obfervations to *Aftronomy* continued; the *Doctrine of the Spheres* explained and illuftrated by the *Armillary Sphere*: *Ptolemy*'s, *Tycho Brahe*'s and *Copernicus*'s different *Syftems* fhewn with a *Planetarium*, and examined with an *Orrery* of four Feet eight Inches Diameter, containing all the *primary* and *fecundary Planets*.
Note, *This Inftrument not only performs the Revolutions of all the Planets in the ufual Method, by a Winch, but by means*

(12)

means of an Eight-Day-Clock, exhibits their periodical Revolutions in their true Times.

XIII. The different *Seafons* of our Year fhewn on the *Orrery*; the *Moon*'s different *Phænomena* confidered; the *Sun* and *Moon*'s *Eclipfes* explained; the Planets *Mercury*, *Venus* and *Mars* enquired into.

XIV. *Jupiter* with his *Belts* and four *Satellites* defcribed; *Saturn* with his *Ring* and five *Moons*; with Illuftrations on the Orrery and Planetarium; *Obfervations* on the moft probable Methods of attaining to the *Difcovery* of a Method of finding *Longitude at Sea*: To conclude, Aftronomy with Thoughts on *Comets* and on the *Tides* according to the *Newtonian Doctrine*.

Twelve LECTURES
On *Hydrostaticks and Pneumaticks*.

XV. HYdroftatical Principles explained and illuftrated by *Experiments*, fuch as *Glafs Figures* floating at the Top, Middle and Bottom of a *Veffel*

(13)

Veffel of Water, a *large leaden Weight* made to *fwim* in Water, &c. the various *Methods* of *leading Water* thro' Pipes, with *practical Rules to cure* fuch as are *Wind-bound* laid down, with *Models of the Edinburgh Water-works*, and many ufeful Obfervations collected from the late Doctor *Defaguliers* and others.

XVI. The *Experiment* of the *Hydroftatick Paradox* tried, and *demonftrated* by the *Hydroftatick Bellows*, &c. *Syphons* examined, their *Effects* explained by a *Syphon Fountain*, different *Syphon Cups*, and *Tantalus*'s *Cup*; alfo fome *Phænomena* of *Springs* and *natural Syphons* explained by thefe Principles.

XVII. On the Conftruction of *fucking, forcing and lifting Pumps*, illuftrated with curious *glafs Models*; the *Pump* invented by *Archimedes*, and the *Dutch forcing Pump* ufed in *Ships*, &c. defcribed; the Model of a *new improved Bucket* for Pumps *ufed in Mines* explained, alfo *a Glafs Air Veffel* applied to the Model of a forcing-Pump, to produce a conftant Stream of Water, to fhew the Effects of *Engines for extinguifhing of Fires*: Some *Experiments* to vary

(14)

vary the Demonſtrations of the *Preſſure* of the *Atmoſphere* on Fluids.

XVIII. The *Preſſure of Fluids* from *different Heights* demonſtrated by an adequate and accurate *Engine*; ſome *Obſervations* on the *Acceleration of Bodies* in their *Fall* further added, and the *Repreſentation* of different *Parabolas* ſhewn on a different *Principle* than the former; the *Conſtruction* of the *Water-Mills du Bazacle at Touluſe*, for *grinding of Corn*, elegantly *modelled* with utmoſt Care, as Doctor *Demainbray* had obtained Leave of ſo doing, as an Academician of that City, of the Magiſtrates; in *this Conſtruction* not only the *Velocity of the Stream* of the River *Garonne*, but the *perpendicular Preſſure and Weight* of the *Water* is employed to *work* the *Engine*: The Neceſſity of a *Counter-Preſſure from above for Fluids to flow*, illuſtrated by various, obvious *Experiments*, ſuch as the *French* and *Engliſh Fountains at Command, Conjurer's Funnel*, &c.

XIX. The *Clepſydra* examined, *old* and *new Conſtruction* ſhewn; the *ſpecific Gravity* of Bodies enquired into; *the Weight of Fluids* ſhewn by the *common Hydrometer*

(15)

Hydrometer or *Water-poiſe*, and by *Clerk's Invention*; the *Uſe* of the *Hydroſtatick Balance* introduced by ſome ſelf-evident *Experiments*; ſeveral *Illuſtrations* with the *Hydroſtatical Balance* to find the relative Weights of Solids and Fluids.

XX. On PNEUMATICKS, the *Air's Preſſure* proved by *Experiments* on *the Air-Pump*, by *the Barometrick Mercury* ſupported in *a Glaſs Tube*: The Application of Barometers to find the Weight of the Atmoſphere, its Extent, and the Height of Mountains; the different *Conſtruction* of Barometers explained, by *common, portable, diagonal, wheel, pendant*, &c. *Barometers*; with the Effects of an *artificial Storm* ſhewn on theſe Inſtruments, the *Model* being after *Haukſbee's* Manner.

XXI. On the Art of living under Water, the *Diving-Bell* illuſtrated with an adequate *Apparatus*; in which *the Air's Elaſticity* will be examined, and its *Preſſure* further illuſtrated with *Braſs Hemiſpheres*, &c. Enquiries on *vitiated Air* after paſſing thro' *the Lungs*; ſome curious *Obſervations* and new *Experiments* made by the Royal Academy of Sciences of *Touluſe, on Damps in Wells, Mines,*

(16)

Mines, &c. Hints taken from the Rev. Doctor *Hales*, and applied to the *fatal Conſequences of Burials in Churches* by Doctor *Haguenot* of the Royal Society of *Montpellier*: Doctor *Halley's Improvements* on the *Diving Bell*; Captain *Rowe's Tub*, with a *Model* of Mr. *Triewald's* new *Campana Urinatoria*.

XXII. *The Air's Elaſticity and Preſſure* further proved by various *Experiments*; *Reſpiration* illuſtrated, with different Inſtruments to repreſent *artificial Lungs* both *in open Air* and *in vacuo*; *Wind-Guns* explained; the *Air's Spring* ſhewn by a beautiful *Hiero's Fountain*; the Conſtruction of *a new Cock* for *anatomical Injections*.

XXIII. Some Obſervations on *Winds*; *Smoak Jacks* illuſtrated with a *Lanthorn with Tin Vanes* turned round by the conſtant Riſe of a Stream of rarified Air; *Thunder, Lightning* and *Earthquakes* examined and illuſtrated, with other Meteors, as *Flying Dragons, Ignes fatui*, &c. explained; *Obſervations* taken from *the Rev. Doctor Hales* on *the Riſe of the Sap in Plants*; the conſtant and alternate Circulation explained; the *Neceſſity of Air* to promote *Vegetation*

(17)

getation proved by *Experiments*; particularly ſome Obſervations illuſtrated in a clear, and not uſually practiſed Manner, to prove the Doctor's Aſſertion; *that Elaſticity is no immutable Property of Air*; in theſe Experiments large *Quantities of Air* will appear to be *generated* and *abſorbed*, and their Meaſures aſſigned.

XXIV. Conſiderations on *Hail* and *Snow*, their Formation laid down according to the Prize Diſſertation crowned by the Royal Academy of *Bordeaux*; the *Doctrine of Sounds* explained, *muſical Sounds* illuſtrated with a *Sonometer*, whoſe Contrivance will ſhew the Grounds on which Harps, Violins, Harpſichords, and all String Inſtruments of Muſick, are conſtructed; *the ſpeaking Trumpet and auricular Tube* explained, with *an Account of Echoes*: A *Machine* ſhewn *to fetch up Water from any Depth at Sea*; and *another Machine* for *taking Depths at Sea*, where the Lead and Line cannot reach.

XXV. On *Fire Engines*, with a curious and compleat *Model*, exhibiting all the Improvements made on that moſt uſeful Contrivance; *a Model* of Captain

(18)

tain *Savery's Fire Engine* will raife Water by means of Fire in this Lecture; fome modern Methods made ufe of for *preferving* the Boiler, and *mending the Cylinder*, if *broken*; being of utmoft Moment to Perfons concerned in Mines, where thefe Engines are wrought.

XXVI. The Conftruction of *Thermometers* explained; various *Hygrometers* fhewn, *Heat* and *Cold*, *Drynefs* and *Moifture* confidered: An Explanation of the *Comet Mill* for grinding Corn, and applicable to other Ufes, as lately contrived by Monfieur *Duboft* on the River *Rhône* at *Lyons*; this ufeful and ingenious Piece of Mechanifm, never before explained in publick, will be illuftrated by a *Model* taken on the Spot by Doctor *Triboudet Demainbray*, according to an exact Scale, in Right of his being a Fellow of the *Lyon's* Royal Society; and one of the Members appointed to examine mechanical Improvements, during his Stay in that City.

One

(19)

One LECTURE

On Magnetifm.

XXVII. THE (*as yet publifhed*) Properties of *Magnetifm* illuftrated experimentally, *viz.*
The Direction;
Attraction and Repulfion;
Magnetic Effluvia, *as fome fuppofe it*;
Communication;
Inclination; And,
Declination of *Loadftones*, and *artificial Magnets*. With feveral Obfervations taken from Doctor *Gowen Knight*, and others.

Seven LECTURES.

On Opticks.

XXVIII. ON *Catoptricks*, in which reflex Vifion will be varioufly illuftrated by a *cylindrical Speculum*, by the *catoptrick Cylinder*, and *catoptrick Cone* with *optically deformed Pictures*; Light's progreffive Motion enquired

(20)

enquired into; fome Illuftrations on *Perfpective*, its different Effects in a *Picture* and in *the Scenes* of a *Theatre* examined and illuftrated by a Sketch of the latter.

XXIX. The famous *Queftion of Opacity* confidered; *Dioptricks* or the *Paffage of Rays of Light* thro' *different Media* fhewn by *Experiments*; the *Direction of Rays of Light* thro' *Lenfes* enquired into; *Methods* of finding the *Focus* of a *Lens* examined; their apparent *Increafe of Magnitude* laid down, and illuftrated by *a diagonal Mirrour and Lens: Paintings on irregular Surfaces* explained, their Effects experimentally fhewn by different *anemorphofes Perfpectives*.

XXX. *Vifion* explained, by tracing the refracted *Rays of Light* thro' the *Humours of the Eye*, with different *Diffections*, and *anatomical Preparations of the Coats* of the Eye. The Ufes of the *Humours of the Eye*, *explained* and proved by *Experiments*; the Methods of *couching a Cataract*; the whole to be illuftrated with an *Apparatus* adapted to the *Doctrine of Vifion*.

XXXI. *Vifion*

(21)

XXXI. *Vifion* further *examined* and *explained* by *Machines* which *fhew the Direction of the Rays of Light* in the *Myopes* and *Prefbytæ*; the *Manner* by which their *Defects* are affifted by *concave* or *convex Lenfes*; the *Objections* fome Authors have made to the *Retina's* being the immediate *Organ of Senfation of Sight*: A further *Explanation of Vifion by the fcioptrick Ball*, and *the Camera Obfcura*; *the Magick Lanthorn* fhewn; *Experiments* on *plane*, *convex* and *concave Mirrours*; the Advantages of the latter in viewing *Perfpective Defigns* will alfo be fhewn; the Ufes of *Scenographical Boxes* for the fame Purpofes.

XXXII. On the different Conftructions of *Microfcopes*, *fingle* and *compound*, illuftrated by *Leeuwenhoeck's*, *Wilfon's*, *Joblot's*, *Clerke's* of *Edinburgh*; *Marfhal's*, *Scarlet's*, *Cuff's*, &c. alfo Mr. *Trembley's* Conftruction of *an aquatick Microfcope* for viewing *Infects in Water*, without difturbing them; with fome remarkable *Experiments* to *explain* their *magnifying* Power; Mr. *Leeuwenhoeck's*, Doctor *Hook's*, and Doctor *Jurin's* Methods of *giving* us the *natural* and *true Magnitude* of exceedingly minute *microfcopical*

I

(22)

crofcopical *Objects* ; Doctor *Lieberkuhn* of *Berlin*'s *Microfcope* for *opake Objects* explained, and the *Ufes* of his *perforated Specula* enquired into. The *Duke de Chaulnes Micrometers*, for the real and apparent Diameter of Objects, made by *Don Noel* the Benedictin.

XXXIII. An *Account* given of fome *microfcopical Objects*, explaining Mr. *Needham*'s and Monfieur *de Buffon*'s capital *Difcoveries* ; *Solar Microfcopes* as Doctor *Lieberkuhn* invented them ; and *Cuff* improved them, *Cuff*'s *Micrometer* ; a *new* contrived *Inftrument* for viewing in the *folar Way* the *Circulation of the Blood* in a *Frog*'s *Mefentery* improved on Mr. *Baker*, Fellow of the Royal Society's Method ; fome Obfervations taken from that ingenious Gentleman ; the Conftruction of a *new Microfcope*, with all its requifite *Apparatus* contained within *the Compafs of a Snuff Box*, to be ufed in the *fingle*, *compound*, and may occafionally be applied in the *folar Way* ; this Contrivance not only contains whatever is ufeful to the various Purpofes of the fore-mentioned different Microfcopes, but is perfectly fteady, and ftrong in all its Parts : On *Burning Glaffes* ; a fhort View of the different

I

(23)

different *Opinions* on *the Doctrine of Fire*, with an Explanation of a *Model* of Monfieur *de Buffon*'s *Specula* in Imitation of thofe by which *Archimedes* burnt the *Roman* Ships at the Siege of *Syracufe*.

XXXIV. On the different Conftructions of *Telefcopes*, illuftrated with *a dioptrick Telefcope* of the aftronomical Kind of *two convex Lenfes* ; its further Ufe applied to a Ball and Socket, as an *Heliofcope* for viewing the Sun's Face ; *a dioptrick Telefcope* with *four convex Lenfes* for viewing Objects on Earth, and a *dioptrick Telefcope with fix Lenfes* to obtain a double Field of Vifion at Sea, this Inftrument was executed by Signor *Cofta*, Optician at *Bordeaux*, *a Galileo*'s *Telefcope*, with a concave Eye-Glafs, and a convex Object Lens ; a *Polemofcope* or *Opera Glafs* defcribed ; various *Machines* to fhew the *refracted Direction* of the *Rays of Light* thro' thefe Inftruments ; *an Apparatus for aftronomical Obfervations* explained : *Cata-dioptrick*, or *reflecting Telefcopes* fhewn, and their Conftructions enquired into ; with an Explanation of the Mechanifm of the *Gregorian* and *Newtonian* Telefcopes : Sir *Ifaac Newton*'s *Doctrine of Light and*

(24)

and Colours enquired into ; *the Direction of colour-making Rays* demonftrated with *proper Machines* ; the Proofs alledged by *Newtonians* in Support of Sir *Ifaac Newton*'s Doctrine, with the Arguments of their Antagonifts delivered in a candid Manner.

F I N I S.

Si quid novifti rectius iftis,
Candidus imperti ; fi non, his utere mecum.

MECHANICS

D1 Model bucket

before 1753

H 101mm (without handle), **D** 86mm

A sketch in Chabrol's lecture notes shows this bucket being used in an experiment on centre of gravity in Demainbray's second lecture. The Queen's Catalogue describes it as a 'pail of water to be supported by the end of a stick with centre of gravity under point of support on shelf'. It is not mentioned specifically in the syllabuses.

The bucket is made of turned applewood with a lead rim and a brass handle which may have originally been painted. It was cracked but has been repaired. If it is filled with water and hung on a stick overhanging the edge of a shelf or table, the stick cannot support it. However, if another stick is wedged between the end of the first and the bottom of the bucket to push it out of the upright and towards the support, its centre of gravity is shifted and therefore it remains suspended. Adams included the same experiment in *Principles of Mechanics* and attributed it to Desaguliers.

Q. C., Item 21; pail of water to be supported by the end of a stick with centre of gravity under point of support on shelf
Desaguliers 1734, Vol. I, p. 62, Pl. 5, Figs 8 and 9
Chabrol 1753, p. 13
Adams 1762, p. 24, Pl. 13, Fig. 59
See also E215
Inventory No. 1927-1240

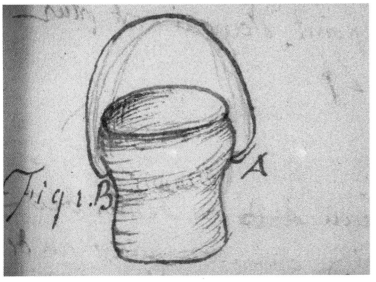

Chabrol 1753

D2 Compound lever

before 1753

L 525mm, **W** 150mm, **H** 310mm

Despite not being mentioned specifically in the syllabuses, this can be identified as belonging to Demainbray, and would have featured in his lecture on levers where 'their combination' was 'illustrated with proper instruments'. It is listed in the Queen's Catalogue and has the characteristic green/blue paint and style of his material.

The compound lever shows a means by which a heavy weight can be supported by a light one. Three beechwood stands on a rectangular beechwood base support three brass levers. The first carries the weight on its short arm while its long arm operates against the short arm of the second lever. The long arm of the second lever depresses the short arm of the third lever and the long arm of the third lever carries the

counterweight or 'power'. In this case the velocity ratio is about 60.

Compound levers were standard items in courses of experimental philosophy throughout the eighteenth century.

Q.C., Item 31; Three compound levers.
Hauksbee 1714, *Mechanics* Pl. 2, Fig. 9
's Gravesande 1747, Vol. I, p. 61, Pl. 11, Fig. 1
Demainbray 1754*b*, p. 6 'The different kinds of Levers enquired into, and their Combinations set in a clear Light by Proper instruments.'
C 33

See also M48   **Inventory No.** 1927-1129

D3 Two frames each with three single pulleys

before 1753

L 170mm, **W** 8mm, **H** 50mm, **D** 40mm (wheel)

This set of pulleys was identified from a sketch in Chabrol's notes which also enabled the 'compleat set of brass pulleys for all the combinations of their power' to be found.

The pulleys are fairly lightweight with four-spoked wheels $1\frac{1}{2}$ inches in diameter. The frames are rectangular; the hooks are of brass wire. This set was used with the frames horizontal, one above the other, in one of the two principal methods of combining pulleys in the early eighteenth century. It was to improve on these methods that Smeaton designed his 'new tackle of pullies' in 1752.

Pulleys were ubiquitous in courses of experimental philosophy during the eighteenth century featuring as one of the five simple machines.

Q.C., Item 32; A compleat set of brass pullies for all the combinations of their power
Hauksbee 1714, Pl. 3, Fig. 5
Smeaton 1752
Chabrol 1753, p. 31
Demainbray 1754*b*, p. 6 'Pullies illustrated with curious Tackles'
C 30

Used with D4, D5, D6, D103.

Inventory No. 1927-1211

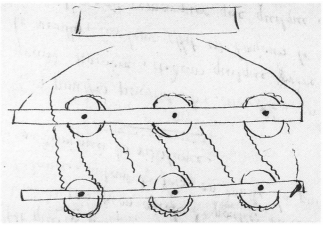

Chabrol 1753

D4 Five single pulley wheels

before 1753

Four pulleys: **D** 40mm, **H** 78mm, **W** 8mm
One pulley: **L** 33mm, **H** 51mm, **W** 8mm

Chabrol sketched the smallest of these pulleys in his notes on Demainbray's lectures. The other four are identical in style and size to D3, D5, and D6, so can be assumed to be part of Demainbray's complete set as listed in the Queen's Catalogue.

As D3

Inventory No. 1927-1236

D5 Two frames each with two single pulleys

before 1753

L 130mm, **W** 8mm, **H** 50mm, **D** 40mm (wheel)

Stylistically these pulleys are identical to those in D3 and can be assumed to be part of Demainbray's 'compleat set of brass pullies . . .' as listed in the Queen's Catalogue. Since one has a hook on a side as well as the ends it is likely that they were used both horizontally, as D3, and vertically. Some hooks are now missing. A sketch in Chabrol's notes shows a similar set of pulleys, with plain rectangular frames, being used vertically.

As D3

Inventory No. 1927-1212

D6 Triple pulley block

before 1753

D 40mm (wheel), **H** 67mm, **W** 26mm

The triple pulley is part of the set of pulleys which includes the above items. It is stylistically identical to the others and would have been used with them in Demainbray's lecture on pulleys, usually the third in the course. Originally there would have been an upper and lower block each containing three pulleys set alongside each other in a frame. The lower block is now missing. According to Adams this type of pulley was 'very compendious and generally used'.

Hauksbee 1714, Pl. 3, Fig. 7
Adams 1762, p. 53

As D3

Inventory No. 1927-1213

D7 Model of water wheel coupled to two beams for working two mine pumps

1753

L 675mm, **W** 340mm, **H** 545mm

This model is mentioned specifically in the Queen's Catalogue but not in Demainbray's syllabuses. It was probably used with the other model water wheels in the lecture on pulleys. An oak wheel is set between two oak beams which each operate pumps. The beams are driven by iron cranks moving connecting rods; they are pivoted on the wooden stand and have chains attached which would carry the pistons of the pumps. The wheel is interchangeable with those in D8; it is probable that experiments were shown using different wheel widths and types since there was some interest in the design of water wheels at this time.

In addition to this model Demainbray had an ordinary water wheel and a model of a great wheel at Newcastle for raising coal out of the mine, neither of which has survived.

Q.C., Item 55; ditto [water wheel] to work two pumps in a mine
Chabrol 1753, p. 33
Demainbray 1753, p. 7
Demainbray 1754*a*, p. 7
Demainbray 1754*b*, p. 7; 'water wheels for draining mines'
Smeaton 1759   **Inventory No.** 1927-1630

D8 Model water wheel and treadmill in common frame

1753

H 535mm, **L** 730mm, **W** 420mm

This apparatus probably belongs with D7 since the wheels are interchangeable. It consists of an oak frame holding an overshot water wheel of the same design and diameter but greater width than that in D7, and a treadmill of the same diameter as the water wheels. The treadmill has two pulleys of different diameters on its axle. It is likely that the wheels were made to show the various powers obtainable using different wheel widths and gearing.

See D7

Inventory No. 1927-1655

D9 Model of Persian wheel with chain of pots

1754   France

H 730mm, **W** 200mm, **L** 410mm

This model was probably made for Demainbray when he was in France in 1754 since it appears in his lecture courses in that year described as 'the Persian wheel or a Model of the Machine with which Gardens Meadows etc are watered in Languedoc and Provence'. A Persian wheel was to be found in the Marquis de Grollier's collection at Lyons which may have been the inspiration.

The model is made of French walnut and the pots are boxwood. It consists of a frame on four legs holding a vertical wheel, over which a chain of buckets passes. The wheel could be turned by a capstan or handle which is now missing. A trough full of water would have been placed between the legs and as the handle was turned the buckets would scoop up water and raise it to a reservoir placed on the axle; from there it could be piped as required.

The Persian wheel or chain of pots is a very primitive device which is unable to raise water much above the centre of the wheel. It was invented in about the second century BC and used extensively in the Middle East and Egypt, where it was known as the *saqiya*. The vertical wheel is geared to a capstan which is turned by animals or people.

Q.C., Item 57; model of Persian wheel with chain of buckets
de Servière 1719, p. 42, Pl. 42
Desaguliers 1744, Vol. 2, Pl. 34, p. 455
Demainbray 1754*a*, p. 7
Demainbray 1754*b*, p. 7; 'the Persian wheel'
Singer *et al.* 1956, Vol. 2, p. 676
Inventory No. 1927-1635

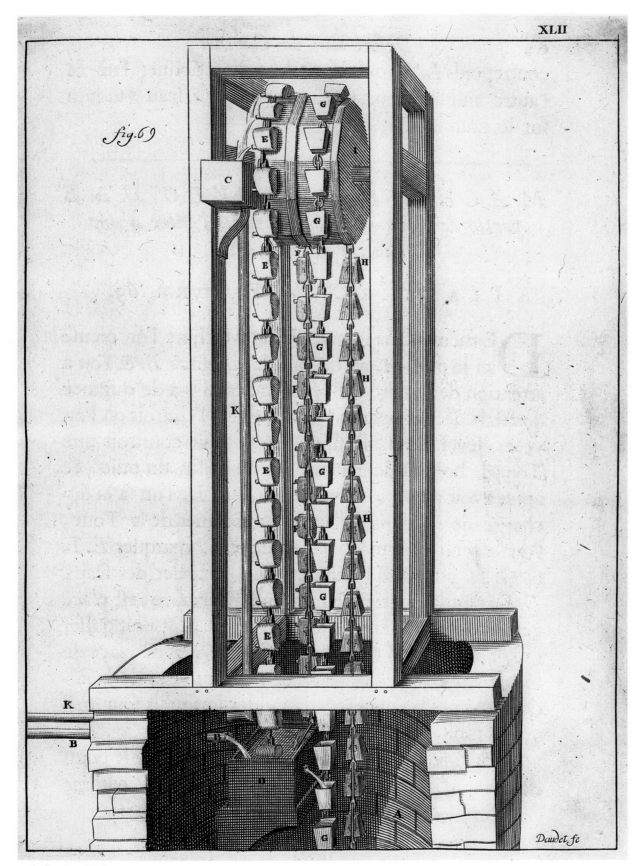

de Servière 1719

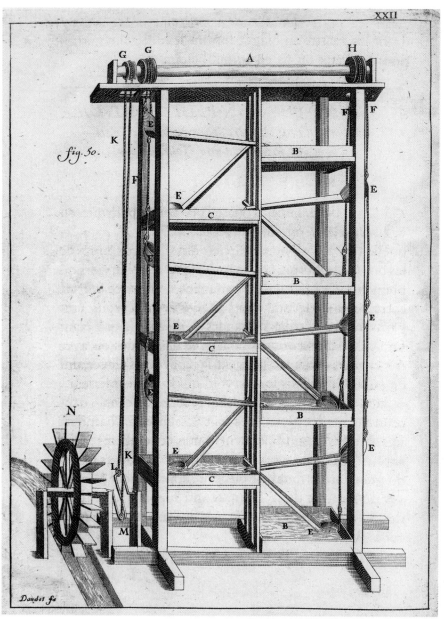

fig. 50.

Doudet fe

de Servière 1719

D10 Model of a ladle machine for raising water

1754   Lyons

H 805mm, **W** 260mm, **L** 700mm, **D** 265mm (waterwheel)

Like the previous item this model has connections with the Marquis de Grollier's collection of curios at Lyons. It was copied from a machine in the Marquis's gardens which stood on a river bank and lifted water up a tower by means of a series of ladles. The Marquis had several machines of this type with slightly differing design features.

A boxwood shaft running across the top of the machine is turned by the water wheel which is operated by means of a handle. As the shaft turns one way the first series of ladles scoops water from each reservoir while the second pours water into the next reservoir. When the shaft turns back the first series empties while the second refills. The ladles and reservoirs are made of painted tinplate, the water wheel is of mahogany, and the axles are of iron.

Demainbray introduced the model into his course when in France in 1754 and brought it back to England later in the same year. He described it as a 'machine of singular construction', using it in his lecture on pulleys.

Q.C., Item 58; model of the Marquis de Grollier's ladle machine to raise water
de Servière 1719, pp. 5–26, Pl. 22
Demainbray 1754*b*, p. 7 'a Machine of Singular Construction copied from the Marquis de Grollier's Collection at Lyons . . . in order to raise water . . . by means of ladles.'
C 71

Inventory No. 1927-1642

D11 Inclined plane

before 1753

SCALE

0 (10) 90

W 160mm, **L** 645mm, **H** 200mm

In the surviving syllabuses the inclined plane demonstration formed part of Demainbray's fourth lecture on mechanics which also included the wedge and the screw. However, in Chabrol's notes it is in the second lecture. Chabrol gives a sketch of this particular apparatus.

This instrument is of a construction similar to that of Hauksbee in that the mechanism to raise the plane is hidden in a curved wooden structure; in this case the wood is beech. There is a small brass quadrant on the side of the plane which would have carried a plumb bob. It is graduated in degrees and numbered in tens from 0 to 90. The inclination is altered by means of a strut on a thick beech screw thread which is turned by the large boxwood handle at the end of the machine. A small brass pulley similar to Demainbray's other pulleys is mounted at the centre of the top of the plane.

The inclined plane was a standard item in a course of experimental philosophy by the mid-eighteenth century.

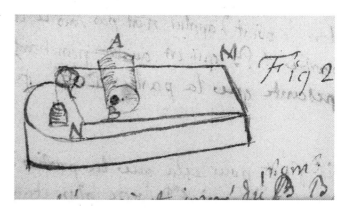

Chabrol 1753

Q.C., Item 39; an inclined plane machine with a brass quadrant for its elevation

Hauksbee 1714, Pl. 5, Fig. 5

Desaguliers 1734, p. 108, lect.3, Pl. 11, Fig. 2

Chabrol 1753, p. 13

Demainbray 1754*b* 'Experiments . . .on the inclined Plane'

Inventory No. 1927-1457

D12 Model of an Archimedean screw

before 1753

L 278mm, **W** 85mm (stand), **H** 55mm

The Archimedean screw model was introduced into the lecture courses on experimental philosophy to illustrate the screw as the fifth simple machine, having a close relationship to the wedge, usually considered the fourth simple machine. The theory of the screw was elucidated by Parent in 1703, although it was probably beyond the scope of most public lecture courses.

This model is rather small and not so well fitted for demonstration purposes as M47, in which the material in the screw is visible. It consists of an enclosed oak screw on an oak stand which can be inclined by means of a prop on a brass hinge under the base. There is an iron axle turned by an ebony handle.

The Archimedean screw became a standard piece of demonstration apparatus in the first half of the eighteenth century, featuring in several collections. Like the scorpion, it illustrated the technology of the ancients although the Archimedean screw was also widely used at the time. According to Diodorus it was invented by Archimedes to irrigate the Nile Delta in the first century BC. Archimedean screws were sometimes used to raise water in gardens, as by the Marquis de Servière at Lyons and by Smeaton at Kew, who in 1761 designed an Archimedean screw worked by two horses to raise water to supply the lakes and basins of Kew Gardens.

Q.C., Item 43; Archimedes screw
de Servière 1719, Figs 76, 77
Belidor 1737, Vol. I, p. 387
Desaguliers 1744, Vol. 2, p. 215, p. 20, Fig. 7
Demainbray 1754*b*, p. 7 'Models of an endless Screw, and an Archimedes' Screw'
Nollet 1754, Vol. 3, p. 134, Pl. 8, Fig. 12
Emerson 1758, p. 228, Fig. 272
Gentleman's Magazine 1763, p. 324

Crommelin 1951, no. GM 89 and A 366
Singer *et al.*, 1956, Vol. 2, p. 677
Turner 1973, p. 374

Inventory No. 1927-1455

D13 Model of a rat's tail crane

before 1753

HANDWRITTEN—UPRIGHT
'perpendicular piece'
HANDWRITTEN—ARM
'brace and ladder'

L 280mm, **W** 110mm, **H** 310mm

Nearly all Demainbray's apparatus for the lecture on compound engines has survived including the model scorpion, Vauloué's pile-driver, de Grollier's pile-driver, and this model of a rat's tail crane. Model cranes were popular demonstration apparatus used to illustrate compound machines, or machines that used combinations of the five simple machines. This particular model is copied directly from Desaguliers's plate in *A Course of Experimental Philosophy*.

The model crane is made of boxwood with the exception of the iron pillar which is held upright by a four-legged pyramidal stand. The main beam is supported by the pillar and carries a small brass pulley at the end of the long arm. Various cross-pieces also carry pulleys over which the rope passes before being wound round the axle of a large capstan acting as a counterweight on the short arm.

The crane was used to unload cargo or move masonry. It was operated by 'men, horses or an ass' inside the capstan or sometimes by means of a rope outside the capstan. Later this type of crane became known as Perronet's crane after the French civil engineer who described a similar machine in 1783.

Examples of crane models can be seen in several eighteenth century collections of instruments including the van Musschenbroek Collection at Leiden and the van Marum collection at Haarlem.

Q.C., Item 51; model of rat's tail crane
van Musschenbroek 1739, p. 186, Pl. 5, Fig. 5
Desaguliers 1744, Vol. 1, p. 127, Pl. 12, Fig. 4
Demainbray 1754*b*, p. 7 'Rat's Tail Cranes'
Nollet 1754, Vol. 3, Pl. 6, Fig. 51, p. 106
Desaguliers 1763, Vol. 1, p. 127, Pl. 12, Fig. 4
Perronet 1783, Vol. 2, Pl. 39, Figs 9-13
C 65
Crommelin 1951, No. G.M.79
Turner 1973, item no. 67
Inventory No. 1927-1929

D14 Model of wharf jib crane

before 1753

HANDWRITTEN—ON ROOF IN INK
'28'

L 480mm, **W** 75mm, **H** 195mm

This model was identified from Chabrol's notes as Demainbray's model capstan or windlass, previously thought to be lost. It illustrated his lecture on compound engines.

The model is made of mahogany and consists of a horizontal beam supported by three uprights with a simple jib crane at one end. A vertical axle set between two of the uprights is turned by a capstan. The string from this passes along the beam over a small vertical pulley and between two horizontal pulleys before passing over the crane to the weight. The small roof keeps the string dry. The horizontal direction of the crane is adjusted by a windlass with iron pins as spokes; this turns a vertical toothed wheel which is geared to a horizontal toothed wheel.

Desaguliers described this type of

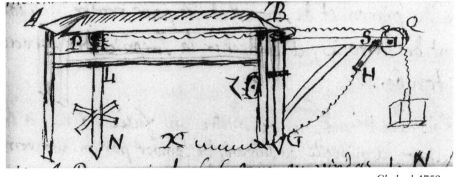

Chabrol 1753

crane in detail in *A Course of Experimental Philosophy* in 1745. The mechanism for turning the lifting arm was due to Padmore who designed the crane for a Mr Allen of Bath.

When the crane is operating at the extremities of its rotation, the length of rope necessary is greater than at its mid-point because the rope has to turn around the horizontal pulleys. This meant that a man operating the crane had to sustain a proportion of the weight when the crane was swung around, placing a limit on its capabilities. The windlass and gearing give a

greatly increased mechanical advantage so that the crane can be turned easily.

Q.C., Item 11; A Gibbet frame
Q.C., Item 52; Model of a Capstane
Desaguliers 1744, p. 125, Pl. 12, Fig. 1, p. 187
Chabrol 1753, p. 39; le modele d'un vindas ou Cabestan qui est applique' a une grue tournante ou gibet . . .
Demainbray 1754*b*, p. 7; 'common Capstanes'
C 66

Inventory No. 1927-1928

D15 Model of scorpion or catapult

before 1753

L 180mm, **W** 74mm, **H** 130mm

Demainbray used a 'scorpion of the antients' together with a capstan, crane, and pile-driver to illustrate his lecture on compound engines. The model is similar to that shown in Desaguliers's *A Course of Experimental Philosophy* 1744, from which it was probably copied. Desaguliers himself copied it from Justus Lipsius, a literary critic specializing in antiquity who lived in the sixteenth century. The original source was Vitruvius. Desaguliers described the scorpion as an instrument for throwing arrows, fire-balls, or great stones. Before firing, the tail has to be hauled to the ground, held by a pin and attached to the ammunition. For this reason and because of its weight and inconvenience, Desaguliers compared it unfavourably with a battery of cannons.

The scorpion uses the principle of the lever. There is a long arm and two short ones which have heavy weights attached. The pivot is situated just above the fork. The machine is loaded by bringing down the long arm and attaching the ammunition, a pin being used to prevent it from firing. When the pin is removed the weights on the short arms drop and the ammunition is thrown into the air. The baseboard is very similar to those on the optical models and the construction similar to that of the rat's tail crane. The arms are of boxwood and the base is of beech. The weights are of lead.

Q.C., Item 53; model of the Scorpion of the antients
Lipsius 1596, Fig. 3, dial 4
Desaguliers 1744, Vol. 1, pp. 72–4, Pl. VI, Figs 1, 2
Demainbray 1754*b*, p. 7 'Scorpion of the Antients'
C 63

See also D13

Inventory No. 1927-1107

D16 Model of Vauloué's pile-driver

before 1753

L 500mm, **H** 900mm, **W** 375mm

The model pile-driver was used by Demainbray in his lecture on compound engines. It is made entirely of mahogany. It consists of a capstan attached to a winding drum, which in turn is attached to a follower, which lifts the ram by means of tongs when the capstan is rotated. When the follower reaches the top the ram is released to fall on the pile. The follower disconnects the drum simultaneously, and a revolving fly geared to the capstan prevents the sudden loss of tension. In the full-scale machine this would prevent the horses from falling.

The pile-driver invented by James Vauloué, a watch-maker, was described by Desaguliers as 'the best of that kind that perhaps was ever seen'. It was invented in about 1737 and was used in the building of Westminster Bridge from September 1738. Piles were sunk around the piers to protect the works from barges. By means of three horses a weight could be lifted to fall on a pile five times in two minutes, driving a standard firwood pile 14 feet in an hour.

Vauloué was paid £150 gratuity by the Bridge Committee in recognition of the worth of his invention. However, ten years later the bridge was in need of extensive repair and the pile-driver was brought out of store. Models of the pile-driver can be found in many collections.

Q.C., Item 59; model of Vauloué's machine for driving piles
Desaguliers 1744, Vol. 2, pp. 417–18; Pl. 26
Demainbray 1754*b*, p. 7 'Vauloué's Machine for driving of Piles used at Westminster Bridge'
C 70
Walker 1979, pp. 89–92, pp. 199–200

See also D17

Inventory No. 1927-1197

D17 Model of a pile-driver driven by a paddle wheel

1754   Lyons

L 360mm, **W** 290mm, **H** 660mm

When in Lyons in 1754 Demainbray was given this model by the Marquis de Grollier, the inheritor of an unusual collection of clocks, machines, and *objets d'art*. The machines were built in the second half of the seventeenth century by Baron Nicolas Grollier de Servière. The Baron had a pile-driver constructed on his estate which was operated by a water wheel held between two boats, although in the model it is attached at one side of a floating structure.

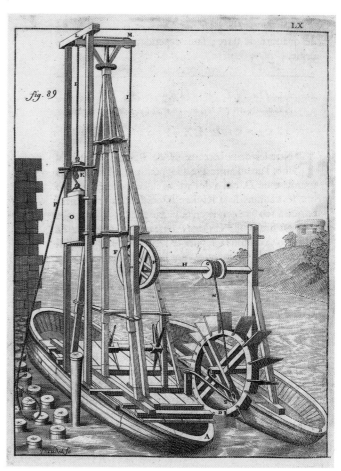

de Servière 1719

    The model is made of oak and consists of a platform supporting a tower with a water wheel on one side. A reverse axis in peritrochio lifted the weight which was then released to drop on the pile. There appears to have been no function for the small windlass under the tower. A swiftly running stream was necessary and so tidal rivers would have been inadequate. Unlike Vauloué's pile-driver it did not require men and horses and so did not need the mechanism to protect them from being thrown when the weight was released. It is a very simple device which probably delivered little power.

Q.C., Item 66; the Marquis de Grollier's pile-driving engine
de Servière 1719, Pl. LX; pp. 59–61
Demainbray 1754*a*, p. 8
Demainbray 1754*b*, p. 8 'a new Engine for driving of Piles without the Assistance of Men or Horses . . . given to the Doctor by the Marquis de Grollier'

See also D16

Inventory No. 1927-1641

D18 Demonstration of friction bearings

Graham, G.

before 1753   London

L 290mm, **W** 250mm, **H** 150mm, **D** 130mm (wheel)

This machine is probably the 'friction engine by Graham' listed in the Queen's Catalogue and certainly 'Dr. Desaguliers' friction engine' mentioned by Demainbray in his syllabuses. It is very similar to M63, the friction engine by Adams, but more closely based on Desaguliers's design described in his *A Course of Experimental Philosophy*. It demonstrates the advantage of friction bearings over other supports.

A spoked brass wheel five inches in diameter is mounted on a brass axle which is half an inch in diameter in the central part and a quarter of an inch in diameter towards the ends. When in use it rests at each end on a pair of brass friction rollers two inches in diameter. The friction rollers have steel axles set on brass screws and points. An additional pair of brass uprights could be used to support the main axle to show the increase in friction by not using the rollers. The machine is mounted on a mahogany base which has a square hole to allow a weight to fall. Another piece of mahogany has been added to the base at a later date.

Like M63 the machine had a spiral spring which was attached both to the base and the axle. When it was wound and released the wheel made oscillations, and the number of these could be counted when the axle rested on the friction wheels, and again when it rested directly on the supports in the brass uprights. Alternatively the spring could have been removed, the wheel spun, and the time taken for it to come to rest be measured both with and without the friction wheels. The revolutions of the friction wheels could be used

to count the revolutions of the main wheel since they will turn eight times as slowly in their present arrangement. The spring is now missing.

Sully, an apprentice of Graham, was one of the first to use friction wheels in a clock mechanism in the early eighteenth century. Gould states that Graham wrote to Sully in 1724 saying that he had seen friction rollers used in an old watch twenty years previously.

Q.C., Item 60; The friction engine by Graham
Desaguliers 1744, Vol. I, pp. 261–2, Pl. 18, Fig. 8
Demainbray 1754*b*, p. 8 'Enquiries on the Resistance of Weights and Surfaces by the late Doctor Desaguliers Machine'.
Adams 1762, pp. 83–5, Pl. 41, Fig. 140
Rees Cyclopaedia (1819) under Friction
Gould 1923, p. 37
See also L24, M63
Inventory No. 1927-1259

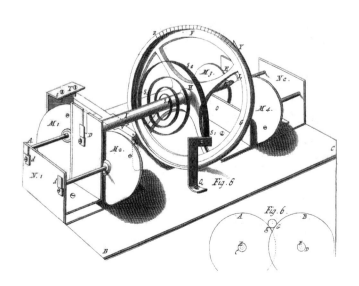

Desaguliers 1744

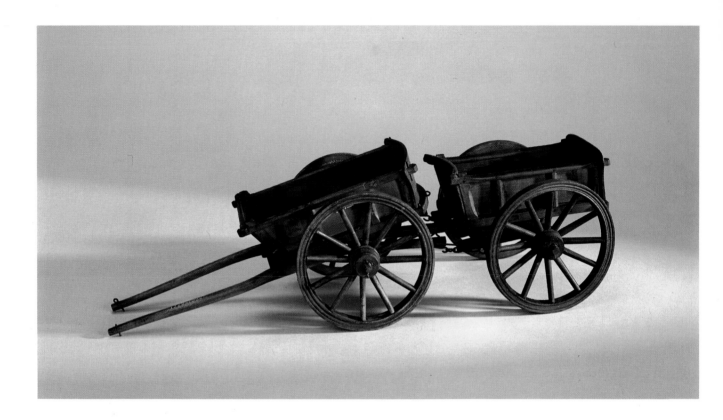

D19 Model of a four-wheeled wagon

before 1754

L 490mm (total), **W** 180mm, **H** 140mm, **D** 125mm (wheel)

A sketch in Chabrol's lecture notes identified this model as the four-wheeled carriage or wagon that was presented to the Dublin Society in about 1751. Demainbray claimed that it had all the advantages of a four-wheeled vehicle, could carry heavy weights, and was easy to turn. The back of the front section and the front of the back form an arc of the circle whose centre is where the shaft from the back is attached to the front. It is made of holly. Chabrol commented that although Demainbray called this his Irish improved carriage, the type had been known for some time in France as the *chariot brise*. The model was rejected by the Dublin Society but unfortunately no records remain there.

Q.C., Item 72; 'Dublin Society's Irish improved Waggon'
Chabrol 1753, p. 55
Demainbray 1753, p. 8
Demainbray 1754*a*, p. 8
Demainbray 1754*b*, p. 8 'Model of an improved four wheeled carriage offered to the Dublin Society, the Reasons why rejected'
Gentleman's Magazine 1754, Vol. 24, pp. 326–9, p. 376, p. 426, p. 473
C 329

Inventory No. 1927-1932

Chabrol 1753

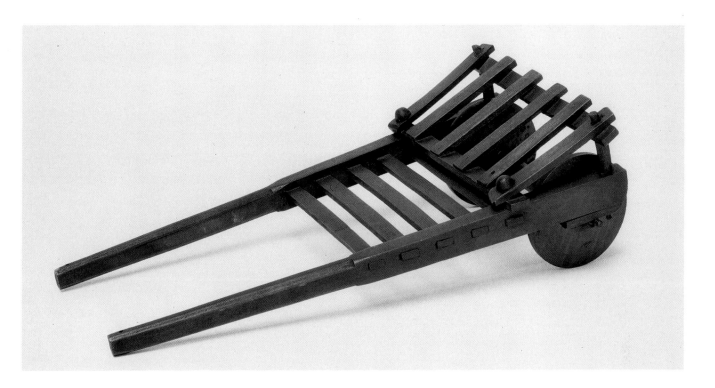

D20 Model Irish low-backed car

before 1754

HANDWRITTEN—INSIDE OF WHEEL
'*2 feet diam/Axis 1 1/2 inch/ratio of friction as 1 to 16/advge as . . .*'

W 117mm, **H** 74mm, **D** 50mm (wheel), **L** 240mm

Like the four-wheeled carriage, this cart was identified from a sketch in Chabrol's lecture notes. The cart is not mentioned specifically in the syllabuses until 1754, although it was being used in 1753 and Demainbray had a collection of vehicle models before this. It was described as being very strong with an axle of iron. Apart from this it is made of holly.

The design of carts and carriages played a significant part in the mechanics sections of experimental philosophy courses in the early and mid-eighteenth century. The width of wheel rims, cladding of wheel rims, and the diameter and dishing of wheels were subjects of debate. This cart was used in calculations of velocity ratio and friction as can be seen from the writing on the wheel.

Q.C., Item 73; A model of the Irish low-backed car
Desaguliers 1734, pp. 201–28
Helsham 1739, pp. 132–48
Chabrol 1753, p. 55
Demainbray 1754*a*, p. 8
Demainbray 1754*b*, p. 8 'the Irish low-back Carr'
Gentleman's Magazine 1754, Vol. 24, pp. 326–9, p. 376, p. 426, p. 473
C 328   **Inventory No.** 1927-1935

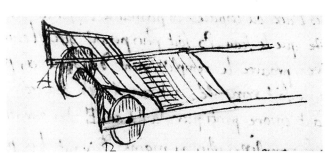

Chabrol 1753

D21 Model experimental cart

before 1753

L 180mm, **D** 35mm (small wheels), **D** 70mm (large wheels),
W 108mm, **H** 55mm (body)

Like the Irish vehicles, D19 and D20, this cart was identified
from a sketch in Chabrol's lecture notes. Demainbray used it
on the inclined plane D11 to compare the performance of
large and small wheel diameters carrying various weights.
Originally he had two pairs of both large and small wheels
but one pair of large wheels is now lost. He stressed the
advantage of large wheels and, according to the syllabuses,
recommended the use of a cart rather than a sledge. The
model cart is made of oak with brass wheels.

Q.C., Item 67; A model of a cart with 4 high brass wheels and 4 ditto
low ones
Chabrol 1753, p. 51
Demainbray 1754*b*, p. 8 'Wheel carriages examined ... with
models'

Inventory No. 1927-1818

Chabrol 1753

D22 Model to show the maximum of a man's power to raise water

before 1753

H 1050mm, **W** 460mm, **L** 660mm

This model was identified from a plate in Desaguliers's *A Course of Experimental Philosophy* of 1744. It is a machine to show the maximum of a man's power to raise water, friction having been reduced as far as possible. Desaguliers states that he made a model of the device which suggests that this may have been his machine before being aquired by Demainbray. The fact that it is referred to as a new model in the earliest syllabus indicates that Demainbray acquired it in about 1750.

The model is made of oak and consists of the framework of a tall building on two floors connected by a ladder. In the base is a well. A square hole in the upper floor allows a lift in the form of a platform to be hoisted by means of a pulley in the rafters. The floor of the platform fits the hole and becomes a trap door when the lift is raised. The lift slides on two brass wires and another four wires at its corners are twisted together to provide a loop for the rope. The rope passed over the pulley in the rafters and then over another pulley at the same height at the other side of the building. Beneath this second pulley is the counter-weight for the platform which takes the form of a 'square bucket' also in oak, which has a valve in its base. When the bucket was lowered into the well the valve opened and water entered. The water was raised in the bucket which emptied itself on being caught on a hook in the upper floor. The water was poured into a cistern which is now missing.

The machine worked as follows: a man stood on the trap door making it heavier than the bucket of water which was therefore raised. When the man reached the bottom he ran up the ladder as fast as he could, during which time the empty bucket descended since it was now heavier than the platform. The bucket then refilled itself and the process was repeated 'all day'. Desaguliers recommended tavern-drawers since they were used to running up and down stairs.

Q.C., Item 105; maximum engine to raise water
Desaguliers 1744, Vol. 2, pp. 503–4, Pl. 43
Demainbray 1750–2, p. 5 'a new Model to shew the Maximum of a Man's Power in raising of Water'.
Demainbray 1754*b*, p. 9 'Model to shew the Maximum of a Man's Power in raising of Water'
Inventory No. 1927-1634

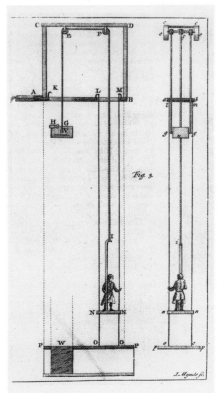

Desaguliers 1744

D23 Diagonal machine

before 1753?

L 325mm, **W** 160mm, **H** 325mm

Motion and mechanics were separate in Demainbray's lecture courses. This instrument probably illustrated the first lecture on motion in which came 'the Law of Compound motion and the diagonal direction of a Body so moved, rendered obvious by an instrument self-evident in its contrivance'.

Although now incomplete, the instrument is similar to Adams's diago-

nal machine. A pulley is held in a frame which can be moved across the top of the instrument. A weight is held by a string passed over the pulley and attached to one corner of the frame. When the frame is moved the weight is lifted and moved horizontally simultaneously, the result being diagonal motion.

Q.C., Item 106; diagonal machine
Desaguliers 1744, Vol. 1, p. 298, Pl. 23; Fig. 12
Demainbray 1754*b*, p. 9 see text

See also M67

Inventory No. 1927-1855

D24 Path of a projectile apparatus

before 1753

L 630mm, **W** 220mm, **H** 370mm, **W** 200mm

The path of projectile apparatuses associated with Adams and Demainbray are remarkably similar; it is possible that Adams made both pieces. They are based on 's Gravesande's illustration and description. This model does not conform to the dimensions given by Adams in the *Mechanics* manuscript nor does it have a brass slope for the balls which he mentions; therefore it has been associated with Demainbray.

The model consists of a mahogany channel in the form of an arc of a circle down which balls can be rolled, their motion being horizontal when they leave it. If they start rolling at the top of the channel and then fall freely, their subsequent paths follow a particular parabola. This can be demonstrated by setting a series of brass hoops in the form of the parabola onto a backing board and observing the balls passing through. A nest of cotton wool in the base should catch the balls but this is missing from both examples in the collection. The instrument stands on three brass levelling screws and has a

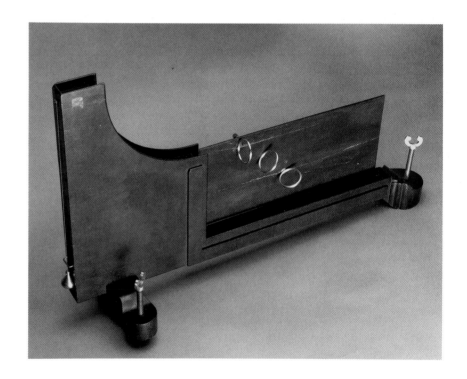

brass plumb bob. There are three balls with this inventory number—two ivory and one brass.

This type of apparatus was probably invented by 's Gravesande, first appearing in a slightly different form in his *Mathematical Elements of Physicks. . .* in 1720. It is attributed to him in all Demainbray's syllabuses and in the Queen's Catalogue.

Q.C., Item 107; 's Gravesande's machine to shew parabolic motion of projectiles
's Gravesande 1720, p. 135, Pl. 15
Adams 1746, No. 133
's Gravesande 1747, p. 123, Pl. 19, Fig. 3
Demainbray 1754*b*, p. 9 'parabolic Motion of Projectiles made evident to the sight by 's Gravesande's Machine'
C 8, 7

See also M75   **Inventory No.** 1927-1112

D25 Tubes for the whirling table

before 1754
L 225mm, **W** 134mm, **H** 65mm

Demainbray used his whirling table to demonstrate central forces at the start of his section on astronomy. This is one of two sets of tubes for the whirling table in the collection; the other is by Adams. Like the previous instrument it is based on 's Gravesande's design from the 1720 edition of *Mathematical Elements of Physicks*. There are four tubes containing respectively water with cork, water with a lead ball, water with mercury, and a white suspension which according to Chabrol was spirit of turpentine. The tubes are on a mahogany base. When the apparatus was whirled the denser substances would rise to the top of the tubes so that cork would appear to sink and lead and mercury to float.

Whirling tables, or central forces machines, enabled experiments to be done on a rapidly revolving plane. The prototype was probably 's Gravesande's which is in his collection at Leiden. A standard experiment involved placing tubes containing mixtures of liquids or solids and liquids on the table inclined to the horizontal, and observing the effects of the central force.

Q.C., Item 153; four glass tubes on inclined plane containing bodies of different weights
Hauksbee 1714, Pl. 5, Fig. 6
's Gravesande 1720, p. 152, Pl. 16, Fig. 8
Desaguliers 1734, p. 289, Pl. 24, Fig. 13
Helsham 1743, p. 42, expt 1
Chabrol 1753, p. 91
Demainbray 1754*b*, p. 11 'instructive Experiments on the Whirling Tables'

Inventory No. 1927-1880

Chabrol 1753

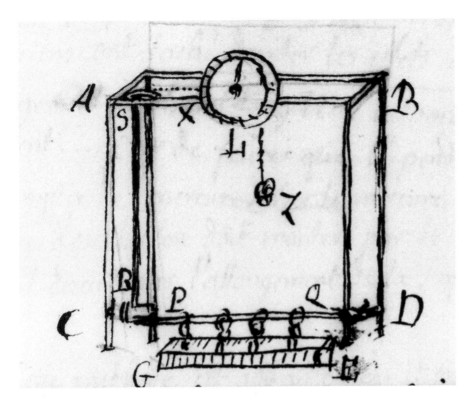

Chabrol 1753

D26 Hindley's pyrometer with three expansion tubes

Hindley, Henry

before 1753   York

SCALE—ENGRAVED—DIAL
0 (10) 100

L 327mm (frame), **W** 125mm (frame), **H** 435mm (frame),
L 285mm (rod), **D** 4mm (rod)

A sketch in Chabrol's notes identified this pyrometer as being the one which Demainbray claimed to have invented, at least in his French syllabuses. It has been linked by Law on stylistic grounds with the entry in the Queen's Catalogue which lists a 'pyrometer made by Hindley with brass, steel and silver rods'.

The pyrometer consists of a rectangular iron frame into which a test bar can be placed horizontally near the base. One end of the bar is fixed while the other is attached to a piece of watch spring wrapped round the barrel on a pivotal arbor which carries a long vertical lever. A thread is attached to the top of the lever and turned around the arbor which carries the hand. A small weight maintains the tension. The dial is marked in 100 divisions with every 10 numbered.

Chabrol's notes show that the bar was heated along its length by a series of wicks, probably floating in an oil bath. The elongation was then made apparent on the dial.

Another connection between Demainbray and Hindley was a certain John Smith who, according to a note by Smeaton, had been brought up with Hindley and was involved with making instruments for Demainbray at Kew in the 1770s. Hindley was known to be interested in temperature compensation so he would have been a logical choice for advice on the design of such an instrument.

Q.C., Item 139; pyrometer made by Hindley with brass, steel and silver rods
Chabrol 1753, p. 81
Demainbray 1754b, p. 10 'a Pyrometer of a new Construction'
Smeaton 1786, p. 27
C 79
Law 1970–2a
Law 1970–2b

Inventory No. 1927-1184

ASTRONOMY

D27 Desaguliers's cometarium

c.1755

'Doctor Demainbray'

L 500mm, **W** 390mm, **H** 115mm

Demainbray does not mention an instrument associated with his lecture on planets and comets so it is possible that the cometarium is later than 1754.

The instrument is a two-dimensional representation of the path of a comet. It forms the top of an oak box and has a glass cover. There are two dials: the upper one gives hours and minutes; the lower one shows the motion of the comet. When the handle is turned a brass pointer moves round each of the dials. The upper dial shows equal arcs being covered in equal times; the lower pointer consists of a brass rod on which slides the ball representing the comet. The tip of the pointer moves over the circle representing the great circle of the heavens, while a groove carries the comet in an elliptical path, giving equal areas in equal times but not equal arcs. The Sun is represented by a larger brass ball at the centre. Strangely, the upper dial is divided into 24 but the lower one is divided into 22.

Ferguson describes the instrument very clearly in *Astronomy Explained Upon Sir Isaac Newton's Principles* in 1757: 'This curious machine shows the motion of a comet or excentric body moving round the Sun, describing equal areas in equal times ... it was invented by the late Dr. Desaguliers'. The mechanism is shown in the illustration. Desaguliers explained that the eccentricity was greater than any planet 'only to make the phenomena more sensible' and less than any comet 'which would have exceeded the bounds of the instrument'.

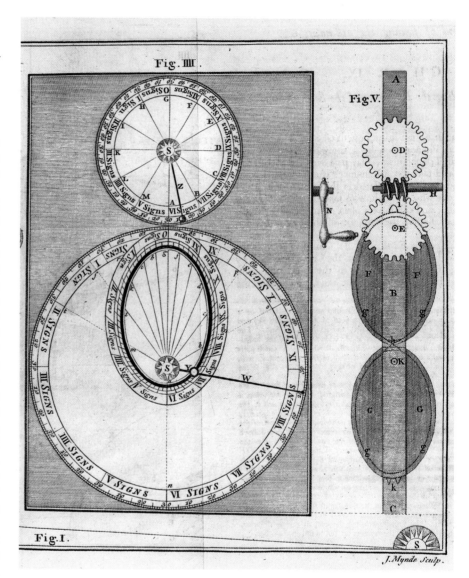

Ferguson 1757

Q.C., Item 154; Cometarium
Desaguliers 1744, p. 464, Pl. 29, Figs 7, 8
Ferguson 1757, p. 270
Desaguliers 1763, Vol. 1, pp. 465–6; Pl. 29,
Figs 7, 8
C 143, 744
King and Millburn 1978, p. 173
Millburn 1981

Inventory No. 1927-1254

D28 Large orrery with eight-day clock

Butcher, Charles

1733   Bedford

ENGRAVED—CASE OF CLOCKWORK
'CHA BUTCHER: BEDFORD EUPHKA.'

H 770mm (base), **H** 650mm (clock mechanism), **D** 1350mm (metal framework), **D** 100mm (Sun ball)

In 1990 the condition of this instrument was so poor as to make it almost unrecognizable. However, it was discovered to have a radius of comparable size to that of the large orrery described by Demainbray in his syllabuses, and the mechanism was compatible with that of an eight-day clock. It was therefore attributed to Demainbray and subsequently other evidence was found to confirm the attribution.

The large orrery was one of Demainbray's most impressive instruments and was used to illustrate four of his lectures on astronomy. He described it as an orrery of four and a half feet diameter, containing all the primary and secondary planets, and adds 'this instrument not only performs the revolutions of all the planets in the usual method, by a winch, but by means of an Eight-Day-Clock, exhibits their periodical Revolutions in their true Times'.

A broadsheet relating to the orrery was found at Oxford which revealed that Butcher had made a very similar instrument for Lord Trevor, the Baron of Bromham, in 1733. It is almost certain that Butcher would only make one such instrument, given its size and complexity, and that the orrery was subsequently acquired by Demainbray. Thomas, the first Lord Trevor, Baron of Bromham, died in 1730, possibly while the orrery was under construction. In an attempt to sell it Butcher may have produced the broadsheet. A more likely alternative is that the orrery was made for the wife of the second Baron who died in 1734 near Grosvenor Square, London. If the orrery was intended for Lady Trevor it might have been sold some time after her death, and if it was in London it would have been more likely to come to the attention of Desaguliers and Demainbray. Certainly Demainbray had the orrery before the second Baron died in 1753.

The orrery is currently being restored. The sphere is made of brass which shows signs of having been painted with decorative floral patterns. It consists of two colures, the polar circles, tropic circles, equator, and ecliptic. The zodiac band is 18 degrees wide and has painted brass figures representing the signs attached around the outside. The names of the signs are painted around the inside. It appears from the broadsheet that the ecliptic was painted in degrees but this has not survived. The sphere is supported on an octagonal wooden pedestal with walnut veneer standing about two feet high. It is inclined for a latitude of 52 degrees; approximately that of Bedford. The clock was fixed to the base and from it a central arbor or sun-stem rises. The arms supporting the plants are attached to collars surrounding the sun-stem. Also on the stem are two horizontal silvered dials, one just above the arm for Venus and one just above the arm for Mercury. Originally the orrery included Mercury, Venus, Earth and Moon, Mars, Jupiter and four moons, and Saturn and five moons. The Earth–Moon system is now lost as are two moons of Saturn, the ring of Saturn, and Venus. The upper small dial plate has a zodiac circle over which an index for the orbit of the planet Mecury could pass to show its position. It also has the orbit of the planet with respect to that of the Earth engraved with the aphelions and perihelions marked. The lower large plate has a zodiac circle, a circle giving the age of the Moon, an equation of time, a further zodiac circle and the orbit of the planet Venus with respect to the Earth, also with aphelions and perihelions marked.

The eight-day clock provided the annual rotation of all the planets. The face showed days as well as hours and minutes. Diurnal rotation of the Earth, Jupiter, and Saturn was provided by means of handles operated manually which turned shafts inside the arms of these planets. The system of gears is all that remains of the Earth–Moon mechanism and this carries the inscription. Apparently the Earth was three inches in diameter and marked with 'a Geographical description of both Sea and Land'. The remaining planets are made of ivory. The clock mechanism is rather crude in parts and consistent with the work of a provincial clock-maker of the period.

The orrery bears a slight resemblance to the Leiden sphere, and rather more resemblance to a print of Hauksbee's orrery published in 1755 which appears in Wynter and Turner's *Scientific Instruments*.

Q.C., Item 161; large orrery with 8 day clock
Butcher 1733
Demainbray 1754*b*, p. 11 'an Orrery of four Feet eight Inches Diameter. . . .'
Universal Magazine of Knowledge and Pleasure 1755
White 1959, Vol. 12, p. 32
Wynter and Turner 1975, p. 51, Fig. 55
King and Millburn 1978
Dekker 1986

See also E34

Inventory No. 1927-1618

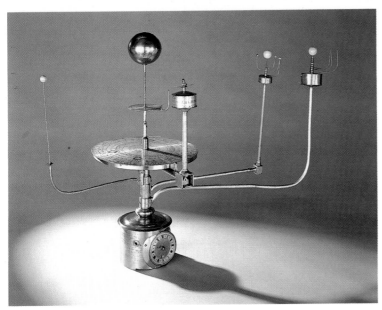

HYDROSTATICS AND PNEUMATICS

D29 Vessels for the hydrostatic paradox

before 1753

H 120mm (all), **D** 80mm (cylinder), **D** 80mm (max double cylinder), **D** 20mm (min double cylinder), **D** 171mm (max funnel), **D** 80mm (min funnel)

Demainbray's six hydrostatics lectures were considerably influenced by Desaguliers's *A Course of Experimental Philosophy*. Having discussed leading water in pipes and the Edinburgh water works in the first lecture, Demainbray moved on to syphons and the hydrostatic bellows in the second.

These vessels were used by Demainbray to illustrate the 'hydrostatic paradox'. They are made of tinplate with brass screw threads at their bases. They have approximately the same base diameter and height but their capacities are 1:2:4. In the Queen's Catalogue they are listed as pint, quart, and half-pint vessels. The vessels were screwed in turn to a stand to form containers which were filled with water. A balance beam was suspended above the apparatus with one arm carrying a piston which could move vertically in the container, and the other carrying a counter-balancing weight in a scale-pan. The weight necessary to maintain the piston was noted and the vessels were interchanged. The pressure on the piston in all three cases was found to be the same, since pressure is dependent on depth and not on the volume of water present.

This is essentially the same demonstration as the nineteenth century 'Pascal's vases' L35.

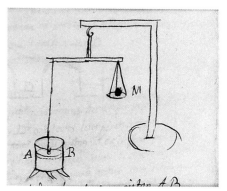

Chabrol 1753

Q.C., Item 174; Hydrostatic paradox machine, pint cylinder, quart vessel, half pint vessel
Van Musschenbroek 1739, Vol. 1, p. 380, Pl. 11, Figs 5, 6
Chabrol 1753, p. 117
Demainbray 1754*b*, p. 13 'the Experiment of the Hydrostatic Paradox tried'
Nollet 1754, Vol. 2, p. 268, Figs 15–17

Inventory No. 1927-1749

D30 Hydrostatic bellows

before 1753

D 410mm, **H** 130mm

In the earliest syllabus Demainbray uses the hydrostatic paradox and bellows in his third lecture on hydrostatics, but later they came into the second lecture. The bellows comprise a further demonstration of the hydrostatic paradox.

They consist of two circular pieces of ash joined by a piece of leather to make a cylindrical vessel. There is a hole in the top to fill the bellows and a brass butterfly nut screwed into the hole. A brass pipe screwed into the side was used to demonstrate the paradox. Weights were placed on the upper surface of the bellows. Demainbray followed 's Gravesande whose instructions are as follows: 'Pour water into this bellow thru the tube, and the water in the Tube will sustain the weights all which, together, weigh 300lbs. These make the water rise into the Funnel, but the height of the water in the Funnel is but small. The weights will even be raised by continuing to pour water into the Tube'. The paradox was that a small volume of water appeared to outweigh the heavy weights until pressure as force per area was considered. 's Gravesande used it to show that 'pressure exerts itself in every way and in every way equally'.

Q.C., Item 175; 'hydrostatic bellows/cock for breath/pipe and receiver for water.
Desaguliers 1744, Vol. 2, pp. 109–10, Pl. 10, Fig. 1
Adams 1746, No. 155.   's Gravesande 1747, p. 350, Pl. 47, Fig. 5
Demainbray 1754*b*, p. 13 'the Hydrostatick Bellows'   **Inventory No.** 1927-1407

D31 Syphon fountain

before 1753

L 445mm, **D** 165mm

The materials and design of this item are consistent with its belonging to Demainbray. It is probably the syphon fountain listed in the Queen's Catalogue, although Demainbray did possess several other fountains. It would have been used in the second lecture on hydrostatics.

The fountain consists of a bell-shaped glass vessel on a brass base having a short central pipe and a longer pipe at one side. The brass cap at the top is permanently closed. A small quantity of water was put into the inverted vessel, the short tube closed, the vessel righted, and the water ran out of the longer tube leaving rarefied air in the vessel. If the base of the short tube was placed in a bowl of water, a fountain would spring into the vessel. A similar version of this type of fountain can be found in Musschenbroek's *Essai de Physique*. Presumably an air pump could also be used to partially evacuate the vessel.

From 1753 Demainbray also had a glass air vessel which he used with a forcing pump to produce a constant flow of water. This apparatus could fulfil that purpose if the vessel was partially filled and then kept topped up by the pump as water flowed out of the longer pipe at a constant rate.

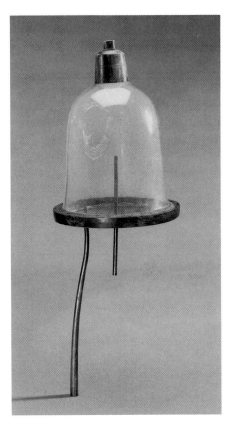

Q.C., Item 177; syphon fountain
Van Musschenbroek 1739, p. 663, Pl. 23, Fig. 17, p. 36, Pl. 3 Fig. 17
Demainbray 1754*b*, p. 13; 'glass air vessel applied to the model of a forcing pump to produce a constant stream of air'
Demainbray 1754*b*, p. 13; 'syphon fountain'
Rees Cyclopaedia (1819), under Fountain
C 129

Inventory No. 1927-1387

D32 Flanged pump plate with pipe and tap

mid-eighteenth century

D 170mm, **L** 254mm

The pump plate has the same dimensions as the base of the fountain above. It would have been used in one of Demainbray's fountain experiments. The plate was first attached to an air pump, and then a tall glass receiver and a wet leather put on it and the air exhausted. The tap was closed and the plate and receiver removed from the pump. The end of the brass pipe was placed in a basin of water. When the tap was opened, the water rose into the receiver making a fountain.

Desaguliers 1744, Vol. 2, p. 384; Pl. 25, Fig. 14
Demainbray 1754b, p. 13 'Syphon Fountain' and 'Syphons examined'

Inventory No. 1927-1403

D33 Syphon or 'Tantalus' cup

before 1753

H 195mm, **D** 100mm

This is possibly the only syphon cup in the collection to survive, although D34 has identical dimensions in glass and so is almost certainly associated in some way. Originally there were several syphons and at least three types of cup. Although there is no Tantalus figure it can be referred to as a Tantalus cup according to Desaguliers.

It is a copper vessel in the shape of a goblet with a handle formed from a copper pipe with an open end. If the water is raised beyond a certain level, e.g. by throwing in an apple or orange, the water runs out through the handle until the cup is empty.

Q.C., Item 178; syphon in handle of cup
Desaguliers 1744, Vol. 2, p. 149, Pl. 13, Fig. 11
Adams 1746, No. 174
's Gravesande 1747, Vol. 2, p. 38, Pl. 74, Fig. 1
Demainbray 1754b, p. 13 'syphon cups'
C 134

See also D34

Inventory No. 1927-1131

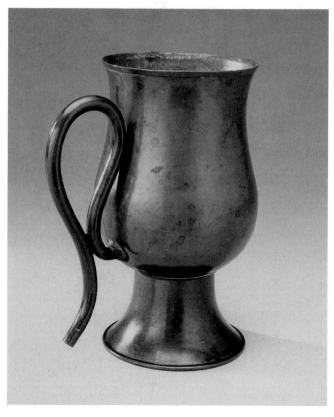

D34 Syphon cup

before 1753

H 195mm, **D** 100mm

Since the dimensions of this item are identical to those of the above it has been assumed that they are related. It is likely that this demonstrated the action of a syphon while the Tantalus cup showed how it could be used. This is a glass goblet which has a J-shaped syphon tube associated with it, although this tube does not appear to be original.

Desaguliers 1744, Vol. 2, p. 149, Pl. 13, Fig. 11
Adams 1746, No. 174
's Gravesande 1747, Vol. 2, p. 38, Pl. 74, Fig. 1
Demainbray 1754*b*, 'syphon cups'
C 129

Inventory No. 1927-1384

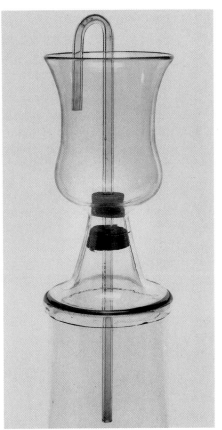

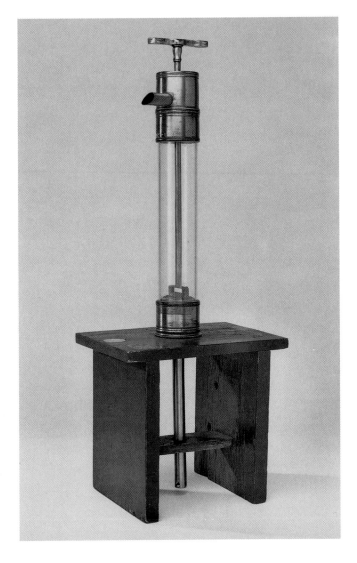

D35 Model sucking pump

before 1753

H 530mm, **L** 210mm, **W** 150mm

The third lecture on hydrostatics was largely devoted to the construction of pumps, two models of which remain. This is believed to be the glass model of a sucking pump mentioned in the syllabuses.

The model consists of a glass barrel with a brass and leather piston worked by a brass handle. There are two valves, one in the base of the pump and the other on the piston; both open upwards. When the piston is raised water enters the barrel of the pump through the brass pipe beneath. If the piston is then lowered water passes through the barrel of the piston so that when it is subsequently raised the water pours through the spout as more water is drawn up. The water would have initially been in a trough at the base of the model. The stained mahogany stand is not original.

Q.C., Item 181; glass sucking pump
Desaguliers 1744, Vol. 2, p. 164, Pl. 15, Fig. 2
's Gravesande 1747, Vol. 2, p. 40, Pl. 74, Fig. 3
Demainbray 1754*b*, p. 13 'On the Construction of sucking, forcing and lifting Pumps, illustrated with curious glass Models'
C 104   **Inventory No.** 1927-1285

D36 Model double-barrelled forcing pump

before 1753

H 520mm, **L** 215mm, **W** 150mm

The model is similar stylistically to the above and features in a sketch in Chabrol's notes. It consists of two glass barrels mounted vertically on a stained mahogany stand which is not original. The barrels are connected by a brass tube in the base. One barrel contains the piston which is worked by a brass handle. It has a valve in the base so that water can be sucked from a trough beneath through the brass pipe when the piston is raised, as in the previous example. However, as the piston is depressed the water is forced through the connecting pipe into the second chamber. This barrel has a glass tube running down its axis extending almost to the bottom. As the water enters, it compresses the air which in turn forces water out upwards via the central tube.

The advantage of this type of pump was that water could be raised to greater heights. The design was often used in engines for putting out fires because it gave a reasonably continuous flow.

Q.C., Item 184; 'glass forcing pump'
Desaguliers 1744, Vol. 2, p. 165, Pl. 15, Fig. 4
Chabrol 1753, p. 135
Demainbray 1754*b*, p. 13 'On the Construction of sucking, forcing and lifting Pumps illustrated with curious glass Models'
C 108

See also L37   **Inventory No.** 1927-1286

D37 Model sucking pump

Sisson, Jeremiah

1756   London

'ıEREMIAH SISSON. LONDON.INVENTOR/*Instrument Maker to late PRINCE OF WALES*'

H 720mm, **L** 202mm, **W** 255mm

In February 1758 Jeremiah Sisson advertised an improvement to the water pump which enabled double the quantity of water to be raised in the same time and with the same force as a common pump. The mechanism also translated the unequal ascending and descending strokes into a uniform circular motion for the benefit of the operation. A working model was displayed at 'Dr de Mainbray's' experiment room at Carey Street and at Sisson's premises in the Strand. This pump is likely to have been the one at Demainbray's since it bears the maker's name so prominently. From an advertisement in the *Daily Advertiser* in December 1756 it seems that Demainbray was exhibiting a model of a Sisson pump the

previous winter in his lecture course. The late Prince of Wales refers to Frederick, George III's father, who died in 1751.

The pump is held vertically in a rectangular mahogany frame. The cylinder is glass in order that the action can be observed, and a valve is set in its base. A brass pipe extends downwards from the cylinder. A thread from the piston passes over a large mahogany pulley in the top of the frame and is attached to a set of three removable counter-weights on one side of the instrument. Turning the ivory handle turns a steel axle to which is attached a brass bar. One end of the bar is fixed to a slider which moves horizontally in a slot in the vertical rod attached to the piston. Thus the rotary motion of the axle is translated into the vertical oscillation of the piston.

If a trough of water is placed between the feet of the stand, as the handle is turned the piston will suck up the water through the valve and into the cylinder. At the next stroke the water will be raised so that it will pour out of the spout in

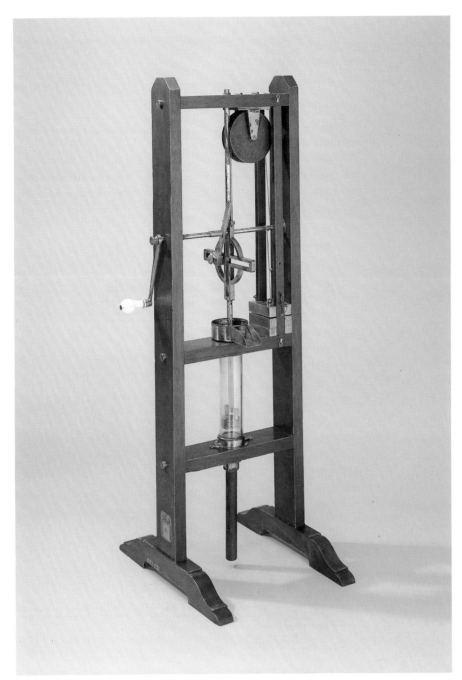

Gentleman's Magazine *1758*

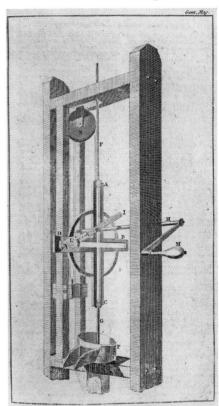

JEREMIAH SISSON. LONDON. INVENTOR.
Instrument Maker to the late PRINCE of WALES.

the top of the cylinder to be collected in a reservoir. The counter-weights enable the action to remain continuous.

Q.C., Item 186; Model of Sissons pump
Daily Advertiser 1756, No. 8083, p. 1
Gentleman's Magazine 1758, Vol. 28, p. 80
C 105

Inventory No. 1927-1251

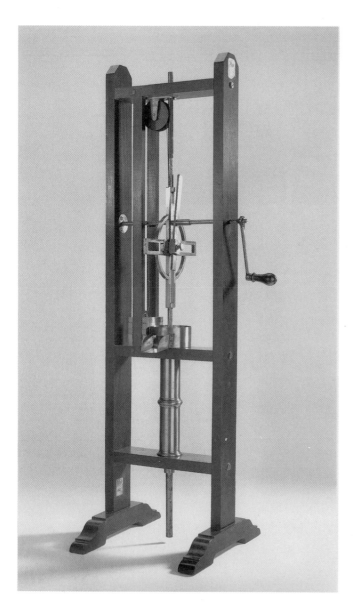

D38 Large model sucking pump

Sisson, Jeremiah

c. 1756   London

H 845mm, **L** 245mm, **W** 284mm (base)

This is a larger version of the above, similar in most respects. The barrel is of brass instead of glass, the handle is ebony, and there are five counter-weights instead of three. Obviously it is more difficult to see the valve in operation. It may have been the model exhibited in Sisson's shop in the Strand in 1758.

Q.C., Item 186; Model of Sissons pump
Daily Advertiser 1756, No. 8083, p. 1
Gentleman's Magazine 1758, Vol. 28, p. 80
C 106

Inventory No. 1927-1252

D39 Model of Comet mill at Lyon

1753

L 810mm (boat), **H** 640mm, **W** 230mm, **L** 600mm (propeller)

Demainbray had this model made while at Lyon in 1753. It has been identified as the model of a 'Comet' mill designed by Dubost for use on the River Rhône. In 1754 Demainbray described it as a 'useful and ingenious Piece of Mechanism, never before explained in publick . . .' although he had shown it in France the previous year.

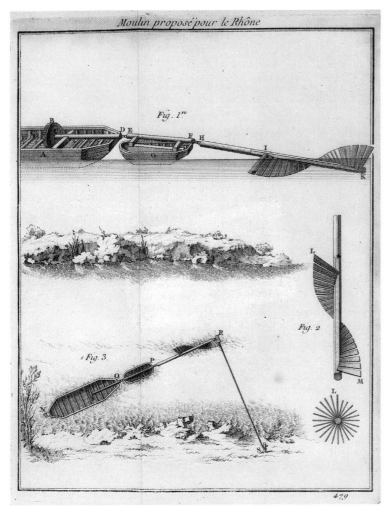

Moulin proposé pour le Rhône

Fig. 1.

Fig. 2

Fig. 3.

479

The model is in walnut in the form of a boat with an Archimedean screw propeller at the stern. The grinding wheels are on a raised platform under a roof. The upper grinding wheel is turned and the lower one is fixed. The flour is collected in a hopper next to the wheels. The axle of the upper grinding wheel runs through the lower one and carries a toothed wheel which meshes with a large vertical toothed wheel in the body of the boat. This in turn meshes with another small horizontal wheel just above the floor. The axle of this wheel passes through the base of the boat where there is a handle. The large wheel can be disengaged from the lower small wheel by a brake operated by a small windlass. The propeller is attached at the stern. Turning this turns a heavy shaft running from the stern to the centre of the boat where it forms the axle of the large vertical wheel. Hence turning the handle operates both the screw and the grinding wheels.

In the original machine the boat would have been moored in the river where it would align itself against the current. The flow of the water would operate the mill. Dubost initially contacted the Academie Royale des Sciences with a new design of mill in 1743. This is a model of a simplified version which he proposed in 1747.

Q.C., Item 195; Comet mill at Lyon
Dubost 1747, in Gallon 1777, Vol. 7, p. 369 and plate facing p. 379
Demainbray 1753, p. 17
Demainbray 1754*a*, p. 84
Demainbray 1754*b*, p. 18; 'the Comet mill for grinding corn'
Inventory No. 1929-125

Dubost 1747

D40 Two specific gravity jars

before 1753

H 195mm, **D** 60mm

The specific gravity of substances was enquired into during Demainbray's last lecture on hydrostatics, which covered hydrometers and the hydrostatic balance. The two jars were used to compare the different specific gravities of various types of wood. Cylinders of the wood in question were immersed in water in the jars and shown to sink to different depths. Cylinders of various sorts of wood are listed in the Queen's Catalogue but are now lost. The glass cylinders stand on painted tinplate bases. One has rough lines marked in black paint.

Q.C., Item 166; glass tube with divisions on tin stand
Q.C., Item 167; glass tube on tin stand
Q.C., Item 199; cylinders of various sorts of wood
Desaguliers 1744, Vol. 2, p. 102, Pl. 8, Fig. 21
Demainbray 1754*b*, p. 14 'the specific gravity of bodies enquired into'
C 131

Inventory No. 1927-1283

D41 Model diving bell with 12 sinking weights

before 1753

D 210mm, **H** 410mm, **L** 85mm (weights)

The 'Art of Living under Water' formed an important part of Demainbray's course, taking up an entire lecture. It was illustrated with a model diving bell, Halley's improvements to the diving bell, Triewald's campana urinatora, and Captain Rowe's tub. This is the first model mentioned, being a simple glass bell which can be weighted by hanging up to 12 lead pieces on the central brass band.

Diving bells had received considerable attention in the first half of the century since the salvaging of wrecks was lucrative. Halley, Triewald, and Rowe all had diving companies with rights to various areas. Unfortunately only two of the four original models survive. A very similar glass diving bell model is in the Gabb collection at present at the National Maritime Museum.

Q.C., Item 216; diving bell and weights with brass hoop
Desaguliers 1744, Vol. 2, p. 212, Pl. 20, Fig. 2
Demainbray 1754*b*, p. 15 'the diving bell illustrated with an adequate apparatus'
Daily Advertiser 1756, No. 8090
Daily Advertiser 1757, No. 8166
Daily Advertiser 1758, No. 8487
C 130

Inventory No. 1927-1288

D41   D42 (part)

D42 Two cages for use with the diving bell

before 1753

H 175mm, **D** 100mm

These brass wire cages fit exactly inside the diving bell above. Although Adams did make cages, his had no bases and the mesh was different. Adams used them for preventing exploding glass phials from damaging his receivers, while Demainbray used these for holding birds or small mammals

in the diving bell to show it in action. Both cages have pine knobs, one has a hook, and the other probably had a hook originally.

Q.C. Item 217; cage to enclose a bird in ditto (diving bell)
Adams 1761, pp. 42–3, Pl. 20, Figs 61, 62
C 118

Inventory No. 1927-1301, -1302

D43 Model of Halley's diving bell

before 1753

H 350mm, **D** 180mm (maximum)

Halley wrote in 1716 'being engaged in an affair that required the skill of continuing under water I found it necessary to obviate these difficulties which attend the use of the common diving bell'. A simple diving bell, such as D41, cannot provide fresh air to sustain the diver beyond a few minutes. Halley devised a system whereby fresh air was sent down in barrels to replenish the bell and the used air was let out via a stopcock. The bell was in the sh___ of a truncated cone and h_____ut clear glass' in t_____. Desaguli___

Air sent down in barrels.

_____
D__
Da_____o December, No. 816___
Daily ___rtiser 1758, 27 March, No. 8487
Heinke and Davis 1873, p. 10
Armitage 1966, pp. 118–121
Ronan 1969, p. 201
Lindqvist 1984, p. 195

Inventory No. 1927-1452

Condensed fountain

___ '753

_, **D** 330mm

___ought to be the 'condensed fountain' advertised in the two earlier ___mainbray syllabuses. Although as such it was dropped from the later courses, from Chabrol's notes it appears to have continued in use in a simpler form in the lecture on the air's elasticity. The spherical chamber was partially filled with water, and then air was pumped in via a pipe and stopcock which are now missing. If the base of the pipe was below the water surface, when the stopcock was opened, the compressed air in the chamber forced up the water to give a fountain. The water collected in the upper basin. Chabrol shows a very similar piece of apparatus being used to provide a jet of water for the 'random of projectiles'. A jet of water was released under pressure from the fountain at a particular angle and its path traced to find the distance covered before it reached the ground. This appears to have been the principal use of the instrument in later courses.

Q.C., Item 108; Condensed copper fountain, ruler and quadrant on moveable cock to show random of projectiles (sic)
Hauksbee 1714, p. 19, Pl. 5, Fig. 8
Demainbray 1750–2, p. 9; a condensed fountain
Chabrol 1753, p. 143
Demainbray 1753, p. 15
Demainbray 1754*b*, p. 9; 'engine to play jets'

Inventory No. 1927-1622

D45 Wind gun

before 1753

L 1175mm, **D** 45mm (barrel), **W** 63mm (maximum)

The wind gun was used by Demainbray in his third lecture on pneumatics which dealt with the pressure of air. The design is probably based on that of Desaguliers which in turn was taken from Petrus van Musschenbroek.

The gun is made of brass and has two barrels, one inside the other. A spring can be operated in the handle which forced the air through a valve into a chamber behind the pellet where it could be retained. The pellet was put into position with a rammer as in other contemporary guns. When the trigger was pulled another valve was opened which allowed the air to escape on to the pellet which in turn was shot out. If the trigger was pulled and released very quickly, several shots could be fired using one valve of air. Alternatively a lock held the trigger open and all the air was used. In this instrument the spring is stuck and a pellet is lodged in the bore of the tube.

Q.C., Item 223; Wind gun
Leybourn 1694, pp. 17, 28; mechanics section
van Musschenbroek 1739, Vol. 2, p. 264, Pl. 23, Fig. 12
Desaguliers 1744, Vol. 2, p. 398, Pl. 24, Figs 14, 15
Demainbray 1750, p. 9; a curious wind gun will be taken to pieces
Demainbray 1754*b*, p. 16
C 111

See also P54

Inventory No. 1927-1472

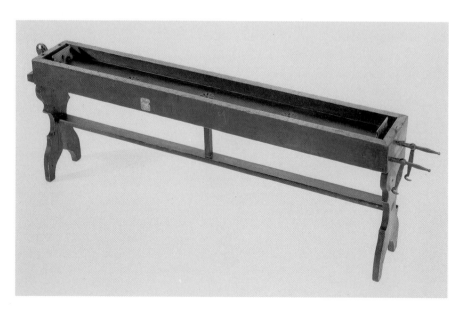

D46 Sonometer

before 1753

L 1050mm, **W** 150mm, **H** 375mm

The sonometer was a central feature of Demainbray's lecture on acoustics. It showed 'the grounds on which harps, violins, harpsichords and all string instruments of musick are constructed', illustrating the 'doctrine of sounds'. It has a pine soundbox one metre long with two fixed hardwood bridges 36 inches apart. Two strings could be stretched over these by tying them to pegs at one end and attaching them to green/blue painted iron hooks at the other on which weights could be hung. There is a scale in inches on one side of the instrument and four marks at 9 inch intervals on the other. 'Octave' is written at the central mark and 'fifth' two thirds along. The experiments performed were probably those described by 's Gravesande in which the relationships between pitch, length, tension, and diameter of the strings are explored. One string was also used to make the other resonate. The brass pulley is probably a later addition.

Q.C., Item 80; sonometer with bridge and strings
Wallis 1677, p. 839
's Gravesande 1747, Vol. 2, p. 57
Demainbray 1754*b*, p. 17 'a Sonometer whose Contrivance will shew the Grounds on which Harps, Violins, Harpsichords and all String Instruments of Musick are constructed.'
C 252

See also L59–66

Inventory No. 1927-1244

D47 Speaking trumpet

before 1753

L 585mm, **D** 184mm (max)

The speaking trumpet demonstration formed part of Demainbray's lecture on acoustics which came under the general heading of pneumatics. The trumpet is made of tinplate and painted in the blue/green colour found on a range of Demainbray's instruments. There was probably a leather or calfskin mouthpiece originally.

Speaking trumpets had received some attention following the publication of Morland's book *Tuba Stentoria-Phonica* in 1671. Here he described instruments he had made and experiments performed on them which included hearing sounds transmitted from Vauxhall at Battersea, and sounds transmitted from Hyde Park at Chelsea. He advocated their use for the issuing of commands on board ship and for overseers to instruct a large workforce. He urged others more mathematically inclined than himself to find the optimum shape for the instrument, but when this was reported to the Royal Society in the following year by Pardies there was little interest in England. Morland used copper or glass for his instruments but in 1678 Conyers suggested tinplate. Speaking trumpets became standard items in courses of experimental philosophy in the mid-eighteenth century.

Q.C., Item 90; speaking trumpet
Morland 1671, pp. 3056, 3058
Pardies 1672, Vol. 9, pp. 57–61
Conyers 1678, p. 1027. Letter about his improvement of Sam Moreland's speaking trumpet
Chabrol 1753, p. 175
Demainbray 1754*b*, p. 17
C 249
Hunt 1978
Gouk 1982, 36. pp. 155–78

Inventory No. 1927-1120

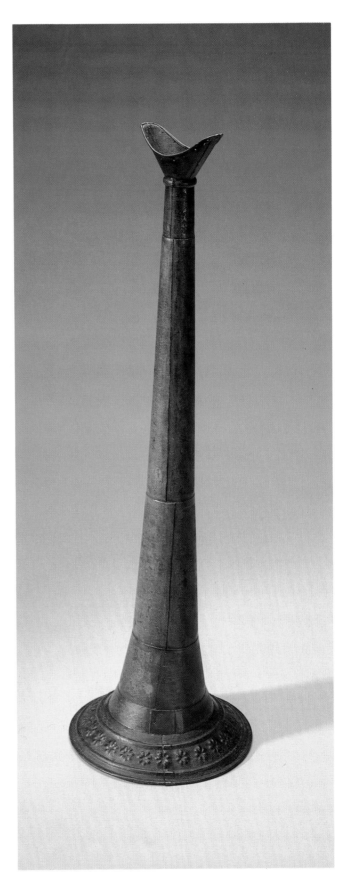

D48 Glass vessel with top, tap, and slipwire

mid-eighteenth century

H 515mm, **D** 190mm (at top), **D** 255mm (base)

This almost certainly forms part of Demainbray's collection of instruments since it has some characteristic features, notably the green/blue paint on tin and the poor quality of the workmanship. However, it is not known to what purpose the apparatus was put. It is possible that it is one of a set of receivers for use with the air pump, since it somewhat resembles a figure in Desaguliers's *A Course of Experimental Philosophy* representing an apparatus for this purpose; however, the glass is not of a strength normally associated with receivers for this purpose. Alternatively it could be a receiver used in the rotting of fruit or the fermentation of fluids, both of which are mentioned in the Queen's Catalogue.

It consists of a tall glass cylinder with a flat tinplate top in which there is a tap and slipwire extending into the cylinder.

Q.C., Item 86; Two compleat sets of different sized receivers for the air pump

Q.C., Item 227; Receiver to measure the quantity of air generated from the rotting of fruit

Q.C., Item 228; note: large glass receiver to measure the quantity of air generated by the fermentation of fluids

Desaguliers 1744, Vol. 2, p. 377, Pl. 24, Fig. 4

Demainbray 1754*b*, p. 16

C 129

Inventory No. 1927-1394

D49 Model sea gauge

before 1754

D 85mm, **H** 310mm (overall)

In his book *Vegetable Staticks*, published in 1727, Hales suggested the use of honey or treacle to mark the maximum point reached by a column of mercury in pressure measurements. The honey or treacle was poured over the mercury so that when the height of mercury in a tube rose and then fell the honey or treacle left a mark at the highest point. Hales hinted at wider applications, and the idea was taken up by Desaguliers who presented the 'sea gage' to the Royal Society in 1728.

The instrument is essentially a barometer and consists of a strong glass vessel which held the mercury, with a glass tube with its closed end uppermost fixed with the lower end in the mercury. The tube is surrounded by a strong glass bulb which holds enough air to enable the instrument to float. Sea water could permeate through the small holes in the brass cover between the lower vessel and the bulb.

Mercury was placed in the lower strong glass vessel and covered with treacle. It rose in the sealed inner tube in proportion to the pressure which is proportional to depth. The gauge was calibrated in two known depths and marks could be made on the tube. Desaguliers suggests a diamond for this purpose: however this model is marked with black paint. When the device was sunk a weight was attached to the base which was released by a catch when it hit the sea bed. The air trapped in the upper vessel carried the device back to the surface where the treacle mark gave the maximum pressure and hence depth. This was a substantial improvement on former depth gauges which had relied on the time an object took to return to the surface from the sea bed.

Q.C., Item 224; Sea Gage
Hales 1727, p. 206
Desaguliers 1728*a*
Hales 1728, p. 559
Hales 1731, pp.
Desa

th of the Sea

s at sea

I.

[handwritten note:] As mercury is 13× heavier than (distilled) water, instrument would measure depths approximately 13× the change in mercury column. Here (upper part) about 100 mm, so would measure water only about 4½ feet deep! (Assuming vacuum above mercury, as in barometer). ↘

D50 Model of Savery's engine

before 1754
ON COCK
'R.RICE'

L 600mm, **W** 300mm, **H** 720mm

Demainbray used this model with the model Newcomen engine in his course on pneumatics. During the lecture this model, but not the Newcomen model, was shown as a working demonstration. Considerable attention was given to methods of preserving the boiler and repairing the cylinder 'being of utmost moment to persons concerned in Mines'.

Savery patented his 'fire engine' or atmospheric engine in 1698 and presented it to the Royal Society the following year. Although it was a major technological breakthrough it did suffer considerable problems, the principal one being that the steam had to be held at very high pressures and temperatures which sometimes melted the solder and blew apart the boiler. According to Desaguliers 'these discouragements stopped the progress and improvement of the Engine . . .' making it useful only for raising water for gentlemen's seats and not in mines where the demand was greatest. Desaguliers and his colleagues studied the design of the early engines and made models and improvements. The Queen's Catalogue lists 'Savery's first fire engine': in fact this model closely resembles that illustrated in Bradley's *New Improvements in Planting and Gardening* which went through several editions in the early eighteenth century. It is a very simple engine with a single boiler and receiver.

The boiler and receiver are made of copper. The boiler sits on a tin cylinder which contained the fire; there is a stoking hole in the side. The water was driven out of the receiver by the pressure of the steam entering. The valve at the base of the water pipe was closed and the upper valve opened so that the water went upwards into the reservoir in the top of the wooden structure. Then the upper valve was closed, the lower one opened, and the cold water was sprinkled on the receiver from the cock. The steam condensed, sucking up water from below into the receiver. The process was then repeated, lifting water from a trough below the model into the reservoir at the top.

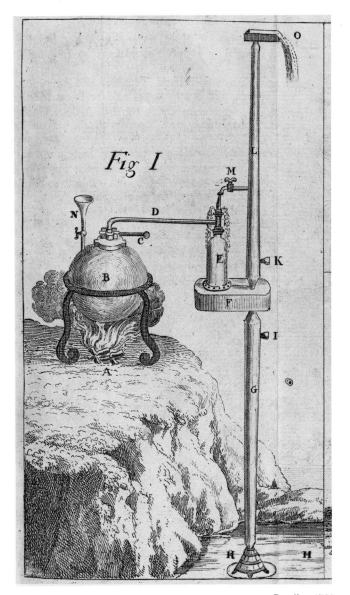

Bradley 1739

Q.C., Item 229; Captain Savery's first fire engine
Savery 1699, p. 228
Savery 1702
Harris 1704
Bradley 1739, Pl. 7, Fig. 1
Desaguliers 1744, Vol. 2, p. 484, Pl. 40
Demainbray 1753, p. 17
Demainbray 1754*a*, p. 18
Demainbray 1754*b*, p. 18; 'Savery's fire engine'
C 325
Buckland 1984, pp. 1–20
See also L36, L37, L38, L39   **Inventory No.** 1927-1620

D51 Model of Newcomen's engine

before 1753

L 1070mm, **W** 400mm, **H** 1570mm

'Fire engines', now known as atmospheric engines, occupied a lecture of Demainbray's course on pneumatics. The lecture was illustrated with 'a curious and compleat model' which he later described as perfect of its type with all the additions and corrections made since the invention of the machine. In the Queen's Catalogue it is described as 'a compleat model of a fire engine with all the modern improvements'. As with several of the pneumatics and hydrostatics items it is based on the design described by Desaguliers in his *A Course of Experimental Philosophy*. Desaguliers related that Newcomen, an ironmonger, and Cauley, a glazier, had brought the fire engine to 'its present form' which had been used 'these 30 years' and had solved some of the problems appertaining to the Savery engine. The actual engine that Desaguliers described was one of those at Griff near Coventry in use in the 1720s to help water out of a coal mine. It is possible that Desaguliers made this model himself or had it made for him, since he did construct a model engine in his garden.

Although the model has many components it is only intended to describe the principal features here. The model works in the same manner as a full-sized engine, although the efficiency would be considerably less, as Desaguliers points out. It is made of beech with a brass cylinder and pipes and a copper housing for the boiler. The fire for the boiler and the housing stand on a painted iron base. Steam from the boiler fills the cylinder as the piston is drawn up by the weight hung on the opposite end of the beam. The steam is cut off and a jet of cold water is admitted to the cylinder causing the steam to condense so that atmospheric

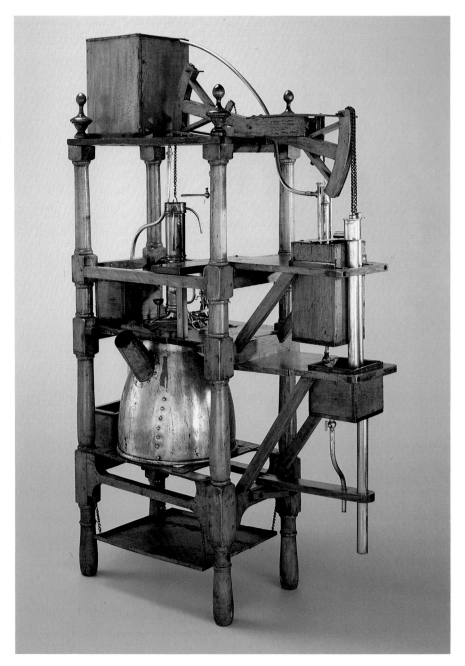

pressure drives the piston down. The steam and water valves are worked automically by pins set in the plug rod that hangs from the beam in front of the cylinder. Water can be pumped from a pail placed on the same level as the foot of the model to a cistern at the level of the cylinder.

The model is complete but has been altered.

Q.C., Item 230; a compleat model of a fire engine with all the modern improvements
Desaguliers 1744, Vol. 2, p. 464
Demainbray 1754*b*, p. 17 'Fire Engines with a curious and compleat Model'
C 326
Chew 1968, item 19

See also L36, L39

Inventory No. 1927-1619

MAGNETISM

D52 Magnetical magazine consisting of twelve magnets in a case

before 1754

L 345mm, **W** 64mm, **H** 50mm

Twelve steel bars are numbered and arranged in two stacks inside the brass casing; the screws are also numbered. One of the bars is broken. Demainbray used the 'magasin magnetique' in the courses he gave in France in 1753 and 1754, but not when he returned to England, possibly because the bar had then been damaged. He gave only one lecture on magnetism, usually placed after hydrostatics and pneumatics.

Artificial magnets were becoming important demonstration items in the 1750s following the improvements in steel-making introduced in 1740 and the publication of the methods of making the magnets by Michell and Canton. Gowin Knight had been making strong artificial magnets during the 1740s but had kept his methods secret. In November 1744 he demonstrated a compound artificial magnet of 12 bars similar to, but smaller than, this example which was said to lift 23 oz Troy.

Q.C., Item 120; Magnetical magazine mounted in brass
Knight 1744, p. 161
Michell 1750, p. 73
Canton 1751, p. 31
Demainbray 1753, p. 18
Demainbray 1754*a*, p. 19; 'un magasin magnetique'
Inventory No. 1929-101

OPTICS

D53 Polyoptric pyramid

before 1753

H 205mm, **D** 136mm (base), **D** 40mm (prism)

Polyoptric pyramids were one of several types of optical amusement popular in the seventeenth and eighteenth centuries. They are similar to multiplying glasses in that light from an object is deflected in various directions to form a compound image. When viewed through the pyramidal prism the original picture appears completely different. Unfortunately the pictures for use with this instrument are now lost, but we know that five demons' heads became a vestal, a hermit became a water fowl and a garland of flowers the head of a Samean sibyl. Although Demainbray described the use of the polyoptric pyramid as 'a very singular experiment' he did not use it after 1753 in his lectures, possibly because it can only be viewed by one person at a time. George Adams the younger described these instruments as 'rather tedious in the making'. This example is on a walnut stand.

Q.C., Item 236; polyoptrick pyramid
Della Porta 1658, p. 369
Adams 1746, No. 267
Demainbray 1750–2, p. 14 'the Polyoptrick Pyramid'
Adams 1799*b*, Pl. IX, Figs 7, 8, 9; p. 257
C 200
Inventory No. 1927-1153

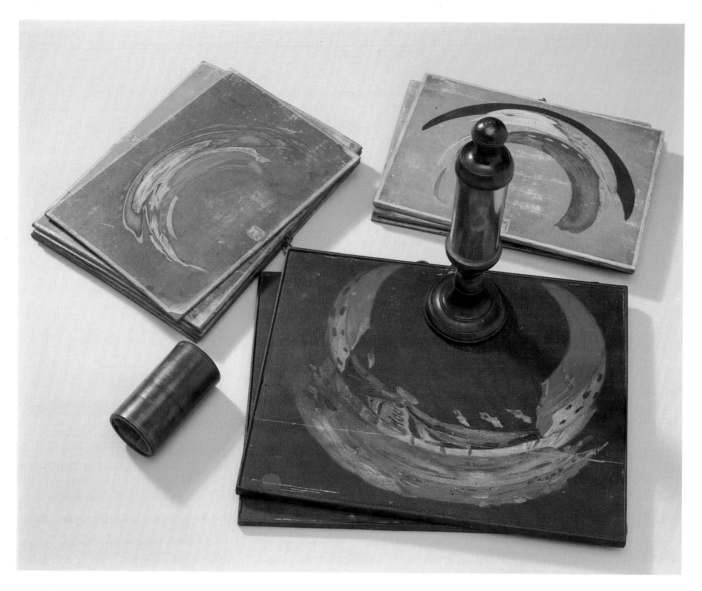

D54 Catoptric cylinder and fourteen pictures

before 1753

H 265mm (cylinder), **D** 86mm (base), **D** 46mm (cylinder),
L 361mm (two pictures), **W** 280mm (two pictures),
L 298mm (four pictures), **W** 235mm (four pictures),
L 285mm (eight pictures), **W** 210mm (eight pictures)

Anamorphic pictures using reflection had become common by the mid-eighteenth century when methods of drawing them were widely known and mirrors of sufficient quality and durability could be made. Although the earlier magical associations of anamorphoses had been lost, they were presumably still an attraction as Demainbray used the 'catoptrick cylinder and deformed pictures' in the first lecture on optics in all the surviving syllabuses.

Twelve pictures form a set with the pictures on paper glued

to oak boards; these are not original. The two larger pictures are painted directly on to the wood which is French walnut; one of these has S.D painted on the back. The cylinder is of speculum metal on a stained boxwood stand. There is a stained boxwood sheath to protect the mirror.

Q.C., Item 234; two cylindric pictures with a polished cylinder
Niceron 1638, 1st edn, title page
Leupold 1713
Martin 1740, p. 2
Adams 1746, No. 265
Demainbray 1754*b*, p. 19 'the catoptrick Cylinder'
C 199
Baltrusaitis 1977
Turner 1987, p. 84, Pl. 68
Inventory No. 1927-1159

D55 Catoptrick cone and two pictures

before 1753

D 300mm (octagonal picture), **H** 100mm (cup height),
D 69mm (max)

The catoptrick cone was used with the catoptrick cylinder described already. The two hexagonal pictures show human heads, possibly portraits, when viewed from above with the cone. Cones were less common than cylinders. This was 'as highly finished as that kind of instrument will admit of'; the pictures are painted directly on to the wood which is oak. The cone is speculum metal and is mounted on an applewood screw which can be screwed into a conical applewood cup when not in use.

Q.C., Item 235; two conic pictures with a polished cone
Niceron 1638, title page
Leupold 1713
Martin 1740, p. 2
Adams 1746, No. 266
Demainbray 1754*b*, p. 19 'catoptrick Cone'
C 199
Baltrusaitis 1977
Turner 1987, p. 84, Pl. 68   **Inventory No.** 1927-1160

D56 Anamorphic picture of a horse

before 1753

L 312mm, **W** 95mm

' horse seen with an oblique view' was listed as one of the demonstrations of perspective in the earliest syllabus of Demainbray that we possess. In later ones he simply gives 'different anamorphic perspectives'. This is a simpler type of anamorphic picture than those used with the cylinder and cone since no reflection is involved. To some extent the reflecting type replaced these during the eighteenth century. Horses were common subjects. This example is painted on beech in a pine frame.

Q.C., Item 255; a painted horse anamorphosis perspective
Demainbray 1750–2, p. 11; 'a horse seen with an oblique view'
C 199
Baltrusaitis 1977   **Inventory No.** 1927-1525

D57 Optical model showing the virtual image formed by a concave lens

1752

HANDWRITTEN—BASE
'Virtual focus. . .'

L 230mm, **W** 80mm, **H** 85mm

Demainbray had some trouble having these optical 'machines' or models made, as was discussed in Chapter III. He commissioned John Cuff the well-known microscope-maker to produce them. However, Cuff did not deliver and Demainbray was forced to try another maker. Although we know that this attempt was successful we do not know who eventually made them.

Since models showing the paths of rays in convex and concave lenses are listed in the Queen's Catalogue, and since this model is so similar to the others, it can be assumed that this is one of Demainbray's collection. In the French syllabuses Demainbray claims to show the paths of rays of light through lenses in an ingenious fashion, which may well refer to these models. However, the one illustrating the convex lens has not survived.

This model consists of a slit, virtual focus, concave lens, and screen, all in ivory on a fruitwood base. It is the only model in which the light passes from left to right. The threads are now missing.

These devices were made commercially by George Adams the Younger in the last decade of the century. He describes them as 'a curious set of optical models where the rays of light are represented by silken strings'.

Q.C., Item 253; a model of the direction of the rays, to shew the virtual focus of a concave lens
Baker 1751, Vol. 4, f. 74
Demainbray 1753, pp. 19–20
Demainbray 1754a, p. 20; 'La direction des rayons de la lumière, se verra d'une façon ingénieuse, dans son passage à travers les lentilles'
Inventory No. 1927-1435

D58 Concave mirror on stand

before 1753

D 440mm (mirror), **H** 750mm, **D** 500mm (stand)

Demainbray described his convex and concave mirrors as 'beautiful large specula' and used them to explain reflection at curved surfaces. This instrument was also used to view perspective prints—prints made to be viewed through 'zograscopes' or mirrors and lenses. The mirror is in a beech mount supported on a brass swivel frame with two butterfly nuts. The stand is mahogany. The instrument is now cracked. Mirrors were very expensive items and a main attraction: in 1757 Demainbray mentioned a model of Buffon's specula which was the only one of its kind in England. Unfortunately this instrument has not survived.

Q.C., Item 305; a large concave mirror and stand
Martin 1740, p. 2
Demainbray 1750–2, p. 12; 'plane, convex and concave mirrors explained with beautiful large specula'
Demainbray 1754*b*, p. 21
Daily Advertiser 1757, April 12th, No. 8190, p. 2
C 217

Inventory No. 1929-119

D59 Convex mirror on stand

before 1753

D 450mm (mirror), **H** 720mm,
D 500mm (stand)

The convex mirror was used by Demainbray to illustrate reflection at curved surfaces. It forms a pair with the concave mirror above.

Q.C., Item 304; a large convex mirror and stand
Martin 1740, p. 2
Demainbray 1750–2, p. 12; 'plane, convex and concave mirrors explained with beautiful large specula'
Demainbray 1754*b*, p. 21
Daily Advertiser 1757, No. 8190
C 216

Inventory No. 1929-120

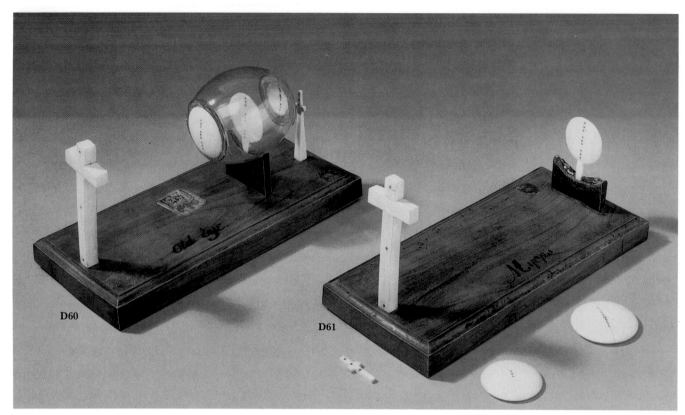

D60 Optical Machine to show the working of an old eye

1752

HANDWRITTEN—BASE
'Old Eye'

L 245mm, **W** 109mm, **H** 130mm

In his early course Demainbray used an artificial eye to explain long and short sight, long sight being the usual affliction of old age. However, in 1753 the artificial eye was replaced by one of a number of 'machines' which showed the direction of the rays of light by threads, as a three-dimensional ray diagram. This 'machine' has an ivory object in the shape of a cross, an ivory eye-lens, a glass eyeball, an ivory retina, and a deformed ivory cross as the image which is formed behind the retina. The threads are now missing. The whole is on a fruitwood base.

Q.C., Item 96; artificial glass eye with silken lines to shew direction of the rays of light in an old flattened eye.
Demainbray 1753, p. 20
Demainbray 1754*a*, p. 21
Demainbray 1754*b*, p. 21 'Machines which shew the Direction of Rays of Light in the Myopes and Presbytae'
Adams 1794, Vol. 2, Pls 3, 4
C 202
Inventory No. 1927-1152

D61 Optical model of the human eye illustrating myopia

1752

HANDWRITTEN—BASE
'Myops'

L 245mm, **W** 80mm, **H** 80mm

Had the glass eyeball survived, this model would have been similar to that demonstrating long sight. An object, represented by a coloured ivory cross, forms an image by means of the eye lens. The image which is also represented by an ivory cross falls short of the retina. The paths of the light rays would have been traced using threads. The model is on a fruitwood base.

Q.C., Item 97; artificial eye with silken lines to shew the direction of the rays of light in the myops
Demainbray 1754*a*, p. 21
Demainbray 1754*b*, p. 21 'Machines which shew the Direction of Rays of Light in the Myopes and Presbytae'.

Inventory No. 1927-1438

As photographed, threads reflecting foot of cross from upper surface of 45° mirror are missing. (see also drawing)

D62 Model demonstrating the principle of a camera obscura

1752

HANDWRITTEN—BASE
'Plan of Camera Obscura with threads'

L 380mm, **W** 70mm, **H** 115mm

Demainbray does not mention this particular optical 'machine' specifically in his syllabuses; however, it is listed in the Queen's Catalogue and was sketched by Chabrol, so it can be identified as his with certainty. The camera obscura was covered in the lecture on long and short sight for which 'machines' were used so it is likely that Demainbray explained the camera obscura by the same means. This machine also demonstrates light passing through a convex lens, so it may be what Demainbray was referring to when he said that the passage of rays of light through lenses would be shown in an ingenious manner.

The light rays, represented by threads, start from the object on the right which is a coloured ivory cross, pass through an ivory convex lens, are reflected by a mirror represented by a wooden screen set at 45° to the horizontal, and form an image on the horizontal screen above. Another

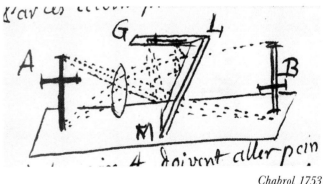

Chabrol 1753

image on the left shows the path of the rays had the mirror not been present. The whole is on a fruitwood base.

Q.C., Item 252; a model of the Direction of the Rays of Light by silken lines in a Camera Obscura
Chabrol 1753, p. 187
C 194

See also D136

Inventory No. 1927-1141

D63 Yarwell type microscope

before 1753

L 180mm, **W** 90mm, **H** 270mm

According to Court this instrument came from the King George III Collection. It is probably the item referred to in the Queen's Catalogue as Yarwell's double microscope and it may also be the Marshall type mentioned by Demainbray in his syllabuses, as Yarwell and Marshall types are very similar.

The microscope is fixed on a pillar mounted on a ball and socket joint which screws into one corner of the box. The stage consists of a glass slide held between brass clips and includes a stage forceps. A jointed arm can be attached to the side of the box for the application of a condenser or mirror; however, both the latter are now missing. The optical system consists of an eyepiece, a field lens, a 'between lens', and an objective, and so corresponds to the four-lens microscope described by Martin in 1759. Focusing is achieved by the use of a steel spring which allows smooth sliding of the arm on the pillar. Since the microscope is entirely of brass, it is unlikely to date from c. 1710 as indicated by Court and Whipple, nor is there evidence of it having been made by Culpeper. The box is made of pine.

Q.C., Item 294; Yarwell's double microscope
Martin 1759, p. 52
Demainbray 1754*b*, p. 21, note: Marshal's [microscope]
Crisp 1925, No. 305?
Whipple 1926, p. 515
Clay and Court 1932, p. 102, Fig. 66
Bradbury 1967, p. 135

Inventory No. 1928-829

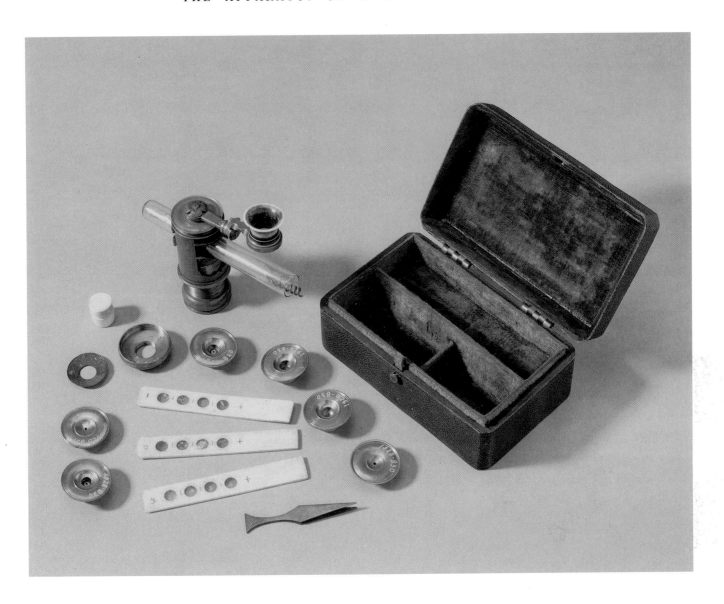

D64 Wilson screw-barrel microscope and hand magnifier

before 1753

L 125mm (box), **H** 50mm (box), **W** 75mm (box), **L** 65mm (screw barrel), **D** 28mm (screw barrel), **L** 70mm (hand magnifier), **W** 20mm (hand magnifier), **T** 15mm (hand magnifier)

There is no reason to doubt Court's claim that this instrument came originally from the King George III Collection. Demainbray had a Wilson screw-barrel microscope which he used in his lectures from 1753.

It is a standard Wilson screw-barrel microscope in brass which occupies a fishskin-covered box, together with a hand magnifier in brass and ivory. There is a choice of six powers and a pair of tongs for specimens. The microscope has a fitting for a stand which is now missing. The complete instrument would probably have resembled the Wilson screw-barrel microscope on a stand which features on Scarlett's trade card.

Q.C., Item 290; Wilson's microscope
Demainbray 1754*b*, p. 21 Wilson's [microscope]
Clay and Court 1932, p. 47, Fig. 21

Inventory No. 1928-830

D65 Culpeper type compound microscope

Loft, Matthew or Scarlett, Benjamin

c. 1745   London

ENGRAVED—BODY
'DOCTr DEMAINBRAY'
H 410mm, **L** 240mm, **W** 240mm

This is likely to be the 'large compound microscope' listed in the Queen's Catalogue since it is above average size for a Culpeper type. Court described it as being made by Matthew Loft for Demainbray in about 1745, but no evidence for this survives apart from stylistic; according to Court the shape of the legs is characteristic of this maker. It is probably the microscope referred to by Demainbray in his syllabuses as Scarlett's, since Scarlett improved the Culpeper type and his name was associated with this particular form.

The instrument is almost entirely of brass but has a French walnut eyepiece. It is mounted on three triangular pillars on an octagonal mahogany base which contains the box for accessories. Focusing was achieved by sliding the body tube and then clamping it into position with a screw, but it is now immovable.

A brass disc with nine specimen cells can be slid along a slot in the stage and rotated. The contents of the cells are listed on a handwritten note with the accessories. In addition there are six ivory sliders, each with four cells, a live box, two glass tubes for specimens, three objectives, forceps, a glass slide, two contrast discs, a mount for a hand magnifier, a hair brush, and an ivory box for talcs.

Q.C., Item 295; large compound microscope
Demainbray 1754*b*, p. 21 Scarlet's [microscope]
Crisp 1925, No. 162
Whipple 1926, p. 515, Fig. 8
Clay and Court 1932, p. 120, Fig. 81
Bradbury 1967, p. 112

Inventory No. 1928-790

D66 Cuff solar microscope attachment

Cuff, John

before 1754   London

'*J. Cuff, London*'

L 310mm, **W** 135mm, **H** 170mm

Again there is no reason to doubt Court's claim that this item originated from the King George III Collection. Demainbray used a Cuff solar microscope in his second lecture on microscopy and appears to have viewed the circulation of a frog with one, using apparatus by Hindley which has not survived. He had 'solar microscopes as Lieberkühn invented them and Cuff improved them'.

This is a standard Cuff solar microscope attachment in brass, set in a square brass plate; there is no microscope with the instrument. The plate can be attached to a shutter or hole in the wall by means of two nuts at opposite corners. The inclination of the mirror can be adjusted using the ring handle and it can be rotated using the knurled screw.

Q.C., Item 299; Cuff's solar microscope
Baker 1742, p. 21
Chabrol 1753, p. 205
Demainbray 1753, p. 22
Demainbray 1754*a*, p. 22
Demainbray 1754*b*, p. 22 'solar microscope . . . as Cuff improved them'

Inventory No. 1928-851

D67 Pocket microscope

Sisson, Jeremiah

before 1753   London

'*Dr. Demainbray invent/J Sisson London*'

L 85mm (box), **W** 65mm (box), **H** 50mm (box), **H** 200mm (microscope and box)

Demainbray described this instrument as follows: 'Apparatus contained within the compass of a snuff box to be used in the single, compound, and may occasionally be used in the solar way, this contrivance not only contains whatever is useful to the various purposes . . . but is perfectly sturdy and strong in all its parts.'

The microscope is made of brass and fits into a small green shagreen box. It can either be used by hand or be fitted on to the top of the box for use with the mirror. In its present form it is possible to use it only as a single instrument; it may at one time have had an additional compound body. The six powers are mounted in a brass bar which slides under the eye cap. A stage forceps can be attached to an arm on the stage which also carried a brass disc containing ten cells for specimens. The brass disc is now missing.

There is a similar instrument by Thomas Bureau at Oxford, but no other Sisson microscope is known to exist. Another microscope with Demainbray's name on can be seen in the Museum of the History of Science at Oxford.

Q.C., Item 307; a pocket microscope by Demainbray
Demainbray 1754*b*, p. 22 note: as text
Crisp 1925, No 21
Whipple 1926, p. 516, Fig. 8

Inventory No. 1928-819

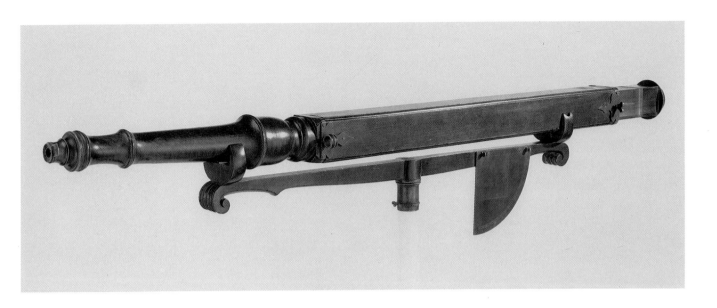

D68 Square-section terrestrial telescope with brass quadrant and mahogany bracket

Robertson, William

c. 1745   Edinburgh

ON BRASS FITTING AT TOP
 'W.R. Fecit'
 Small circles stamped on brass fittings

L 1080mm, **D** 60mm, **L** 700mm (bracket), **H** 125mm (bracket)

It is likely that this instrument is the one described by Demainbray as 'a curious dioptrick with a large apparatus for astronomical observation' which he used in his lecture on the construction of telescopes. However, another instrument now at Armagh Observatory could also fit the description, as could D140. The Queen's Catalogue lists both 'a large 4 lens Telescope with a Stand and a Brass Quadrant' and 'a lesser Ditto'. This is probably the lesser version while the one at Armagh could be the large one.

The telescope is an early example of the use of mahogany. It has a square-section body with a single draw tube which contains the objective. An X-ray photograph shows the telescope to have four lenses; the eyepiece and erectors are in a turned mahogany eyepiece section which is fixed. The telesope is mounted on a mahogany stand which would have screwed into a pillar. A brass quadrant is attached to the stand and is graduated in degrees from 0 to 90. The upper surface of the body tube carries a pair of sights. The brass mounts for the sights are marked with a circular stamp.

William Robertson was primarily a microscope maker who signed his instruments 'W.R. Fecit' and often used a circular stamp. Judging by advertisements he was fairly prominent in Edinburgh at the time Demainbray was there in the late 1740s.

Q.C., Item 277; a large 4 lens telescope with a stand and quadrant
Q.C., Item 278; a lesser ditto
Demainbray 1750–2, p. 13; a curious dioptick with a large apparatus for astronomical apparatus
Demainbray 1753, p. 23
Demainbray 1754*a*, p. 24
Demainbray 1754*b*, p. 23
Whipple 1926, p. 512, Fig. 6
Disney *et al.* 1928, Pl. 8, p. 179
C 176
Turner 1967*a*, pp. 213–42
Bryden 1972*b*, pp. 167–9

Inventory No. 1927-1282

D69 Optical model of a Galilean telescope

1752

HANDWRITTEN—BASE
  'Galilean telescope with threads'

L 275mm, **W** 55mm, **H** 120mm

This model is one of three optical 'machines' Demainbray had for showing the paths of light rays inside telescopes. It consists of an object, a virtual image, a concave lens, a convex lens, and a final image all in ivory; the threads are now missing. The two images are represented by coloured crosses. As usual in these machines the light travels from right to left taking the writing on the base as a reference.

Q.C., Item 274; model of ditto with silk lines [Galileo's telescope]
Demainbray 1753, p. 23   Demainbray 1754*a*, p. 23
Demainbray 1754*b*, p. 23; 'various machines to shew the refracted direction of light through these instruments'   **Inventory No.** 1927-1431

D70 Optical model of a two-lens telescope

1752

HANDWRITTEN—BASE
  'Plan of/Dioptrick Telescope with 2 lenses/by/Threads'

L 350mm, **W** 110mm, **H** 70mm

Demainbray illustrated his lecture on telescopes with models of a two-lens telescope, a four-lens telescope, and a Galilean telescope. This model has an object represented by an ivory cross, two ivory lenses, an ivory cross as an image, and a hole for the eye. Like most of the other optical models the base is fruitwood and the threads are missing.

Q.C., Item 270; a model of ditto with silken threads [2 lens telescope]
Demainbray 1753, p. 23   Demainbray 1754*a*, p. 23
Demainbray 1754*b*, p. 23; 'various machines to shew the refracted direction of the rays of light through these instruments' [telescopes]   **Inventory No.** 1927-1434

D71 Optical model of a four-lens telescope

1752

HANDWRITTEN—BASE
  'Plan of/Four Lens Telescope/with Threads'

L 350mm, **W** 110mm, **H** 70mm

This model is the most complex of the three telescope models owned by Demainbray. It consists of an object, primary and a final image, four convex lenses, and an eye hole. It shows the arrangement of the standard four-lens telescope of the day. The optical parts are made of ivory and the base is fruitwood. The threads are missing.

Q.C., Item 279; a model of the direction of the rays through ditto [a four lens telescope]
Demainbray 1753, p. 23   Demainbray 1754*a*, p. 23   Demainbray 1754*b*, p. 23
Inventory No. 1927-1437

D72 Terrestrial telescope

Costa

1752   Bordeaux

HANDWRITTEN IN INK ON DRAW TUBE
  'Dr. Demainbray 1752'

L 647mm, **D** 61mm

This is one of a number of instruments which Demainbray acquired while travelling in France. He introduced it into his course of lectures in 1754, describing it in the English syllabus as 'a dioptrick telescope with six lenses for obtaining a double field of image at sea, by Costa of Bordeaux'. However, in the earlier French syllabus he states merely that the lenses were made in the manner of Costa of Bordeaux. Christiaan Huygens discusses telescopes with additional lenses to increase the field of view but warns against using them without good cause because of chromatic aberration. An X-ray photograph revealed that this instrument does in fact have six lenses; otherwise it is a fairly standard instrument of the period with a red shagreen-covered body tube, green vellum draw tube, and brass mounts.

Dollond 1753, p. 103
Demainbray 1754*a*, p. 24; 'un Telescope dioptrique à six lentilles convexes qui double le champ de vision . . . ces lentilles font de la façon du Signeur Costa de Bordeaux'
Demainbray 1754*b*, p. 24
Huygens 1888, Vol. 13, Pl. 468   **Inventory No.** 1927-1419

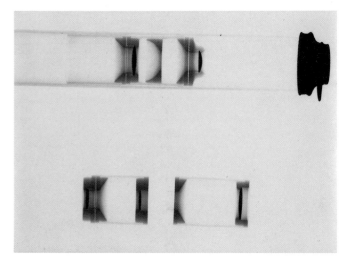

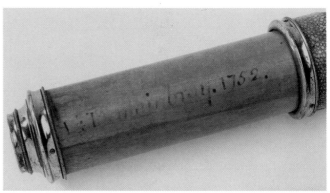

D73 Polemoscope and case

before 1753

L 95mm, **D** 30mm, **L** 115mm (cap), **D** 32mm (cap)

In all the syllabuses Demainbray advertises a 'polemoscope or opera glass' which according to Smith were both names used to describe what was properly known as a reflecting perspective. This gives rise to some confusion since the standard opera glass does not use reflection and does not have the same function as this instrument.

The polemoscope superficially resembles an ordinary perspective glass but allowed the user to view the audience while appearing to view the performance on stage. Even if people were not fooled, it was difficult to know whether the instrument was directed at any particular person. The best-known maker was Edward Scarlett and a polemoscope can be seen on his trade card.

The instrument consists of two lenses, one convex and one concave, and an oblique mirror at 45° to the axis of the tube.

A hole in the side of the instrument allows light to enter. The cap has tooling consistent with a date of 1700 to 1725 but it may not belong with the instrument.

Q.C., Item 275; a polemoscope
Smith 1738, Vol. 2, p. 377, Fig. 625
van Musschenbroek 1739, Vol. 2, Pl. 19, Fig. 16
Adams 1746, No. 233
Demainbray 1754*b*, p. 23 'a Polemoscope or Opera Glass described'
C 201
Turner 1966
Calvert 1971, No. 339
Inventory No. 1927-1158

D74 Optical model showing the direction of rays through a prism

1752

HANDWRITTEN—BASE
'Prism, 2 Screens and Convex Lens/ from single ray'
L 230mm, **W** 80mm, **W** 80mm

Three of the original four optical 'machines' demonstrating spectra have survived. Newton's experiments on spectra could only be performed 'weather permitting', so these machines were particularly useful. From 1753 until 1758 we know Demainbray ended his courses with optics and the final lecture contained 'Sir Isaac Newton's doctrine of light and colours enquired into; the direction of colour-making rays demonstrated with proper machines, the proofs alleged by Newtonians in support of Sir Isaac Newton's doctrine, and with the arguments of their antagonists delivered in a candid manner'.

As the model is constructed, the rays, which were represented by threads which are now missing, pass through the prism to form a spectrum on the first screen. If separate colours are allowed to pass through this screen they are brought to a focus on the second screen by the convex lens. Each colour will be brought to a different place. The convex lens and second screen are not well aligned and only one colour is shown reaching the second screen. The materials are ivory and fruitwood.

Q.C., Item 263; a plan with threads to shew the direction of rays of light through a prism
Demainbray 1753, p. 23
Demainbray 1754*a*, p. 23
Demainbray 1754*b*, p. 23

Inventory No. 1927-1436

D75 Optical model of a spectrum with inverted colours

1752

HANDWRITTEN—BASE
'Prism screen & convex lens/Colours inverted' '4'
L 230mm, **W** 80mm, **H** 80mm

This machine is the last and most complex of the series; it consists of a pinhole in wood with a prism, a convex lens, a second pinhole at the focus of the lens, and a screen, all of them in ivory. The resultant spectrum would have red at the top. '4' has been written on the side of the model opposite to the inscription. The base is fruitwood and the threads are now missing.

Q.C., Item 267; a plan with threads to shew the colours of a spectrum inverted
Demainbray 1753, p. 23
Demainbray 1754*a*, p. 23
Demainbray 1754*b*, p. 23; 'the direction of colour making rays demonstrated with proper machines'

Inventory No. 1927-1432

D76 Optical model of a spectrum brought to a focus

1752

HANDWRITTEN
'Prism, Screen & Convex Lens/. . ./Focus' '3'
L 230mm, **W** 80mm, **H** 80mm

This model is the third of four optical 'machines' Demainbray used to demonstrate spectrum experiments. It consists of a pinhole in wood and a prism, a convex lens, and a screen in ivory, all on a fruitwood base. The threads are now missing. An improved spectrum can be obtained by the use of a lens to bring the coloured rays to a focus. '3' has been written on the side of the model opposite to the inscription. The threads are now missing.

Q.C., Item 266; a plan to shew rays after retraction thro a prism, brought to a focus by a convex lens
Demainbray 1753, p. 23
Demainbray 1754*a*, p. 23
Demainbray 1754*b*, p. 23

Inventory No. 1927-1433

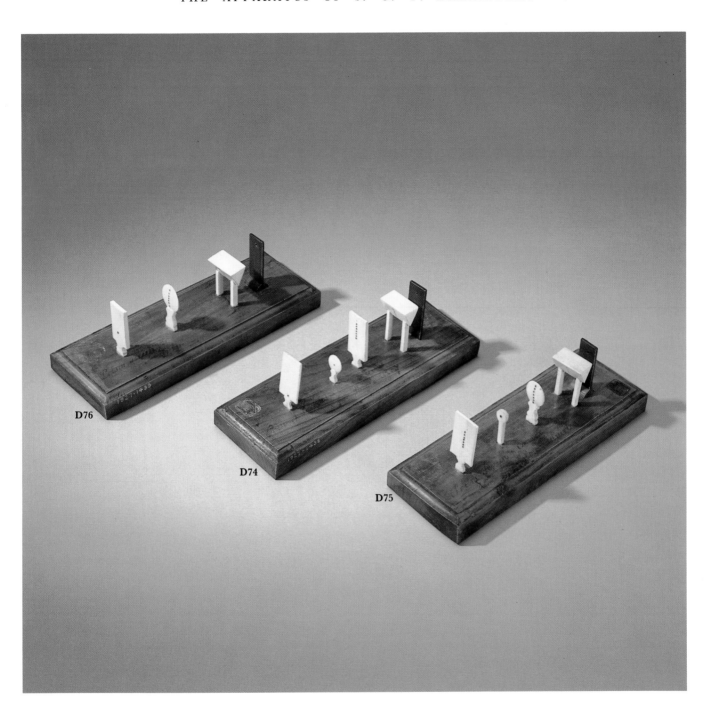

D76

D74

D75

The following items are probably from Demainbray's collection but the evidence is not as conclusive as for those listed above.

GEOMETRICAL FIGURES

D77 Hollow skew pyramid

mid-eighteenth century?

ON BASE
'7.14'

H 180mm, **L** 175mm (sides of base)

The inscription '7.14' on the pyramid refers to the height in inches.

This object belongs to a set of 21 two- and three-dimensional geometric figures with similar features: they are made of applewood painted cream which has darkened to grey over the years. The colour is similar to that on the inside of the vessels for the hydrostatical paradox D29. Many of the plane figures have been used in experiments to find their centres of gravity. Most of the solid figures have inscriptions which have been handwritten in ink in the first instance and then painted over at a later date.

It is quite possible that the set belonged to Demainbray since a collection of geometrical figures is listed in the Queen's Catalogue, and we know that Demainbray covered the topic in his lecture course, although Chabrol's notes do not confirm that this particular set was used. The mathematics required to find minimum areas is rather advanced for this period. It seems likely that while the collection was at King's College the figures were used again and the inscriptions painted.

Q.C., Item 15; centers of motion, of gravity in regular and irregular bodies
Demainbray 1750–2, p. 3
Demainbray 1753, pp. 5–6
Demainbray 1754*a*, pp. 5–6
Demainbray 1754*b*, pp. 5–6; 'Experiments to find these centers [gravity & motion] in all kinds of bodies . . .'
C 247

See also D79, D80, D82, D83, D84, D86, D90, D92, D93, D95, D96

Inventory No. 1927-1645

D78 Cone

mid-eighteenth century?

PAINTED
'Cone of Min: Surface'
ON BASE
'9.8'

H 255mm, **D** 176mm

The inscription '9.8' on the base of the cone refers to the height in inches. A cone of this angle gives the minimum surface area per unit volume.

As D77

Inventory No. 1927-1650

D79 Cone

mid-eighteenth century?

PAINTED
'Cone of Minm Curved Surface'

H 150mm, **D** 212mm (base)

This angle of cone will give the minimum surface area per unit volume if the area of the base is excluded.

As D77

Inventory No. 1927-1651

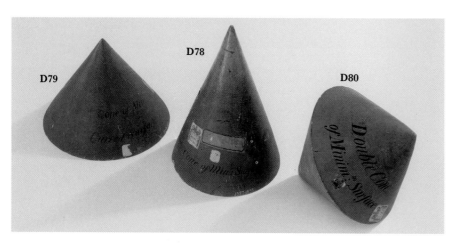

D80 Double cone

mid-eighteenth century?

PAINTED
  '*Double cone/of Minim surface*'

H 230mm, **D** 230mm (max)

The double cone has the minimum possible surface area per unit volume.

As D77

Inventory No. 1927-1751

D81 Semi-solid cylinder

mid-eighteenth century?

HANDWRITTEN
  '*116 1/4*'

D 210mm, **H** 65mm

The number refers to the surface area in square inches.

As D77

Inventory No. 1927-1757

D82 Semi-solid cylinder

mid-eighteenth century?

HANDWRITTEN
  '*Open Cylinder of/Min:m Surface*'
  '*109 3/4*'

H 88mm, **D** 170mm

Again the number refers to the surface area in square inches.

As D77

Inventory No. 1927-1756

D83 Semi-solid cylinder with lid

mid-eighteenth century?

HANDWRITTEN
  '*Cylinder of/Minimum Surface/138 1/4*'

H 140mm, **D** 135mm

The number refers to the surface area in square inches.

As D77

Inventory No. 1927-1753

D84 Hollow square pyramid

mid-eighteenth century?

PAINTED—BASE
  '*7.14*'
HANDWRITTEN
  '*7.14*'

L 230mm (side), **H** 180mm

The number refers to the height in inches.

As D77

Inventory No. 1927-1649

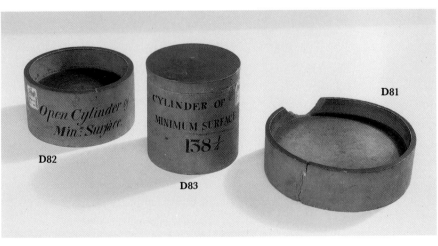

D85 Hollow rectangular prism

mid-eighteenth century?

L 163mm, **W** 130mm, **H** 80mm

As D77

Inventory No. 1927-1647

D86 Rhomboidal parallelepiped

mid-eighteenth century?

PAINTED
'182 3/4'

L 128mm (side), **L** 128mm (side), **L** 148mm (side)

The number refers to the surface area in square inches.

As D77

Inventory No. 1927-1754

D87 Hexagonal prism

mid-eighteenth century?

H 205mm, **L** 72mm (side)

Three triangular segments have been cut from the prism in order that the upper surface becomes a triangle. The angle of the cut is shallower than that in D88.

C 247

As D77

Inventory No. 1927-1648

D88 Hexagonal prism

mid-eighteenth century?

H 205mm, **L** 70mm (side), **W** 125mm

As with the above item, three triangular segments have been cut off the top of the hexagonal prism so that the upper surface becomes triangular.

As D77

Inventory No. 1927-1755

D89 Hexagonal prism

mid-eighteenth century?

'Hex Prism of Min: Surface'
'3.03' '6.03'

H 150mm, 77mm (side)

The numbers refer to the length of the sides and the height in inches. For this particular volume the hexagon has the least possible surface area.

As D77

Inventory No. 1927-1752

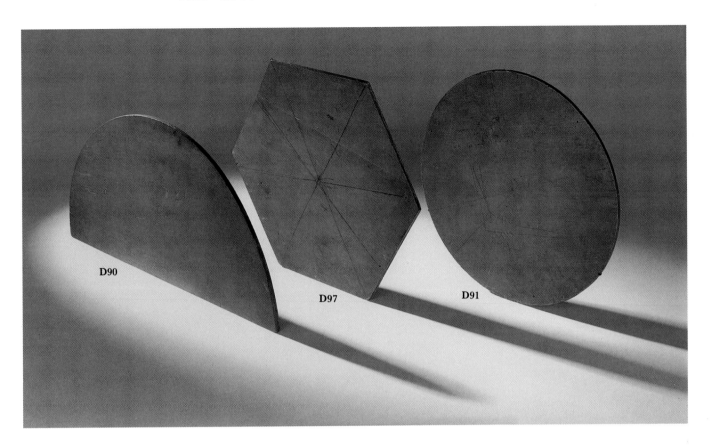

D90 Semicircle

mid-eighteenth century?
D 505mm

C 246

As D77

Inventory No. 1927-1683

D91 Circle

mid-eighteenth century?
D 340mm, **T** 4mm

Two small iron hooks are embedded in the circumference of the circle for the location of the centre of gravity. The item appears to be well used.

C 246

As D77

Inventory No. 1927-1685

D92 Square

mid-eighteenth century?
L 303mm (sides), **T** 4mm

C 246

As D77

Inventory No. 1927-1684

D93 Quadrilateral

mid-eighteenth century?
L 505mm, **W** 240mm, **T** 4mm

C 246

As D77

Inventory No. 1927-1681

D94 Quadrilateral

mid-eighteenth century?
W 350mm (max), **H** 505mm (max), **T** 4mm, **L** 165mm (side)

This figure has a small iron hook attached to each of two corners for centre of gravity experiments

C 246

As D77

Inventory No. 1927-1686

D95 Rectangle

mid-eighteenth century?
L 430mm, **W** 212mm, **T** 4mm

C 246

As D77

Inventory No. 1927-1687

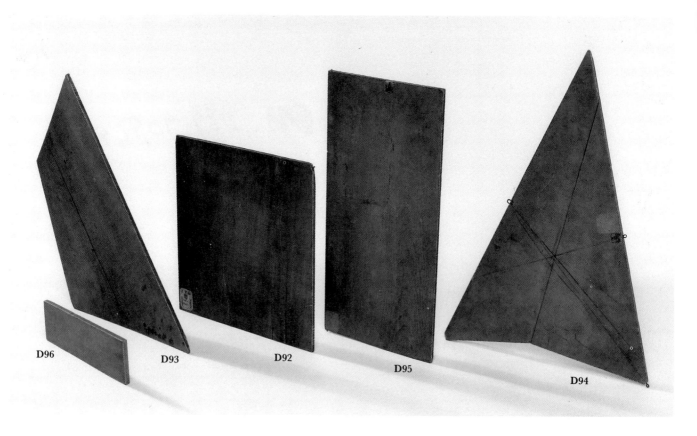

D96 Rectangle

mid-eighteenth century?

L 235mm, **W** 78mm, **T** 4mm

C 246

As D77

Inventory No. 1927-1688

D97 Hexagon

mid-eighteenth century?

L 187mm (side), **T** 4mm

One small iron hook and some lines remain to show that the hexagon was used for experiments on the centre of gravity

C 246   *As* D77   **Inventory No.** 1927-1682

MECHANICS

D98 Oblique cylinder

c. 1750

L 206mm, **D** 57mm

Because this cylinder is made of beech rather than boxwood, it does not belong in Adams's sets of solids for centre of gravity experiments. It is likely to have been Demainbray's since he began his mechanics course with centre of gravity

experiments and often used beech as a material.

Q.C., Item 15; centers of motion, of gravity in regular and irregular bodies
Demainbray 1754*b*, p. 5, 'centers of gravity and motion made clear'
Adams 1762, p. 22, p. 12, Figs 49, 50
C 247

Inventory No. 1927-1207/2

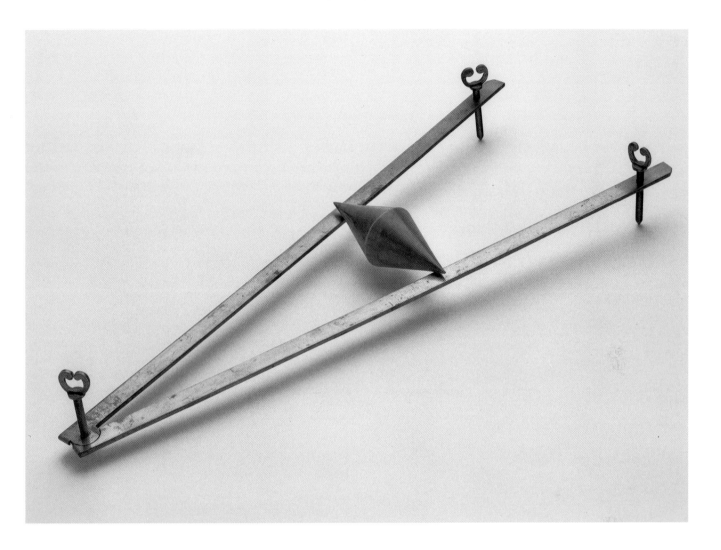

D99 Rolling double cone and arms

mid-eighteenth century

L 120mm (cone), **D** 45mm (cone), **L** 490mm (arm), **H** 65mm (screw)

Of the three rolling double cones in the King George III Collection, only this model is in brass. It is therefore almost certainly the brass double cone listed in the Queen's Catalogue and, as such, is likely to have been Demainbray's. It is not mentioned specifically in the syllabuses which suggests that it was acquired after 1755.

It is entirely of brass and rolls on a pair of flat arms which are inclined by means of three levelling screws and jointed at one end. If the arms are kept horizontal, the double cone will roll towards the open end because its centre of gravity descends as the arms increase in separation. If the arms are slightly lifted at the open end, the double cone will appear to roll uphill, thus creating the paradox.

The double cone was common in courses of experimental philosophy by the mid-eighteenth century. It was first described in 1694 by a certain J.P. in Leybourn's *Recreations of Divers Kinds* where it was referred to as a 'New and Diverting experiment'.

Q.C., Item 47; brass double cone with 2 brass inclined rulers
Leybourn 1694, Mechanical Recreations, pp. 12–13
Desaguliers 1744, Vol. 1, p. 56; Pl. 4, Fig. 14
Adams 1746, No. 114
's Gravesande 1747, Vol. 1, p. 41, Pl. 7, Fig. 1
C 37

See also E55, M22

Inventory No. 1927-1104

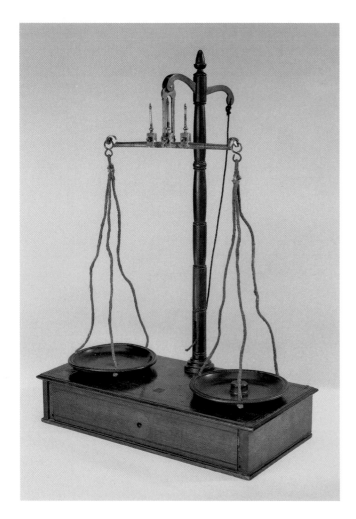

D100 Balance

second or third quarter eighteenth century

L 400mm (box), **W** 200mm (box), **H** 70mm (box), **H** 600mm, **L** 255mm (beam), **D** 145mm (pans)

Since this device does not reach equilibrium when held centrally, it could be the false balance listed in the Queen's Catalogue. It appears to be a balance primarily intended for demonstration purposes which also suggests that it belonged to Demainbray.

The balance consists of an iron beam about ten inches long, engraved with rings around its circumference in a manner slightly similar to E17. There are three fulcra with pointers, one at the centre of the beam and the others about one and a half inches along each arm. The beam hangs from an iron suspension unit attached to one arm of a brass lever which is fixed at the top of a mahogany pillar screwed into the lid of the box. To operate the beam the other arm of the lever is lowered by means of a piece of cord to raise the balance; otherwise the pans rest on the lid of the box. The walnut pans are attached by three thick cords to the swan-neck ends of the beam. One of the pans has a hole in the centre and a raised portion around the hole which makes it slightly heavier than the other. The box contains a half guinea piece dating from before 1772 and some other items such as paper, glass, and pieces of metal. The box is mahogany.

Q.C. Item 25; false balance

Inventory No. 1927-1453

D101 Balance beam

mid-eighteenth century

L 610mm, **H** 155mm, **W** 28mm

It is possible that this is the 'beam of a balance' referred to in the Queen's Catalogue and that it was used by Demainbray to demonstrate the balance or law of the lever.

The boxwood balance beam has a steel fulcrum and a steel hook one foot from the fulcrum on each arm. The suspension unit is boxwood with the pointer in the form of a fleur de lys. One arm is marked with a scale in inches and tenths from 1 to 11.

Q.C., Item 22; Beam of a balance
C 53

Inventory No. 1927-1238

D102 Chinese dotchin in case

c. 1760

L 285mm, **W** 58mm, **T** 15mm

Although Chinese dotchins were not mentioned specifically by Demainbray in his lecture syllabuses, the listing in the Queen's Catalogue suggests that it may have belonged to him.

The instrument is a type of small steelyard suspended from a silk thread. The arm consists of an ivory stick similar to a chopstick. The small brass scale-pan which carries the weight is suspended on three threads at a fixed distance from the fulcrum and a fixed weight can be hung on a silk thread and moved along the arm which is marked with a series of dots. The instrument is housed in a mahogany case in the shape of a guitar which can be kept closed by a straw collar.

Dotchins were used to weigh coins, jewels, opium, and medicine, and were recorded by Hakluyt between 1598 and 1600. Their origin is unknown but it has been assumed that they derive from Roman steelyards. They are not commonly found in lecture courses in the early and mid-eighteenth century, but at least one other lecturer used them: this was Griffiss who was lecturing in London in the mid-1750s. 'Chinese Dodgins' were highlighted as a special feature of his course.

Q.C., Item 24; Chinese dotchins
Griffiss 1755, p. 4
C 50
Thurkow 1978, pp. 55–9   **Inventory No.** 1927-1130

D103 Two tapered pulley blocks each with three pulleys

before 1753   **H** 115mm, **L** 25mm, **W** 8mm

These brass pulleys are similar to but not identical with the others in Demainbray's complete set of pulleys. They are of similar construction and size but the wheels are solid and the frames taper. Tapering frames were usual in pulleys set vertically, and since D3 and D5 were primarily intended for horizontal use this is not a significant difference. It is therefore very probable that these also formed part of Demainbray's collection of brass pulleys. Both blocks have two hooks.

Q.C., Item 32; A compleat set of brass pullies for all the combinations of their power
Hauksbee 1714, Pl. 3, Fig. 6
Smeaton 1752
Chabrol 1753, p. 31
Demainbray 1754*b*, p. 6 'on Pullies illustrated with curious Tackles'

Inventory No. 1927-1235

D104 Axis in peritrochio, or wheel and axle

before 1753   **L** 610mm, **W** 310mm, **H** 385mm

This is certainly not Adams's axis in peritrochio and is likely to have been Demainbray's although it may have been modified subsequently. It is similar to that in a sketch in Chabrol's lecture notes.

The wheel is made of mahogany and is set on a steel axle which is supported on two brass pillars, the whole being on a mahogany base. There are three choices of axle diameter for the weight, namely the steel axle, the brass cylinder, and the larger wooden cylinder. There are also three choices for the 'power', each with a groove for the rope. The wheel probably carried twelve spokes at one time but these are now missing. A ratchet mechanism can stop the wheel turning.

Axes in peritrochio were universally incorporated into courses of experimental philosophy as examples of one of the simple machines. A small 'power' can lift a large weight in proportion to the diameters of the axle and the wheel.

Q.C., Item 34; The axis in peritrochio machine
Hauksbee 1714, Pl. 14, Pl. 4, Fig. 3
Desaguliers 1734, p. 108, Pl. 10, Figs 10, 11
's Gravesande 1747, Vol. I, p. 50, Pl. 8, Fig. 6
Chabrol 1753, p. 27
Demainbray 1754*b*, p. 6 'The Axis in Peritrochio, simple and compound'

Inventory No. 1927-1831

D105 Compound axis in peritrochio

Graham, G.

before 1752   London

L 253mm (overall), **W** 200mm (overall), **H** 505mm (overall), **L** 85mm (mechanism), **W** 65mm (mechanism), **H** 183mm (mechanism)

This machine is likely to be the 'brass model of the compound axis in peritrochio by Graham' listed in the Queen's Catalogue, and therefore associated with Demainbray. We know that Demainbray used such a machine and no similar apparatus is described in the Adams *Mechanics* manuscript.

Like several other instruments in Demainbray's collection, it closely follows Desaguliers's description from *A Course of Experimental Philosophy*. It essentially consists of two steel axles of different diameters linked by a wheel and pinion. The upper axle bears a brass drum of two inches diameter which carries the 'power' or applied force, probably a weight of one ounce in this model. This axle also has a steel pinion of eight teeth which meshes with a larger steel wheel with 40 teeth mounted on the lower axle. The lower axle has a diameter of a quarter of an inch and carries the weight. The combined velocity ratio of the machine is the ratio of the diameters of the two axles times the ratio of the number of revolutions, or velocity, of the two wheels carried by them: in this case 40. Desaguliers points out that this relationship always holds no matter how many wheels and axles come between the power and the weight. Hauksbee and Whiston called their similar machine 'only a train of wheel work'. The brass and steel mechanism is on a platform with a rectangular hole cut in it to allow the weights to fall. It is held on four pillars on a rectangular base.

Graham was one of the leading precision instrument-makers of his time, specializing in clocks and astronomical instruments.

Q.C., Item 38; a brass model of the compound axis in peritrochio by Graham
Hauksbee 1714, Pl. 4, Fig. 4
Desaguliers 1734, p. 104, Pl. 10, Fig. 13
Desaguliers 1744, Vol. 1, pp. 109–10, Pl. 10, Fig. 13
Demainbray 1754*b*, p. 6: 'axis in peritrochio simple and compound'
C 35
Taylor 1966, pp. 120–1
Inventory No. 1927-1873

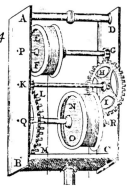

Hauksbee 1714

D106 Folding inclined plane

mid-eighteenth century?
L 900mm, **W** 161mm, **H** 260mm

This is very likely to be one of the two wooden inclined planes listed in the Queen's Catalogue and as such is probably from Demainbray's collection. It is made of very heavy mahogany but folds for ease of carrying. The plane is 3 feet long marked in quarter inches with each half inch numbered in ink.

Q.C., Item 151; Two wooden inclined planes
Desaguliers 1734, p. 108, Pl. 11, Fig. 2
Inventory No. 1927-1841

D107 Model tipping cart

mid-eighteenth century   France?

L 270mm, **W** 130mm, **H** 133mm

It is possible that this was one of Demainbray's collection of model carts, particularly since the use of iron pins in the wheel rims was a French fashion at the time. It is a simple tipping cart made of mahogany with the axle of the body slightly above and in front of the wheels' axle. The wheels are also made of mahogany and are dished. The body can be fastened by pinning two pieces of mahogany between the shafts.

Gentleman's Magazine 1754, Vol. 24, pp. 326–9, p. 376, p. 426, p. 473

Inventory No. 1927-1937

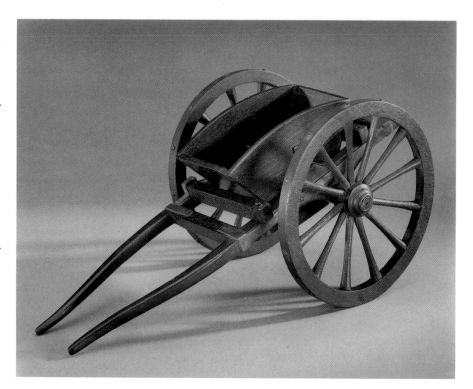

D108 Model tipping cart

mid-eighteenth century

L 468mm, **W** 430mm, **H** 202mm (without wheels), **D** 250mm (wheels)

Since Demainbray had a number of model carts it is possible that this is one of his objects, but there is no particular evidence. It is a deep-bodied tipping cart made of French walnut with an iron axle running through the centre on which the body pivots. The axle is attached to the rectangular frame at either side. An iron loop on the upper surface of each side of the frame indicates the position in which a cross-bar would have been inserted to prevent the cart from tipping when carrying a load. The wheels are entirely of French walnut and have 12 spokes.

As D107

Inventory No. 1927-1938

D109 Model tipping cart

mid-eighteenth century

L 280mm, **W** 175mm, **D** 192mm (wheel), **H** 192mm

Like the two previous items, this could have been one of Demainbray's model carts but there is no particular evidence available. The wheel rims are an inch wide and are lined at each edge with a strip of brass. The wheels have a double set of spokes with 16 in each set. The iron axle passes through the rectangular pearwood body of the cart. The front and back panels are removable and are not attached to the sides or base, allowing a load to be tipped out when the body is tilted.

It is likely that the cart was made to demonstrate a design of wide-rimmed wheel since this was a subject of debate at the time.

Gentleman's Magazine 1754, Vol. 24, pp. 326–9, p. 376, p. 426, p. 473

Inventory No. 1927-1931

D110 Model three-wheel heavy wagon

mid-eighteenth century

L 730mm, **W** 290mm, **H** 245mm, **D** 200mm (two wheels), **D** 150mm (one wheel)

Demainbray does not mention this model in his syllabuses but it is likely that it belonged to him. It is listed in the Queen's Catalogue with the Irish low-backed cart and the four-wheeled car-riage, and it was painted with the characteristic green/blue paint, although it appears to have been repainted more recently.

The ash body of the wagon, sup-ported by the two rear wheels, is attached to a large horizontal ash ring which turns above another ring by means of a steel rim. The front wheel is slung on the lower ring which also carries the shafts; hence turning is made easier than in the standard four-wheeled vehicle.

Q.C., Item 74; A model of a three-wheeled cart improved
Gentleman's Magazine 1754, Vol. 24, pp. 326–9, p. 376, p. 426, p. 473
C 331

Inventory No. 1927-1936

D111 Model axle and pair of wheels

Pease

c. 1750   London

D 270mm (axle), **D** 155mm (wheels)

This is almost certainly the model axle by Pease listed in the Queen's Catalogue. It was probably used by Demainbray to illustrate his seventh lecture, which dealt largely with wheeled carriages. In particular it may have been used to illustrate 'the manner in which wheels should be shod with iron'.

The axle is made of iron embedded in carved mahogany. The hubs are of brass. Each wheel has eight carved spokes and the rims are clad with iron. One wheel is broken.

Q.C., Item 69; Model of Mr Pease's improved axle
Demainbray 1754*b*, p. 8 'the Manner in which Wheels should be shod with Iron'
Demainbray 1761

Inventory No. 1927-1939

D112 Model carriage spring

Pease

c. 1760   London

L 310mm, **W** 185mm, **H** 105mm

It appears from the fact that it is listed in the Queen's Catalogue that this model could have been presented to the collection by Pease, or that it formed part of Demainbray's collection. In 1761, Demainbray wrote to Bute sending him a model by Pease whom he described as 'the most notable spring maker of London'.

The model is made of walnut and iron with a lacquered steel spring. The wood parts are painted in black and gold. The base consists of a pair of shafts with two cross-pieces and sits on five studs. Above this is an iron rod which would form the wheel axle. At each end of the rod is pivoted an iron bracket in the 'whip' shape which is attached at one end to the spring and at the other to another iron piece which is attached to the other end of the spring forming a triangle. The carriage body would hang from the top of the bracket.

Q.C. Item 4; new springs for carriages . . . Mr. Pease

Q.C., Item 70; ditto (model of Mr. Pease's) of improved springs
Demainbray 1761

Inventory No. 1927-1933

D113 Hooke's universal joint

Pease

c1760   London

L 415mm, **W** 120mm, **H** 145mm (max)

This is very likely to be the 'Hook's Universal Joint' listed in the Queen's Catalogue as being presented by Mr Pease. Since Mr Pease was a spring-maker who made the model carriage springs above, it is probable that this joint was also made by him.

Using two double joints, rotary motion can be transferred from one shaft via a second shaft to a third lying in the same direction as the first but at a level two inches lower. The components can be rearranged to show transferrence of the rotary motion when the third shaft lies in a direction at right-angles to that of the first, as in the illustration. The shafts are brass rods of square cross-section separated by two joints, each made up of two ball-type joints on either side of a rigid chain. The central portion of the model is supported on an iron and steel bracket near the centre of the baseboard. The first section is held on a bracket about one inch higher at one end and the third on a bracket about one inch lower at the other. Alternative brackets for the first and third section enable the rods to be held in a straight line or to turn through a right-angle. A walnut handle is provided to rotate the rods. The base is walnut with a rosewood inlay.

This type of joint has been attributed to Hooke since he described it in detail in his Cutlerian Lectures of 1679. He had already introduced similar joints in *Animadversions . . .* in 1674 and in *A Description of Helioscopes* in 1676. They were 'for communicating a round motion through any irregular bent way.'

Q.C., Item 11; Hook's Universal Joint . . . Mr. Pease
Q.C., Item 71; ditto of universal joint
Hooke 1674, Fig. 22
Hooke 1676
Hooke 1679, pp. 133–41, illust. opp. p. 152
Demainbray 1761
C 64

See also D112

Inventory No. 1927-1109

D114 Model of mill at the Bazacle, Toulouse

1753   Toulouse

H 1050mm, **L** 730mm, **W** 285mm

Demainbray had this model made while at Toulouse in 1753. It is described in his syllabus of that year as a model of the water mills of the Bazacle at Toulouse for grinding corn. Demainbray states that in this construction not only the velocity of the stream of the River Garonne but the perpendicular pressure and weight of the water is employed to work the engine. In fact this is a very primitive device demonstrating the backwardness of the region. In 1757 and 1758 Demainbray advertised his lectures by citing the model of the water mill at the Bazacle at Toulouse as 'being the only one brought to England'.

The model is principally of oak on curved legs. The water is directed through a narrowing oak channel on to the wheel which has curved ash paddles. The axle passes through the lower grinding wheel to turn the upper grinding wheel above. The grinding wheels are made of French walnut.

Q.C., Item 193; Mill du Bazacle at Toulouse
Q.C., Item 194; horizontal wheel to ditto
Belidor 1737, Vol. 1, pp. 302–3, p. 1, Fig. 5
Demainbray 1753, p. 13
Demainbray 1754*a*, p. 14
Demainbray 1754*b*, p. 14; 'model of water mills at Bazacle for grinding of corn'
Daily Advertiser 1757, 7 March, No. 8157
Daily Advertiser 1758, 20 March, No. 8483
Singer *et al.* 1956, Vol. 2, pp. 594–5

Inventory No. 1927-1633

D115 Bridge frame

c. 1750

L 386mm, **H** 230mm, **W** 160mm, **D** 55mm (roller)

This is almost certainly an item from Demainbray's collection since it has butterfly nuts identical with those on the inclined plane D11. Its function is unknown but it is probably connected with the machine listed in the Queen's Catalogue to show friction over round bodies, which consisted of 'a brass cylinder on steel axis and two iron supporters'. This has two brass cylinders on steel axles which can roll on two brass supports set in a mahogany frame. Four small pulley wheels are attached to the internal surface of the frame.

Q.C., Item 61; friction over round bodies with a machine consisting of a brass cylinder on steel axis and iron supporters

Inventory No. 1927-1204

D116 Model signal cannon

mid-eighteenth century?

L 500mm, **W** 200mm, **H** 220mm

This may be associated with Demainbray although it could be later. Demainbray's second lecture on motion, explaining Newton's second and third laws, was chiefly concerned with gunnery. Having considered the parabolic path of a projectile, he went on to illustrate action and reaction with experiments to show the disadvantages of the 'new forged and other lighter cannons proposed for shipping'.

This model is very heavy, being made of solid brass on cast iron trunnions and base. The cascable, or large cap, screws into the breech end, or rear end, and there is a long protrusion on the front of the screw extending almost the length of the bore.

Q.C., Item 109; Brass cannon
Demainbray 1754*b*, p. 10; 'new forged and other lighter cannon'
C 341

Inventory No. 1927-1614

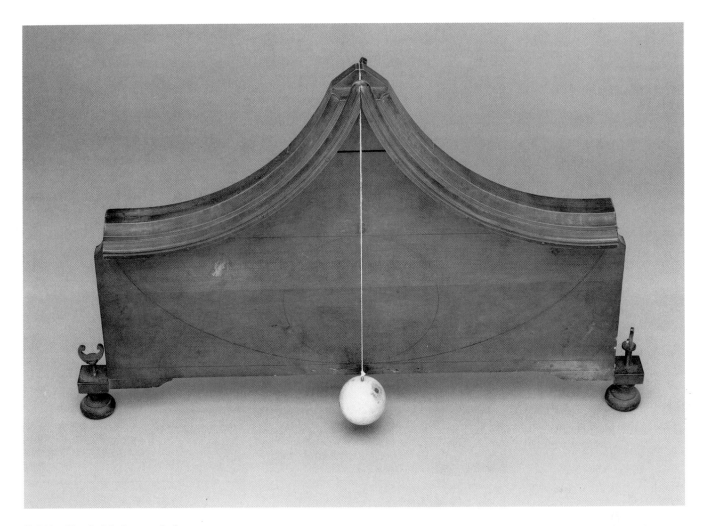

D117 Cycloidal pendulum

mid-eighteenth century

ENGRAVED
'*A,B,C,D*'
'*line of cycloid and inscribed circle*'

H 313mm (stand), **L** 470mm (stand), **W** 150mm (stand)

The appearance of a cycloidal pendulum in the Queen's Catalogue beside those of Graham and Julien le Roy indicates that this item belonged to Demainbray. In each syllabus he advertised the third lecture on motion as dealing with 'the doctrine of pendulums' and named the above specifically. It is likely that he used the isochronous pendulum since he dealt with temperature compensation, although the mathematics would have been omitted.

The pendulum backboard is made of pearwood. It stands on three feet, the one behind having a lead weight. The plumb bob is made of brass. The positioning of the engraved letters does not correspond to that in any of the standard sources of the time so possibly Demainbray explained the

working of the instrument without their help.

Pendulums were standard items in courses of experimental philosophy, often featuring together with gunnery, collisions, and central forces in sections on motion. The principle of the isochronous pendulum was discovered by Huygens in 1659.

Q.C., Item 114; pendulum swinging in cycloidal cheeks
's Gravesande 1747, Vol. I, p. 103, Pl. 16, Fig. 3
Huygens 1888, Vol. 14, pp. 404–6
C 4
Mahoney 1980, pp. 234–270
See also M73

Inventory No. 1927-1199

ASTRONOMY

D118 Orrery

Wright, Thomas

1720–48   London

H 190mm, **D** 480mm

The orrery closely resembles that shown in the plate in Thomas Wright's *A Description of an Astronomical Instrument being the Orrery Reduced* and has been attributed to him. Wright constructed these small machines for 'Ladies and Gentlemen rather than noblemen or Princes', which might suggest that it belonged to Demainbray rather than the royal family. An entry in the Queen's Catalogue strengthens this view.

Although Wright states that the orreries could be made of silver, brass, or wood he mentions only a wooden frame in the text. This model is on a painted and lacquered oak base on three small feet. Twelve pillars support the oak horizon ring. Paper pasted on the ring gives the degrees of each sign of the zodiac from 0° to 30°, the signs of the zodiac, the Julian calendar or 'old stile used in England', together with dates in the church calendar, the Gregorian calendar, and, lastly, the points of the compass.

The gilded ball in the centre of the orrery represents the Sun. The presence of Mercury and Venus on spindles is a small departure from Wright's description in the pamphlet. The Earth–Moon system is on a circular brass base which is inclined at an angle of 5° to the horizontal in order to show the difference between the Moon's orbit and the ecliptic. The Earth is represented by a small ivory ball inclined at 23°, which is marked with a crude map of the world with a black spot on London. Wright states that if the instrument is intended for places other than London, a further black spot could be added. The equator, two tropics, two polar circles, and two colures are also marked. The equator is numbered in hours from London. Diurnal motion is provided manually; a brass pointer indicates the place where the Sun is directly

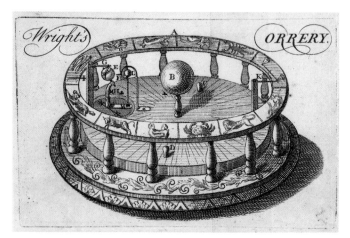

Wright 1720

overhead. A brass semicircle over the Earth separates the half in sunlight from that in darkness. The Moon's motion is also manual: it can be rotated about its axis so as to keep the dark side away from the Sun and also rotated about the Earth.

A crank handle, the second departure from Wright's description, can be used to turn the central platform in relation to the horizon ring thus providing annual motion. The wheel work is hidden under the platform which is painted with the Sun's rays in gold on a blue background.

The instrument could be used to find the time of the Sun's rising and setting at various points on Earth at various times of the year, the eclipses of the Sun and Moon, and the parallelism of the Earth's axis.

Q.C., Item 162; small orrery with Earth, Moon, Venus and Mercury
Wright 1720
C 141
King and Millburn 1978, p. 160, Fig. 99

Inventory No. 1927-1414

D119 Mechanized glass celestial globe with terrestrial globe

Senex, J. (terrestrial globe)

1742–50   London

ON TERRESTRIAL GLOBE
'A New and Correct/GLOBE/of the Earth/by In Senex'

D 330mm (globe), **D** 370mm (base), **H** 570mm

This globe is almost certainly the 'glass globe on a stand' listed in the Queen's Catalogue and therefore came into the collection via Demainbray. However, apart from a mention of terrestrial and celestial globes in his syllabuses, there is nothing to indicate that he used this particular instrument in his lecture courses, possibly because it was too fragile.

The instrument closely resembles Long's 'glass sphere' illustrated in the frontispiece of his *Astronomy*, published in 1742. It comprises a glass sphere of about 13 inches in diameter, unsigned, and very crudely engraved with the principal stars and figures of the constellations, the celestial equator, polar circles, and tropic circles. The ecliptic is marked with two calendars, the Gregorian and the Julian, indicating a date of construction shortly before 1752. Set inside the glass sphere is a terrestrial globe of three inches diameter by Senex, similar to that of E40 but with a steel band round the equator engraved with the hours.

The globe is on a steel axis which can be rotated by means of a winch acting on one of the arbors of a gear mechanism. The gear mechanism is an oval brass box at the lower end of the axis. Turning the second arbor causes the glass sphere to rotate so that both real and apparent motion can be viewed. The terrestrial globe has a steel horizon ring and meridian circle associated with it: the horizon ring is marked in degrees from 0 to 90 each way from the equinoxes, and with the points of the compass. The meridian circle is marked with the degrees of latitude. It is likely that the horizon circle could have been moved with respect to the meridian circle but the mechanism for this is not now apparent. The globes can be tilted by moving the instrument on its brass central axis.

The sphere was a useful teaching aid as Long states, '. . . by this machine, the real motion of the Earth round its axis within the sphere of the heaven, or the apparent motion of the heaven round the earth, may be represented.' A similar instrument by Cowley, with a later terrestrial globe, is in the Astronomy Collection at the Science Museum and another Cowley sphere with a Hill globe was sold by Trevor Philips & Son Ltd in 1988.

Long 1742

Q.C., Item 115; a glass globe on a stand
Long 1742, Frontispiece and p. 71
Adams 1746, No. 331
C 161
King and Millburn 1978, p. 175
Philip & Sons 1987*a*
Philip & Sons 1987*b*, Lot 35

See also 1913-531 (Astronomy Collection)

Inventory No. 1927-1412

HYDROSTATICS AND PNEUMATICS

D120 Curved glass tube

second half eighteenth century or first half nineteenth century

H 300mm, **D** 30mm, **L** 1400mm

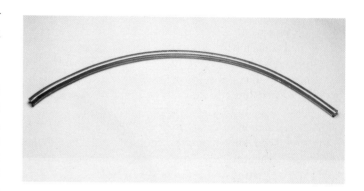

This is possibly the 'glass bent tube mounted to explain wind bound pipe' listed in the Queen's Catalogue, although it could be much later. Demainbray, following Desaguliers, paid great attention to the flow of water through pipes and in his later syllabuses included a model of the Edinburgh water works. 'Practical rules to cure such as are wind bound', or having an air lock, figured in his first lecture on hydrostatics. It is simply a glass tube bent in an arc of a four foot chord.

Q.C., Item 172; glass bent tube mounted to explain wind bound pipe
Desaguliers 1744, p. 124, Pl. 11, Fig. 13
Demainbray 1754*b*, p. 13; 'practical rules to cure such [pipes] as are wind bound'
C 129

Inventory No. 1927-1395

D121 Syphon tube

mid-eighteenth century

L 520mm, **W** 180mm, **D** 20mm (main pipe)

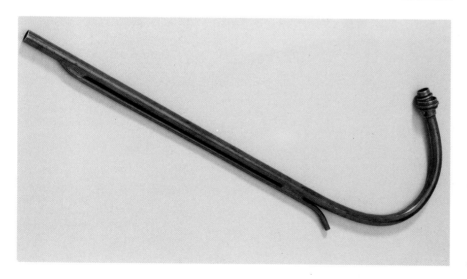

Judging by the material and workmanship this is almost certainly one of Demainbray's objects. It is possibly part of one of the fountains at command listed in the Queen's Catalogue. Alternatively, it is part of Demainbray's apparatus used in his lecture on leading water through pipes which by 1753 included a model of Desaguliers's water works at Edinburgh. It is a J-shaped brass pipe with a screw collar at one end and a smaller syphon tube, or rider, adjacent to it.

Q.C., Item 171; Model of main with rider
Q.C., Item 172; Model of main with rider
Q.C., Item 173; Model of main with rider at delivery
Q.C., Item 189; French fountain at command
Q.C., Item 190; English fountain at command
Desaguliers 1744, Vol. 2, Pl. 11, Figs 13–16
Demainbray 1754*b*, p. 13 'Methods of leading Water through Pipes'

Inventory No. 1927-1771

D122 Pair of marble adhesion discs with band and ring handle

before 1753

D 70mm, **H** 70mm (each disc)

Demainbray lists the relevant experiment in his earliest course, and a sketch by Chabrol shows a similar piece of apparatus. It is almost certain that these are Demainbray's adhesion discs. The discs are made of solid marble and to each is attached brass handles. One has a brass ring with clips, presumably to keep them aligned during adhesion.

The phenomenon of the adhesion or cohesion of smooth bodies had been known and discussed in relation to the existence of a vacuum for many centuries before it became a standard demonstration. In about 1659 Boyle used an air pump to perform experiments on the cohesion of marble planes hoping to make them a centrepiece of experimental philosophy. He expected that they would adhere in the atmosphere but fall apart in the near vacuum of the exhausted air pump; however, he had difficulty in producing the desired result. When he redesigned his pump and added a greater weight to the lower disc he did achieve success. The experiment was performed by Hauksbee and Vream and mentioned by Desaguliers. Demainbray did not mention it specifically in later lecture courses, but Chabrol provides the evidence that it was continued.

Boyle 1660, pp. 69–70
Vream 1717, p. 15
Desaguliers 1744, Vol. 2, p. 386
Demainbray 1750, p. 8; marble planes supported by [air pressure]
C 121
Shapin and Schaffer 1985, p. 48, p. 157, pp. 191–201

Inventory No. 1927-1304

D123 Ear trumpet

before 1753?

L 180mm, **W** 160mm, **H** 54mm

The auricular tube, sonometer, and speaking trumpet remained central items in Demainbray's lecture on acoustics. The description of the auricular tube is not sufficient to be able to identify this item with absolute certainty.

This type of auricular tube or ear trumpet appears to have been invented by du Quet in 1706. It was intended for use with a hat; the ear-pieces could be put through holes in the hat with the instrument suspended under the chin. The sound was guided to the ears by the shape of the interior. It is made of painted tinplate covered in silk with ivory ear-pieces.

Q.C., Item 91; auricular tube
du Quet 1706, in Gallon 1777, Vol. 2, p. 121, Pl. 2, Fig. 2
Demainbray 1750, p. 10; auricular tube of a singular construction
Demainbray 1754b, p. 17
C 248

Inventory No. 1927-2021

D124 Diagonal barometer

Finney, Joseph

1760–72   Liverpool

'Finney LIVERPOOL'
scale 28–31 inches in 1/20ths and 0 to 60 in 10s
Stormy, Rain, CHANGEABLE, Fair, Drought

H 890mm, **L** 635mm, **DE** 75mm

A diagonal barometer is listed in the Queen's Catalogue but it is not certain that the reference is to this instrument.

The barometer is more domestic in character than E114. The main tube is concealed in a mahogany case in the form of a pillar. The cistern is a late example of the open type and has a piece of leather covering to exclude dirt. The scale runs from 28 to 31 inches in the space of 18 inches, giving a magnification of 6. It is marked from 0 to 60 above the tube. There is a brass shield at the end of the tube to prevent the glass breaking.

The instrument is similar to one by Whitehurst of Derby who was probably Finney's brother-in-law. Finney appears to have made his scientific instruments in the last decade of his life. It is likely that only a few were made, most of which are now in the King George III Collection.

Q.C. Item 210; diagonal barometer
Chew 1968, No. 4
Goodison 1968, p. 63, Pl. 25
Fairclough 1975
Banfield 1985, p. 152, Fig. 199

See also D148, E21, E114

Inventory No. 1927-1911

D125 Two glass cylinders

mid-eighteenth century

L 960mm (1), **D** 95mm (max), **L** 670mm, **D** 195mm (2)

The larger cylinder tapers slightly towards one end. The quality of the glass is very poor. It is probable that this is one of the 'two glass tubes of great bores and lengths' mentioned in the Queen's Catalogue and, as such, it is likely to have been part of Demainbray's apparatus. The smaller tube is similar and has a waxed-on glass cover plate.

Q.C., Item 82; Two glass tubes of great bores and lengths C 129

Inventory Nos 1927-1399, -1353

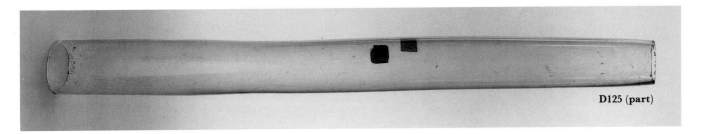

D125 (part)

MAGNETISM

D126 Lodestone and keeper

before 1753?

L 68mm **W** 46mm **H** 90mm

This is probably a 'loadstone' used by Demainbray although it does not appear in the Queen's Catalogue. Magnetism, although only taking one lecture, came under a separate heading in his courses. He demonstrated 'the properties incident to magnets viz direction, attraction and repulsion, magnetic effluvia (as some suppose it), communication, inclination and declination' by means of natural lodestones and artificial magnets. A popular experiment was to weigh the attractive force of the lodestone against weights on the balance, first performed by van Musschenbroek in 1725. This example is bound in brass with a steel keeper.

van Musschenbroek 1725, p. 370, Vol. 33
Michell 1750, pp. 4–5
Demainbray 1750–2, p. 10, 'natural loadstones'
Demainbray 1754*b*, p. 19
C 262

See also M6

Inventory No. 1929-112

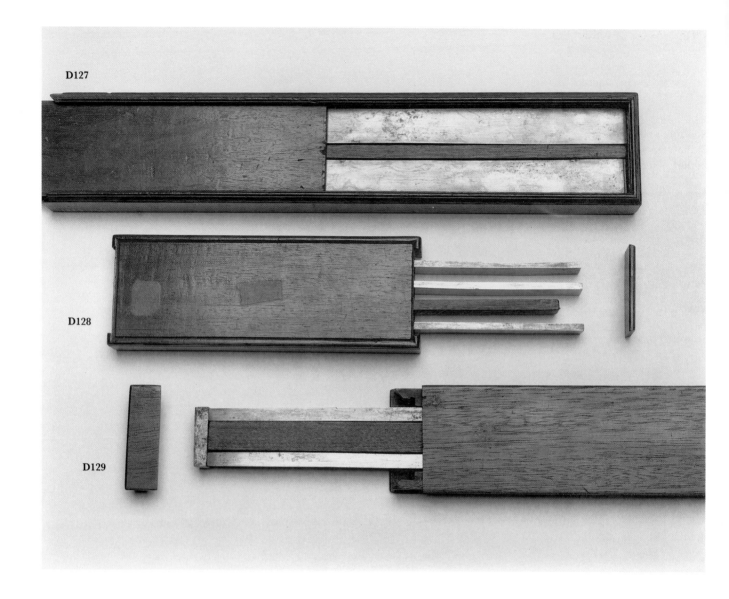

D127 Pair of bar magnets in box

c. 1750

L 310mm (box), **W** 60mm (box), **DE** 23mm (box), **L** 265mm (magnets)

The Queen's Catalogue lists magnetic bars by Dr Knight, Hindley of York and Morgan, and a collection of magnetic bars by different hands. The five anonymous eighteenth century sets of magnets in the collection could be any of these, or could possibly have come into the collection via another route. In his syllabuses Demainbray advertised the use of Gowin Knight's magnetic bars together with his apparatus for magnetic experiments, which has not survived.

These steel bars are about ten and a half inches long. They are in a mahogany box made to house them and two keepers which are now missing. The north poles are marked with a line. Artificial magnets were a recent invention in the 1750s, following the work of Canton, Knight, and Michell, and improvements in steel-making. Knight suggested housing bar magnets in pairs arranged with the north pole of one at the same end as the south pole of the other, a piece of wood between them, and keepers across their ends.

Q.C., Item 122; magnetical bars by Dr. Knight.
Q.C., Item 123; magnetical bars by Morgan
Q.C., Item 124; magnetical bars by Hindley of York
Q.C., Item 127; A collection of magnetical bars by different hands
Knight 1744, p. 162
Demainbray 1750–2, p. 10, 'with Gowin Knight's curious magnetic bars'
Michell 1750, p. 73
Canton 1751, p. 31
Knight 1766
C 265

Inventory No. 1929-102

D128 Three bar magnets in a box

c. 1750

L 150mm (bars), **W** 13mm (bars), **L** 160mm (box), **W** 60mm (box) **H** 17mm (box)

Again, these are likely to have been magnets used by Demainbray and listed in the Queen's Catalogue as being by Knight, Hindley, Morgan, or 'different hands'. They are the only bars left from a set of six or possibly eight which fitted into the box. They are about six inches long with the north poles marked by lines. They fitted on top of each other in two stacks so as to form a pair of larger laminated magnets, an arrangement suggested by Gowin Knight in 1744.

Another similar set of magnets, three inches long, is now missing. These magnets would have been fairly popular by mid-century since in 1747 they were referred to as 'common small magnetic bars'.

Q.C., Item 122; magnetical bars by Dr Knight
Q.C., Item 123; magnetical bars by Morgan
Q.C., Item 124; magnetical bars by Hindley of York
Q.C., Item 127; A collection of magnetical bars by different hands
Knight 1744, p. 162
Demainbray 1750–2, p. 10, 'with Gowin Knight's curious magnetic bars.'
Michell 1750, p. 73
Canton 1751, p. 31
Knight 1766
C 265

As D127

Inventory No. 1929-109

D129 Pair of bar magnets with keepers in box

c. 1750

'*N*' *(one magnet)*
'*S*' *(other magnet)*

L 265mm, **W** 52mm, **DE** 18mm

As for the previous items, these magnets could have formed part of Demainbray's collection and could be by Knight, Hindley, Morgan or 'different hands'. They are nine inches long with 'N' and 'S' engraved at the poles. They fit into a mahogany box and have two steel keepers.

Q.C., Item 122; magnetical bars by Dr Knight
Q.C., Item 123; magnetical bars by Morgan
Q.C., Item 124; magnetical bars by Hindley of York
Q.C., Item 127; A collection of magnetical bars by different hands
Knight 1744, p. 162
Demainbray 1750–2, p. 10, 'with Gowin Knight's curious magnetic bars'
Michell 1750, p. 73
Canton 1751, p. 31
Knight 1766
C 265

As D127

Inventory No. 1929-107

D130
Pair of bar magnets with keepers in box

second half eighteenth century

L 460mm (box), **W** 100mm (box), **DE** 20mm (box),
L 370mm (magnets)

As for the previous items, these magnets could have been used by Demainbray in his lecture on magnetism and could be one of the several magnetic bars listed in the Queen's Catalogue. It is possible, however, that they are slightly later and were made by Adams since they are stylistically similar to some of his instruments.

The steel bars are fifteen inches long with lines at the north poles. They have keepers across their ends and are kept apart by a shaped piece of mahogany. The box is of polished mahogany with the end piece removable.

Q.C., Item 122; magnetical bars by Dr Knight.
Q.C., Item 123; magnetical bars by Morgan
Q.C., Item 124; magnetical bars by Hindley of York
Q.C., Item 127; A collection of magnetical bars by different hands
Knight 1744, p. 162
Demainbray 1750–2, p. 10, 'with Gowin Knight's curious magnetic bars'

Michell 1750, p. 73   Knight 1766
Canton 1751, p. 31   C 265

As D127   **Inventory No.** 1929-106

OPTICS

D131 Plano-convex lens in stand

mid-eighteenth century

H 255mm, **D** 108mm (lens)

From the many convex lenses in the collection this is the most likely to have been used by Demainbray. It could either be the 'convex lens on stand' or the 'plano-convex lens' listed in the Queen's Catalogue. The materials are consistent with those in Demainbray's collection and the style indicates an earlier date than the majority of the other mounted lenses. It is a plano-convex lens of crown glass with a focal length of about two feet. The stand is made of boxwood and the lens is set in a walnut mount on a brass swivel holder similar to those of the large mirrors D58 and D59.

Q.C., Item 241; a plano-convex lens
Q.C., Item 302; a convex lens on stand
C 206

Inventory No. 1927-1250

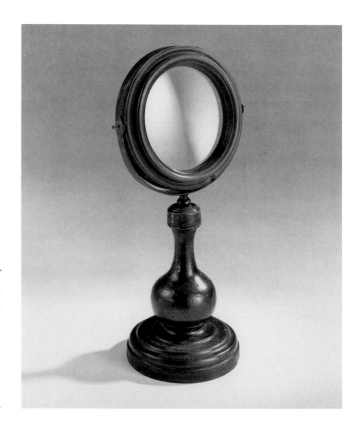

D132 Plano-convex lens in stand

mid-eighteenth century

D 180mm (lens), **D** 143mm (base), **H** 345mm, **L** 240mm (overall)

This could be the plano-convex lens listed in the Queen's Catalogue, but since no stand is mentioned this is uncertain. It could also be 'a convex lens on a stand' but D131 is more likely to fit this description.

The lens is mounted in a brass swivel frame which is held by butterfly screws similar to others featuring on Demainbray's instruments. The stand is made of mahogany. The focal length is about one foot.

Q.C., Item 241; a plano-convex lens
Q.C., Item 302; a convex lens in stand
C 305

Inventory No. 1927-1249

D133 Scioptic ball

mid-eighteenth century

D 130mm, **D** 140mm (overall), **D** 80mm (ball)

This is possibly Demainbray's 'lesser scioptric ball' listed in the Queen's Catalogue. It is probably not large enough to be the large one listed. It consists of a turned sycamore outer ring with a rosewood ball containing one lens. There is a turned sycamore lens cap.

Scioptic balls were devices used to direct and focus light for various experiments. They were often fixed into window shutters.

Q.C., Item 249; a lesser scioptric ball
C 192

See also D133, E152

Inventory No. 1927-1166

D134 Scioptic ball

mid-eighteenth century

D 125mm (frame), **D** 80mm (ball)

This is probably Demainbray's 'scioptic ball with two lenses'. It is in a square frame which is entirely of boxwood.

Q.C., Item 250; scioptic ball with 2 lenses
C 192

See also E152

Inventory No. 1927-1167

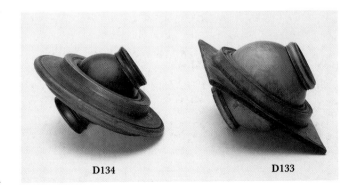

D134   D133

D135 Model eye

mid-eighteenth century

L 50mm (overall), **D** 35mm (ball), **D** 12mm (lenses), **L** 35mm (screw), **D** 12mm (screw)

This could possibly be Demainbray's 'artificial' eye but it seems very small for demonstration purposes. It is not mentioned in the latest syllabus. It has previously been referred to as a pocket microscope but this is not its function.

It consists of a hollow ivory ball with a walnut surround to the pupil which screws into one side, giving an aperture of about a quarter of an inch. Inside are two small lenses with different magnifications and thicknesses, presumably to be used one at a time as eye lenses. At the other side of the ball there is a hollow walnut screw with a ground glass screen representing the retina at the end which enters the lens.

Q.C., Item 98; an artificial eye with lenses to show how the defects of sight are cured by glasses
Demainbray 1750–2, p. 12, 'an artificial Eye'
Demainbray 1753, p. 20
Demainbray 1754a, p. 21
C 190
Inventory No. 1927-1568

D136 Camera obscura

before 1753

L 430mm, **W** 210mm, **H** 195mm

This is a common design of box camera obscura which featured on many instrument-maker's trade cards in the mid-eighteenth century, including those of Edward Scarlett and Oliver Combs. Smith describes the instrument in *A Compleat System of Opticks* as the standard type of camera obscura. The uncertainty with which this instrument can be associated with Demainbray arises from the listing of two camera obscuras in the Queen's Catalogue. One belonged to Demainbray, but the other was 'placed by her Majesty' and is of unknown origin. Demainbray used a camera obscura to illustrate his lecture on vision along with a scioptic ball and magic lantern. He also had an 'optical machine' showing a camera obscura by threads which has survived.

The instrument is almost entirely made of oak. The lens is held in a square draw tube which has a brass knob. A similar knob allows the cover to be opened. The back panel slides out so that the instrument can be used without the mirror; however, in the more usual arrangement the ground glass screen would have been placed horizontally below the cover with the mirror at 45°. Neither the mirror nor the ground glass screen have survived.

Q.C., Item 251 and p. 20 item 15; a camera obscura
Smith 1738, p. 386, Fig. 634
Demainbray 1754b, p. 21, 'the Camera Obscura'
C 194
Hammond 1981, pp. 82–8

See also D62

Inventory No. 1927-1139

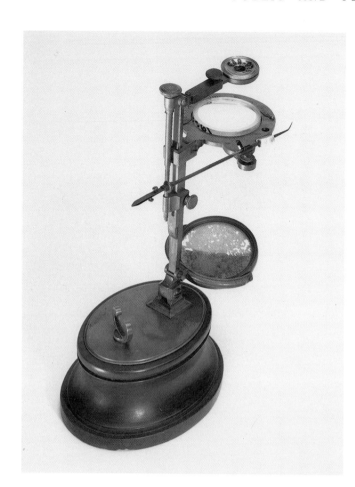

D137 Aquatic microscope

Cuff, John

c. 1745   London

'J Cuff Londini Inv & Fec. No.34'

L 130mm, **W** 70mm, **H** 165mm

There is some uncertainty as to whether this instrument can be associated with Demainbray. A Cuff aquatic microscope was listed in the Queen's Catalogue and Court asserted that this item originated in the King George III Collection; however, Crisp states that the microscope was bought in Lucca. Whipple assumes Court's opinion to be correct and describes it as item 291 from the Queen's Catalogue.

The microscope consists of a lens with a Leberkühn mirror and a stage containing a concave glass dish and reflecting mirror. It is attached to a pillar which stands on a weighted base. Focusing is achieved by sliding the lens attachment on the pillar and fine focusing by means of a screw; the lens can also be moved across the stage. There is a choice of four powers, stage forceps, specimen slides, and tongs. The microscope is made of brass and the base is mahogany.

Q.C., Item 291; Cuff's aquatic microscope
Crisp 1925, No. 94?
Whipple 1926, p. 515, Fig. 7

Inventory No. 1928-803

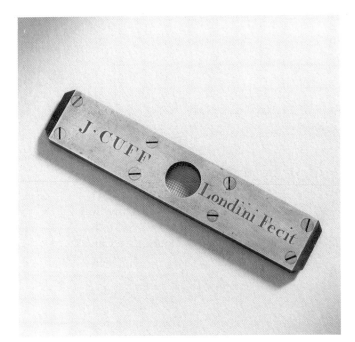 

D138 Micrometer in slider

Cuff, John

1745   London

SCRATCHED—BACK
'J. CUFF Londini Fecit 1/50 in 1745'
L 62mm, **W** 12mm, **T** 3mm, **D** 8mm (aperture)

Since this micrometer came from Court's collection it is possible that it belonged to Demainbray and is the one referred to in the Queen's Catalogue, although it is not listed as a separate item in the catalogue of the Crisp sale. Demainbray used the micrometer, or one like it, with the solar microscope in his lecture on microscopy.

It is a brass slider with an aperture about $\frac{1}{3}$ inch in diameter. It has a 12×12 lattice in silver wire, each division giving a linear measure of $\frac{1}{50}$ inch. It could be placed in the focal plane of the eyepiece of a microscope to give the size of the specimen when the magnifying power of the objective was known.

Baker states that Cuff developed these micrometers in 1747. The 1745 scratched on the back of this item could indicate an earlier date or could be a later addition, although Court does not mention it in *The History of the Microscope*.

Q.C., Item 293; Ditto [Cuff's micrometer] in slider
Baker 1753, p. 425
Demainbray 1753, p. 22
Demainbray 1754*a*, p. 22
Demainbray 1754*b*, p. 22 'Cuff's micrometer'
Clay and Court 1932, pp. 139–40

Inventory No. 1938-706

D139 Gregorian telescope

Robertson, W.?

c. 1745   Scotland

L 245mm, **D** 40mm, **H** 135mm, **L** 117mm (box), **W** 90mm (box), **H** 65mm (box)

The telescope has small circles stamped on the brass mounts, very similar to those on D68, so it is possible that it is by William Robertson of Edinburgh and came into the collection through Demainbray. The Queen's Catalogue lists a 'small reflecting telescope by the late James Short' but this is not signed in Short's characteristic manner. Whipple believed the Short telescope to have been at Kew Observatory in 1926.

The telescope is entirely of brass with a 6-inch body tube and an aperture of $1\frac{1}{2}$ inches. The mirrors are in good condition. Focusing is obtained by turning a screw which runs down the side of the instrument. When in use the telescope and stand screw into the top of the mahogany box for stability. A ball and socket joint allows the inclination of the body to be varied both vertically and horizontally.

James Gregory invented this design of reflecting telescope in 1663. It comprises a primary concave mirror with a central hole and a smaller secondary concave mirror which share a common focus. The image is viewed through the hole in the primary mirror using an eyepiece lens.

Q.C., Item 282; A small reflecting telescope by the late James Short
Gregory 1663, p. 94
Whipple 1926, p. 510
C 173
King 1955, pp. 70–1
Turner 1969
Bryden 1972*a*

See also D68, E186, E187

Inventory No. 1927-1415

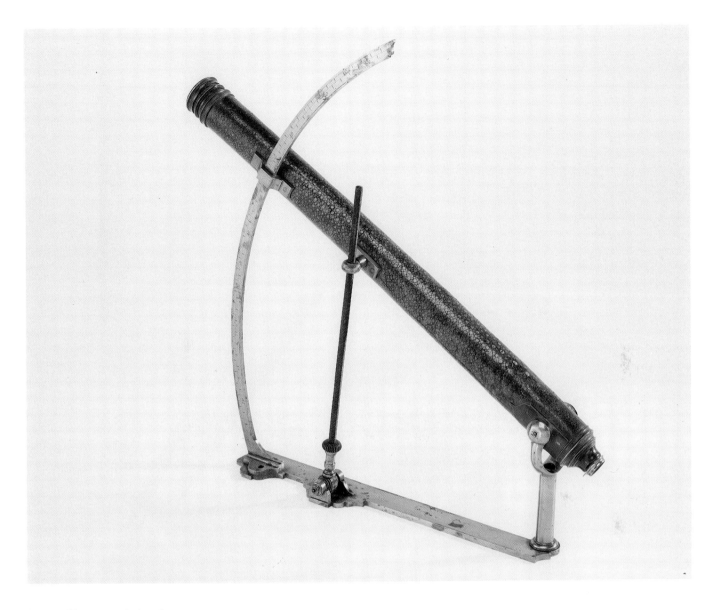

D140 Terrestrial telescope

mid-eighteenth century

L 595mm (closed), **H** 495mm, **D** 50mm

This is a possible contender for the entry in the Queen's Catalogue 'a large 4 lens Telescope with a stand and a Brass Quadrant', and also for Demainbray's description in his syllabuses of 'a curious dioptrick with a large apparatus for astronomical observation'. However, D68 is more likely to be the instrument concerned; this telescope is probably not large enough.

The body of the telescope is covered in green shagreen. There are two vellum draw tubes, one at each end. The objective is non-achromatic with remains of a cross-wire in evidence. The erector lenses are in a lignum vitae and cardboard cell which was probably removable. The mounts and objective lens cap are made of horn while the eyepiece cap is brass with a slider. The telescope is mounted on a brass pivot near the eyepiece and on a short brass pillar. Its inclination can be adjusted by turning a long steel screw which runs from the base to the centre of the instrument. The inclination or altitude can be read off a brass arc over which the objective end of the telescope passes. The scale is marked from 10° through 0° to 65° in 20′ intervals to give the inclination of the telescope but it is not clear against what index the scale was read.

Q.C., Item 277; A large 4 lens Telescope with a stand and a brass Quadrant
Demainbray 1754*b*, p. 23 as text
C 177

See also D68
Inventory No. 1927-1424

D141 Heliostat

Sisson, Jeremiah

before 1753?   London

L 325mm, **W** 325mm, **H** 264mm

In the earlier syllabuses Demainbray was advertising 'a curious new and improved heliostate'; however, it does not appear to have been included on his return to London nor does it feature in the Queen's Catalogue, unless it formed part of the helioscope. Nevertheless, Demainbray did use instruments by Sisson so it is possible that it belonged to him. Like the other Sisson instruments in the collection it is an example of high quality workmanship.

Chaldecott gives a very comprehensive description of this object. It is a complex device, almost all in brass. The mainframe of the instrument consists of two brass squares held at right angles to one another by two supports. The vertical square holds the fixed mirror in a hood at an angle of about $22\frac{1}{2}°$ to the horizontal. Behind this is a circular disc containing four sizes of hole which can be rotated to give the required aperture. The vertical square is hinged vertically to another square frame which is hinged horizontally to another which could be attached to a window shutter or the objective of a telescope. In the former case the working parts of the instrument would be on the outside of the room.

The horizontal brass square holds the movable mirror on a smaller square which is hinged above it. The angle of this smaller square would be set at the co-latitude (90° minus the latitude) so that it would lie in the plane of the celestial equator. The horizontal square has four levelling screws and a spirit level, now empty, in order to improve accuracy. Set inside the small square is the hour ring which can be moved by means of a handle on a rod from inside the room or while operating the telescope. The movable mirror is held between two supports on pillars and can be rotated vertically according to the Sun's declination; this is done by means of a screw on the outside. The co-latitude is marked up to 47° with a vernier giving an accuracy to 1', the declination is marked from 90°N to 90°S, also with a vernier, giving an accuracy to 5' of arc, and the time is given to an accuracy of 4 minutes, numbered every 20 minutes.

This instrument enables the Sun's rays to be reflected from both mirrors in such a way that the resulting beam is maintained in a constant direction. This is particularly important when making solar observations.

Demainbray 1750–2, p. 13; curious new and improved heliostate
Chabrol 1753, p. 205
Demainbray 1753, p. 23
Demainbray 1754*a*, p. 24
C 145

See also D37   **Inventory No.** 1927-1195

D142 Pair of mounted coloured glass discs

c. 1750

'Num 1: yellow' 'Num 2: green'

D 46mm, **H** 125mm

Although these glasses were referred to as blue and yellow in the Queen's Catalogue, 'num 1: yellow' and 'num 2: green' are handwritten on their respective bases. They were probably used by Demainbray to illustrate his lecture on Newton's theory of colour. Each disc of coloured glass is mounted in ivory on a boxwood stand. One of the ivory mounts is broken.

Q.C., Item 148; blue glass mounted in ivory stand
Q.C., Item 149; yellow glass mounted in ivory stand
C 212

Inventory No. 1927-1156

D143 Three prism stands

c. 1750

L 170mm, **W** 155mm, **H** 720mm (max)

Although not mentioned specifically in the syllabuses, it is likely that these prism stands were used by Demainbray in his experiments on Newton's theory of colour. They are probably parts of the three prisms and frames listed in the Queen's Catalogue.

The frames are made of beech, a wood frequently used in Demainbray's collection. One holds three prisms, while the other two hold one each. The heights can be adjusted by sliding the frames holding the prisms inside the outer frame. They are a standard design; similar models can be found in the 's Gravesande and Van Marum collections.

Q.C., Items 259, 260, 261; a prism and frame, ditto, ditto
's Gravesande 1747, Vol. 2, p. 242, Pl. 110, Fig. 7
C 215
Crommelin 1951, No. GM72a
Turner 1973, No. 248

Inventory No. 1927-1137

D144 Glass sphere on a stand

mid-eighteenth century

D 240mm, **D** 125mm (base), **D** 50mm (sphere)

The entry in the Queen's Catalogue together with the materials and style of this object suggest that it was part of Demainbray's collection. It was probably used to enhance illumination during his two lectures on microscopy.

Q.C., Item 155; a glass globe on a stand

Inventory No. 1927-1427

MISCELLANEOUS

D145 Globe on stand for electrical experiments

second half eighteenth century or first half nineteenth century

H 530mm, **D** 150mm (globe), **D** 180mm (base)

This item may have been Demainbray's since it is stylistically similar to some of his instruments. The glass globe is nearly cylindrical in shape. A feather is suspended on a thread from the top of the interior, which indicates that it was used for electrical experiments. The mahogany stand has a brass collar and the globe is sealed with wax.

C 129

Inventory No. 1927-1386

D146 Channel

mid-eighteenth century

L 600mm, **W** 128mm, **H** 67mm

This channel is thought to have been in Demainbray's collection because it is made of the characteristic green/grey painted tinplate found in several other items known to have belonged to him. It is not known to what use it was put.

Inventory No. 1927-1758

D147 Tin box

mid-eighteenth century

D 37mm, **H** 30mm

The box is made of tinplate which has been painted in the characteristic blue/grey colour of Demainbray's collection. It has therefore been associated with him despite having been assigned to Adams's Archimedean screw when it came into the Museum. It was probably made to hold iron filings for magnetic experiments or balls for various mechanical demonstrations.

Inventory No. 1927-1106/2

D148 Straight cistern barometer

mid-eighteenth century

H 930mm, **W** 100mm, **DE** 35mm

This is likely to be the portable barometer listed in the Queen's Catalogue. It is now in very poor condition and the screw is missing. It has a pine cistern with a leather base; the back board is mahogany. The scale, over which an index could be moved, is now illegible.

Q.C., Item 209; portable barometer
Desaguliers 1744, Vol. 2, p. 287, Pl. 21, Fig. 19

See also D124, E114

Inventory No. 1929-129

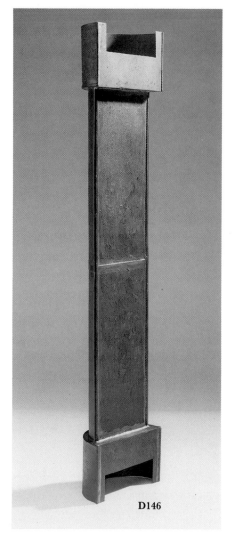

D146

PART 2

The apparatus supplied for King George III by George Adams in 1761–2

About one-fifth of the present-day King George III Collection consists of items supplied in 1761–2 by George Adams, one of the best known instrument-makers of his time. Adams was born in early 1709 in London.[1] From 1724 he was the apprentice first of James Parker and then of Thomas Heath, receiving his freedom in 1733. From 1734 he had his own shop in Fleet Street where, though he moved premises, his business remained for the rest of his life. Adams was a very successful maker and supplied many instruments for the Office of Ordnance, for example.[2] He was also a prominent member of one of the City Livery Companies, the Grocers. After Adams's death in 1772 the business was carried on by his widow, Ann, and his son, George.[3]

George III had appointed Adams to be his Mathematical Instrument-Maker on 15 December 1760.[4] By the end of the year it seems that he had commissioned Adams to make the pneumatics and mechanics apparatus for a sketch of one item, the Philosophical Table, is dated 'Jan. 5 1761'.[5]

The bulk of this material from Adams can be identified from the illustrated manuscripts on pneumatics and mechanics he provided with the apparatus. These manuscripts exist in three versions; a rough draft and an incomplete fair copy, both in the Science

---

[1] See the forthcoming work by John Millburn, *The Adams Family: Mathematical Instrument Makers to George III*, London, 1993.

[2] See Brown (1979) p. 36 for details of Adams. See Millburn (1988) for Adams and the Office of Ordnance.

[3] George Adams junior was also apprenticed to his father at the time of his father's death. He carried on the business until his own death in 1795. He took his own brother, Dudley, as an apprentice and Dudley later set up in business himself. See Millburn (1993) forthcoming.

[4] See Public Record Office LC/3/67 p. 30. Mr Dolland was appointed Optician and Dr Robert Smith, Master of the Mechanicks at the same time. LC/3/67 pp. 31–2.

[5] Science Museum MS 203, plate 7. It is possible that one of the people who carried out the work was Nikolai Chizhov who worked for Adams at this time and who later returned to Russia to become head of the Academy of Sciences workshop in the 1760s. See Boss (1972) pp. 204–5 and Cross (1979) p. 35.

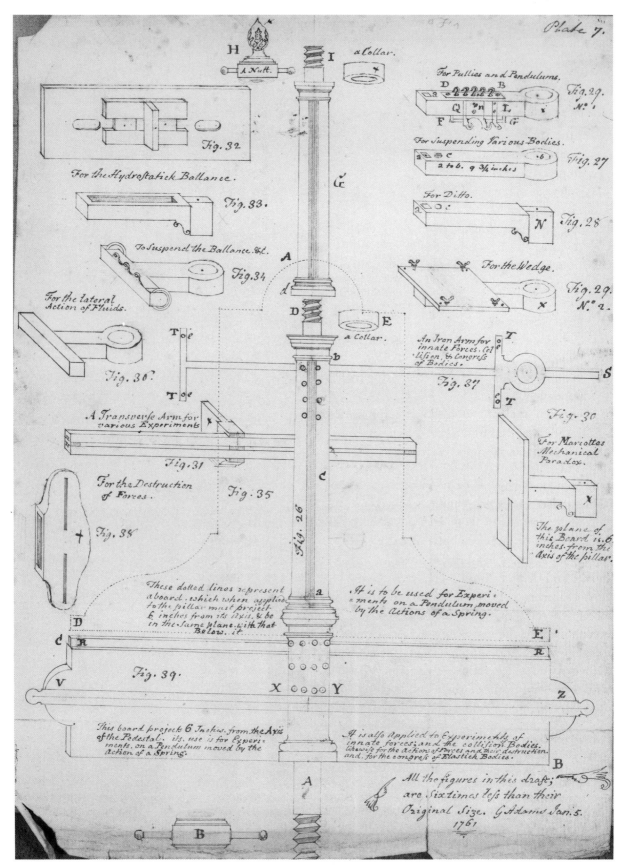

A sketch of the Philosophical Table by George Adams dated January 5, 1761.

Museum, and a complete finished copy in the Royal Library, Windsor Castle.[6] The *Pneumatics* manuscript is dated 1761 and the *Mechanics* manuscript, 1762. The rough draft contains only sections of the final version; the other sections, presumably drawn up in the same way, were later lost. The draft contains instructions for carrying out the various demonstrations together with sketches of the apparatus used. These sketches were also used in designing the apparatus itself for in several cases details only relevant to constructing the apparatus are included. This dual purpose, providing details of how to make the apparatus as well as to use it, is carried over into the final version for there Adams describes the air pump in terms which he thinks sufficient for 'intelligent Workmen' to make a copy.[7]

As his source for many illustrations and much of the text, Adams used the sixth (or third in English) edition of 's Gravesande's, *Mathematical Elements of Natural Philosophy*, translated by J. T. Desaguliers, though details such as the dimensions of the apparatus Adams supplied were incorporated in his version of the text.

The *Pneumatics* manuscript states that the apparatus was made for the King in 1761 and so it is assumed that the apparatus described in the manuscript can be dated to that year.

The *Pneumatics* manuscript

Though the main source used by Adams for the *Pneumatics* manuscript is the work by 's Gravesande, Adams also drew on an article by Nollet and the work of both Hauksbees and Whiston.[8]

The apparatus supplied by Adams illustrates that by the mid-eighteenth century an instrument-maker could supply quite a wide range of accessories to carry out what were the standard experiments with an air pump. Many of these demonstrations originated in the work of Boyle, von Guericke, Huyghens, Papin, Hooke, and

others in the seventeenth century. A little later Hauksbee senior introduced several electrical experiments carried out on his air pump. In 1717 William Vream published a book of pneumatical experiments, a work which was copied with little alteration by Desaguliers in various editions of his published lectures.[9] Vream's example of selling both pumps and a work describing the demonstrations was continued by Davenport in 1737 and Martin in 1766, amongst others.[10]

The *Mechanics* manuscript

In the case of the *Mechanics* manuscript, Adams also drew on parts of Helsham, Nollet, Emmerson, and Desaguliers. Though the basic design of the apparatus was usually copied from descriptions and illustrations in these sources, Adams often improved the designs both in function and appearance. By the 1760s Adams had been selling examples of many of the items used for the mechanics demonstrations for over 20 years so that his designs were well established.

The course of experiments described by Adams combined topics known to classical authors, such as demonstrations showing the significance of the centre of gravity, simple machines, quadrature of figures, with relatively new discoveries, such as the isochronous pendulum, the cycloid, and Marriott's paradox. Magnetism was also included as part of mechanics. Curiously, however, there is little discussion of contemporary machines.

In style the apparatus is restrained and almost austere when compared with other contemporary collections such as those of Nollet and 's Gravesande. The equipment is made almost exclusively from mahogany and brass but ivory was used occasionally for scales and balls.

As the *Mechanics* manuscript is dated 1762 and there

6   The copy in the Royal Library was bought in the mid-nineteenth century at a sale of books belonging to a son of Lord Bute. It is possible this copy was made for Lord Bute though nothing in the manuscript suggests that. It seems more likely that this copy was made to accompany the King's apparatus and the close connection between Bute and King George III in 1761–2, with their mutual interest in natural philosophy,

allowed this copy to find its way into the possession of Bute's son.
7   Adams (1761*a*) p. 8.
8   Nollet (1740), Hauksbee (1714).
9   See Vream (1717) and Desaguliers (1744), Vol. 2, pp. 375–90. See Anderson (1978) pp. 67–70 for a discussion of the history of the air-pump in the early eighteenth century.
10 See Davenport (1737) and Martin (1766).

is the evidence quoted above that the Philosophical Table at least was designed around January 1761 we have concluded that the mechanics apparatus by Adams was constructed in 1761–2. In this period Adams was working in Fleet Street, at what became No. 60 when the street was numbered a few years later.

PNEUMATICS

P1 Double-barrelled air pump

*'Geo. Adams/Mathematical Ins*ᵗ *Maker to his/ MAJESTY/Fleet Street LONDON'*

H 1230mm, **W** 790mm, **L** 700mm (approx. with handles), **L** 330mm (barrel), **D** 64mm (barrel)

This magnificent air pump is the centrepiece of the pneumatics apparatus made for King George III by George Adams in 1761. In the manuscript he provided to accompany the apparatus, Adams explained that the design was based on a pump constructed by John Smeaton some years before except that Adams's version had two barrels where Smeaton's had one. As in Smeaton's design, the pump was also provided with a cock to allow it to be used to compress or 'condense' the air as well as to 'rarefy' it.

The main similarity with Smeaton's own pump lay in the design and arrangement of the valves. The valve at the bottom of each cylinder consisted of a piece of bladder covering a grating of seven hexagonal holes. Each hexagon was one tenth of an inch across and its sides were filed 'almost sharp' so the area of the grating in contact with the bladder was as small as possible. There was a valve in each piston and a third pair near the top but to the rear of each cylinder. The purpose of the last of these valves was to prevent the external air from pressing on the pistons in order to make the pump easier to operate.

To remove air from a receiver on the plate on top of the pump, the handle on the front of the pump was turned. This moved the two pistons in opposite directions, raising one while lowering the other; both pistons went through the same sequence but out of step with each other. The rising piston sucked air from the receiver into the lower part of the cylinder through the valve at the bottom. When the motion of the handle was reversed, the piston descended and the air now passed through the valve in the piston into the upper part of

the cylinder. On the next up-stroke the air above the piston was compressed until it was at a high enough pressure to pass through the valve near the top of the cylinder to the open air.

A further advantage of Smeaton's design was that the pump could be used to compress air in a 'condenser'. To do this a key was turned so that the upper part of each cylinder was connected to the condenser while the lower part communicated with the open air. On an up-stroke, air passed into the cylinder through the valve in its base. The subsequent down-stroke allowed that air to pass through the piston into the upper part of the cylinder. The following up-stroke forced the air through the topmost valve into the condenser. Water was used to keep the joints at the bottom of the cylinders airtight.

There is a wheel and pulley mechanism at the back so that bodies could be rotated in a receiver on top of the pump to show the effects of friction.

In 1763 the King arranged for George Adams to demonstrate the pump for Lalande.

Adams 1761*a*, pp. 1–8, Pls 1–3
C 101, Pl. 8
Lalande 1980, p. 72

See also p. 19 for colour illustration

Inventory No. 1927-1624/1

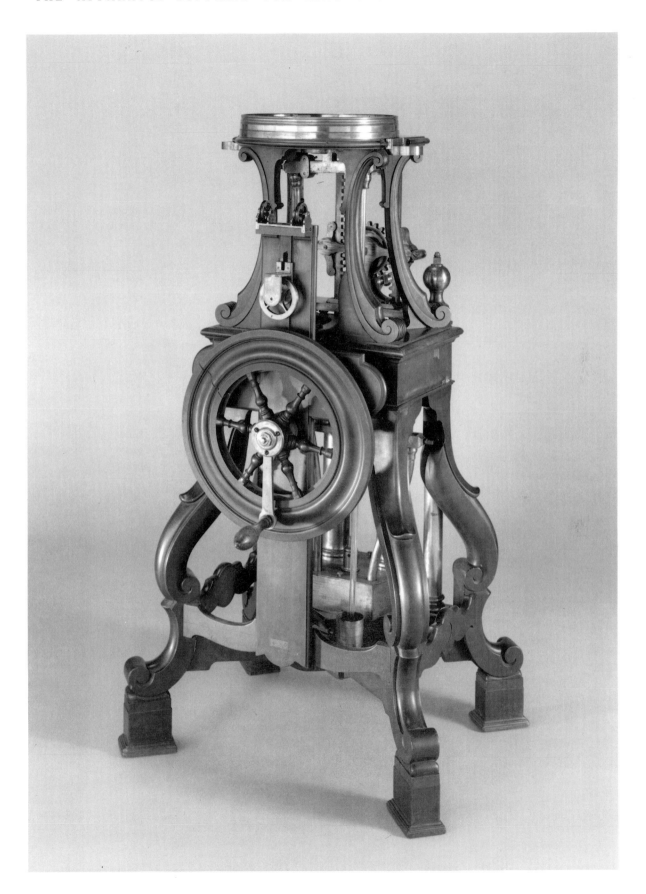

P2 Cork cube

L 75mm

This apparatus, now incomplete, consisted of a cork cube suspended from one arm of a balance and a brass weight from the other. This was placed in a receiver on an air pump. In air the two weights were in balance, but when the air was removed from the receiver, the cork fell. This was because the cork cube displaced a greater volume of air than the brass weight.

's Gravesande used a piece of wax rather than cork in his version of the demonstration.

Birch 1744, Vol. 1, p. 50
Desaguliers 1744, Vol. 2, pp. 236–7, Pl. 20, Fig. 15
's Gravesande 1747, p. 20, Pl. 67, Fig. 2
Adams 1761a, p 11, Pl. 4, Fig. 8
Adams 1799a, Vol. I, p. 68, Pl. 2, Fig. 9

Used with P3

Inventory No. 1927-1315

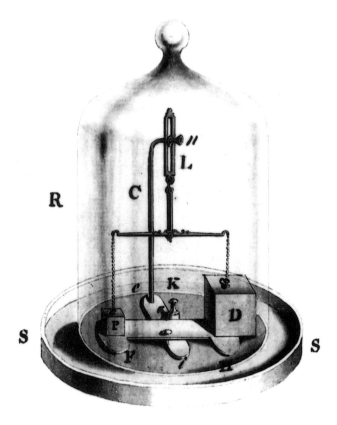

Adams 1761a

P3 Brass bracket

H 275mm, **L** 110mm, **W** 30mm

This was used to support a balance to suspend the cork cube P2.

's Gravesande 1747, Vol. 2, pp. 19–20, Pl. 67, Fig. 2
Adams 1761a, pp. 10–11, Pl. 4, Fig. 8

Inventory No. 1927-2054

P4 Cylindrical box with four weights

H 285mm, **D** 170mm

This apparatus was used to demonstrate the spring of the air. A bladder, not quite full of air, was placed in the cylindrical mahogany base, the wooden top replaced, and lead weights fitted over the rod on the top of the box. If this apparatus was put under a receiver and the air removed, the air in the bladder raised both the lid of the box and the weights.

's Gravesande suggested a weight of 40 lb in his version of the demonstration.

Hauksbee 1714, p. 19, Pl. 5, Fig. 7
's Gravesande 1747, Vol. 2, p. 20, Pl. 67, Fig. 3
Adams 1761*a*, p. 12, Pl. 4, Fig. 10
Adams 1799*b*, Vol. 1, p. 85, Pl. 2, Fig. 6

Inventory No. 1927-1292

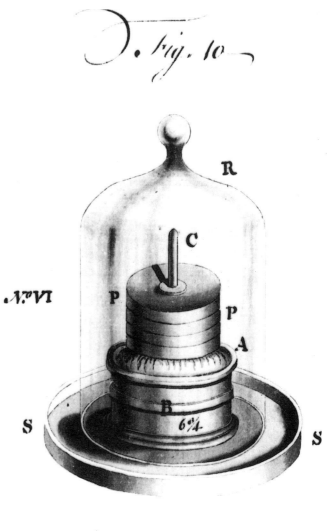

*Adams 1761*a

P5 Glass tube with base

D 102mm (base), **L** 600mm

This glass tube was used in a demonstration of the effects of air pressure. The tube was placed on top of a receiver on the plate of the air pump so that both vessels covered a barometer tube filled with mercury. As the air was removed from the receiver, the mercury in the barometer tube fell. The receiver was also connected to the top of a second vertical tube beneath it. The other end of that tube was in a reservoir of mercury. The mercury rose in this lower tube as much as it fell in the tube on the receiver, showing that the heights of both these columns of mercury depended on the air pressure in the receiver.

The illustration in the *Pneumatics* manuscript shows an inverted U-tube over the receiver. However, the sketch in the *Pneumatics Scrapbook* shows one tube inside the other which suggests that the version in the *Pneumatics* manuscript is based on a misinterpretation.

Hauksbee 1714, p. 15, Pl. 1, Fig. 6
Birch 1744, Pl. 13, pp. 6–26
Adams 1761*a*, p. 12, Pl. 5, Fig. 11

Inventory No. 1927-1332

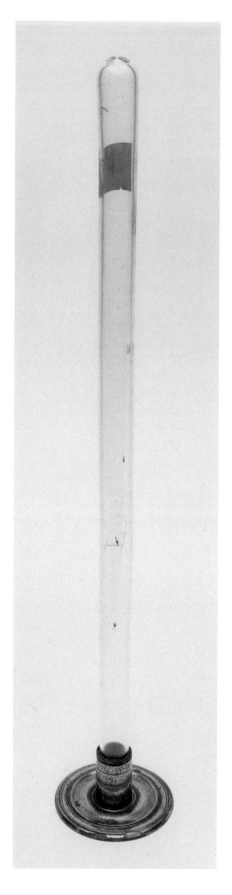

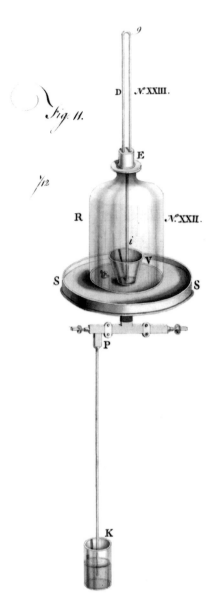

*Adams 1761*a

P6 Mercury manometer

SCALE
0 ($\frac{1}{10}$) 3 inches

H 190mm, **D** 41mm (base)

Adams explained that this gauge consisted of a glass syphon tube, 'J-shaped', $\frac{2}{10}$ inch bore, closed at one end, and filled with mercury. The difference in height of the mercury in the two limbs was measured on a scale divided into inches and tenths.

According to Adams the design of this gauge was taken from one contrived by M. de Mairan and described by Nollet. However, the account by Nollet explains that de Mairan had used a reservoir for the mercury on the shorter limb of his J-shaped tube. Nollet suggested improvements to this design: his tube was wider than the one de Mairan had used; it was also the same diameter throughout and did not have a reservoir.

Nollet 1741, pp. 345–6, Figs 3, 4
Birch 1744, Vol. 3, pp. 23–5
Desaguliers 1744, Vol. 2, pp. 395–6, Pl. 24, Fig. 11
Adams 1761*a*, pp. 13–14, Pl. 5, Fig. 12, Pl. 18, Fig. 56

Inventory No. 1927-1313

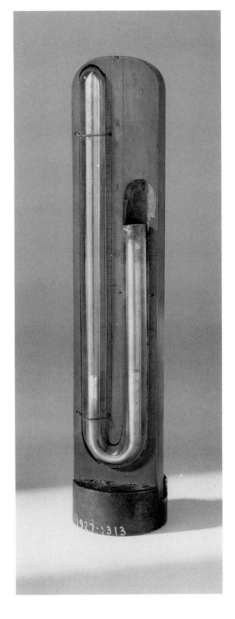

P7 Glass tube with bulb

L 540mm, **D** 55mm (bulb)

This spherical glass vessel with a long neck, or 'bolt head', was filled with water. The open end of the tube was put in a glass of water and placed in a receiver on an air pump. When the pump was operated the level of the water in the tube fell. From this the expansion of the matter in the tube could be estimated.

Desaguliers 1744, Vol. 2, p. 379, Pl. 29, Fig. 2
's Gravesande 1747, Vol. II, pp. 24–5
Adams 1761*a*, Pl. 4, Fig. 14, Pl. 5, Fig. 15
C 126

Used with P60

Inventory No. 1927-1689/2

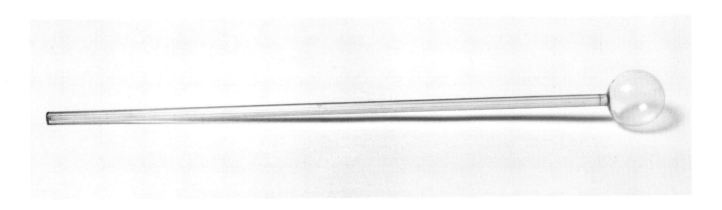

P8 Receiver with brass collar

H 810mm, **D** 150mm

This tall glass receiver was used in an experiment on air pressure. A syringe was mounted above the receiver with a glass tube going from the syringe through the receiver to a jar of water on the pump plate. If the piston of the syringe was raised when the air had been removed from the receiver, the level of the water in the glass tube did not rise. This showed that air pressure had to act on the water before the syringe could raise it.

Hauksbee 1714, p. 16, Pl. 2, Fig. 4
Adams 1761*a*, pp. 15–16, Pl. 5, Fig. 16, p. 35, Pl. 17, Fig. 52 (possibly)
C 127

Inventory No. 1927-1338

P9 Two truncated cones with glass plates

D 118mm (large cone), **H** 63mm (large cone), **D** 70mm (small cone), **H** 56mm (small cone)

One of these truncated brass cones is placed, wider end up, on the plate of an air pump and covered with a glass plate. When the air is removed from underneath the glass, the pressure of the air on one side of the glass will eventually break it, as Adams explained, into a kind of powder. With the narrow end of the cone upwards, however, it was more difficult to break the plate, showing that it was the pressure of the air that caused the damage.

Desaguliers 1744, Vol. II, p. 390, Pl. 25, Fig. 29
's Gravesande 1747, Vol. II, p. 26, Pl. 68, Fig. 2
Adams 1761a, p. 16, Pl. 5, Fig. 17
C 116   **Inventory No.** 1927-1291

P10 Magdeburg hemispheres

H 317mm (stand), **D** 188mm (stand),
D 103mm (hemisphere), **L** 240mm
(both hemispheres)

Following the experiment of Otto von Guericke in 1654 this demonstration became standard in courses of pneumatics. Adams described how the pump was used to remove the air between these brass hemispheres. The tap was closed and the hemispheres removed from the pump. Hooks and eyes were attached to the hemispheres so that one could be held fixed and the other attached to a steelyard which was used to apply a force large enough to pull the hemispheres apart. Another version of the experiment was carried out in the brass condenser P13; now air was left between the hemispheres and the air pressure in the condenser was doubled; in this case roughly the same force was needed to separate the hemispheres as in the first experiment.

von Guericke 1672, p. 104
Hauksbee 1714, p. 19, Pl. 5, Figs 1, 3
Hauksbee 1719, pp. 89–93, Pl. 4
Desaguliers 1744, Vol. 2, pp. 383–4, Pl. 25, Fig. 28, Fig. 13, pp. 396–7, Pl. 24, Fig. 12
's Gravesande 1747, pp. 27–8, Pl. 69, Figs 2, 4, pp. 30–1, Pl. 70, Fig. 3
Adams 1761a, pp. 17–18, p. 20, Pl. 6, Figs 21, 22; No. 1, Pl. 7, Fig. 22, No. 2, Pl. 8, Fig. 2
C 122

Used with P11, P12

Inventory No. 1929-115

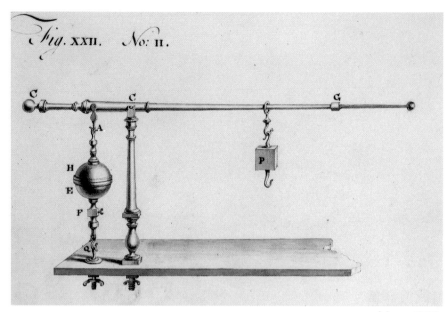

Adams 1761a

P11 Steelyard

SCALE
10 to 100

L 1200mm

This steelyard was used in a demonstration to show the weight needed to pull apart a pair of Magdeburg hemispheres (such as P10). Adams describes two versions of the demonstration. In the first the air was removed from the hemispheres using a pump. The hemispheres were then suspended from the shorter arm of the steelyard and the lower hemisphere was fixed to a table. A weight was then moved along the longer arm of the steelyard until the hemispheres came apart.

In the second version of the demonstration, the hemispheres, with air between them, were placed in the brass condenser. As the amount of air in the condenser increased, a greater weight was needed on the steelyard to separate the hemispheres.

Adams modified 's Gravesande's demonstration to use a steelyard instead of a scale-pan with weights totalling 140 lb, which must have been inconvenient to use. However, though Adams took the trouble to show the steelyard in the diagram for the experiment, he did not alter the description given in the text from that given by 's Gravesande so that the diagram and text are inconsistent. In using a steelyard Adams was following the version of the demonstration commonly used by other lecturers.

Hauksbee 1714, p. 19, Pl. 5, Fig. 3
Desaguliers 1744, Vol. 2, pp. 396–7, Pl. 24, Fig. 12
Adams 1761a, pp. 18, 20, Pl. 7, Fig. 22, No. 2, Pl. 9, Fig. 26
C 51   *See also* P10, P12   **Inventory No.** 1927-1337

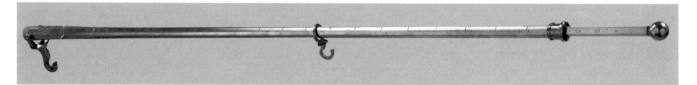

P12 Pillar

L 230mm, **D** 45mm (max)

This walnut pillar with a brass fulcrum supported the steelyard P11.

Adams 1761a, p. 20, Pl. 9, Fig. 26

Inventory No. 1927-1330

P13 Cylindrical condenser

L 290mm (overall), **D** 260mm (overall), **H** 520mm

This condenser was used for demonstrations carried out in air at greater than atmospheric pressure on, for example, animals or the clockwork bell. The body of the cylinder is made of brass; the two ends are of brass with plano-convex windows of glass.

This is the more elaborate of the two condensers supplied by Adams, the other being P11. Adams described this 'machine' as ''s Gravesand's Brass Condenser'. The design, together with the description given by Adams, comes from 's Gravesande except that Adams's version was slightly larger; 7.8 inch diameter rather than 6 inch.

Desaguliers 1744, Vol. 2, pp. 396–8, Pl. 24, Figs 12, 13
's Gravesande 1747, Vol. 2, pp. 28–30, Pl. 70, Figs 1, 2, 4
Adams 1761a, pp. 18–20, Pl. 8, Figs 23, 24, Pl. 9, Fig. 26, p. 24a, Pl. 12, Fig. 35, pp. 45–6; Pl. 23, Fig. 66, Pl. 24, Figs 67, 68

Used with P10, P11, P14, P51, P53, P54, P55, P56, P57, P58, P59

Inventory No. 1927-1326/1

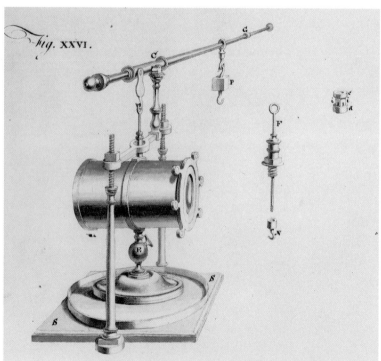

*Adams 1761*a

P14 Hook and eye

L 215mm

This brass hook was used to attach the steelyard P11 to the Magdeburg hemispheres P10 when these were set up inside the condenser P13. With air inside the hemispheres, if there was air at twice normal pressure in the condenser, the force needed to separate the hemispheres was about the force required to carry out the demonstration with the hemispheres standing in the open air but with the air between them removed.

Hauksbee 1714, p. 19, Pl. 5, Fig. 3
Adams 1761a, p. 20, Pl. 9, Fig. 26

Used with P11

Inventory No. 1927-1326/2

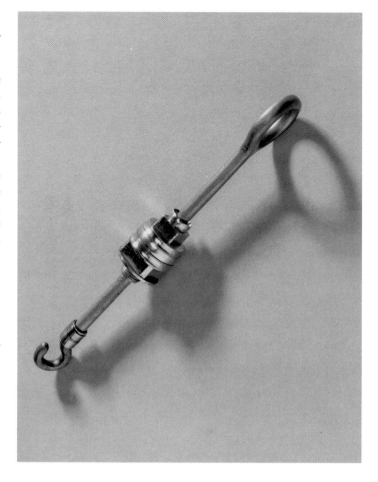

P15 Guinea and feather apparatus

H 210mm (release device), **L** 440mm (wooden collar), **H** 730mm (glass barrel)

'A Machine Whereby two bodies are let down at the same time either in Common, Rarified or Condensed Air'. The bodies were a guinea and a feather and they dropped down through two glass cylinders set one above the other. The join of the cylinders was sealed by a combination of a brass and a leather ring. Two wooden struts and a cross-piece were used to hold the glass cylinders in place over the air pump. The brass release mechanism allowed up to five pairs of guineas and feathers to be dropped in succession. One of the glass cylinders is now broken.

Adams suggested that the experiment should be carried out once in air, twice after the air pump had removed air from the glass column, and twice more when the pump had been used to increase the air pressure to greater than normal. The point of repeating experiments twice was to confirm better the result for the audience. When there was little air to slow the guineas and feathers, they fell at the same rate. In air, however, the feather fell more slowly than the guinea.

This demonstration was shown to King George III on one occasion by John Miller, who was apprenticed to George Adams at the time.

Mr Miller . . . used to tell that he was desired to explain the airpump experiment of the guinea and feather to Geo: III. In performing the experiment the young optician provided the feather the King supplied the guinea and at the conclusion the King complimented the young man on his skill as an experimenter but frugally returned the guinea to his waistcoat pocket.

's Gravesande 1747, Vol. 2, pp. 34–8, Pl. 72, Fig. 4, Pl. 73
Adams 1761a, p. 21, Pl. 10, Fig. 27 (parts), pp. 22–4, Pl. 11, Fig. 31, Pl. 12, Fig. 32
C 128

Inventory No. 1927-1308

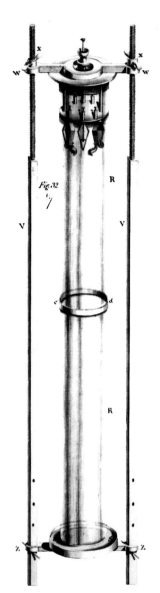

*Adams 1761*a

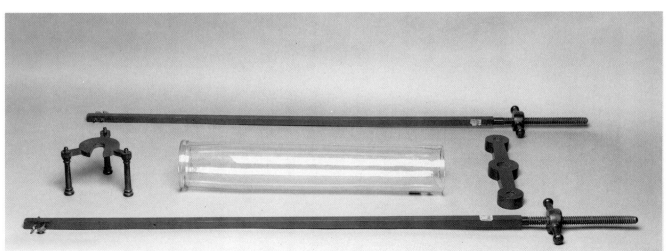

P16 Clockwork bell

'Geo. Adams Mathematical/Instrument Maker to His/
MAJESTY/Fleet Street LONDON'

H 168mm, **D** 109mm (base)

According to Adams the brass clockwork bell strikes 24 times on each occasion the button is depressed, and will do so 46 times before the spring unwinds fully.

This mechanism was used in demonstrations of the propagation of sound in air. If the bell was placed under a receiver on an air pump and the air was removed, the bell became fainter. In a condenser, with air at greater than atmospheric pressure, the strokes of the hammers are slowed down.

Hauksbee 1719, pp. 125–30
Desaguliers 1744, Vol. 2, pp. 387, 398
Adams 1761a, pp. 24–5, Pl. 12, Figs 33, 34
C 124
Turner 1987, p. 129, Pl. 117

Used with P17

Inventory No. 1927-1293

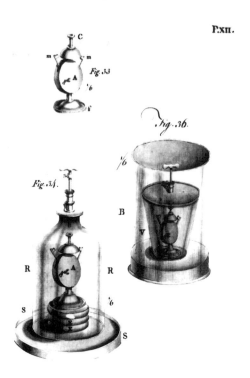

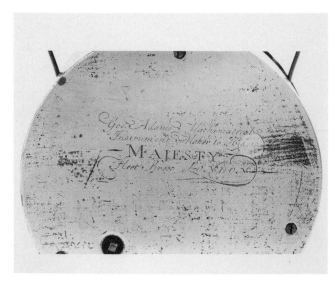

*Adams 1761*a

P17 Plate with three feet

D 205mm, **H** 43mm

This paper-covered lead plate was used to support the clockwork bell P16. Adams suggests two cushions made of several thicknesses of 'Old Hatts' between the bell and this plate. The purpose of the cushions was to prevent the sound of the bell being transmitted through the plate in the bell *in vacuo* experiment.

Desaguliers 1744, Vol. 2, pp. 387, 398
Adams 1761a, p. 24, Pl. 12, Fig. 34

Used with P16
See also P18

Inventory No. 1927-1678

P18 Plate with three feet

D 130mm

This is similar to P17. This lead plate has one hole and one socket and is covered with marbled paper.

Adams 1761a, p. 24, Pl. 12, Fig. 34

See P17

Inventory No. 1927-1677

P19 Apparatus to demonstrate chemical action

L 475mm (assembled), **W** 130mm (assembled)

The *Pneumatics* manuscript explains that this apparatus was used to show the effect of dropping sal ammoniac (or ammonium chloride) into diluted oil of vitriol (or sulphuric acid). The brass groove was filled with sal ammoniac and the brass arm turned to push some of this salt into a jar containing the acid. This demonstration was carried out in a receiver from which air had been removed. Thermometers (P20, P21) were used to measure the changes of temperature of the acid and of the 'air' in the receiver.

's Gravesande 1747, Vol. 2, pp. 70–1, Pl. 77, Fig. 5
Adams 1761*a*, pp. 25–6, Pl. 13, Fig. 37
C 135

Used with P20, P21, P22, P36, P61

Inventory No. 1927-1322

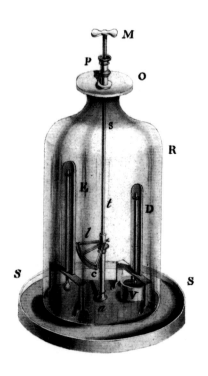

*Adams 1761*a

P20 Mercury-in-glass thermometer

'*Freezg*|*Tempe*|*rate*|*Sumr*|*Heat*|*Blood*|
Heat|*Coaguln*|*Heat*|*Boilg*|*Alcohl*|
Boilg|*Water*'
ENGRAVED
'*ADAMS*|*Fleet Street*|*London*'
SCALE—ENGRAVED
20(10)210

H 265mm (thermometer stand),
W 75mm (thermometer stand),
L 168mm (thermometer tube)

This mercury-in-glass thermometer was used in the demonstration in which sal ammoniac (ammonium chloride) was dropped in dilute sulphuric acid. The whole apparatus was in a receiver on the air pump with the air removed. This thermometer showed that the temperature of the 'Efervescence' given off during the reaction was lower than that of the surroundings. The thermometer has a spherical bulb and a silvered brass back mounted on a brass stand. The thermometer is graduated from −20 to 210°F in two scales. The scale on the left of the tube is marked in 2° intervals. A similar scale on the right is offset by a degree to enable the temperature to be estimated to the nearest degree.

's Gravesande 1747, Vol. 2, pp. 70–1, Pl. 77, Fig. 5
Adams 1761*a*, p. 26, Pl. 13, Fig. 37 left
C 93

Used with P19, P21

Inventory No. 1927-1810

P21 Mercury-in-glass thermometer

1770

ENGRAVED
'G Adams London'
'Just/Freez, Temper/ate, Sumr/Heat, Human/Heat, Fever/Heat'
SCALE
12(10)130

H 265mm (thermometer), **L** 72mm, **W** 75mm, **L** 93mm (thermometer tube)

This mercury-in-glass thermometer was also used in the demonstration where sal ammoniac (ammonium chloride) was dropped in dilute sulphuric acid.

Compared with P20, the bulb thermometer proje~~~~~
brass st~~~~~

[handwritten note:]
Errata:
H 265 (thermometer)
W 75mm L 193 (tube)
(L 72 mm)

cf. P 20
Both same height, width, but this tube extends well below, and ends higher up.

m~~~~~en list of ten~~~~~his list is in Demainbray's handwriting and records two temperatures of particular interest to him: the 'Heat at Toulouse 7. July. 1753' when he is likely to have been there, and 'Myrtles', a plant he had experimented with, were listed under 'Green house Thermometer'. The list of temperatures also records that Mr Brown at St Petersburg 'fixed' quicksilver at −245°, a reference to experiments by I. A. Braun in 1760, published by the St Petersburg Academy in 1767.

's Gravesande 1747, Vol. 2, pp. 70–1,

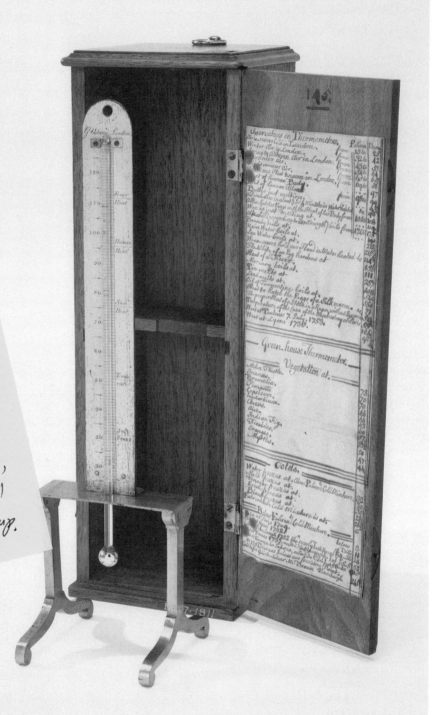

Pl. 77, Fig. 5
Adams 1761a, p. 26; Pl. 13, Fig. 37 right
C 94, Pl. II
See P19, P20

Inventory No. 1927-1811

P22 Brass fork

L 55mm

Adams 1761*a*, p. 41, Pl. 19, Fig. 59

Used with P19, P42.

See also illustration for P42

Inventory No. 1927-1326/3

P23 Connecting pipe

L 305mm

This copper pipe connected the air pump to a pump plate on which stood a glass receiver. The experiments using the pipe, illustrated by Adams, were the mercury fountain and the 'Mercurial Phosphori', where mercury was made to fall down through an exhausted receiver. In a darkened room flashes of light were seen as droplets of mercury hit the sides of the receiver.

Hauksbee 1719, pp. 8–9
's Gravesande 1747, Vol. 2, pp. 80–1, Pl. 79, Fig. 5
Adams 1761*a*, p. 26, Pl. 13, Fig. 38, p. 41, Pl. 15, Fig. 53

Used with P1, P44

Inventory No. 1927-1323

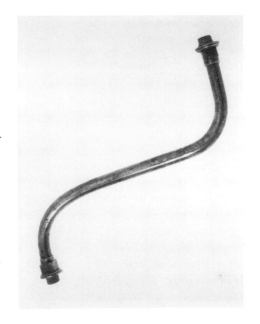

P24 Iron cube in an iron box

L 26mm (cube), **W** 41mm (box), **L** 50mm (box)

This is one of a pair of identical iron cubes used in a demonstration on heat; the second is no longer extant. This cube was heated to red heat; its iron box prevented ashes from adhering to the cube. The cube was then put in a scale-pan of a small balance on the plate of the air pump. The other cube, unheated, was placed in the other scale-pan so that the two cubes balanced. The apparatus was covered with a receiver and the air removed. The balance remained in equilibrium as the heated cube cooled, showing that the heat of the cube had no effect on its weight. Though Adams makes no reference to it, this demonstration had a bearing on contemporary discussions about phlogiston.

's Gravesande 1747, Vol. 2, p. 91
Adams 1761*a*, pp. 28–9, Pl. 4, Fig. 8

Inventory No. 1927-1346

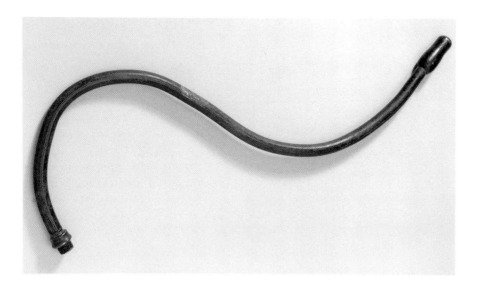

P25 S-shaped tube

L 520mm, **D** 11mm

One end of this iron tube was attached to a stopcock on a glass receiver from which the air had been removed. The other end of the tube was placed in red-hot coals. The stopcock was opened and the 'air', as Adams called it, from the coals passed into the receiver. A lighted taper put in the receiver was extinguished.

Desaguliers 1744, Vol. 2, p. 389, Pl. 25, Fig. 27
's Gravesande 1747, Vol. 2, p. 93
Adams 1761a, pp. 30–1, Pl. 15, Fig. 44

Inventory No. 1927-2053

P26 Iron cylinder

L 46mm, **D** 50mm

This iron cylinder was heated to red heat and then placed on a fire-stone on a plate on the air pump. The two demonstrations suggested by Adams were to cover the cylinder with a receiver, remove the air, and then let small pieces of wood, or a few grains of gunpowder, drop on to the cylinder. In the case of the wood, though it was consumed, no flame was emitted.

Hauksbee 1714, p. 16, Pl. 2, Fig. 3
Birch 1744, Vol. III, pp. 250, 252
Desaguliers 1744, Vol. 2, pp. 386–7, Pl. 25, Fig. 23
's Gravesande 1747, Vol. 2, p. 91, Pl. 82, Fig. 3
Adams 1761a, p. 29; Pl. 14, Fig. 41, p. 32, Pl. 16, Fig. 47

Inventory No. 1927-1347

P27 Flask with stopper

H 240mm (overall), **D** 100mm

In the experiment described by Adams, this glass flask was placed in the receiver of an air pump and the air was removed. The stopper was attached by soft wax to a wire which passed through a plate on the top of the receiver. The wire was used to place the stopper in the mouth of the flask. When the air was let back into the receiver, the stopper was held tightly, provided that it had been ground to fit the neck well.

This flask and stopper seem to be a decanter with a metal cap fixed over the cut glass stopper.

's Gravesande 1747, Vol. II, p. 80, Pl. 81, Fig. 2
Adams 1761a, pp. 32–3, Pl. 16, Fig. 48
Adams 1799b, Vol. 1, p. 175, Pl. 3, Fig. 7
C 119

Inventory No. 1927-1375

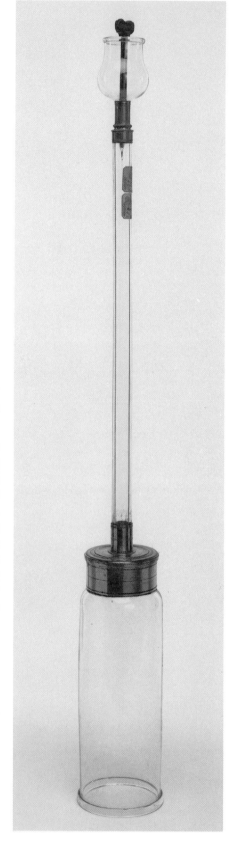

P28 Pair of cohesion plates

D 120mm (assembled), **D** 103mm (disc face), **H** 110mm (one disc)

The *Pneumatics* manuscript mentions 'two brass plates full four inches in diameter' with 'three side buttons', a description which fits these brass plates. However, the manuscript illustration does not match, implying that the drawing was not made from these particular plates. This may be a case where both the design of the apparatus and the illustration were based on the same source—and not on each other—so that differences could arise.

The plates were heated and were held in the smoke of a candle before being pressed together. It then required a large force to separate them, showing the effect of cohesion. Adams suggested that the steelyard P11 might be used to separate the plates.

Hauksbee 1714, p. 19, Pl. 5, Fig. 6
Birch 1744, Vol. 1, p. 45
Desaguliers 1744, Vol. 2, p. 386, Pl. 25, Fig. 13
Adams 1761*a*, p. 33, Pl. 16, Fig. 49
C 120

Inventory No. 1927-1303

P29 Apparatus for filling a barometer tube

H 355mm (glass jar), **D** 124mm (glass jar)

This apparatus was assembled on the plate of an air pump with a funnel at the top of the apparatus over an inverted barometer tube. The funnel was filled with mercury. The air was removed and then the plug was opened so that the mercury was forced through the funnel into the barometer tube. The advantage of this procedure, according to Adams, was that there were fewer air bubbles in the mercury filling the barometer tube than otherwise.

Adams 1761*a*, pp. 33–4, Pl. 16, Fig. 50
C 125

Used with P30, P31

Inventory No. 1927-1309

P30 Bell jar

H 325mm, **D** 130mm

This may have been the bell jar originally supplied by Adams for the apparatus described in the previous entry, P29, for the dimensions given by Adams in his manuscript match this item more accurately than those of the bell jar shown in that entry.

Adams 1761a, p. 33, Pl. 16, Fig. 50
C 129

See P29

Inventory No. 1927-1367

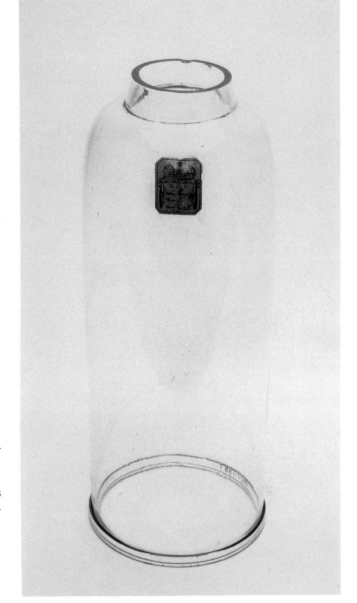

P31 Glass tube with closed end

L 780mm

This is probably the barometer tube used with the apparatus for demonstrating how to fill such a tube, or for demonstrations involving a barometer inside a receiver on the air pump.

's Gravesande 1747, Vol. 2, pp. 24–5
Adams 1761a, Pl. 16, Fig. 50
C 125, 126

Used with P29

Inventory No. 1927-1689/1

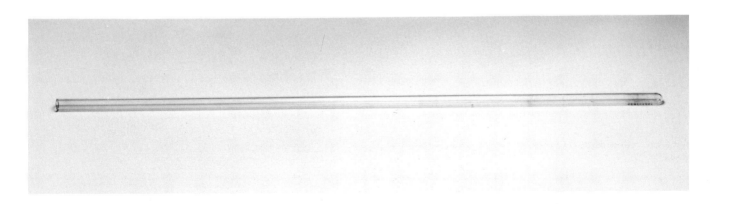

P32 Grooved wheel with five spherical beads

HANDWRITTEN
'50'

D 80mm (disc), **DE** 18mm (disc), **D** 19mm (beads)

This boxwood wheel was used in one of the experiments done on whirling bodies *in vacuo*. The wheel was fixed on a vertical steel spindle which could rotate. As it did so, the amber beads round the wheel were made to rub against two flannel-covered cork pads supported on flexible metal plates. This apparatus was set up on the air pump P1 and covered with a receiver. The air was then removed. The spindle and wheel were made to rotate by a cord driven by a wheel and pulley mechanism fixed to the back of the pump. The rubbing of the amber beads on the flannel pads produced light which could be seen 4 or 5 feet away.

Hauksbee 1719, pp. 24–6, Pl. 5, Fig. 2
Adams 1761*a*, p. 35, Pl. 14, Fig. 43
Adams 1761*b*, Fig. 43

Used with P1

Inventory No. 1927-1766

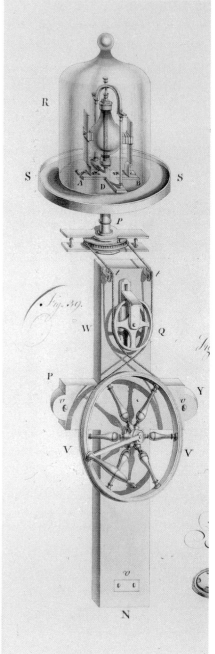

*Adams 1761*a

*Adams 1761*a

P33 Apparatus for comparing different airs

H 395mm (1), **D** 130mm, **H** 355mm (2), **D** 130mm

This apparatus makes up most of what Adams called 'A Convenient Apparatus to examine a portion of Air, taken by chance or choice either in the Atmosphere or in a place filled with vapours or known Exhalations'.

The account given by Adams of the experiments with this apparatus is an abridged translation of a paper by Nollet, the source also for the design of the apparatus.

The central part of the apparatus consists of two glass receivers on a double stand P34 connected to the air pump. These receivers are connected by copper pipes, one to a large receiver and the second to two spherical glass flasks arranged one above the other; the upper flask is now missing. The lower flask contains a glass tube with a perforated glass ball at one end; the other end passes through the brass cap of the flask to the open air. A mahogany board on two brass pillars supports the whole apparatus on the air pump.

One use of this apparatus was to compare airs from different sources. For example, burning matter could provide combustion products to fill the large receiver. The two glass globes were arranged to introduce air from the atmosphere. The lower globe contained distilled rainwater and the other contained cotton or linen which had first been soaked in a strong solution of salt of tartar (potassium tartarate) and then dried. Operating the pump sucked air through the water in the lower globe via the perforated ball and water and then into the upper globe where the air lost any moisture as it passed through the potassium tartarate. In these processes, according to Adams, the air would dispose of many of its

'strange particles' and become purer than if it had been taken straight from the atmosphere.

Nollet 1741, p. 355, Pl. 13, Fig. 13.
Adams 1761a, pp. 35–40, Pl. 19, Figs 55, 56, 57
C 138
Used with P34, P35, P36, P37
Inventory No. 1927-1311/1, 2

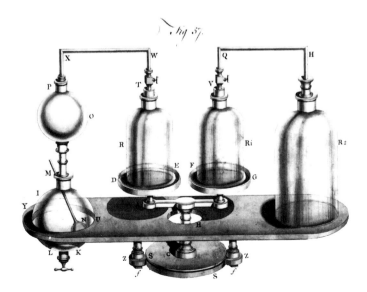

Adams 1761a

P34 Double stand for bell jars

H 180mm (overall height), **W** 335mm (across stand), **D** 215mm (base)

This double stand was referred to as a 'transferror' by Adams.

Hauksbee 1714, p. 18, Pl. 4, Fig. 3
Nollet 1741, p. 355, Pl. 12, Figs 10, 11, 12, Pl. 13, Fig. 13
Adams 1761*a*, pp. 36–40, Pl. 18, Figs 54, 55, 56, Pl. 19, Fig. 57
C 138

See P33

Inventory No. 1927-1316

P35 Spherical flask

D 204mm (sphere), **L** 330mm

Adams 1761*a*, p. 39, Pl. 19, Fig. 57
C 138

See P33

Inventory No. 1927-1333

P36 Bell jar

H 350mm, **D** 230mm (base)

Adams 1761*a*, pp. 35–40, Pl. 19, Fig. 57, Pl. 13, Fig. 37
C 138

See P19, P33

Inventory No. 1927-1334

P37 Mahogany board

L 910mm, **W** 290mm, **H** 126mm

Adams 1761*a*, pp. 38–9, Pl. 19, Fig. 57, Pls 13, 17, Figs 38, 53
C 138

See P33

Inventory No. 1927-1335

P38 Pump plate

D 165mm

This and other pump plates were to be used in many demonstrations to support bell jars and other apparatus.

Adams 1761*a*, p. 37, Pl. 18, Fig. 54

See P39, P40

Inventory No. 1927-1318

P39 Brass disc for air pump

D 132mm (disc)

See P38

Inventory No. 1927-1319

P40 Brass disc for air pump

D 120mm, **H** 5mm

See P38

Inventory No. 1927-1320

P41 Pear-shaped glass bulb

L 365mm

This glass bulb was fixed on top of a receiver and contained a liquid to be introduced into a bowl or other vessel in the receiver. As with other items, Adams took the design from Nollet.

Nollet 1741, pp. 359–60, Pl. 13, Fig. 15.
Adams 1761a, pp. 40–1, Pl. 19, Fig. 58
Adams 1799b, Vol. 1, pp. 175–6, Pl. III, Fig. 21

Inventory No. 1927-1314/1

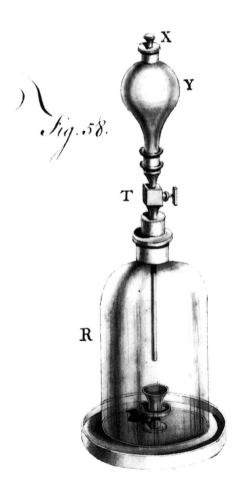

*Adams 1761*a

P42 Stand with two tilting flasks

H 178mm, **L** 176mm, **W** 80mm

This apparatus was used to mix two liquids under a glass receiver from which the air had been removed. Two glass vessels, described by Adams as 'cruetts', are supported in an adjustable brass framework so that they can be tilted to empty out any liquid they contain. The bent brass fork P22 was used to tilt the flasks. This fork was manipulated by a thin brass rod which passed through a 'collar of leathers', part of P19, which acted as a seal. A small brass plate P43 supported a basin to hold the mixed liquids.

Again, Adams's design follows Nollet who had modified an earlier design of Musschenbroek's.

Nollet 1741, pp. 360–1, Pl. 12, Figs 16, 17, 18, 19, Pl. 13, Fig. 21
Adams 1761*a*, p. 41, Pl. 19, Fig. 59
C 137

Used with P19, P22, P43

Inventory No. 1927-1324

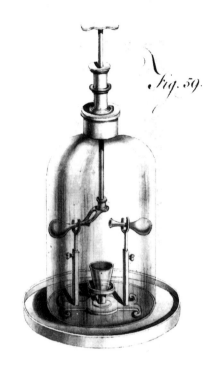

*Adams 1761*a

P43 Circular plate

HANDWRITTEN
'58'

D 72mm

This brass plate was used to support a small glass jar in which liquids or other materials could be mixed in the receiver.

Adams 1761*a*, p. 41, Pl. 19, Figs 58, 59
Adams 1799*b*, Vol. I, Pl. III, Fig. 21

Used with P42

Inventory No. 1927-1314/2

P44 Glass mushroom cover

H 65mm, **D** 115mm

This is perhaps the 'hemispherical glass cover' used in an experiment originally described by Hauksbee. Mercury is dropped into an evacuated receiver on to this glass cover, the mercury '. . . will be broke into very small particles and give a surprising appearance of a shower of Fire . . .'.

Hauksbee 1719, pp. 8–9, Pl. 3, Fig. 3.
Desaguliers 1744, Vol. 2, pp. 388–9, Pl. 25, Fig. 25
Adams 1761*a*, pp. 41, 42, Pl. 17, Fig. 53

Used with P23

Inventory No. 1927-1339

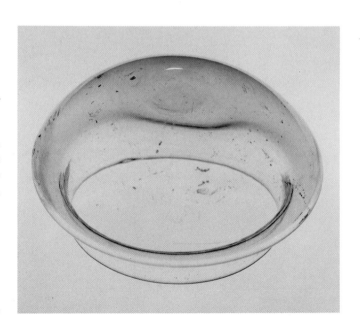

P45 Vacuum fountain

H 176mm (overall), **D** 96mm

This glass globe could be used for demonstrating a fountain. If it was partially filled with water and then air was forced in using a syringe, the air in the globe would be above atmospheric pressure. If the syringe was removed and the tap opened, the water would be forced out of the globe in a fountain.

Hauksbee 1714, p. 6, Pl. 6, Fig. 7
Desaguliers 1744, Vol. 2, pp. 118–19, Pl. 11, Figs 4, 5, 6, 7
Adams 1761a, p. 42, Pl. 20, Fig. 60
C 115

See also P46, L50

Inventory No. 1927-1298/1

P46 Vacuum fountain

H 176mm (overall), **D** 96mm

See P45

Inventory No. 1927-1298/2

P45

P47 Glass globe for collapsing bladder

H 190mm

Adams followed Desaguliers in describing this apparatus. It consisted of a bladder, which is now missing, attached to a small pipe fixed inside a glass globe. Air can enter or leave the bladder via this pipe and a nozzle on top of the globe. When the globe is placed under a receiver and the air is removed, the bladder is also emptied of air. However, because air is trapped between the bladder and the glass globe, the bladder contracts. Similarly, if air is let back into the receiver and so into the bladder, the bladder expands.

There is a small hole in the foot of the apparatus to let the air escape which would otherwise be trapped there. This suggests that the apparatus was made to be used with an air pump as Adams describes and not with a syringe which was an alternative arrangement.

Adams describes this device as demonstrating what happens to the lungs of an animal when one is put under a receiver on an air pump and the air removed. Years later, George Adams junior stated that this apparatus was 'improperly' called a lungs glass because it was not a demonstration of the way lungs work.

Birch 1744, Vol. 1, p. 75
Desaguliers 1744, Vol. 2, pp. 380–1, Pl. 25, Fig. 7, p. 405
Adams 1761*a*, p. 42, Pl. 20, Fig. 60
Adams 1799*b*, Vol. 1, pp. 171–2, Pl. 3, Fig. 9
C 129

See E77, E78

Inventory No. 1927-1381

P48 Thin-walled bottles

H 80mm (approx.), **W** 52mm (approx.), **H** 100mm (with brass caps)

These phials are made of glass in the form of square-sided or 'case' bottles with burnt-off necks. They are designed for demonstrations showing the effect of pressure. Most of the bottles are sealed. If one of these was placed on the pump plate under a receiver and the air was pumped out, the bottle eventually exploded, an effect of the pressure of the air inside the bottle. A similar demonstration was carried out with the bottle under water. A third demonstration employed a bottle with a valve arranged to let out air as the receiver was exhausted. If air was let back in the receiver, it could not enter the bottle which would eventually implode.

The bottles were enclosed in wire cages to prevent fragments of them hitting the glass receiver and causing it to break. Such wire cages, though illustrated by Adams, have not survived. However, D42 are similar.

Desaguliers 1744, Vol. 2, p. 395, Pl. 24, Figs 8, 9
Adams 1761*a*, pp. 42–3, Pl. 20, Figs 61, 62
C 117

Inventory No. 1927-1294

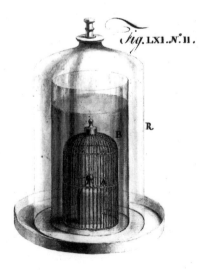

Adams 1761a

P49 T-shaped brass stand with valve

ENGRAVED—ARMS—ENDS
  'A', 'B'
ENGRAVED—STOPCOCK
  'A', 'B', 'Open'
L 480mm, **H** 370mm

This apparatus used bladders (now missing) to illustrate muscular motion. One arm was connected to a chain of three bladders with a weight at the end. The other arm had one large bladder and the same weight attached. Fixed to a plate on the air pump, the pump could remove air from the single bladder, or from the chain, or from both together by turning the stopcock. Adams explained that a smaller quantity of air in the chain of bladders was able to raise the weight through the same distance as a larger quantity of air in the single bladder. In his account of this experiment, Adams refers to both Desaguliers's and Hauksbee and Whiston's descriptions. However, Desaguliers was uncertain that comparing muscular motion to a chain of bladders was valid, saying that it illustrated this motion but was not a 'Demonstration' of it.

Hauksbee 1714, p. 6, Pl. 6, Fig. 5
Birch 1744, Vol. 3, p. 13
Desaguliers 1744, Vol. 2, p. 392, Pl. 24, Fig. 7
Adams 1761a, pp. 43–4, Pl. 21, Fig. 63

Inventory No. 1927-1328

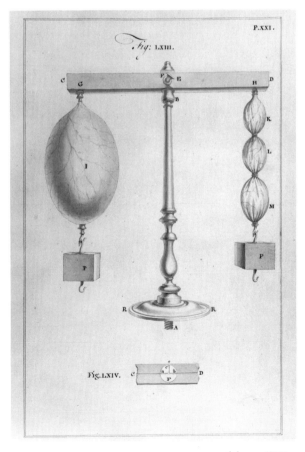

Adams 1761a

P50 Glass condenser

D 210mm (pump plate), **H** 205mm

This seems to be the 'Glass condenser' mentioned but not illustrated by Adams in the *Pneumatics* manuscript. It consists of a cylindrical vessel of thick colourless glass. One end has a metal collar to which a brass base plate can be screwed. A leather ring helps to improve the seal.

Adams suggests that either this or the brass condenser P13 could be used for experiments such as the air fountain or air gun which Adams called wind fountain and wind gun. A basin of water was placed in the condenser and then the air pump would be used to compress more air into the condenser. The cock P53 was used to retain the air. The water joint P51 was attached and the cock opened to release the water.

Desaguliers 1744, Vol. 2, pp. 396–8, Pl. 24, Figs 8, 12
Adams 1761*a*, pp. 44–5
C 129

Used with P51, P53

Inventory No. 1927-1397

P51 Arm for air fountain

L 290mm, **W** 40mm, **DE** 27mm

This arm was used for an air or 'Wind' fountain. A quadrant for use with this arm is attached to P54. The arm was attached to one of the condensers P13, P50. If the condensers contained water and air under pressure, a jet of water would be forced out through this arm. The arm is jointed to allow the angle of the jet to be varied to demonstrate the motion of projectiles.

Adams 1761a, pp. 44–5, Pl. 22, Fig. 65
C 111

Inventory No. 1927-1340

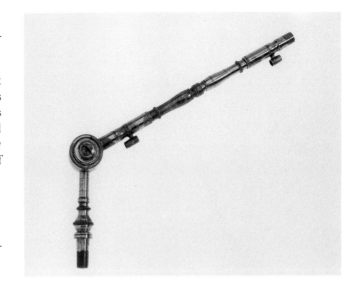

P52 Pipe with tap

L 1200mm

One end of this copper pipe was attached to the underside of a pump plate on which stood a tall glass receiver. The air was removed from that receiver. The lower end of the pipe was placed in a vessel full of water and the brass tap opened. The water then rose through the pipe into the receiver with great force. The pipe is hinged to make it more convenient to use.

Desaguliers 1744, Vol. 2, p. 384, Pl. 25, Fig. 14
's Gravesande 1747, Vol. 2, p. 31, Pl. 71, Fig. 2
Adams 1761a, p. 20, p. 44, Pl. 10, Fig. 27

Used with parts of P15, P60

Inventory No. 1927-1341

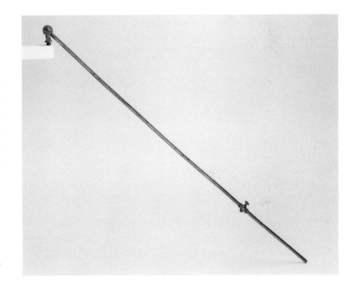

P53 Six sections of brass pipe with one plug

L 270mm, **L** 145mm

These are accessories for use with the rest of the pneumatics apparatus.

Adams 1761a, p. 44, Pl. 22, Fig. 65

Used with P50

Inventory No. 1927-1906

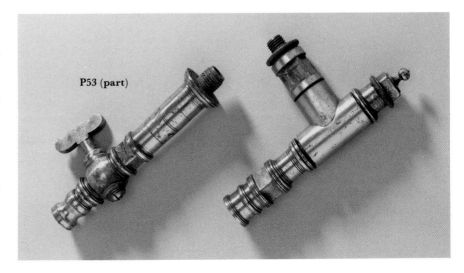

P53 (part)

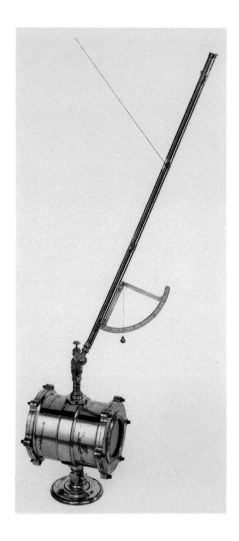

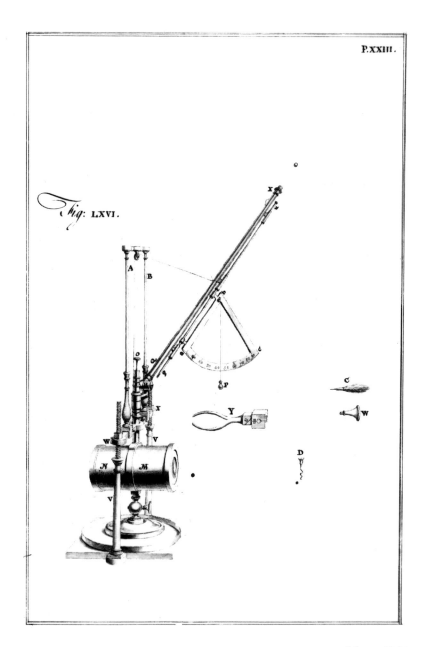

*Adams 1761*a

P54 Barrel of air gun

SCALE
15 (1) 90°

L 945mm, **H** 220mm, **D** 16mm

This tube was attached through a valve to the brass condenser P13. The tube could be set at a particular elevation using the quadrant and plumb-line and was held in position by a string fixed to a cross-piece P55 on the pump.

In use, the air in the condenser was released by pushing down the spring-loaded valve. A pipe-clay or lead pellet, or a dart P59 could be fired in this way.

Adams 1761*a*, pp. 45–6, Pl. 23, Fig. 66
C 111

See P13, P55, P56, P57, P58, P59. The quadrant was also used with P51

Inventory No. 1927-1343

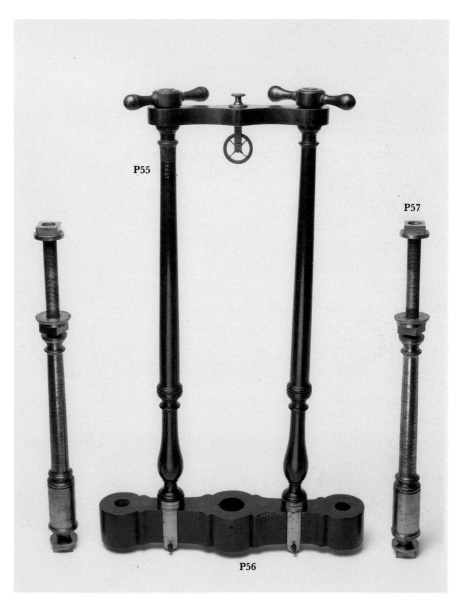

P55 Two pillars

L 735mm, **W** 150mm

These are to support the air gun, P54.

Adams 1761*a*, p. 45, Pl. 23, Fig. 66

Inventory No. 1927-2037

P56 Cross-piece for an air pump

L 440mm, **DE** 60mm, **W** 90mm

Though not identical to the cross-piece illustrated by Adams, the mahogany cross-piece is certainly meant as part of the air gun apparatus, though it may be a replacement.

Adams 1761*a*, p. 45, Pl. 23, Fig. 66
C 111

Used with P45

Inventory No. 1927-2058

P57 Two pillars

L 530mm, **D** 45mm

These brass pillars supported the air gun apparatus, P54, above the brass condenser, P13.

Adams 1761*a*, pp. 44–6, Pl. 1, Fig. 1, Pl. 23, Fig. 66
C 111

Inventory No. 1927-2038

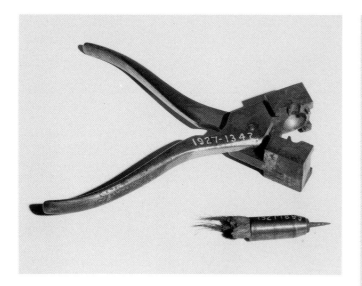

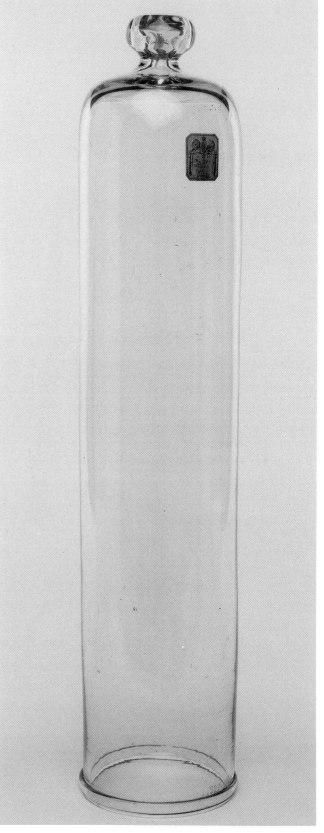

P58 Mould for clay or lead pellets

L 147mm, **W** 28mm

Adams 1761*a*, p. 46, Pl. 23, Fig. 66

See P54

Inventory No. 1927-1342

P59 Dart

L 65mm, **D** 11mm

This dart has a brass body, a steel point, and a fur flight. It was fired from air gun P54.

Adams 1761*a*, p. 46, Pl. 23, Fig. 66
C 111

Used with P54

Inventory No. 1927-1899

P60 Bell jar

H 605mm, **D** 127mm

This tall glass receiver was used for demonstrations of a water fountain and mercury fountain, and to measure the elasticity of 'matter'.

Adams 1761*a*, p. 15, Pl. 4, Fig. 14, Pl. 5, Fig. 15, p. 20, Pl. 10, Fig. 27, p. 21, Pl. 10, Fig. 28
C 126

See P7, P23, P31, P52, P61

Inventory No. 1927-1390

P61 Pump plate

D 220mm

This flanged brass pump plate has a hollow central pipe, threaded at its end to connect it to the air pump P1. A second threaded hole allowed a pipe such as P23 to be attached also. This pump plate has a turned mahogany stand.

C 135

Used with P19, P60

Inventory No. 1927-1329

P62 Disc

D 145mm, **T** 5mm, **D** 20mm (hole)

This brass disc has a central hole. It is an accessory for pneumatics experiments.

Adams 1761a, p. 25, Pl. 12, Fig. 36

Inventory No. 1927-1317

P63 Four plugs

L 75mm, 80mm, 80mm, 45mm

These brass plugs are accessories for the pneumatics apparatus.

C 127

Inventory No. 1927-2048

MECHANICS

M1 Syringe

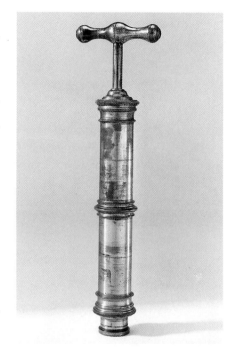

L 345mm, **D** 24mm

The syringe is the first instrument to be specified in the *Mechanics* manuscript. Before introducing the Philosophical Table, Adams presents a small section on 'Body', closely following 's Gravesande and the ideas set out by Locke. Extension, solidity, divisibility, and mobility are discussed, as well as the action of gravity upon bodies and their attraction to each other, a property we would now call capillarity. Although 'the air in which we live almost always escapes our sight and touch', the resistance of the piston can be made to demonstrate its 'solidity'.

The simple brass syringe is almost identical to that in the first illustration in the *Mechanics* manuscript, although the handle is slightly different. Adams does use a syringe in the *Pneumatics* manuscript but the description is not compatible with this instrument.

Adams 1761*a*, pp. 15–17, Pl. 5, Fig. 16
Adams 1762, p. 2, Pl. 1, Fig. 1

Inventory No. 1927-1325

M2 Apparatus to demonstrate 'attraction of bodies'

L 232mm, **W** 168mm, **H** 215mm

The experiment demonstrates what is now known as surface tension. Adams's apparatus is similar to that of 's Gravesande. The two plates of glass are held vertically in a mahogany frame which is placed in a basin of coloured water. The plates touch at one side but are held slightly apart at the opposite side by a mahogany wedge, the original ivory wedge having been lost. The water rises between the plates to form a hyperbola. It is likely that the basin M3 was used with this apparatus.

The use of vertical glass plates to demonstrate the 'attraction of bodies by capillarity' was common during the eighteenth century. Together with capillarity tubes and the horizontal 'oil of oranges' experiment they formed part of the introduction of many lecture courses which usually concerned 'Body in General'. Whiston attributes this particular experiment to Hauksbee who, he claims, devised it as an improvement on the tubes. Adams included all three experiments in the *Mechanics* manuscript but the tubes have not survived.

Hauksbee 1714, p. 18
Desaguliers 1734, p. 13, Pl. 2, Fig. 2
Helsham 1739, p. 4, expt. 3, Fig. 1

's Gravesande 1747, Vol. I, p. 19, Pl. 3, Fig. 4
Adams 1762, p. 3, Pl. 1, Fig. 5
Crommelin 1951, No. G.M.1
Turner 1973, No. 124

Inventory No. 1929-116

M3 Glass bowl on whirling stand

second half eighteenth century or first half nineteenth century

D 276mm (overall), **H** 110mm, **D** 240mm (bowl), **DE** 58mm (bowl)

This is very likely to be the glass vessel used by Adams in the above experiment since the proportions and dimensions are exactly right. It would have been filled with coloured water. However, the bowl has been set in a mahogany frame at a later date. The stand can be rotated by hand and the apparatus could have been used to show the effects of centrifugal force on the surface of liquids.

Adams 1762, p. 3, Pl. 1, Fig. 5

See also M2   **Inventory No.** 1927-1875

M4 Apparatus for 'oil of oranges' experiment

L 250mm, **W** 185mm, **H** 140mm

Adams copied this apparatus directly from 's Gravesande who featured it in his first chapter 'Of Body in General' Although 's Gravesande used water or oil, Adams stipulated oil of oranges; hence the name given to the apparatus in our collection.

A glass plate is placed on a horizontal glass plate and held away from it at one end by a coin or similar object. A drop of oil of oranges placed between the plates moves towards the end where the plates are together. This end can be raised by a mahogany screw so that the force due to capillarity is balanced by the force due to gravity and the drop is brought to a standstill. The experiment showed 'attraction in bodies' since the drop of oil was attracted to the glass plates. The glass plates are not original. The stand is made of mahogany.

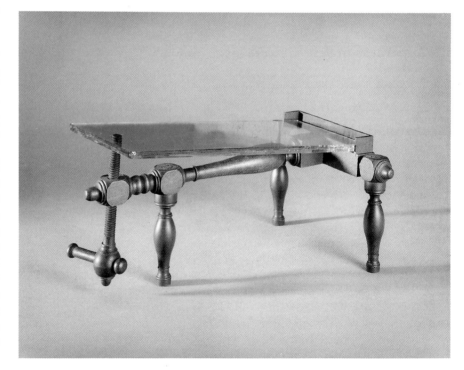

Desaguliers 1744, Vol. I, p. 11, Pl. 1, Fig. 5; p. 37, note 16
Adams 1746, No. 104
's Gravesande 1747, p. 19, Pl. 1, Fig. 4
Adams 1762, p. 3, Pl. 2, Fig. 6
C 43   **Inventory No.** 1927-1105

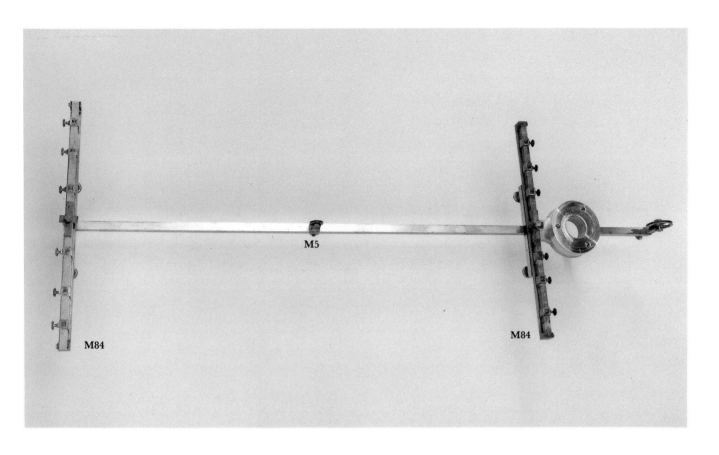

M5 Cross-bar attachment for the pillar of the Philosophical Table

L 975mm (bar), **W** 38mm (central hole), **W** 152mm

Adams described this apparatus as the 'iron arm of the great pillar'. It consists of an iron rod which extends in both directions from the pillar and carries two cross-pieces, both on the longer arm, one close to the pillar and one further removed. Two brass pulleys are attached to the upper surface, one at the end of the shorter arm and one halfway along the longer arm. The two cross-pieces each carry two hooks. Cradles containing magnets were suspended between them for experiments on magnetic 'effluvia'. It was also used for the collision of bodies and 'innate' forces. In the latter experiment the two rulers M84 were attached to the cross-pieces, as shown in the photograph.

's Gravesande 1747, Vol. 1, p. 179, Pl. 25, Fig. 4, p. 116, Figs 194–5
Adams 1762, p. 6, Pl. 3, Fig. 11, p. 17, Pl. 7, Fig. 37, p. 93, Pl. 46, Fig. 151. p. [117], Pl. 60, Figs 194, 195
C 24

Inventory No. 1927-1221/1

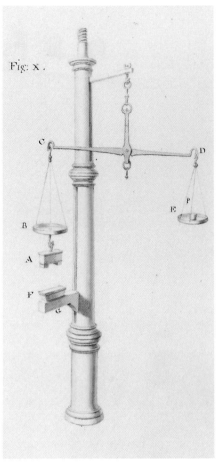

Adams 1762

M6 Lodestone in silver case with keeper

ON MAGNETS
'N', *'S'*
CREST ON CASE
'G III R'

L 83mm (lodestone), **W** 50mm (lodestone), **H** 120mm (lodestone), **L** 99mm (keeper), **W** 21mm (keeper), **H** 85mm (keeper)

At the end of an introductory section based on 's Gravesande's first chapter, Adams inserted a few pages on magnetism as a further example of the attraction of bodies. He does not specifically describe a silver-cased lodestone in the text of the *Mechanics* manuscript, which may indicate that it was made after 1762. However, a sketch for the illustrations shows a similar instrument. The lodestone was used to demonstrate and weigh magnetic attraction following van Musschenbroek, and to attempt to verify the inverse square law of distance for magnetic attraction, about which there was some controversy at the time.

The silver lodestone was suspended from one arm of a balance which was attached to an arm of the pillar of the Philosophical Table. The keeper was screwed to another arm of the Philosophical Table below. Weights were placed in a scale-pan at the other end of the balance to counteract the magnetic attraction. Lodestones were still used in courses of experimental philosophy, although their use in making compass needles and experimenting with magnetism was being superseded by artificial magnets by the middle of the century. There is a large lodestone in the Oxford Museum for the History of Science weighing about 170 lb and capable of lifting the same.

Brook-Taylor 1721, p. 206; pp. 204–8
Desaguliers 1734, p. 40
Helsham 1739, p. 19, lect. 2, expt. 6
Adams 1762, p. 6, Pl. 3, Fig. 10, p. 94
C 261
Chew 1968, Item 11
Heilbron 1979, pp. 87–90

See also D126

Inventory No. 1929-113

M7 Six bar magnets in case with keepers

L 269mm (each magnet), **W** 19mm (each magnet), **H** 12mm (each magnet), **L** 318mm (box), **W** 155mm (box), **H** 33mm (box)

After the above demonstration Adams describes 14 experiments to be undertaken with a 'magnetical apparatus', the central feature of which were these bar magnets. Since he does not give a reference for the experiments, it is possible that he devised them himself.

There are six steel bars each $10\frac{1}{2}$ inches long with two keepers half their length placed across their ends in a mahogany box. The north poles are marked with a double line. Adams gives Canton's method for restoring the magnetic 'effluvia' by stroking from the middle to the ends of each bar with opposite poles of two other bars. The magnets were used to show magnetic attraction and repulsion, demonstrate the 'effluvia' using iron filings on a plate of glass, influence the magnetism of a small lodestone, support a chain of soft iron balls, and magnetize both 'hard' steel and 'soft' iron compass needles.

The last was a significant demonstration since the methods of making artificial magnets had been kept secret by Gowin Knight in order to protect the trade in compass needles. Only when the methods were revealed in 1750 could artificial magnets of reasonable strength become readily available. As Michell pointed out, they had many advantages over lodestones: cheapness, abundance, strength, shape, and restorability. When Gowin Knight obtained a patent for compass needles in 1766, Adams made them for him. Adams made some magnetic compasses for the Office of Ordnance between 1748 and 1773.

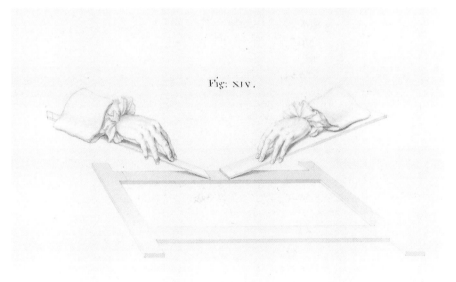

Fig: XIV.

Adams 1762

Knight 1744
Michell 1750, pp. 4–7
Canton 1751, pp. 35–38, tab. 2, Figs 2, 3, 4, 5

Adams 1762, p. 6, Pl. 3, Fig. 11; p. 7, Pl. 3, Fig. 12; Pl. 4, Figs 13, 14
C 265
Knight 1766
Hitchins and May 1955, pp. 27–30
Millburn 1988, pp. 258, 261

Inventory No. 1929-98

M8 Two cradles for magnets

L 203mm, **W** 23mm, **H** 36mm

These cradles were used with the iron arm of the great pillar M5, and a pair of bar magnets from the set M7. They are simply rectangular pieces of mahogany with brass straps and hooks at each end.

The magnets were suspended and shown to attract one another and a piece of unmagnetized iron. The repulsion of like poles was also demonstrated.

Adams 1762, p. 6, Pl. 3, Fig. 11

Inventory No. 1927-1234

M9 Single pulley

H 133mm, **D** 44mm, **W** 36mm

This is the pulley 'whose box turns round its axis'. It is made of brass with a steel axle. The tail fits through square openings such as that at the end of the arm M71 and the collar is pressed against the arm by the screw. When fixed vertically the pulley can have 'any vertical situation'. It is impossible to be precise concerning its use because the type of pulley required is not usually specified in the text.

's Gravesande 1747, Vol. I, p. 33, Pl. 4, Figs 1, 2
Adams 1762, p. 14, Pl. 15, Fig. 15

Used with M42, M43, M44, M71

Inventory No. 1927-1800/1

M10 Single pulley

H 98mm, **D** 43mm, **W** 35mm

Originally there were four 'pullies with square tails' but this appears to be the only one to have survived. It is simply a single pulley in brass with a square tail, collar, and screw. As with all the Adams pulleys, the axle is steel. This type of pulley had many uses: on the transverse arm M38, with the bent lever M60, with the inclined 'horse-way' M54, and with the experiment on oblique powers M18/53.

's Gravesande 1747, Vol. I, p. 34, Pl. 4, Figs 3, 4
Adams 1762, p. 15, Pl. 5, Fig. 16; p. 51, Pl. 29, Figs 105, 109

Used with M42, M43, M44, M71

Inventory No. 1927-1800/2

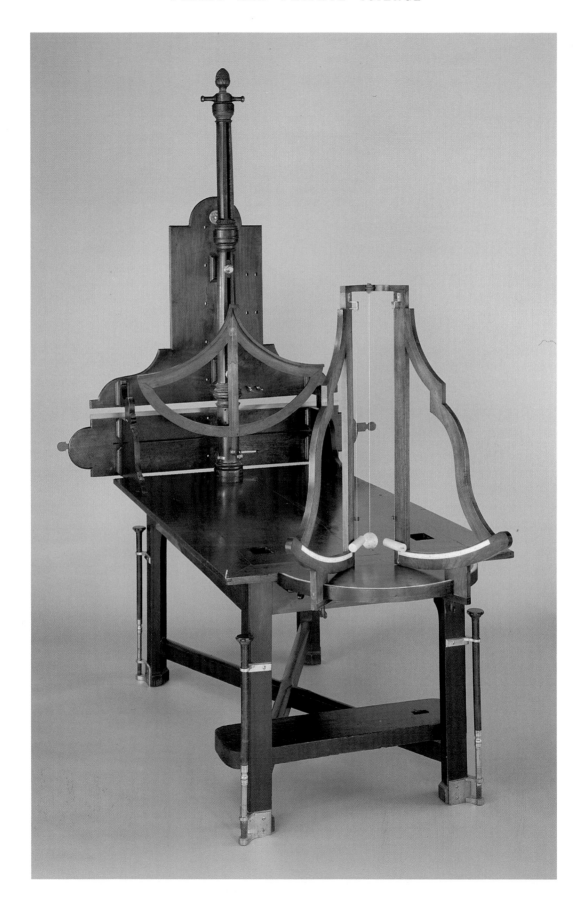

M11 Philosophical Table

SCALE
24(1)0(1)21

L 1510mm, **W** 910mm, **H** 751mm (not including pillar),
H 975mm (backboard (upper)), **W** 1070mm (backboard
(upper)), **W** 1350mm (backboard (lower)), **H** 240mm
(backboard (lower)), **H** 2350mm (including pillar),
D 120mm (pillar)

The 'Philosophical Table' is the centrepiece of the collection
of mechanical instruments made by Adams, as the air pump
is the centrepiece of the pneumatical instruments. It was
designed to carry many experiments: for example, the
inclined horse-way, the path of a projectile, the cycloidal
channels, and the collision of bodies. The pillar provided a
means of suspending magnets, balances, and pulleys, and the
backing board was used for experiments on impact and
pendulums. Although Adams copied extensively the instruments made by van Musschenbroek for 's Gravesande,
including the pillar and backing board, the table was a major
departure. It allowed many experiments to be brought
together which were separate pieces of apparatus in
's Gravesande's collection, notably the central forces table
and the stand for pendulums.

Adams described it as a table fitted for making many
experiments or the 'great table'. He gave the dimensions
including the fact that the wood for the table top is one inch
thick. Each leg has a levelling screw in brass, otherwise the
table and pillar are entirely of mahogany. There is a
decorative 'nutt' resembling a pineapple on the top of the
pillar. The backing board which is 20mm thick was used for
experiments on pendulums, while the 'mahogany plane'
below was used for the collision of bodies and 'innate' forces.
There is a brass strip along the top of this plane and an ivory
scale, marked in inches each way from the midpoint, which
slides horizontally across the centre. Various table accessories
appear under other inventory numbers. Plate 7 in the
Mechanics manuscript, in which the pillar and its accessories
are illustrated, is signed and dated 'G. Adams. Jan 5 1761'.

The table was probably the first of its kind to have been
made since it was drawn in such detail in the *Mechanics*
manuscript. Adams described a 'moveable table for various
experiments' in about 1746 but this was probably a less
ambitious piece. There also appears to have been a similar if
smaller table in Lord Bute's collection which was also used
with apparatus designed by 's Gravesande. Neither George
Adams nor his son advertised a table such as this in their
catalogues during the second half of the eighteenth century.

Adams 1746, No. 149
's Gravesande 1747, Vol. I, Pl. 4
Adams 1762, pp. 15–17, Pl. 6, Fig. 25; Pl. 7, Figs 26, 35, 39

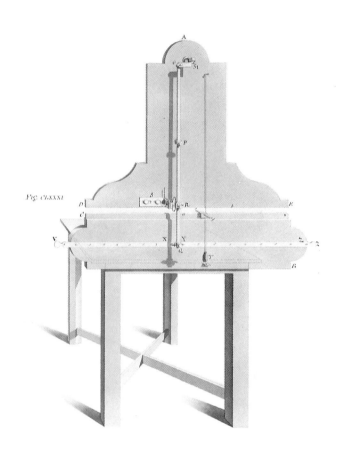

Adams 1762

C 1
Turner 1967, Item No. 237
Chew 1968, Item 1

Used with M54, M68, M70, M73, M75, M77

Inventory No. 1927-1101

M13 Balance beam and fulcrum, on plinth base

'Geo Adams Fleet Street London'
SCALE
100 (10) 0 (10) 100

L 955mm (arm), **D** 150mm (base),
H 280mm

This is the second instrument introduced by Adams to deal with the principles of the balance. The first was a wooden balance with two scale-pans held suspended in the conventional manner which does not appear to have survived.

This balance is of brass; each arm is 19 inches long and divided into 100 parts with every 10 numbered from the centre. It was placed on a sliding arm of the pillar of the Philosophical Table rather than suspended. Weights could be attached by means of brass loops with sharp edges which could be placed on the divisions, or by means of scale-pans. A set of 10 cubical weights with loops and scale-pans were provided to be used with the balance. Each weight together with the loop or pan weighed 1oz; hence the law of the balance could be demonstrated with minimum calculation, following 's Gravesande. The balance was also used to demonstrate that acting powers at right-angles are most effective, Helsham's property of the balance, the 'quadrature' or area of a parabola mechanically, and the action of a screw. Sometimes it was referred to as 'the great balance'.

M12 Attachment for the pillar of the Philosophical Table

L 315mm, **W** 90mm (wooden bracket), **DE** 28mm (wooden bracket), **H** 113mm (with hook)

The attachment simply consists of a mahogany ring for fixing to the pillar and an arm with a large brass hook. The attachment was used to suspend the balance in the experiment to weigh magnetic attraction, to suspend the steelyard, and for experiments on inclined forces. It could be placed at the top of the pillar under the 'nutt' or at the joint part way down.

Adams 1762, Pl. 3, Fig. 10; p. 16, Pl. 7, Fig. 27; Pl. 37, Fig. 124, No. 2

Used with M11

Inventory No. 1927-1219

By the mid-eighteenth century the law of the balance was ubiquitous in lecture courses on experimental philosophy, usually preceding the five simple machines.

Desaguliers 1744, Vol. I, pp. 148–51, Pl. 14
's Gravesande 1747, Vol. 1, p. 37, Pl. 5, Fig. 1

Adams 1762, p. 17, Pl. 9, Fig. 41, Pl. 11, Fig. 46; p. 20, Pl. 10, Fig. 47; p. 58, Pl. 31, Fig. 113, p. 27, Pls. 15–18
C 16, 19

Used with M15, M19, M30, M46

Inventory No. 1927-1119

M14 Four cubical weights with loops

ENGRAVED
'1 oz T'

L 15mm (side of weight)

These are four out of the ten cubical brass weights and loops which Adams describes for use with the 'great balance' M13. Together each loop and weight weighs 1 oz Troy. They were used to demonstrate the law of the balance with minimum calculation and in Helsham's property of the balance.

Adams 1762, p. 18, Pl. 9, Fig. 41; p. 21, Pl. 11. Fig. 47
C 62

Inventory No. 1927-1863

M15 Ten circular scale-pans

D 41mm (each)

These are the ten scale-pans designed to be used with the balance above. They are made of brass.

Adams 1762, p. 18, Pl. 9, Fig. 41

Used with M13, M14, M37

Inventory No. 1927-1896

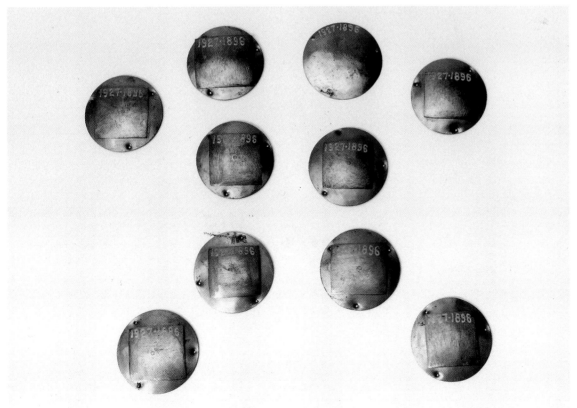

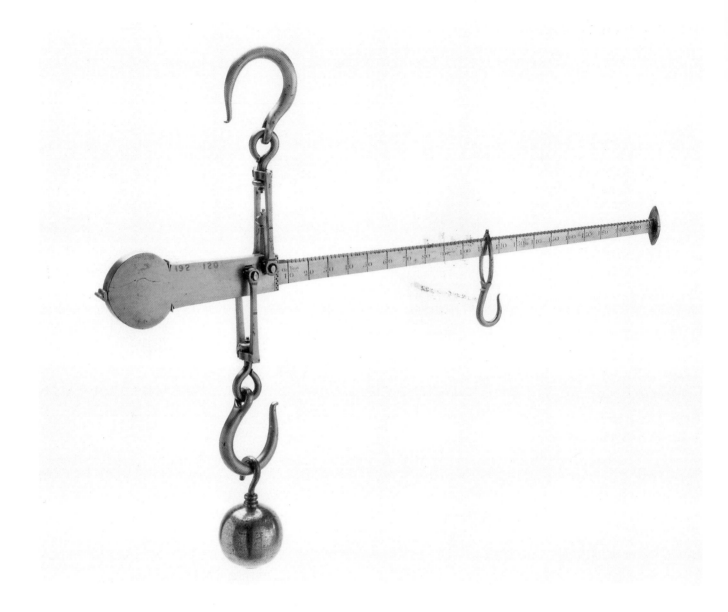

M16 Roman type steelyard

SCALE
10 (2) 210

L 515mm (steelyard), **W** 34mm, **H** 345mm, **D** 53mm (weight), **H** 87mm(weight)

The demonstration of the Roman balance or steelyard followed the law of the balance in Adams's mechanics course, as in 's Gravesande's. However, Adams departed from 's Gravesande's dimensions and made an instrument which measured to 2 oz rather than 1 oz accuracy and could not weigh below 10 oz. The steel arm is marked from 10 oz to 210 oz in 2 oz divisions; the brass counterweight can be hung at each division. The scale-pan which hung from the short arm is now missing, so the instrument is not in equilibrium.

Steelyards were 'of frequent use among us', according to Whiston, and often used in courses of experimental philosophy in addition to the demonstration of the balance.

Hauksbee 1714, Pl. 2, Fig. 2
Desaguliers 1734, p. 92, lect. 3, Pl. 9, Fig. 7
Helsham 1739, p. 88, lect. 6, expt. 8
's Gravesande 1747, Vol. I, p. 38, Pl. 6, Fig. 4
Adams 1762, p. 18, Pl. 9, Fig. 42
C 49
Turner 1973, Item No. 3

Inventory No. 1927-1205

M17 Fixed weights for experiments with the balance

L 101mm, **W** 10mm, **H** 143mm

This is a fixed weight which can be attached to the balance beam M13. Adams described it as 'a mahogany parallelo-piped . . . 4 1/2 inches square and 1/2 inch thick having a hollow tail with a screw to make it fast to the beam'. The tail has a slot through which the beam can be passed and a small boxwood screw holds it in place. It was used to show how the turning moment of a force varied according to its position and the position of the beam with respect to the horizontal. The weight can be attached above or below the beam.

Desaguliers 1744, Vol. I, p. 149, Pl. 14, Figs 6, 7
Adams 1762, p. 19, Pl. 9, Figs 41, 43

Used with M13

Inventory No. 1927-1242

M18 Transverse arm for use on the Philosophical Table

L 1095mm, **W** 13mm, **H** 50mm

There are two transverse arms mentioned in the *Mechanics* manuscript. This one is simply referred to as the 'longest transverse arm' and is not described in detail or illustrated separately. It consists of two pieces of mahogany about $\frac{1}{2}$ inch thick held $\frac{1}{4}$ inch apart and joined at the ends. It can be fixed to the pillar of the Philosophical Table with a mahogany attachment into which it fits. The attachment can be screwed into the pillar with a brass collar and screw. This apparatus was used for experiments on oblique forces when pulleys with flat tails were screwed into the slit between the mahogany pieces. These pulleys may not have been made, since they were drawn, but not mentioned, in the text.

Desaguliers 1744, Vol. 1, p. 143, Pl. 14, Fig. 5
Adams 1762, p. 20, Pl. 10, Fig. 46

See also M38

Inventory No. 1927–1912/1

M19 Apparatus for Helsham's property of the balance

L 35mm (plate), **W** 22mm (plate), **H** 180mm (plate), **L** 465mm (pipe), **D** 10mm (pipe), **L** 400mm (rod), **W** 30mm (rod)

Having discussed the law of the balance, the steelyard, and the false balance, Adams treats Helsham's property of the balance with this apparatus. Helsham reported 'a property somewhat singular and surprising, though it has not, that I can find, been taken notice of by any of the mechanick writers'. He noticed that if a man stood on one scale-pan of a balance he could make himself heavier by pressing upwards on the arm of the balance between the fulcrum and the point of his suspension. Also he could make himself lighter if he pressed upwards on the arm further from the fulcrum than his point of suspension. In order to illustrate and account for this, Adams used the brass balance beam M13, the stand M20, and these three pieces of apparatus.

They are a brass rod with two wheels at one end, a flat piece of brass which can be screwed into the stand, and a 'pushing pipe' which is a long brass helical spring. The man is represented by the brass rod which is suspended from the balance. Its weight is counteracted by the cubical weights. The 'pushing pipe' is attached to a hole at the base of the rod and a hole under the '50' division of the balance arm. To stop the pipe pushing the rod out of its perpendicular position the vertical plate is fitted into the top of the stand so that the rod can move freely up and down against it using the wheels. The larger cubical weights M37 were placed on the base of the stand to keep it stable.

Equilibrium was first reached with the rod alone; then the pushing pipe was attached and equilibrium re-established by adding extra weights. A force upwards halfway along the balance produces an equal and opposite force downwards at the other end of the spring. This downward force is further

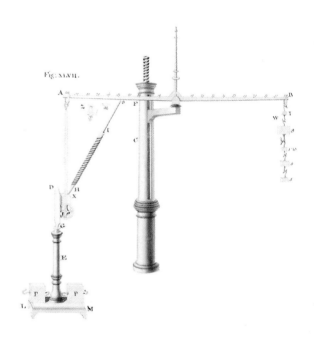

Adams 1762

from the fulcrum than the upward force and therefore has greater effect.

Helsham 1743, p. 91, Pl. 2, Fig. 10
Desaguliers 1744, Vol. I, p. 152, Pl. 14, Fig. 10
Adams 1762, pp. 20–1; Pl. 11, Fig. 47
C 18

Used with M13, M20, M37

Inventory No. 1927-1217

M20 Stand

H 355mm, **W** 149mm, **L** 203mm

The 'small pillar' was intended for use in a variety of experiments in Adams's mechanics course. It appears only in the illustrations not copied from 's Gravesande in which Adams draws instruments of his own design. It is made of lignum vitae on a rectangular mahogany base. There is a flat brass top with an inside screw thread which took a number of fittings. It was used to balance plane figures for centre of gravity experiments, hold the vertical board in Helsham's property of the balance, hold the fulcrum in experiments on the bent balance, and in experiments on oblique forces and on Marriott's mechanical paradox.

Adams 1762, p. 21, Pl. 11, Fig. 47; p. 22, Pl. 11, Fig. 48; p. 38.5, Pl. 22, Figs 85, 86
C 12

See also J17

Used with M36

Inventory No. 1927-1121

M21 Four solids for centre of gravity experiments

L 75mm (parallelepiped), **W** 25mm (parallelepiped), **H** 25mm (parallelepiped), **L** 115mm (long cylinder), **D** 38mm (long cylinder), **L** 57mm (short cylinder), **D** 38mm (short cylinder), **H** 77mm (octagon), **L** 20mm (side octagon)

This group of four boxwood solids does not quite correspond to the group of solids for centre of gravity experiments which Adams described: a cube of 2 inch sides is now missing and the cylinders which survive were not mentioned. However, the octagonal prism is illustrated, and the parallelepiped is both illustrated and described as being 1 inch square and 3 inches long. The other solids are very similar and were probably also used in centre of gravity experiments. In particular, Adams used the cube and octagonal prism to show that the former slid but the latter rolled when placed on a plane inclined at $21\frac{1}{2}^{\circ}$ to the horizontal.

Similar experiments on the centre of gravity were performed by many lecturers in the mid-eighteenth century.

Desaguliers 1744, Vol. I, p. 46, Pl. 4, Fig. 1; Vol. I, p. 62, Pl. 5, Fig. 7

's Gravesande 1747, Vol. I, p. 40, Pl. 5, Figs 3, 4
Adams 1762, p. 22, Pl. 12, Figs 49, 50; p. 87, Pl. 42, Fig. 142

Inventory No. 1927-1207/1

M22 Rolling double cone and arms

L 910mm, **H** 70mm (max), **T** 7mm (rulers), **L** 278mm (brass rod), **L** 300mm (cone), **D** 122mm (cone max)

The dimensions and material given by Adams for his rolling double cone correspond to those of this example. The double cone is made of lignum vitae; it is 1 foot long and has a maximum diameter of 5 inches. The two arms are made of mahogany and are hinged together at one end so that the horizontal angle between them can be varied. The inclination of the arms can also be varied by means of the brass levelling screws at the open ends. A steel pin supports the closed end. A brass bar can be fitted between the arms or 'rulers' to maintain an angle necessary for the paradox to be effective. Two small brass stops prevent the cone from falling off and being damaged.

If the open ends of the arms are slightly raised and the double cone is placed at the closed end, the cone will appear to move uphill. As the point of support of the cone rises its centre of gravity descends. The paradox of the rolling double cone was probably first described in Leybourn's *Recreations of Divers Kinds* in 1694 by a certain J.P. The mathematics is quite advanced and would not have been treated.

Leybourn 1694, Mechanical recreations, pp. 12, 13
Desaguliers 1744, Vol. I, p. 56, Pl. 4, Fig. 13
Adams 1746, No. 114
's Gravesande 1747, Vol. I, p. 41, Pl. 7, Fig. 1
Adams 1762, p. 22, Pl. 12, Fig. 52
C 38

See also D99, E55

Inventory No. 1927-1103

M23 Weighted cylinder in frame

L 173mm, **W** 118mm, **D** 80mm

Adams describes this object as a loaded cylinder and uses it as a further example of a centre of gravity paradox after the rolling double cone.

A walnut cylinder, which is stated to be 6 inches long with a diameter of $3\frac{2}{10}$ inches, has a smaller lead cylinder lying along its length towards one side. It is held in a square brass frame by which it can be drawn or held by its centre. The lead cylinder moves the centre of gravity of the walnut cylinder towards itself, and so the cylinder will roll up an inclined plane if by doing so it lowers its centre of gravity.

Desaguliers 1744, Vol. I, p. 57, Pl. 4, Fig. 15
's Gravesande 1747, Vol. I, p. 41, Pl. 5, Fig. 5
Adams 1762, p. 23, Pl. 12, Fig. 51

Used with M54   **Inventory No.** 1927-1824

M24 Cylinder in frame

L 173mm, **W** 118mm, **D** 85mm

This item is similar in almost every respect to the previous one, except that it is not weighted. It has therefore been assumed, although it is not mentioned by Adams, that it is a matching pair and was made in order to compare the behaviour of an unweighted cylinder with that of the weighted example.

Inventory No. 1927-1825

M25 Three boxwood discs, one weighted, with bracket for suspension

FIRST DISC
  'centre of gravity/ centre of motion,'
SECOND DISC
  'centre of gravity/ centre of motion'
THIRD DISC
  Centre of gravity/Centre of magnitude'

L 102mm (bracket), **W** 17mm, **H** 194mm, **D** 153mm (discs), **T** 15mm (discs)

Adams used these discs in an experiment taken from Desaguliers which compared the centre of gravity, centre of magnitude, and centre of motion.

Each boxwood disc is 6 inches in diameter. The first has a mark at the centre to take the callipers of the bracket so that when it is suspended from an arm of the pillar and rotated, a point on the disc moves in a circle round the centre of motion; two such circles are marked. 'Centre of gravity and centre of motion' are inscribed at the centre. The second disc has

'centre of motion' inscribed at a mark to take the callipers, which is away from the centre of gravity, which is also marked. When the disc is rotated a point makes a circle around the centre of motion; three such circles are marked. The disc comes to rest when the centre of gravity is at the lowest point. The third disc has a square piece of lead at one side which moves the centre of gravity away from the centre of magnitude. Both these are marked to take the callipers, and 'centre of gravity' and 'centre of magnitude' are inscribed. Adams does not mention in the text that one disc was weighted, but it is shown in the illustrations. The third experiment showed the disc in what we would now call 'unstable equilibrium' when the centre of gravity is directly above the point of suspension.

Desaguliers 1744, Vol. I, p. 52, Pl. 4, Figs 4, 5, 6
Adams 1762, pp. 23–4; Pl. 13, Figs 53, 54, 55
Inventory No. 1927-1237

M26 Centre of gravity paradox

L 360mm (awl), **D** 37mm (head)

Like the boxwood discs, this is a demonstration taken from Desaguliers. It consisted of a 'stick 12 inches long $\frac{1}{2}$ broad and $\frac{3}{4}$ thick' and two steel awls with solid brass heads. The stick was placed on one of the sliding arms of the great pillar. It falls if its centre of gravity lies beyond the end of the arm, but if the awls are attached as in the illustration, they shift the centre of gravity so that the stick will remain supported. The

experiment works to a lesser extent with one awl as is shown in the second illustration in the manuscript. The original stick is now missing; the piece of mahogany now with the awls was made in the Museum.

Desaguliers 1744, Vol. I, p. 59, Pl. 5, Figs 5, 6
Adams 1746, No. 110
Adams 1762, p. 24; Pl. 13, Figs 57, 58
C 40   **Inventory No.** 1927-1110

M27 Quadrilateral for centre of gravity experiments

L 205mm (diagonal), **L** 170mm (other diagonal), **T** 15mm

This piece of mahogany was used in the standard experiment to find the centre of gravity of a body mechanically. It was suspended from a point by the bracket and callipers in M25 and a plumb line was taken and drawn. The process was repeated from another point and the centre of gravity found from where the lines crossed. Two such lines are marked on the quadrilateral. The plumb bob was possibly that now with M54.

Desaguliers 1744, Vol. I, p. 52, Pl. 4, Figs 8, 9
Adams 1762, p. 24, Pl. 13, Fig. 56
C 39

Inventory No. 1927-1794

M28 Two inclined cylinders for centre of gravity demonstrations

D 43mm, **H** 75mm (slant)

These are a pair of oblique boxwood cylinders, according to Adams 'about 3 inches long 1½ diameter'. They do not fall if they stand alone because vertical lines from their centres of gravity lie within their bases. However, if they are placed one above the other they both fall, since a vertical line from their combined centre of gravity lies outside the base. Another example of an oblique cylinder is D98.

Desaguliers 1744, Vol. I, p. 57, Pl. 5, Fig. 2
Adams 1762, p. 25, Pl. 13, Fig. 60

Inventory No. 1927-1792

M29 Prism with a knife edge for centre of gravity experiments

L 236mm, **W** 75mm (side), **H** 66mm

The mahogany triangular prism has a steel rim along its upper edge. Adams states that it is about 9 inches long with each side 3 inches long. It could be used to find the centre of gravity of a two- or three-dimensional body by laying the body on the steel rim, marking the position of balance, changing the orientation of the body, and repeating the process. A flat body requires two lines to be drawn which will intersect to give the centre of gravity. A three-dimensional body requires three planes to be drawn to give the centre of gravity at the intersection, although it is difficult to see how this would be done in practice.

Desaguliers 1744, Vol. I, p. 75, Pl. 7, Figs 4, 5
Adams 1746, No. 109
Adams 1762, p. 25, Pl. 13, Fig. 61

Inventory No. 1927-1639

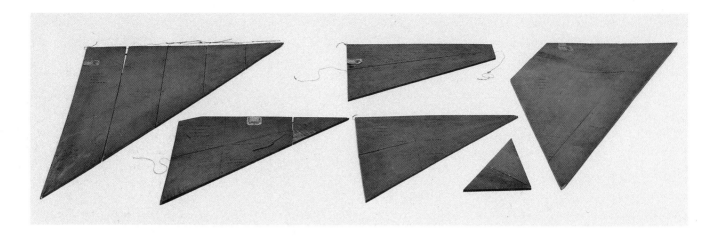

M30 Boards for centre of gravity and Archimedes' quadrature of the parabola demonstrations

PIECE FOR PROP 8 & 9
  '*Archimedes/de   Quadratura/Parabola/
  Prop. 8 & 9 Center of Gravity EDG Centre
  of Gravity CDG*'
PIECE FOR PROP 10 & 11
  '*Archimedes/de   Quadratura/Parabola/
  Prop. 10 & 11 Center of Gravity of the
  Trapez BRHT Center of Gravity of the
  Trapezium BDCH Center of Gravity of
  the Trapezium RDCT*'
PIECE FOR PROP 12 & 13
  '*Archimedes de Quadratura Parabola Prop
  12 & 13 Center of Gravity of the Trap.
  DCHE Center of Gravity of the Trapez
  DTRC*'
  *Also numbers along the top and letters from
  Adams 1762 Pl. 19 Fig. 75*
PIECE FOR PROP 14 & 16
  '*Archimedes de Quadratura Parabola prop
  14 & 16*'
PIECE FOR PROP 17
  '*Archimedes de Quadratura de Parabola
  prop 17*'
  *Also letters from Adams 1762 Pl. 20 Fig.
  78*

L 333mm (prop 8 and 9), **H** 333mm,
W 5mm, **L** 338mm (prop 10 and 11),
H 295mm, **W** 5mm, **L** 333mm (prop 12
and 13), **H** 333mm, **W** 5mm, **L** 445mm
(prop 14 and 16), **H** 668mm, **W** 5mm,
L 445mm (prop 17), **H** 668mm,
W 5mm.

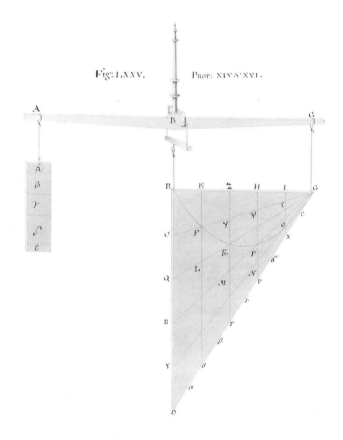

Adams 1762

A major set of experiments in Adams's *Principles of Mechanicks* concerned Archimedes' 'Quadrature of the Parabola' in which Archimedes found the area bounded by a parabola and an intersecting straight line. In this work Archimedes used his 'mechanical' method in several propositions before demonstrating the results geometrically. The 'mechanical' method involved an imaginary balance on which were suspended the figures under consideration and counterweights. Using centres of gravity and the law of the balance the area of the figures could be calculated. The underlying postulates were set out in another work, 'On the Equilibrium of Planes Part I'. Archimedes proved that the area bounded by a parabola and a straight line is four-thirds that of a triangle with the same base and height as the segment.

To realize Archimedes' mechanical method, Adams supplied a balance on which weights could be hung at known distances along each arm and a set of plane figures of constant thickness. Originally propositions 6–17, excluding 15, were intended to be demonstrated, but unfortunately the pieces for propositions 6 and 7 and the counterweights have been lost. The pieces are made of boxwood with the inscriptions stamped on each piece. They were suspended by threads from the balance M13. Adams included a translation of 'Quadrature of the Parabola' by Isaac Barrow, Newton's teacher, to which were added instructions for performing the demonstrations. This was inserted after the section on centres of gravity since it exploited this knowledge.

The mathematics involved in this work is the most demanding in Adams's course, although Archimedes and Barrow have been simplified by using the special case of a right-angled triangle throughout. A knowledge of Euclid is assumed and Apollonius's 'On conics' is referred to.

Other models of geometrical constructions by Adams survive, although they are much smaller. Models for performing the propositions in Euclid's Books 11 and 12 can be seen at the Museum of the History of Science at Oxford and at the Whipple Museum at Cambridge.

Barrow 1675
Adams 1762, pp. 27–38 1/2b, Pls 14–21, p. 22, P. 11, Fig. 48
Heath 1897, Quadrature of the Parabola
C 39
Porter *et al.* 1985, No. 59

Inventory No. 1927-1793

M31 Attachment for the suspension of planes for the quadrature of the parabola

L 90mm, **W** 11mm, **H** 25mm

This small brass suspension unit fitted over the balance M13 and, together with a brass bar on two threads which is missing, held the plane trapezia and triangles for the experiments on the quadrature of the parabola.

Adams 1762, Pls. 15–18

Used with M13, M30

Inventory No. 1927-1870

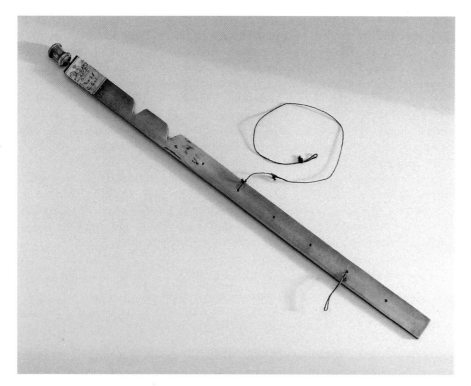

M32 Lever

L 397mm, **W** 19mm, **T** 5mm

This apparatus was used to illustrate the three different kinds of lever. It is a boxwood rule which Adams gives as 15 inches long, $\frac{3}{4}$ inch broad and $\frac{2}{10}$ inch thick. Two notches have been cut for the fulcrum 3 and $4\frac{1}{2}$ inches from one end. A choice from two brass cylinders can be screwed into this end in order to balance the ruler according to whether the fulcrum was at the nearer or further notch. Only one now remains. Holes have been drilled every $1\frac{1}{2}$ inches to hang weights or suspend the ruler by means of weights over a pulley on the great pillar.

The simple machines were routine in courses of experimental philosophy. Levers were often treated after the balance as the second simple machine, but Adams follows 's Gravesande in discussing the balance separately and starting the simple machines with the lever. Adams copied 's Gravesande's apparatus in detail for these experiments.

's Gravesande 1747, Vol. I, p. 45, Pl. 7, Figs 2, 3, 4
Adams 1762, p. 38 1/2b, Pl. 22, Figs 85, 86, 87
C 10, 11

Used with M20, M33, M37, M61
Inventory No. 1927-1241

M33 Fulcrum for various experiments

L 28mm, **W** 20mm, **H** 70mm

The fulcrum screws into the top of the stand M20. It was used to balance the levers M32 and M60 and, in Marriott's mechanical paradox, M61. It is simply a brass knife edge in a rectangular brass frame. Adams describes it as being 'of brass with an oblong perforation one inch long $\frac{4}{10}$ inch broad, the upper and lower parts are chamfered almost to a sharp edge that the levers may have the least friction possible'.

Adams 1762, p. 39, Pl. 22, Figs 85, 86, 87, p. 39, Pl. 23, Fig. 89, Pl. 37, Fig. 119
Inventory No. 1927-1232/1

M34 Right-angled lever and fulcrum

L 420mm, **W** 191mm, **W** 19mm (arm), **T** 5mm (arm)

Adams departs from 's Gravesande and follows Desaguliers in this experiment on the right-angled lever. The lever is of boxwood with a fulcrum of brass which fits onto the stand M20. In this position the long arm is vertical and the short arm is counterbalanced by a brass cylinder screwed into one end. The long arm is 15 inches and the short arm is 5 inches in length according to Adams, so if a weight is hung over a pulley on the great pillar and attached horizontally to the end of the long arm, it can be balanced by a weight three times as heavy on the end of the short arm. There are holes at fixed intervals to attach the strings for weights. Adams commented that in this manner a hammer was used to draw nails.

Desaguliers 1744, Vol. I, p. 98
Adams 1762, p. 39; Pl. 23, Fig. 89
C 12

Inventory No. 1927-1856

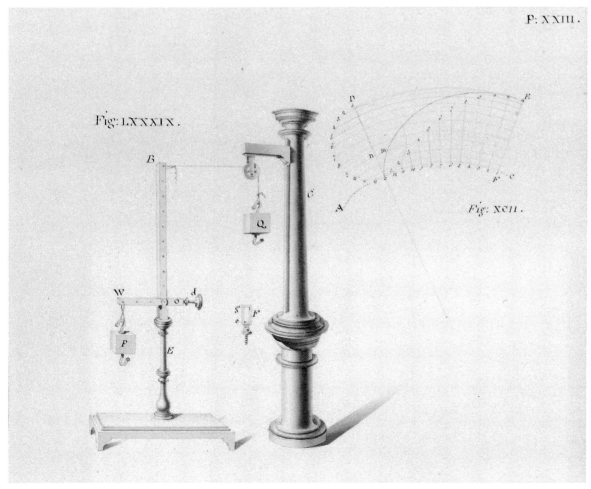

Adams 1762

M35 Cycloidal wheel and axle and epicycloidal teeth

L 200mm, **W** 175mm, **H** 525mm

This piece of apparatus demonstrates two separate but related experiments. The first is a comparison of an epicycloidal and a straight lever. The brass levers have their fulcrums on the two steel axles which each also carry spoked brass wheels on which can be hung weights. The straight lever is 2 inches long and the corresponding wheel radius is also 2 inches. The epicycloidal lever is straight for $1\frac{1}{2}$ inches, the diameter of its corresponding wheel, and then continues along the line of a portion of an epicycloid described with its base as the circumference of the smaller wheel and the larger wheel as its generating circle. When the levers operate against each other they will be in equilibrium when equal weights are hung on each wheel, despite the different lengths, because of the angles of force applied.

The second experiment occupies the centre of the apparatus. This consists of one brass wheel with epicycloidal teeth and a second brass wheel with straight teeth which can be slid along the axles to be meshed together. The levers can be removed for easy access. The figure of the epicycloidal teeth is determined by an epicycloid generated by the circle of the wheel on which the teeth project, on a base given by the circumference of the circle of the tips of the straight teeth. Again, the same weight on each corresponding spoked wheel will bring the toothed wheels into equilibrium. The ratio of the generating circle to the base circle is 4:3 in both cases. The whole is set on a brass base on a mahogany twin pillared stand. The cubical weights M37 were used in these experiments.

Although Adams demonstrated the principle of the epicycloidal lever in these experiments, it is unlikely that the mathematics would have been treated in detail. However, Fig. 92 in the *Mechanics* manuscript is taken from

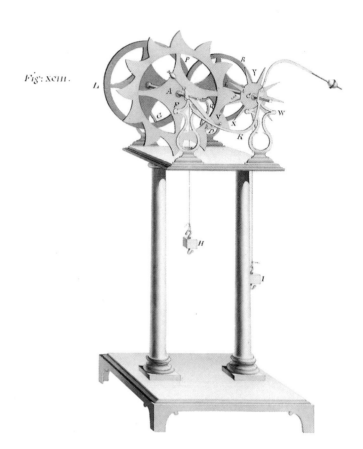

Fig: xciii.

Adams 1762

Emerson and shows how a cycloid can be drawn.

Newton 1729, pp. 100–201
Emerson 1758, p. 35, prop. 23, Fig. 24
Adams 1762, p. 40, Pl. 23, Fig. 92, Pl. 24, Figs 93, 94, 95
C 34

Inventory No. 1927-1850

M36
Fitting for stand

H 77mm, **W** 35mm, **L** 35mm

Like M33 this is a fitting for the top of the stand M20. It provided a fulcrum for the lever M39. There is a small hole in each of the brass cheeks to carry the axis for the sliding socket on which the lever was supported.

Nollet 1754, Vol. 3, p. 20, Pl. 1, Fig. 7   Adams 1762, p. 40, p. 23, Fig. 90

Used with M20, M39   **Inventory No.** 1927-1866

M37 Set of cubical weights

'200 oz T' '100 oz T' '20 oz T' '10 oz T' '8 oz T' '4 oz T'

L 83mm (side, largest), **H** 153mm (overall largest) **L** 12mm (side, smallest), **H** 36mm (overall smallest)

Adams made a large set of brass cubical weights, each with two steel hooks, apart from the heaviest which had one. They ranged from 500 oz Troy to ½ oz Troy and were listed in the *Mechanics* manuscript. The 500 oz and 400 oz weight are now lost, as are some of the others, but the majority have survived. The set now comprises: 1 × 200 oz, 2 × 100 oz, 1 × 50 oz, 1 × 20 oz, 1 × 10 oz, 2 × 8 oz, 1 × 7 oz, 2 × 6 oz, 2 × 5 oz, 2 × 4 oz, 2 × 3 oz, 4 × 1 oz, 2 × ½ oz. Each weight is engraved with the number and 'oz T'. Some of the hooks are missing.

The weights were used in many of the experiments in the mechanics course: for example, properties of the balance, oblique forces, bent and cycloidal lever, the axis in peritrochio, screw machine, compound engine, parallelogram of forces, compound lever, and projection of bodies. The 10 oz weight is drawn to scale and the length of side of each of the other weights is also drawn to give a geometric progression.

Adams 1762, p. 46, Pl. 25, Figs 97, 98
C 62

Used with M13, M32, M35, M40, M41/101, M48, M50, M60, M67, M76

Inventory No. 1927-1806, -1807

M38 Transverse arm

L 760mm, **H** 280mm, **W** 95mm

Adams returned to 's Gravesande for an experiment on the lever for which this transverse arm was needed. In the *Mechanics* manuscript it is drawn with the accessories for the Philosophical Table and it is described in detail later in the text.

The transverse arm consists of two pieces of mahogany $2\frac{1}{2}$ feet long, joined at their ends in a similar manner to M18/53. The slot between them is $\frac{3}{10}$ inch wide, according to Adams. The upper piece has an inlaid ivory scale marked in inches from the centre up to 14 inches each way. A mahogany fixture can attach the arm to the great pillar. The pulleys with tails, M10, were intended for the slot, as were the two bronze rods which acted as stops in the experiments on pendulums; they have been retained with the apparatus on this inventory number. The arm was used to suspend the lever below. It was attached horizontally to the pillar of the table, the pulleys were screwed into the slot at various points, and the straight lever was attached to counterweights by threads passing over the pulleys. The arm was also used with a simple pendulum when the rods could be used as stops, and was probably used to suspend the compound steelyard.

's Gravesande used the same apparatus with the lever below to show that a weight must be held between two powers that support it, and the distances from the powers to the weight will be in inverse ratio to the proportion of the weight carried by the power.

's Gravesande 1747, Vol. I, p. 47, Pl. 8, Fig. 3
Adams 1762, Pl. 5, Fig. 23 (bronze rods), p. 44, Pl. 26, Fig. 96, Nos. 1, 2, 3, Pl. 7, Fig. 31. p. 62. p. 94, Pl. 47, Fig. 152
C 20

See also M18/53

Used with M9/10, M39, M49   **Inventory No.** 1927-1216

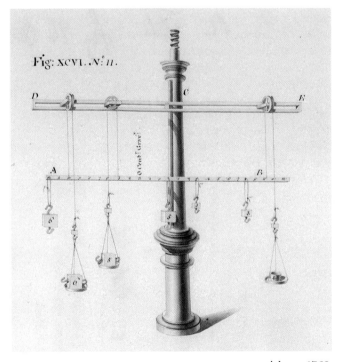

Adams 1762

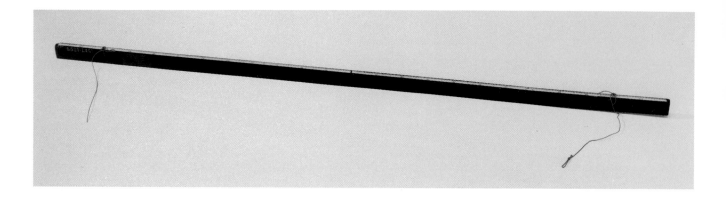

M39 Lever with two fixed and three movable threads

L 610mm, **W** 17mm, **T** 5mm

This is probably the item that Adams describes as 'another straight lever . . . which will always resemble an inflexible mathematical line', meaning that its weight could be ignored. The lever had a sliding fulcrum, a sliding socket for a counterweight, and a hook at either end which hung on pins through holes in the wood. The holes are present, but all the accessories are now missing. It corresponds to both the dimensions and materials specified: it is 2 feet long, about $\frac{8}{10}$ inches wide and $\frac{2}{10}$ inches thick, made of mahogany with an ivory scale. The scale runs along one edge and is numbered in inches up to 12 both ways from the centre. The fulcrum of the lever was placed on the stand M20, which was fitted with the socket M36. Wherever the fulcrum was placed along the lever, if it was balanced by weights it would resemble a mathematical line. Alternatively the lever was suspended horizontally from the transverse arm above by means of weights passing over pulleys. Equilibrium was reached and the relationship between powers and weights was observed.

Nollet 1754, Vol. 3, p. 20, Pl. 1, Fig. 7
Adams 1762, pp. 44–6, Pl. 26, Fig. 96; p. 39, Pl. 23, Fig. 90
C 15

Inventory No. 1927-1799

M40 Square frame and ruler for experiments on oblique forces and the horizontal lever

L 467mm, **W** 467mm, **H** 75mm, **L** 405mm (ruler)

The square frame is made of mahogany with sides $18\frac{1}{2}$ inches long. Each side piece is $1\frac{1}{4}$ inches wide and $\frac{8}{10}$ inches thick. There is an inlaid ivory scale marked in inches and tenths starting from the mid-point of each side, and also marked in degrees from 0° in the mid-points of a pair of opposite sides to 90° in the mid-points of the other pair of opposite sides. A slit runs along the length of each side to hold the pulleys. Three pulleys survived with the apparatus and two were made in the Museum. The frame stands on four mahogany feet and extensions to these allow the frame to be raised a further inch from the stand. The stand is simply a square topped table which was made in the Museum since the original is now lost. The

original was a frame rather than a table since it had a large square hole in the centre through which weights could fall.

The lever or 'ruler' for the experiment to demonstrate the horizontal lever is a piece of mahogany just over 15 inches long with holes every $1\frac{1}{2}$ inches. It does not correspond to the ruler in the text and so may not be original, although it fulfils the purpose and matches the frame. The lever was held stationary by various weights hung over pulleys on two opposite sides and attached by threads through the holes. It is essentially the same experiment as that of the lever suspended on the transverse arm above but horizontal.

For the experiment on oblique forces three pieces of thread were knotted together at one point and their ends taken over pulleys hung with weights. The position and size of the weights can be adjusted until equilibrium is reached. Lines drawn perpendicular to the threads in lengths proportional to the weights will form a triangle.

The frame and stand were copied in detail from 's Gravesande, who appears to have invented the apparatus.

's Gravesande 1747, Vol. I, p. 49, Pl. 9, Fig. 1, p. 68, Pl. 12, Fig. 3
Adams 1762, pp. 47–8, Pl. 27, Figs 99, 101, Pl. 5, Fig. 20, p. 69, Pl. 36, Fig. 118, No. 4, No. 5

Inventory No. 1927-1857

M41 Axis in peritrochio or wheel and axle

L 326mm, **W** 212mm, **H** 365mm

The axis in peritrochio 'vulgarly called the wheel and axle' was treated by Adams as the second of the simple machines. A steel axle is slung between two mahogany pillars on a mahogany base. On it turn four brass co-axial wheels and a mahogany wheel with eight walnut spokes or handles. Each handle has three notches. The axle has two diameters: $\frac{1}{4}$ and $\frac{1}{2}$ inch and the brass wheels have diameters of 1, 2, 3, and 4 inches respectively. The mahogany wheel is 6 inches in diameter and the notches are 8, 10, and 12 inches from the centre of the wheel. Ratios of up to 1:48 can thus be had for the power to the weight. A hole is cut in the base to allow the weights to hang freely when the instrument is placed on a stand. The heavier cubical weights would have been placed on the base to keep the machine stable.

Adams elaborated on 's Gravesande's machine which had only three brass wheels and no notches in the handles. Adams stated that the axis in peritrochio was more useful than the lever because weights could be lifted to a greater height.

Hauksbee 1714, Pl. 4, Fig. 3
Desaguliers 1734, p. 102, Pl. 10, Figs 10, 11
's Gravesande 1747, Vol. I, p. 50, Pl. 8, Fig. 6
Adams 1762, p. 49, Pl. 28, Fig. 104

Inventory No. 1927-1860/1

M42 Two pulley blocks, one with three pulleys and one with two

H 167mm (triple pulley), **H** 147mm (double pulley), **W** 9mm (both), **D** 50mm (wheel max)

These blocks constitute Adams's 'Tackle of three fixed and two moveable pullies'. The brass spoked wheels are set in brass tapering frames with steel hooks and axles. The triple pulley would have been fixed on to the arm of the pillar M71, and the double pulley would have been slung beneath. Adams explains, following 's Gravesande, that the lack of verticality of the ropes is neglected because to achieve it the wheel diameters would have to be in the ratio 1:2:3 for the movable block and 2:4:6 for the fixed, making them cumbersome.

's Gravesande 1747, Vol. 1, p. 53, Pl. 9, Fig. 5
Adams 1762, Pl. 5, Fig. 19, p. 52, Pl. 29, Fig. 107

See also M9/10, M43, M44

Inventory No. 1927-1233

M43 Two triple pulley blocks and a lead ball

D 41mm (ball), **L** 57mm (pulley blocks), **W** 42mm (pulley blocks), **H** 99mm (pulley blocks)

Adams described these pulleys as 'A Tackle of equal sheaves parallel to each other three fixed and three moveable'. They are made of brass with steel hooks and axles as recommended by 's Gravesande. One block was attached to the fixed arm M71 while the other held the weight, in this case a lead ball. Although the pulleys are drawn in detail in the illustration concerning pulleys in the *Mechanics* manuscript, when they are later referred to in the text Adams reverts to 's Gravesande for an illustration. Adams stated that this type of pulley block was very compendious and generally used.

's Gravesande 1747, Vol. I, p. 54, Pl. 9, Fig. 6
Adams 1762, Pl. 5, Fig. 21, p. 53, Pl. 29, Fig. 108
C 30

See also D6

Used with M9/10, M42, M44, M71

Inventory No. 1927-1802

M44 Tackle of pulleys

ON BOTH BLOCKS
'ADAMS/Fecit'

L 225mm (one), **L** 215mm (two), **D** 57mm (one), **D** 61mm (two)

Adams made various pulleys for various experiments in the *Mechanics* manuscript. This tackle is the most distinctive, being larger, stronger, and more complex than the others. It was a departure from 's Gravesande in design, materials, and motivation, for while 's Gravesande was concerned with demonstrating the principle of the pulley, Adams also wanted to lift as large a weight as possible. It is described as a 'New Tackle of Pullies' and is based on Smeaton's design which appeared in 1752. To avoid the problems of having several pulleys in a line beside each other, or attached vertically on one shaft, Smeaton made two sets of sheaves in which each pair was arranged one above the other in compact blocks. Thus the tackle had a high mechanical advantage while retaining its convenience. Adams modified the design by setting the lower sheaves in each block at right-angles to the upper ones, thus separating the ropes and making the blocks nearly equal in size. There are eleven sheaves in the upper block and ten in the lower; hence '25oz will sustain 525oz', although Adams was not taking account of friction. The entire instrument is made of steel.

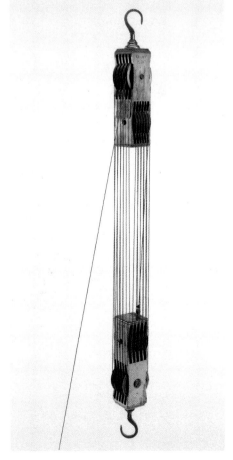

Hauksbee 1714, Pl. 3, Fig. 5
Smeaton 1752
Gentleman's Magazine 1754, Vol. 24, p. 84
Adams 1762, pp. 54–5; Pl. 29, Fig. 110
C 2

Inventory No. 1927-1230

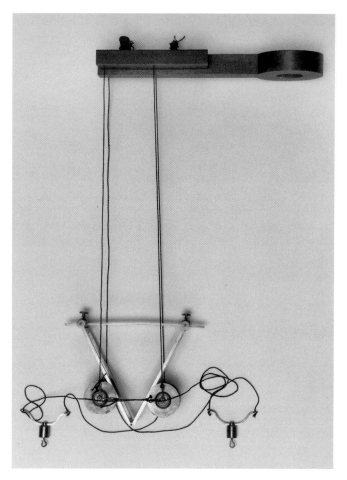

Adams 1762

M45 Apparatus for demonstrating properties of the wedge

SCALE—ENGRAVED ON THE BRASS RULER
'1/4', '1/3', '1/2', '1', '1', '1/2', '1/3', '1/4'

L 312mm (bracket), **L** 160mm (wedge), **T** 22mm (bracket),
D 54mm (cylinder), **L** 81mm (cylinder), **L** 200mm (ruler)

Adams introduced the wedge after the pulley as the fourth simple machine. He advocated 's Gravesande's machine as being the best for demonstrating the properties of the wedge, but since he disagreed with 's Gravesande's interpretation of the experiment he used a different method of drawing the cylinders together. He worked from Rutherforth's principles of the wedge which are as follows: in a wedge, the 'velocity of the resistance' or the downward force of the wedge is to the 'velocity of the power' or applied force acting horizontally to support it as the base of the wedge is to the sum of the length of its other two sides.

The demonstration compares the weight of the wedge with the force required to support it by horizontally pressing the sides together. The apparatus is as follows. A small maho-

gany board the dimensions of which Adams gives as 6 inches × 4½ inches is attached to an arm of the pillar and fixed horizontally at the top. From its corners hang four threads about 3 feet long. These threads are attached to each end of the axles of two brass cylinders by means of small brass brackets. The axles are free to rotate. The threads do not carry the weight of the wedge and cylinders but can be adjusted to keep the wedge vertically aligned, using brass pegs in the board. The two cylinders are 2 inches long with diameters of 1½ inches. At each end they have a rim with a larger diameter in order to keep the wedge in place. The wedge rests on the cylinders and is free to move vertically. The wedge itself consists of two rectangular brass frames about 6 inches long hinged at one end. Adams explained that this was an improvement on 's Gravesande's solid-sided wedge because the resulting apparatus was much lighter.

The square frame on four legs, from M40, was placed on the Philosophical Table and the wedge was lowered into the central space. Two pulleys were screwed to the frame at the two opposite sides and weights were hung from the pulleys which were then attached to the opposing cylinder. Thus the weights hanging at each side acted to bring the cylinders together and support the wedge. Adams made two connecting side-pieces so that only one weight was necessary for the two pulleys at each side. A weight was hung from the vertex of the wedge to give the total downward force.

The remaining piece of apparatus is a brass ruler which can be slotted across the top of the wedge to form its base. The angle of the wedge can be set so that the base is equal to the sides, half the sides, one third, or one quarter. Marks on the ruler show these positions. If the base is set on the marks 1,1, it will be shown that a total downward force of 16 oz will be counterbalanced by a total horizontal force of 32 oz; thus Rutherforth's principle holds.

The basic mechanism was probably copied from Hauksbee.

Hauksbee 1714, Pl. 4, Fig. 5
Adams 1746, No. 120
's Gravesande 1747, Vol. I, p. 56, Pl. 10, Fig. 2
Rutherforth 1748, Vol. I, p. 57, Pl. 3, Figs 4,5,6
Adams 1762, Pl. 7, Fig. 29; pp. 55–8, Pl. 30, Fig. 111
Inventory No. 1927-1231

M46 Apparatus for demonstrating the action of a screw

H 128mm, **L** 177mm, **W** 110mm

The screw demonstration follows the wedge as the last of the simple machines. Adams appears to have devised this apparatus himself: there are no references to 's Gravesande or any other work after the description, and the illustration is his own.

The apparatus is a heavy steel screw of five turns to the inch with a pointed end which forms the axle of a mahogany wheel 4 inches in diameter. The wheel and screw are held in a brass bracket which can be attached to one of the transverse arms of the great pillar. A pulley is fixed at the other end of the transverse arm which carries a weight which is attached by a thread to the mahogany wheel. The force of this weight acts to turn the screw so that the screw moves downwards.

Beneath the transverse arm on the pillar is the balance M13. It is supported at its centre and a weight is placed at one end. The other end is situated just below the screw so that the screw acts to depress the arm of the balance and so raise the weight. Adams found that a weight of 1 oz acting to turn the screw was exactly counteracted by a weight of 38 oz on the balance. In theory this should have been $62\frac{1}{2}$ oz, the circumference of the mahogany wheel multiplied by 5 for five turns of the screw per inch. However, friction accounted for the shortfall. The experiment showed how a small power can overcome a great resistance with the help of a screw.

Adams 1762, p. 58, Pl. 31, Fig. 113
C 19

Used with M11, M12, M13

Inventory No. 1927-1243

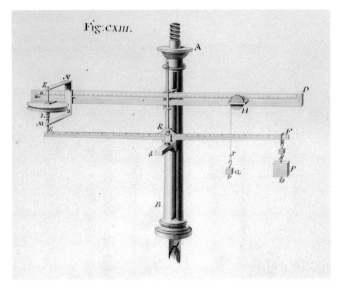

Adams 1762

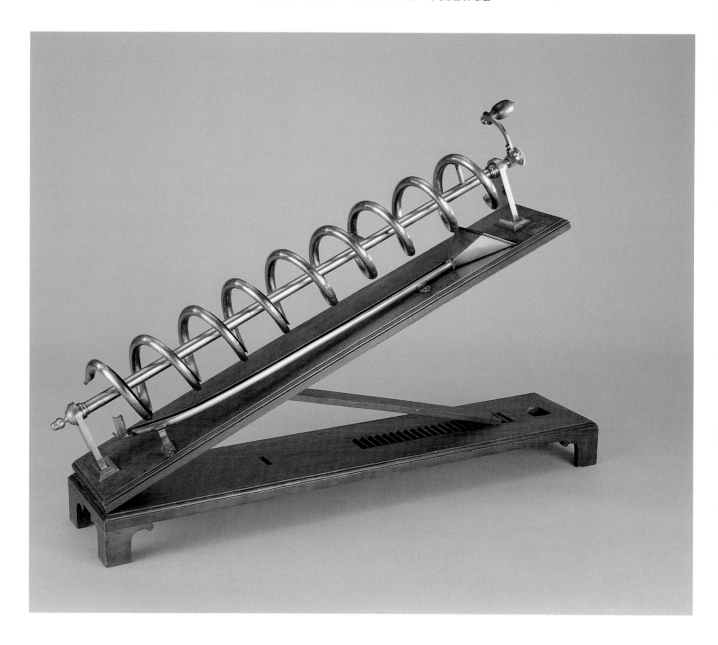

M47 Archimedean screw

ENGRAVED—END
'*Made by|Geo Adams|Mathematical|Instrument Maker|to His|Majesty*'
L 848mm, **W** 165mm, **H** 880mm (max), **D** 8mm (balls)

Following on from the screw demonstration above, Adams describes 'two kinds of screw which are distinguished from all the rest': the endless screw and the Archimedean screw. He does not appear to have made a model of the endless screw but the Archimedean screw is one of the most distinctive items in the collection. The design of this instrument is probably original.

A copper pipe with part of its circumference cut away forms a helix. A brass rod runs down the centre and rotates on its axis when the walnut handle is rotated, thereby turning the screw to which it is connected by a brass rod. Although water is the usual medium to be raised by an Archimedean screw, this model uses two balls, one of copper and one of ivory, to demonstrate the principle. When the balls reach the top of the screw, they drop into a copper funnel and are channelled back to the bottom in a copper pipe. They are scooped up by the next turn of the screw. This makes the action continuous and also prevents damage to the balls. The screw can be inclined by lifting the mahogany base and propping it on one of a series of slots in the stand.

The Archimedean screw was a popular demonstration in courses of experimental philosophy. Having been treated theoretically by Parent in 1703, it featured in several texts of the period.

Barrow 1675
Parent 1703
de Servière 1719, Figs 76, 77
Belidor 1737, Vol. I, p. 387
Adams 1746, No. 122
Saverien 1753, Under 'vis et vis Archimede'
Nollet 1754, Vol. 3, p. 134, Pl. 8, Fig. 12
Emerson 1758, p. 228, Fig. 272
Adams 1762, p. 59, Pl. 31, Fig. 114
Chambers 1763, Pl. 'the section of the water engine'
Gentleman's Magazine 1763, Vol. 33, p. 324
C 44
Crommelin 1951, No. GM 89, A36.
Singer et al. 1956, p. 677
Chew 1968, Item 3
Turner 1973, Appendix 1, p. 378

See also D12

Inventory No. 1927-1106/1

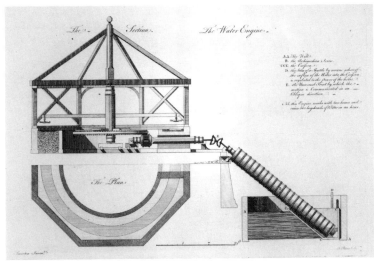

Chambers 1763

M48 Compound lever

L 650mm, **W** 165mm, **H** 358mm

Having covered the simple machines Adams moved on to compound machines with this instrument. With very slight modification it is copied from 's Gravesande's compound lever. Upright walnut pillars carry three fulcrums for three boxwood levers. The first lever is supported by two pillars connected by a brass bracket for extra strength. The levers are each made to balance by a small brass counterweight on the short arms which are all 1 inch long. The long arms are 5, 4, and 6 inches long. Each arm is marked in inches. When the compound lever is in operation a large weight on the short arm of the first lever raises the long arm of the first lever which presses up the short arm of the second lever. The long arm of

the second lever presses down the short arm of the third whose long arm carries the power. The combined power to weight ratio is 1:120. Adams used $\frac{1}{2}$ oz and 60 oz cubical weights for this demonstration.

Hauksbee 1714, Pl. 2, Fig. 9
Desaguliers 1744, Vol. I, p. 101, Pl. 9, Fig. 14
Adams 1746, No. 112
's Gravesande 1747, Vol. I, p. 61, Pl. 11, Fig. 1
Adams 1762, p. 60, Pl. 32, Fig. 115

See also D136

Used with M37

Inventory No. 1927-1102

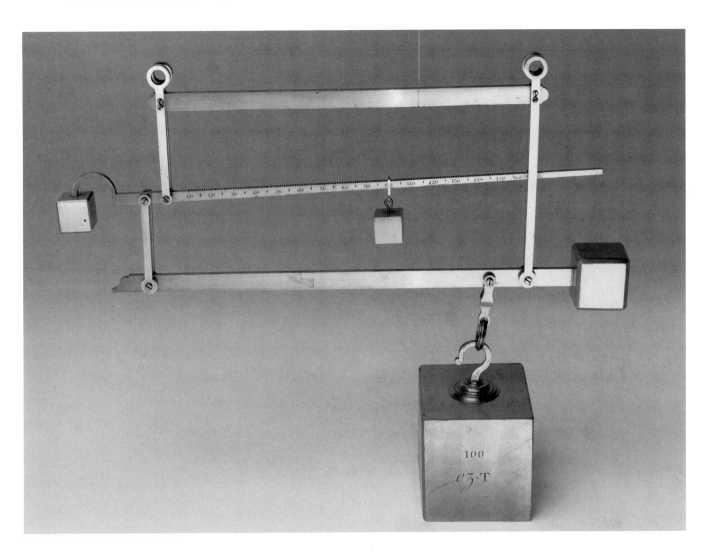

M49 Compound steelyard

L 352mm, **W** 30mm, **H** 170mm

This apparatus demonstrates the same principle as the previous one, but the levers are suspended rather than supported. It appears to have been copied from 's Gravesande.

A brass lever of the first kind is suspended from its fulcrum. On the upper surface of the long arm are 150 incisions which are numbered in tens on a vertical surface. They can support weights hung on loops as in M14. The short arm is counterbalanced by a cubical weight and from this arm is suspended the long arm of the heavier unmarked brass lever of the second kind. The point of suspension of the second lever is equal to a distance of ten incisions from the fulcrum of the first lever. The second lever is also suspended from its fulcrum and counterbalanced. The weight is suspended from a point on the second lever so that the ratio of the distances from the fulcrum of the weight and the power is fixed at 1:10. Theoretically the two levers or 'steelyards' combined give a power-to-weight ratio of 1:100 if the power is placed at the '100' mark of the graduated lever. The maximum power-to-weight ratio would be 150. The weights used are the 100 oz from M37 and a 1 oz weight with a loop from M14. The demonstration was probably performed on the transverse arm M38 when the holes in the brass suspension would have been put over the bronze rods.

's Gravesande 1747, p. 61, Pl. 11, Fig. 2
Adams 1762, p. 61, Pl. 32, Fig. 116

Used with M14, M37, M38

See also P11

Inventory No. 1927-1111

M50 Compound engine

'ADAMS/FECIT'

L 295mm, **W** 310mm, **H** 350mm

Adams combined two of 's Gravesande's machines in this apparatus: the compound axis in peritrochio and the endless screw. He claimed 'it was one of the simplest and most elegant compound engines I have ever seen'.

Essentially the machine consists of a horizontal axis in peritrochio, or wheel and axle, which acts upon a vertical axis in peritrochio by means of toothed wheels. An endless screw is turned by the second axis and can be fitted to a large spoked wheel. The machine rests on a rectangular brass base which, according to Adams, measures $6\frac{8}{10}$ inches $\times 4\frac{2}{10}$ inches. The middle is cut away to allow room for the weight to fall. In the illustration for the *Mechanics* manuscript at Windsor it is shown on an ornamental stand; however, it would also have fitted the frame table which is now missing from M40.

The moving parts are held in a strong brass frame consisting of two main pillars and a cross-piece at the top. The principal axis runs horizontally across the base of the instrument. It is made of steel and has diameters of $\frac{1}{4}$ inch, $\frac{1}{2}$ inch, and 1 inch. A brass contrate wheel of 3 inches diameter with 50 teeth meshes with a pinion with 10 teeth on the vertical axis. Above this pinion is another wheel with 25 teeth which can be made to engage with an endless screw. The screw makes one revolution for each tooth of the wheel, giving a velocity ratio of 1:25. The large spoked wheel has a diameter of 12 inches. It can be attached to the end of the horizontal axis carrying the screw or on to the vertical axis. The machine can be used as a compound axis in peritrochio without the screw if the secondary supports holding the screw in place are moved out of the vertical. Alternatively it can be used without the large spoked wheel.

There are two dials, one on the horizontal axis and one on the vertical axis; they are supported on small pillars on the frame. Pointers can be moved to the desired position on the dials by means of a spring and will rotate with the axes to show the velocity of the rotation more easily. Each dial is numbered 10 to 100 in 10s with each division marked.

The maximum power-to-weight ratio which the machine can achieve is 1:6000. This represents the combination of the toothed wheels 1:5, the endless screw 1:25, and the ratio of the axle diameter of the weight to the diameter of the large wheel which carries the power 1:48. Adams warned against friction considerably reducing this figure.

's Gravesande 1747, Vol. I, pp. 62–3, Pl. 11, Figs 3, 5
Adams 1762, pp. 62–7; Pl. 33, Fig. 117, Pl. 34, Fig. 118, No. 1, Pl. 27, Fig. 100
C 36

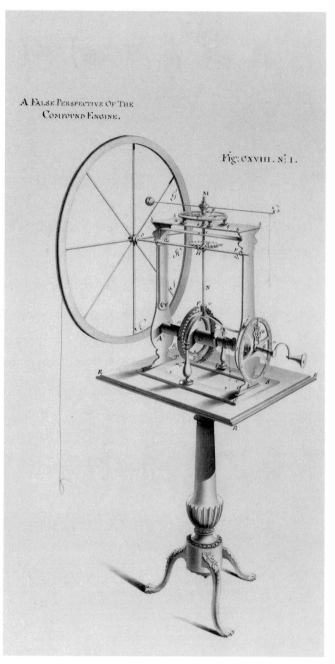

Adams 1762

Chew 1968, Item 2

See also M40

Inventory No. 1927-1851

M51 Model of windlass

L 210mm, **W** 210mm, **H** 270mm

In the mechanics course the windlass forms a pair with the capstan below. This model has the dimensions Adams gives for his windlass: length of the cross rail 7 inches; length of the bottom rail 8 inches; length of the levers 5 inches. It is made entirely of mahogany.

Windlasses and capstans were often cited as examples of the application of mechanical powers in courses of experimental philosophy. Nollet gives their uses in wells, quarries, on ships to raise anchors, at ports to lift loads, and in numerous other situations.

Adams 1746, No. 145
Nollet 1754, Vol. 3, p. 103, Pl. 6, Fig. 48
Emerson 1758, p. 38, cor. 4
Adams 1762, pp. 67–8; Pl. 35, Fig. 118, No. 1, 2
Keill 1776, 2nd edn, p. 120, Pl. 2, Fig. 34
C 68   **Inventory No.** 1927-1126

M52 Model of capstan

H 195mm, **L** 270mm, **W** 260mm

Although Demainbray used a model 'capstane' which is listed in the Queen's Catalogue, this model corresponds to the description given by Adams of his 'capstain' in 1762. It is entirely of mahogany. The height of the upright rail is 5 inches, the side rails are 7 inches and the levers 5 inches.

Adams inserted the examples of the windlass and capstan after considering 's Gravesande's compound engines. The capstan had a vertical axis and the windlass a horizontal axis but otherwise they are the same machine. In the case of the capstan, Adams suggested that four men turn the handles while another guards against slippage of the rope, only two or three turns being taken on the axle. There is a model capstan in the van Musschenbroek collection at Leiden.

Adams 1746, No. 145
Nollet 1754, Vol. 3, p. 103, Pl. 6, Fig. 49
Adams 1762, pp. 67–8; Pl. 35, Fig. 118, No. 3
van Musschenbroek 1762, 1st edn 2 vols., Tab. VII, Fig. 3
C 68

Inventory No. 1927-1125

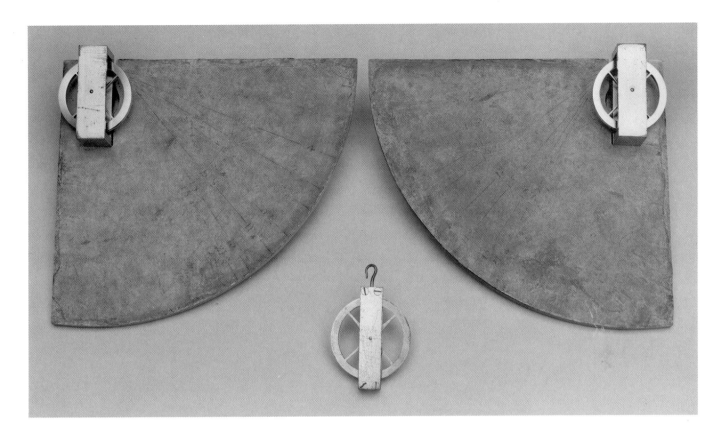

M53 Two quadrants for use in the demonstration of oblique powers

L 175mm (side of quadrant)

These mahogany quadrants are used with the transverse arm M18 for experiments on oblique forces. They are clamped to the ends of the transverse arm by the brass pulleys with square tails set perpendicularly to the pulley so that the pulleys are slightly in front of the quadrants. The illustration shows conical weights being hung in scale-pans on threads between the pulleys and down each side. The central weight is supported by the two 'powers' or applied forces at each side. Lines are drawn on the white paper on the quadrants which correspond to the direction of the thread when various combinations of weights are used. Each line is marked at the middle with the secant × 10 which the angle makes with the horizontal and also at the end with the tangent × 10 which the angle makes with the horizontal. This means that each power holding the central weight is to the central weight as the secant is to the sum of the two tangents.

The experiment was taken from 's Gravesande.

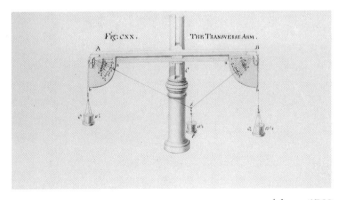

Adams 1762

's Gravesande 1747, Vol. 1, p. 72, Pl. 13, Fig. 3
Adams 1762, p. 70, Pl. 37, Fig. 120, Pl. 5, Fig. 18

Inventory No. 1927-1912/2

M54 Inclined plane and horse-way with quadrant and plumb bob

L 1140mm, **W** 335mm, **H** 98mm (max)

This apparatus can be used in an abstract manner to demonstrate weights and powers on an inclined plane, but it can also be used as a representation of travel on a model road with model obstacles and ruts.

The inclined plane is $3\frac{1}{2}$ feet long by 1 foot wide, with a small rim around the edge. It stands on a mahogany base which has two supports, one at each end. It can be inclined on a hinge at one end while a wooden prop is fitted into a channel and is held by a long brass screw to raise the other. Turning the screw adjusts the inclination of the plane. A brass quadrant, originally with a plumb bob, is attached to the side of the plane to measure the angle of elevation; it is marked from 0° to 90°. The plane was intended for use on the Philosophical Table when a weight would be hung over a pillar on an arm of the Philosophical Table to act as a power for the load on the plane.

In addition to the plane surface, a cobbled surface 'with which the horse ways of our streets are paved' is available. A piece of mahogany with one side cobbled fits exactly into the tray formed by the surface of the plane. A strip of the cobbling 1 inch wide and stretching across the horse-way can be removed to give a 'kennel' and this 'kennel' represents 'a real kennel one foot wide' since the scale of the whole is 1:12. A model carriage and model cart were made especially for use on the horse-way, and the cylinder and weighted cylinder were also demonstrated at this point.

Hauksbee 1714, Pl. 5, Fig. 5
Adams 1762, pp. 73–75, Pl. 38, Fig. 125
C 22, 42

See also D11

Used with M23, M24, M55

Inventory No. 1927-1830

M55 Obstacles for the inclined plane

L 258mm, **W** 47mm, **H** 24mm (one obstacle), **H** 13mm (two obstacles)

Originally there were two pairs of obstacles for use on the inclined horse-way above, which represented 'small hills or risings often to be met with in several Roads'. Three remain: the pair $\frac{1}{2}$ inch high and one of the pair 1 inch high. They are mahogany double wedges which can be placed under the wheels of the carriage M57 to modify the inclination of the plane.

Adams 1762, p. 74 1/2, Pl. 38, Fig. 125
C 22

Inventory No. 1927-1820

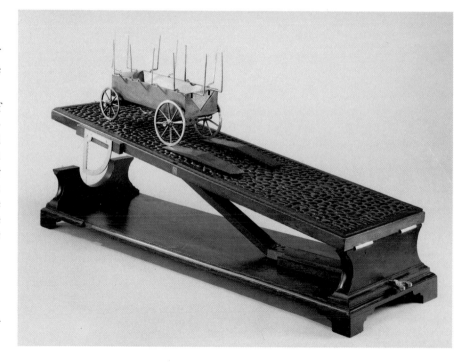

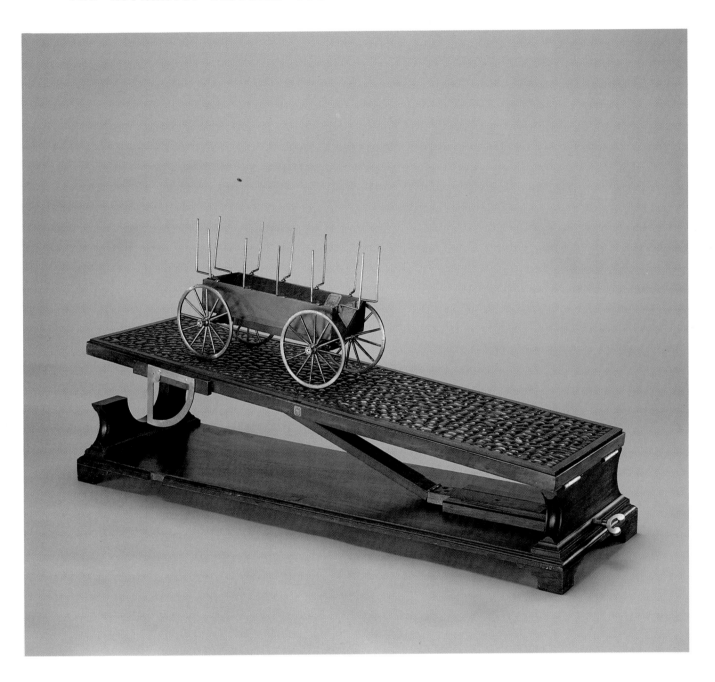

M56 Model carriage chassis with two axles

L 275mm, **W** 165mm, **T** 13mm

The chassis of the model fits beneath the body of the carriage M57. It is a mahogany cross-piece with conical brass axle housing and a steel axle. Various pairs of wheels can be attached by means of screw nuts. The front axle housing carries two hooks for the attachment of threads and is removable to become a two-wheeled cart. The model carriage was used for a variety of experiments on the inclined plane and horse-way: the relation of power to weight needed to hold the carriage, the effect of wheel diameter, width, and type, and the effect of various obstacles on the motion.

Hauksbee 1714, Pl. 5, Fig. 5
Adams 1746, No. 117,
Adams 1762, p. 75; Pl. 39; p. 75, Figs 126, 127, 128
C 22

Inventory No. 1927-1827

M57 Model carriage body

ENGRAVED—ONE END
    'Geo Adams/Fleet Street/London'
ENGRAVED—OTHER END
    '50 ounces'

L 403mm, **W** 170mm, **H** 172mm

This is the body of the carriage for use with the inclined horse-way M54. It consists of a rectangular mahogany block 1 foot by $4\frac{1}{2}$ inches by $2\frac{1}{2}$ inches with a triangular groove cut out of the top surface to receive the cubical weights M37. The ends are brass plates, one signed by Adams and one marked '50 ounces', which is the combined weight of the body, chassis, and axle trees. Twelve brass rods can be fitted into holes around the body to hold a bulky load such as hay. 'If it [hay] be wrapped in a sheet of paper and put between these wires and then a cubical weight laid thereon . . . we can see how easily carriages loaded with hay or any other light substances are liable to be overturned'.

Adams 1746, No. 117
Adams 1762, p. 75, Pl. 39, Figs 126, 128
C 22, 42

Inventory No. 1927-1821

M58 Seven pairs of wheels for use with the model carriage

ENGRAVED—ON WHEELS
'15', '11', '8', '8', '5', '4', '4'

D 147mm (three largest), **D** 114mm (two middle-sized), **D** 92mm (two smallest), **W** 19mm (rim, two largest), **W** 5mm (rim, five smallest)

All the pairs of wheels Adams made for the experiments with the carriage on the inclined plane have survived. They are interchangeable and can also be used on the model cart M59.

They are entirely of brass and are all spoked and dished. The weight in ounces is engraved on each wheel. The weights, diameters, and rim widths vary and one small pair is 'with nails'; it has small protrusions around the rim.

The subject of wheels and wheeled carriages received a great deal of attention in the early and mid-eighteenth century. Desaguliers and others devoted a considerable amount of space to the subject of wheel diameters and dishing.

Desaguliers 1744, Vol. I, pp. 210–39
Adams 1746, No. 117
Adams 1762, pp. 76–7, Pl. 39
Albert 1972, p. 81

Inventory No. 1927-1828

M59 Model experimental cart

HOOK
'4'

L 126mm, **W** 76mm, **H** 28mm

This piece of mahogany can be fixed to the detached fore axle tree of the model carriage M56 to form a model two-wheeled cart. The brass spring clip is fitted into a hole in the centre of the axle tree. The cart can then be used with any of the pairs of wheels in the above for experiments similar to those with the model carriage. The number 4 is engraved on the clip because, together with the axle tree, the cart weighs 4 oz.

Adams 1762, p. 77, Pl. 39, Figs 135, 136

Inventory No. 1927-1819

M60 Bent lever

L 467mm, **W** 86mm, **T** 5mm (arm), **W** 19mm (arm)

This is the boxwood 'bended' lever described by Adams whose arms are 12 inches and 4 inches long respectively, and $\frac{2}{10}$ inches thick, with an angle of 150° at the fulcrum. It was placed on the knife edge M33, on the stand M20, at the edge of the Philosophical Table. Large cubical weights were placed on the base of the stand to keep it stable. A pulley attachment, now lost, was fixed to the edge of the table and a weight was hung from the lever over the pulley in such a position as to make the force act perpendicularly to the long arm while the short arm was horizontal. This weight was counterbalanced by a weight three times as heavy on the short arm, showing that the law of the lever was obeyed. As in the other lever experiments, the apparatus was copied from 's Gravesande.

's Gravesande 1747, p. 46, Pl. 8, Fig. 2
Adams 1762, p. 38 1/2b, Pl. 22, Fig. 88, p. 83, Pl. 40, Figs 137, 139
C 13, 14

Inventory No. 1927-1804

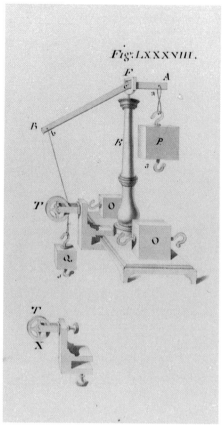

Adams 1762

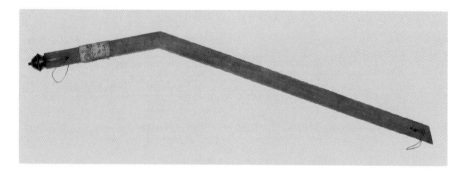

M61 Apparatus for Marriott's paradox

L 50mm (pulleys), **W** 45mm (pulleys), **H** 70mm (pulleys), **L** 45mm (bracket)

Following 's Gravesande, Adams used Marriott's paradox as an example of the equilibrium of three forces acting at a point. A bent lever, now missing, was balanced on the fulcrum M33 on the stand M20. One arm held horizontal carried the apparatus described below while the other carried a weight twice as large at half the horizontal distance from the fulcrum. The paradox is that under a certain condition the weights can be interchanged and the balance is retained. The condition is that a vertical board be held against the end of the sloping arm of the lever so that this point is subject to a horizontal force from the board, a downward force from the weight of the apparatus, and an upward force perpendicular to the line of the lever from the weight on the other arm. When the forces are balanced the point becomes motionless. 's Gravesande states that other ratios can be used and that the experiment is difficult to perform. He gives an explanation in terms of similar triangles.

The apparatus referred to above consists of three coaxial brass pulleys which are attached by threads to each end of a brass bracket. Weights can be hung from the bracket. The whole weighs 5 oz. In operation, the two outer pulley wheels run on the board on either side of a slot while the inner wheel is in contact with the lever. M62 is the vertical board for this experiment.

Marriott 1718, p. 61
's Gravesande 1747, Vol. I, pp. 76–7, Pl. 14, Figs 2, 3, 4
Adams 1762, pp. 82–3; Pl. 40, Figs 137, 138, 139
C 13, 14.

Inventory No. 1927-1232/2, -1805

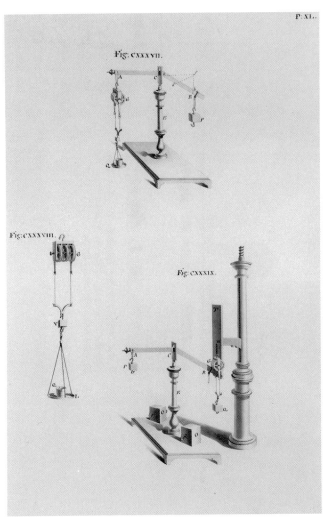

Adams 1762

M62 Pillar attachment for Marriott's paradox

H 440mm, **L** 185mm, **W** 70mm

The mahogany board is attached to the pillar for the experiment on Marriott's mechanical paradox. In this position it has a vertical slot down the centre of its lower part to accommodate the central pulley wheel of M61. A brass plug has been inserted into the bracket to receive a screw to clamp this item to the pillar.

's Gravesande 1747, Vol. I, p. 76, Pl. 14, Figs 2, 3, 4
Adams 1762, pp. 82–3, Pl. 7, Fig. 30, Pl. 40, Figs 137, 138, 139
C 13, 14

Used with M61

Inventory No. 1927-1805

M63 Friction machine

ENGRAVED—ON LARGE WHEEL
'A', 'B', 'C', 'D'

L 165mm, **W** 125mm, **H** 213mm

Adams used this machine to begin the section on motion in the *Mechanics* manuscript. It served to show that friction is due to weight on the moving parts, not the area in contact with them, and that friction wheels are an advantage. 'Friction wheels' are wheels which reduce friction.

A spoked brass wheel of diameter $4\frac{1}{2}$ inches turns on a steel axle of $\frac{1}{4}$ inch diameter with pivots at each end of $\frac{1}{30}$ inch diameter. The pivots each rest on a pair of brass friction wheels 2 inches in diameter. If the wheel is put into motion 'its circumference will go more than the space of a mile before the wheel stops'. Alternatively a spiral watch spring is attached to a strong stud in the base of the instrument and also to the axle. It can be wound by turning the wheel and holding it with a brass catch. When the catch is released the number of 'vibrations' or oscillations the wheel makes before coming to rest gives the relative friction acting, and is smaller the greater the friction. The friction is provided by a small brass lever which can be brought into contact with the axle and hung with weights. It can be held in either of two positions which give different areas of contact with the axle. A third alternative is to fix the pivots into holes above the friction wheels when it is seen that the number of vibrations is greatly reduced, thus proving their benefit. The friction wheels can also be used to count the revolutions of the larger wheel since they turn once for every 60 turns of the large wheel. The whole is set on a rectangular brass base on a mahogany stand.

The machine is a close copy of one described by Desaguliers for similar purposes.

Desaguliers 1744, Vol. I, pp. 261–2, Pl. 18, Fig. 8
Adams 1746, No. 124
Adams 1762, pp. 83–5; Pl. 41, Fig. 140
Rees Cyclopaedia (1819), under Friction
Gould 1923, p. 37
C 41

See also D18, L24

Inventory No. 1927-1116

M64 Instrument to show what is meant by motion and velocity

L 866mm, **W** 120mm, **H** 210mm

Adams added after the above title 'which some authors have confounded'. This refers to the contemporary 'vis viva' controversy which centred on the measurement of motion. Adams took the view that motion was mass times velocity and used this experiment from Desaguliers 'rather as an illustration than a proof'.

The apparatus is simply a pair of mahogany channels with a spring at one end. The spring can be tied back by means of a thread which passes through a hole in the spring and a hole in the backboard to fasten around a brass catch. When the spring is released it knocks two different cylindrical 'weights' or masses along the channels. The distance travelled along the channels is inversely proportional to the 'weight' or mass and so the argument is as follows: the quantity of motion is the same in both cases since it was imparted by the spring. The mass of one is twice the other and the velocity of one is half the other because it travelled half the distance in the same time. Hence the quantity of motion is mass times velocity.

The cylindrical weights for this experiment are probably the two cylinders from M21 since they correspond closely to the dimensions given by Adams.

Desaguliers 1744, Vol. I, p. 44, Pl. 4, Fig. 1
Adams 1746, No. 127
Adams 1762, pp. 87–8, Pl. 42, Fig. 142

Use with M21

Inventory No. 1927-1853

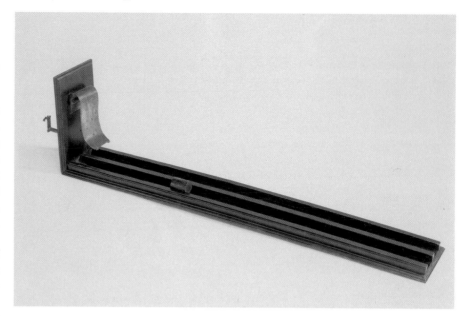

M65 Whirling speculum

L 137mm (whirler), **W** 44mm (whirler), **H** 37mm (whirler), **L** 140mm (mahogany box), **W** 140mm (mahogany box), **H** 100mm (mahogany box), **D** 87mm (speculum), **D** 95mm (tin box), **H** 50mm (tin box)

Adams continued his course on motion with this apparatus invented by Serson which represented a potentially important application of minimizing friction. It is an artificial horizon which spins on its axis in order to maintain a constant plane on a ship which is in motion. It would have been used with an octant or sextant.

A circular mirror of speculum metal forms the top surface of a heavy brass cylinder. Its axis terminates in a cone at the lower end which is slightly rounded off and it is filed square at the upper end. The lower end of the axis, upon which the mirror rotates, lies just above the centre of gravity of the whole piece. The mirror fits into a square mahogany box which has a boxwood pedestal in the centre, upon which is a polished steel concave surface to take the axis. Adams suggests using a few drops of oil of oranges to reduce friction further. The mahogany box has a glass sloping lid so that the mirror can be viewed while being protected from the wind. The mirror is whirled using the whirler which fits on top of the square axle. The whirler is a complex spring-loaded bridge mechanism. The tape is pulled to start the rotation, the whirler is removed, and the lid replaced. When the speculum is not in use it is stored in a tin box.

The instrument represents one of a series of attempts to make an artificial horizon using the gyroscopic principle. The idea was taken up by Short, Shelton, and Graham in the 1740s and '50s. Serson himself died at sea on the sixth HMS *Victory* in 1744.

Adams 1748
Short 1751, p. 352
B.J. 1754, pp. 446–8
Adams 1762, pp. 88–91, Pl. 42, Figs 144–9
Robertson 1805, Vol. 2. p. 253
C 225
Hewson 1983, pp. 86–8
Inventory No. 1927-1123

M66 MS description of artificial horizon

after 1751
L 240mm, **W** 150mm

The manuscript relates to the instrument above. The first part is a copy of a paper read to the Royal Society by James Short in 1752 entitled 'An account of an horizontal top, invented by Mr. Serson, by Mr. James Short FRS, Read February 6th 1752'. Then follows in the same hand a copy of Newton's first law of motion from Motte's English edition of the *Principia*, Vol. 1, p. 19. Following this, two paragraphs describe the artificial horizon referring to 'a pamphlet on the Sea Quadrant published in the year 1748'. The whole is five pages long.

Adams 1748
Short 1752
B.J. 1754, pp. 446–8
Adams 1762, pp. 88–91; Figs 144–9
Robertson 1805, Vol. 2, p. 253
C 323
Hewson 1983, p. 352
Inventory No. 1927-1124

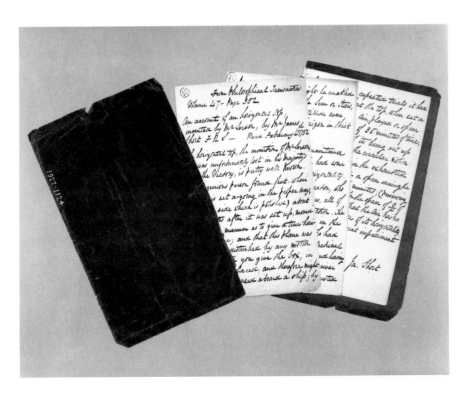

M67 Diagonal machine, or parallelogram of velocities

L 410mm, **H** 420mm, **W** 90mm

Under the titles 'Compound Motion' and Sir Isaac Newton's 'Second Law of Motion', Adams writes 'If a body be acted upon by two forces making any angle with each other the body will move in the diagonal of a parallelogram'. This apparatus demonstrates the principle.

A mahogany square frame with inside dimensions of one foot is supported vertically on a mahogany stand. A brass pulley in a bracket can move along two brass wires running across the top of the instrument between two brass supports. A thread, originally silk, is attached to one of the supports, passes over the movable pulley, and carries a lead ball,

originally a brass ball. Another thread is attached to the bracket, passes over a pulley on the other support and also carries a weight. This thread can be pulled to move the bracket and hence move the ball diagonally across the frame. A lattice of threads helps to mark the path of the ball.

Desaguliers 1744, Vol. 2, p. 289
Adams 1746, No. 131
Nollet 1754, Vol. 2, p. 13
Adams 1762, p. 91; Pl. 42, Fig. 143
C 27

See also D25

Inventory No. 1927-1115

M68 Apparatus for oblique and compound collision

SCALE—TABLE END
80 (1) 0 (1) 80
SCALE—SIDE FRAMES
0 (1) 30

H 680mm (table end), **W** 380mm (table end), **T** 35mm (table end), **H** 840mm (frame), **L** 980mm (frame, max)

This apparatus was made to fit on the base of the Philosophical Table. It is in two parts: a mahogany semicircular extension to the table and a frame on which are supported the 'mallets' and ball. The semicircular extension is fitted onto the end of the table by three 'tennants' which slot into place beneath the surface. It is held firm by two mahogany screws. An ivory scale is inlaid around the circumference of the semicircle marked in degrees from 0° at the centre to 80° each way. The lower hinge of the frame fits under the semicircle and can be clamped by a further two mahogany screws. The degree of opening of the frame can be varied to allow the mallets to hit the ball from different directions. The arcs of the frame have ivory scales marked from 0° to 30° so as to measure the arc through which the mallets fall. The ivory cylindrical mallets are suspended from brass brackets; their radius is 28 inches. The ball, which is also of ivory, has a diameter of $1\frac{1}{2}$ inches and sits in a small hollow on the table surface to keep it still until it is struck.

Adams does not give any particular experiments to be performed on this apparatus but it can be assumed that he would have used it as 's Gravesande did. The mallets were used together or independently to impart their force to the ball whose resulting motion was observed. Adams's apparatus was very much more elegant than 's Gravesande's in this case.

's Gravesande 1747, Vol. I, p. 277, Pl. 39, Fig. 1
Nollet 1754, Vol. 2, p. 19

Adams 1762, p. 92; Pl. 44, Fig. 150; Pl. 45, Fig. 150, No. 2
C 9

Used with M11

Inventory No. 1927-1223

M69 Brass plate for projectile experiment

L 201mm, **W** 199mm, **T** 4mm

This brass plate was used to show that a 'weight thrown up from a body in motion will fall down on the same point that it falls on when the body is at rest'. However, the experiment described does not show this, as the weight is not thrown up but lifted and the 'body' below, which is the brass plate, does not have uniform velocity.

The plate is hung from the iron arm M5 by strings attached to the holes in the four corners. The iron arm is attached to the top of the pillar and the plate is held just above the floor. A weight is placed on the plate and a string is passed from it over the pulley at the centre of the iron arm and then over the pulley at the end of the arm nearest the pillar. By means of this string the weight can be raised or lowered as the plate is swung beneath it. In effect the weight can be considered a pendulum which maintains its swing independently of the plate.

This experiment was taken from Desaguliers.

Desaguliers 1744, Vol. I, p. 300, Pl. 24, Fig. 1
Adams 1762, p. 93, Pl. 46, Fig. 151

Inventory No. 1927-1695

M70 Cycloidal channels

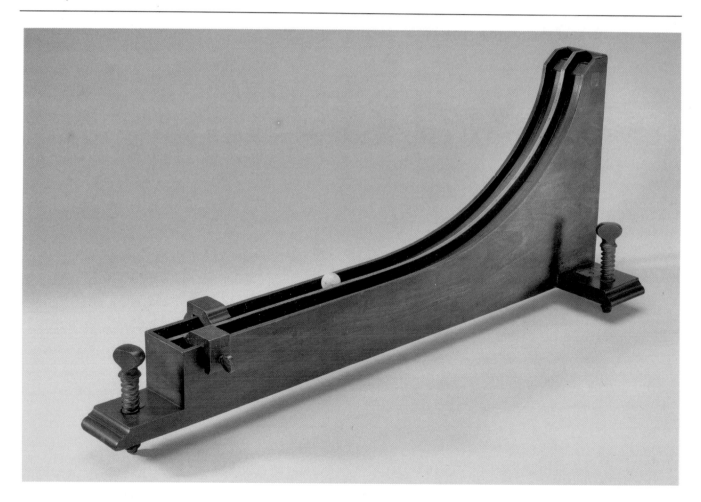

SCALE
9 (1), F, 1 (1) 14

L 760mm, **H** 315mm, **W** 245mm, **D** 33mm (ivory ball)

Adams used this apparatus to compare the velocity of falling bodies. The particular property of a cycloid is that no matter where a body starts it will always take the same length of time to reach the lowest point.

Ivory balls about half an inch in diameter are placed in the two mahogany cycloidal channels. A strip of holly between the channels carries a scale marked from 1 to 9 ascending the cycloidal section and from 1 to 14 along the horizontal section. The lowest point of the cycloid is marked with the letter F. Two mahogany buffers can be moved along the channels to stop the balls at any required position. A brass plumb bob hangs at the back of the instrument and the whole can be kept horizontal by means of the three mahogany levelling screws. The ball now with the apparatus is probably not original because it is too large; it is more likely that the balls in J24 were used.

's Gravesande described two experiments to be performed with the apparatus. Firstly the buffers were clamped at F and balls were let roll from various points on the cycloid. It was observed that the balls always reached F at the same time. Then the clamps were attached at points marked four and six from F on the horizontal section. The balls were set to roll from points four and nine respectively on the cycloid and it is observed that again they reached the buffers at the same time. This demonstated that the starting heights were proportional to the squares of the velocities attained.

Adams copied this apparatus directly from 's Gravesande, who appears to have invented it.

's Gravesande 1747, Vol. I, p. 88, Pl. 15, Fig. 7
Adams 1762, p. 96, Pl. 47, Fig. 153
C 5

Inventory No. 1927-1127

M71 Attachment for pillar

L 315mm, **W** 91mm, **H** 81mm

Adams made several attachments for the pillar which he illustrated on Plate 7 of the *Mechanics* manuscript. This is the one marked 'for pulleys and pendulums'.

It is a mahogany arm with a ring to fit the pillar at one end and a square hole at the other for the pulleys with square tails. It is fitted with several brass plates: one carries five steel hooks, another has three small holes and is slightly raised from the arm, and another has a larger hook and hole. Four brass keys are screwed into four holes in the mahogany. The apparatus was copied with slight modification from 's Gravesande who described it in detail.

's Gravesande 1747, Vol. I, p. 36, Pl. 4, Fig. 9
Adams 1762, Pl. 7, Fig. 29, Pl. 29, Figs 105, 106, 107, 108, 109, p. 96, Pl. 48, Figs 154, 155, 156

Inventory No. 1927-1801

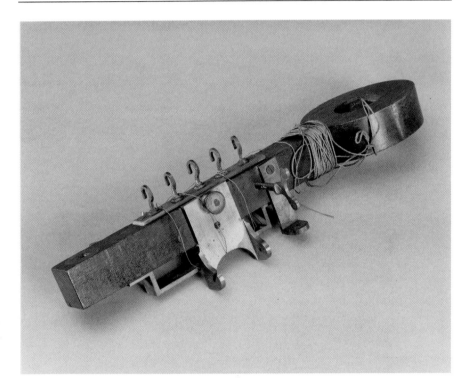

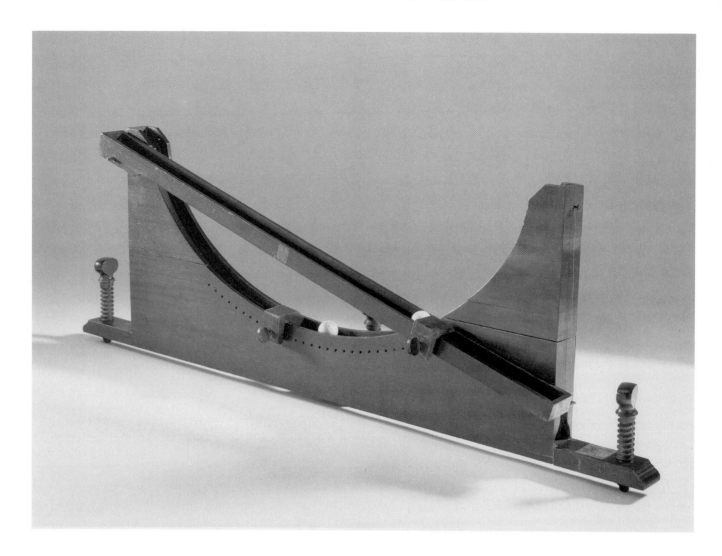

M72 Straight and cycloidal channels

L 880mm, **H** 310mm, **H** 190mm

This is a similar piece of apparatus to M70. Ivory or brass balls of $\frac{1}{2}$ inch diameter can be rolled in a mahogany cycloidal channel and their descents compared with those of similar balls rolling in a straight channel. The cycloidal channel is on three feet and can be made horizontal by means of levelling screws. The straight channel is pivoted at one end by a screw which attaches it to the top of one side of the cycloid. The other end can be attached by means of a steel pin which can be fixed into one of a series of holes around the lower part of the cycloid; thus the inclination of the straight channel is variable. The balls can be stopped by two mahogany buffers. A brass plumb bob hangs at one end of the instrument.

's Gravesande describes only one simple experiment with this apparatus. The channels are placed along the edge of a table in such a manner that the straight one can be freely raised or lowered while the third leg provides a counterbalance on the opposite side. The straight channel is pinned to the side of the cycloidal channel at any angle made available; it can be inclined between about 45° and 20°. The buffers are clamped at the lower point of intersection of the two channels so that when two balls are rolled simultaneously from the top of each channel it will be seen and heard that the one in the cycloid takes less time.

Like M70, this apparatus was copied directly from 's Gravesande.

Adams 1746, No. 129
's Gravesande 1747, Vol. I, p. 97, Pl. 17, Fig. 3
Adams 1762, p. 97, Pl. 48, Fig. 157
C 6

Inventory No. 1927-1128

M73 Cycloidal pendulum

H 510mm, **L** 780mm, **W** 107mm, **D** 25mm (ball)

After studying the oscillation of pendulums on the back board of the Philosophical Table, Adams considers the pendulum with cycloidal cheeks. He copies the description from 's Gravesande which does not correspond to his own instrument and gives only a sketch of the cycloid as an illustration.

The frame is entirely of mahogany and consists of the cycloidal cheeks and a cycloidal arc with a central vertical piece. The pendulum bob is a brass ball which may not be original since Adams quotes four balls of $1\frac{1}{2}$ inches diameter—two brass and two ivory. The instrument is clamped to the pillar of the table and the motion of the cycloidal pendulum is compared with the unconstrained pendulum suspended just in front of it from an arm on the pillar described above. The lengths of the pendulums are half that necessary to give a period of one second.

Adams observes merely that the periods of oscillation are equal when the lengths are equal, and that the weight of the bob does not affect the period; the latter is demonstrated by the interchanging of the brass and ivory balls. Adams does not describe any experiments beyond these and does not discuss amplitude of swing.

's Gravesande 1747, Vol. I, p. 91, Pl. 16, Fig. 3
Adams 1762, pp. 96–7, Pl. 50, Fig. 158, No. 1
C 3

Used with M11, M71   **Inventory No.** 1927-1200

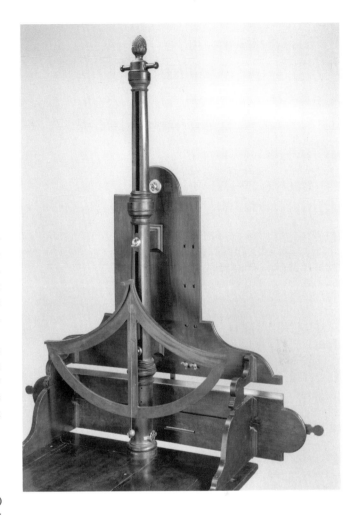

M74 Clinometer with spirit level

L 183mm, **H** 183mm, **T** 15mm

In the *Mechanics* manuscript Adams mentions very briefly a pocket spirit level and this quadrant, but the corresponding illustrations were not provided or have not survived. The pocket spirit level has not been traced.

Adams described the instrument as 'a quadrant six inches radius with a spirit tube to elevate or depress any plane to any inclination'. The quadrant is in brass with the sides of the right-angled frame having flat edges for placing on the plane in question. The index, which is about six inches long, slides over a scale marked from 0° to 90° each way. A vernier allows readings to be taken to tenths of a degree. A spirit level is attached to the index so that when it is horizontal the angle of elevation of the plane in contact with the side of the frame can be read off the scale.

Adams 1762, p. 98
C 231   **Inventory No.** 1927-1917

M75 Path of projectile apparatus

L 615mm, **H** 430mm, **W** 198mm, **D** 25mm (ball)

The two projectile apparatuses in the collection are very similar, both being based on 's Gravesande's design. Fortunately Adams gave a detailed description of his instrument which enabled it to be distinguished from Demainbray's.

The apparatus is in three parts: an upright piece containing a channel, a backing board, and the base. The upright piece is of solid mahogany and is weighted at the lower end for stability. The channel is lined with a brass strip about 1 inch wide and was intended for a marble ball $\frac{3}{4}$ inch in diameter. The upright piece fits into the base which is a mahogany trough with a cross-piece at one end and three levelling screws. The backing board, which is also mahogany, slots into the side of the upright piece and has a pin which attaches it to the base. It is marked with a parabola and has four brass rings screwed at intervals along the curve.

If the ball is allowed to roll down the channel, it leaves it with a horizontal velocity. Gravity acting on the ball makes the resulting motion parabolic, which is confirmed if the ball starts from the highest point by it falling through the rings. A nest of cotton wool was originally provided to avoid damaging the ball, but this is now missing, as is the plumb bob.

's Gravesande appears to have been the inventor of the apparatus.

's Gravesande 1747, Vol. I, p. 123, Pl. 19, Fig. 3
Adams 1762, p. 6, Pl. 3, Fig. 10, p. 94
C 7

See also D24

Inventory No. 1929-114

M76 Trolley for projection of bodies experiments

L 312mm, **W** 182mm, **H** 155mm

Adams uses this apparatus to show that a body thrown vertically upwards from a body in uniform motion, although describing a curve, falls upon the same point on the body as it would fall if the body had been at rest.

The trolley was made to roll along two mahogany boards $\frac{3}{4}$ inch wide and 10 feet 3 inches long, forming two sides of a trough which rested on the Philosophical Table at one end and on a stand at the other. The boards are now missing. The trolley is a rectangular piece of mahogany set on four rollers. At the centre is a tube and funnel containing a movable wooden cylinder. Under the cylinder a brass hammer is held against two long brass springs by a peg at the back of the trolley. An ivory ball was placed in the tube and the trolley was put on the boards at the end furthest from the pillar of the Philosophical Table. The peg was attached by thread to the end of the trough. A weight was used to draw the trolley along the boards using a system of pulleys. When the trolley reached the point where the thread from the peg became taut, the peg released the hammer which struck the wooden cylinder and ejected the ball. The sketch shows the desired outcome: the trolley moves at a uniform speed and the ball returns into the tube having described a parabola. However, it is difficult to see how uniform motion was achieved.

Adams devised the apparatus but the concept of the experiment was taken from Nollet.

Nollet 1754, Vol. 2, p. 31, Pl. 3, Fig. 13   Adams 1762, p. 101, Pl. 51, Fig. 160–2
Inventory No. 1927-1795

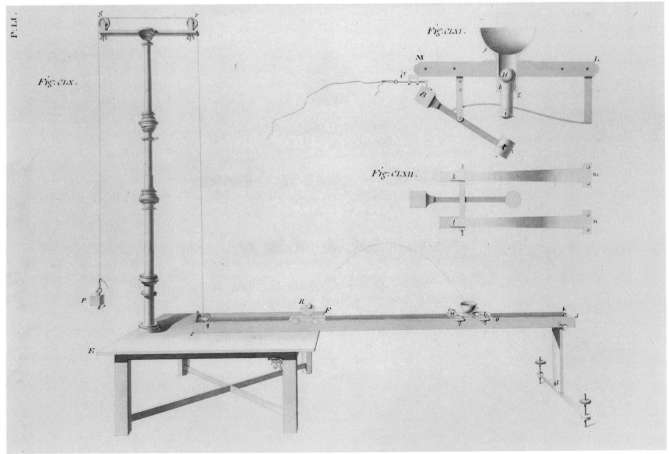

Adams 1762

M77 Central forces machine

L 1200mm (variable), **H** 500mm, **W** 900mm (variable), **L** 755mm (arm)

This is the most intricate piece of apparatus that Adams made for the mechanics course. The use of the Philosophical Table enabled him to improve substantially on 's Gravesande's design for a central forces machine; instead of being a cumbersome piece of furniture it is relatively compact and portable.

Like 's Gravesande's machine it consists essentially of two vertical 'axes' or axles which are made to rotate and a wheel which is used to turn them. The 'axes' and wheel are each clamped to the surface of the Philosophical Table by two screws acting on strong brass discs. The discs are connected to similar discs above by three small brass pillars. The iron axes pass through the upper disc and terminate in a cone which fits into a conical socket in the lower disc. Thus the axes are kept upright. The 'axes' are squared above the upper discs to take the mahogany wheels which are placed over them and firmly screwed down. The wheels, which take the rope, have diameters of 4, 5, 6, and 6 inches diameter on one axis and 3, 4, 4, and 4 inches on the other.

A strong spring is fastened to each axis above the mahogany wheels and the planet-bearing apparatus can be attached to this. The turning wheel is mounted similarly using the brass discs and pillars described. It is a horizontal mahogany spoked wheel 16 inches in diameter which is turned by a steel and mahogany handle from above. The motion is communicated to the axes by means of a catgut 'rope'. This is the basic mechanism which provided whirling for a number of experiments.

The principal experiment involved the use of the two iron 'rulers' or bars to measure the centripetal forces associated with whirling various weights. Numerous accessories are included. Small brass plates screw into each spring from above. Into each of these are screwed two small brass pillars with plates across their tops. These pillars and plates provide frames to hold small vertical brass cylinders in a similar manner to the brass pillars and discs holding the main axes. The small cylinders, which screw into the main axes, weigh 2 oz Troy but the weight can be increased by the addition of cylindrical lead weights with holes in their centres which can be slotted over the cylinders. These have weights of 1, 2, 4, 8, and 16 oz Troy. The top plates are attached to the iron 'rulers' which are $\frac{1}{2}$ inch thick and 'scarce exceeding a quarter of an inch wide', except above the plates where they are $\frac{3}{4}$ inch wide. Each ruler has a small pulley situated above the plate so that a thread from the small brass cylinder can pass above the ruler over the pulley to the weight which can slide along the ruler. The movement of the weights along the ruler can lift the small brass cylinders, but only slightly. Since the moment the cylinder is lifted is crucial to the experiment, Adams provides a bell and hammer on top of the ruler. By means of a spring mechanism, a thread from the base of the cylinder holds the hammer from striking the bell, but when the cylinder rises a little the hammer is released and the bell sounds.

A cylindrical brass weight, which can slide along the iron 'ruler', is connected by a thread passing over a pulley to the vertical brass cylinder. A mill-headed nut is screwed to the end of the 'ruler' to prevent the weight from falling off. The 'ruler' is marked so that the position of the base of the cylinder nearest the end gives the distance of the centre of gravity of the weight from its centre of rotation. It is marked from 10 to 36 in half inches. Each brass cylindrical weight is 2 inches long, $1\frac{3}{4}$ inches diameter and weighs 9 oz Troy. Outer rings weighing 3, 6, and 9 oz can be added. Brass tails with screw nuts can be attached to the opposite ends of the

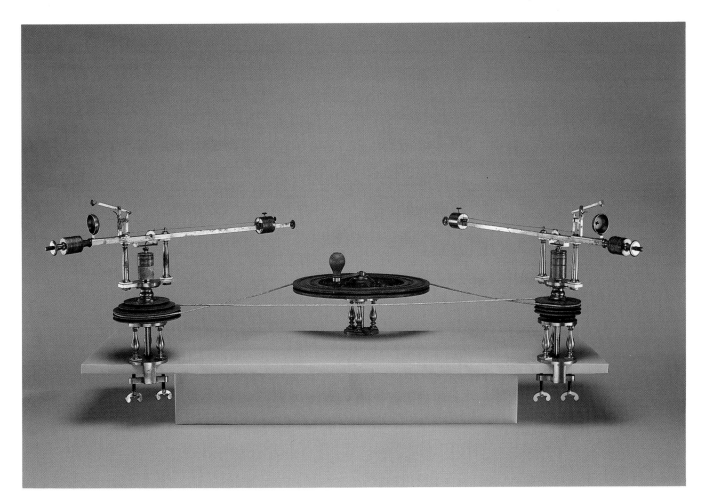

rulers to act as counterweights and extra lead rings can also be added depending on the weight being used in the experiment.

Adams does not give an account of the experiments intended with this apparatus but presumably they correspond to the series described by 's Gravesande, although the latter's apparatus is rather different. For example, if cylindrical weights are placed at equal distances along the two iron rulers, when the machine is in operation they will lift weights on the small brass cylinder in the same ratio to their own weights simultaneously. Thus it is shown that the central force depends on the 'quantity of matter', or mass, if the speed of rotation and distance from the centre are equal. Other experiments involving various combinations of the qualities mass, speed, and radius could be performed. The iron 'rulers' were also used independently for experiments on oblique forces when they were attached to the stand M20.

Besides 's Gravesande's own machine similar apparatus exists in the Van Marum collection at Haarlem and in the Poleni collection at Padua.

Hauksbee 1714, Mechanics, Pl. 5, Fig. 6.
Adams 1746, No. 135
's Gravesande 1747, Vol. I, pp. 129–33, Pls 20, 22, 23, pp. 137–44
Nollet 1754, Vol. 2, Pl. 6, Figs 27, 28
Adams 1762, pp. 102–9, Pl. 52, Fig. 163; p. 71, Pl. 37, Figs 123, 124, No. 1, Pl. 53, Figs 164–78

C 47
Bonelli 1968, Nos 840, 857
Turner 1973, Item 49
Saladin 1986, No. 189

Used with M78, M79, M80

Inventory No. 1927-1118

M78 Five glass tubes for experiments on the whirling table

L 265mm, **W** 217mm, **H** 141mm

As for the central forces machine, Adams describes this apparatus without giving details of the experiments to be performed. The five tubes are designed to be attached to a circular mahogany table, the 'whirling table', so that they are all directed to the centre of revolution. The whirling table would have been placed on one axle of the central forces machine. The tubes are set at an angle of 25° to the horizontal. The tubes originally contained mercury with water, oil of tartar per diliquum and water, water with a cork cylinder, and water and a lead ball. One tube was empty. It is now difficult to distinguish them except for the third and the last, which has a glass stopper. Two contain liquid; the lead ball is missing. The effect of the central force was observed. At high speeds the heavier objects would move to the top of the tubes, apparently defying gravity.

Although Adams's design is his own, the principle of the experiment was taken from 's Gravesande. The whirling table has not survived.

's Gravesande 1747, Vol. I, p. 136, Pl. 21, Fig. 4
Adams 1762, pp. 19–110; Pl. 53, Fig. 178
C 47

See also D25

Used with M77, M79

Inventory No. 1927-1854

M79 Hinged slot and two rectangular pieces for experiments on the whirling table

L 177mm, **W** 15mm, **L** 45mm (base), **L** 150mm (pieces), **W** 50mm (pieces), **H** 20mm (pieces)

Although the whirling table mentioned in the previous entry has not survived, some small pieces of apparatus intended for it have done. These brass pieces were used in experiments on central forces.

The T-shaped piece of brass has a slot at the end of the long arm across which passes a brass peg. The base was screwed onto the surface of the whirling table so that the arm made an angle of about 70° with the horizontal. An ivory ball was attached by thread to the centre of the table and hung over the slot. The peg prevented the thread from rising. When the table was stationary, the ball hung with the thread vertical, but as the speed of rotation was increased, the angle of the thread approached the horizontal.

The other 'two little rectangular pieces' of brass were also attached to the table. Two ivory balls, one 4 oz and one 2 oz, were joined by a thread which passed through the central pipe of the table, a small protrusion with holes for threads, and allowed them freedom to move radially. It was found that they would reach equilibrium when the table was in motion when the smaller ball was twice as far from the centre as the larger ball. The rectangular pieces were placed so that their shorter arms prevented the balls from flying off the table and the longer arms prevented them from being left behind by the motion of the table.

Desaguliers 1744, Vol. I, pp. 302–3
Adams 1762, p. 108, Pl. 53, Fig. 176; p. 109, Pl. 53, Fig. 177

Inventory No. 1927-1868

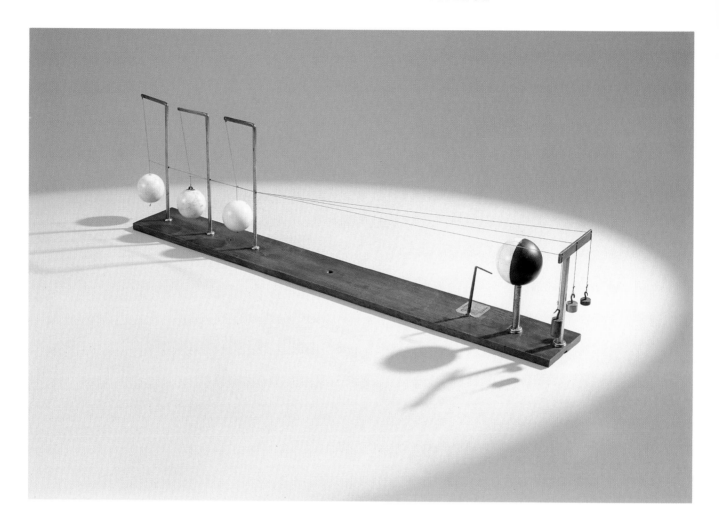

M80 Apparatus to demonstrate tidal motion

L 645mm, **W** 90mm, **H** 165mm

The apparatus in the collection does not quite correspond to that described by Adams in the *Mechanics* manuscript but it is essentially similar. It demonstrates that water rises on both sides of the Earth simultaneously to give high tides, and not just on the side of the Earth nearest the Moon.

It is set on a mahogany rectangle 2 feet by 3 inches with the model of the Moon on a brass stand at one end and three ivory balls hanging from brass supports at the other. The central ball represents the solid Earth and the outer balls the sea on each side of it. The Moon is represented by a ball 2 inches in diameter, half ivory and half ebony to give a white full moon from the position of the balls. The balls are each held by a thread passing through small holes in their stands and small holes in a brass T-shaped stand behind the Moon. Weights are hung from the threads, the nearest ball has the largest weight and the farthest the smallest, to represent the differing pull of the Moon according to distance.

When the apparatus is stationary the differing pulls of the weights act to separate the balls, although they do not hang vertically. To make the central ball hang vertically the apparatus was placed on the whirling table. At a certain speed the centrifugal force of the central ball just balanced the pull of the weight representing the Moon. The outer ball will have a centrifugal force in excess of the pull of the Moon and the inner ball will have a centrifugal force less than the pull of the Moon. Hence the balls will separate further and show the tides rising each side of the Earth more clearly.

Adams does not attribute this apparatus to anyone and it appears to be original.

Adams 1762, pp. 110–2; Pl. 54, Fig. 179
C 144

Inventory No. 1927-1852

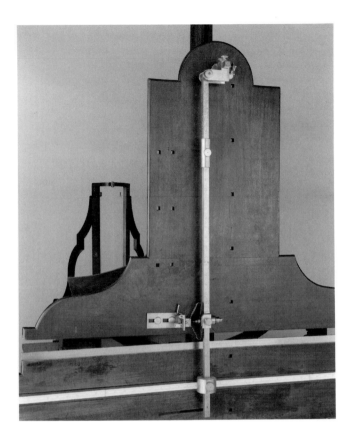

M81 Pendulum moved by a spring, with accessories

SCALE

$0 \left(\frac{1}{10}\right) 36; 36 \left(\frac{1}{10}\right) 0$

L 920mm, **W** 97mm (brass bracket), **D** 62mm (weight), **L** 112mm (suspension), **L** 90mm (iron plate)

The pendulum set in motion by an impulse from a spring is another intricate piece of apparatus Adams provided for the mechanics course. Although he does not give any experiments, the text and apparatus are copied from 's Gravesande and so it is clear that it would have been used to perform the experiments described by 's Gravesande on the relationship between velocity, force, and 'quantity of matter', or mass.

The pendulum is made of iron, and is just over 3 feet long with a square cross-section and sides of just over $\frac{1}{2}$ inch. It is marked in inches and tenths down its length. It is attached to the back board of the Philosophical Table by means of a brass bracket, which fits onto a square-section iron rod which is fixed perpendicularly to the top of the back board. The pendulum is held on conical bearings between screws to eliminate any vibration. Only one of the three original brass cursors now remains, but it is the one that carries the spring. It can slide along the pendulum and is fixed by a screw. On one side it carries the steel spring and on the other it can carry the impact head M83. There is a central iron tongue which

protrudes between the jaws of the spring and is used to attach the spring to the back board from where it can be released. This was done by means of two iron plates, only one of which now remains. The plate screws into one of the double holes in the back board. The piece that is perpendicular to the back board contains a mechanism for holding the tongue. A weak spring clamps two arms together and can be released by a small catch. The design of the tongue enables it to be pushed into the plate in one of three positions, compressing the main spring to differing degrees. In addition to the cursor, spring, and plate with catch, the accessories include two cylindrical weights which can be moved along the pendulum. One is $1\frac{1}{2}$ inches long and weighs 24 oz, and the other is 3 inches long and weighs 6 oz.

The ruler on the lower backing board of M11 is used in these experiments to indicate the height of the pendulum bob before it is released.

's Gravesande 1747, Vol. I, p. 172, Pl. 25, Fig. 2
Adams 1762, pp. 112–6; Pl. 55, Figs 181, 184; Pl. 58, Figs 186, 188; Pl. 59, Figs 190, 191, 192
C 21

Used with M11, M82/3/5/7/8, M113

Inventory No. 1927-1220

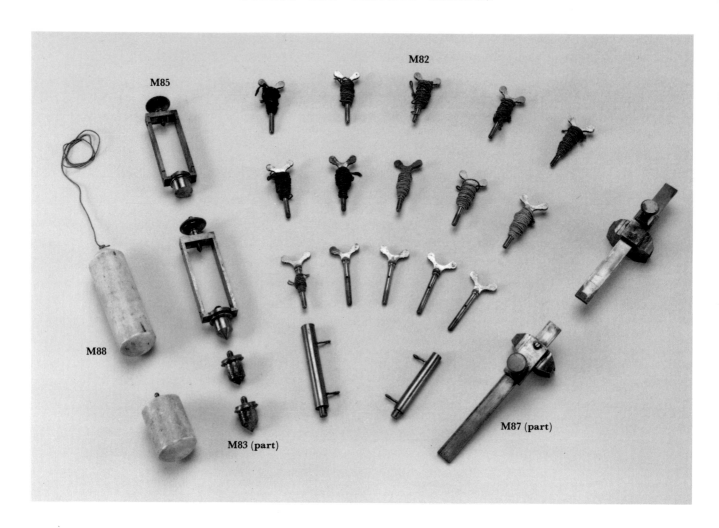

M82 Fifteen buffer screws

L 63mm, **W** 27mm, **D** 5mm

These long brass screws are illustrated on plate 5 of the *Mechanics* manuscript with the pulleys; however, they belong with the impact apparatus for use on the Philosophical Table. They fitted into plates which could be inserted into the fifteen holes in the back board of the table, and acted as buffers for the pendulum. 'By a small turn of one screw the tremendous motion of the axis is hindered.'

's Gravesande 1747, Vol. I, p. 173, Pl. 25, Fig. 1
Adams 1762, Pl. 5, Fig. 21, p. 114, Pl. 56, Fig. 194, No. 2

Inventory No. 1927-1222/1

M83 Five impact heads for pendulum experiments and the collision of bodies

L 33mm, **D** 15mm (max)

There were originally six steel impact heads although only five now remain. One is flat, one is hemispherical, and three are conical with vertices of 85°, 85°, and 102°. They can be attached to the cursors on the pendulum M81 or to the fronts of the brass cradles provided for the collision of bodies experiments. They are 'equally prominent and of the same weight': $1\frac{1}{4}$ oz. They were used to make an impression in clay and also in the collision between cradles in the experiments on forces, following 's Gravesande.

's Gravesande 1747, p. 196, Pl. 28, Fig. 7
Adams 1762, p. 114, Pl. 59, Fig. 189

Inventory No. 1927-1222/2

M84 Two double rulers with hooks, for experiments on the collision of bodies

L 360mm, **W** 30mm, **H** 60mm

These brass rulers were attached to the cross bars of the 'iron arm of the great table' M5, for experiments on the collision of bodies. The upper rulers are attached to the cross bars and carry the lower rulers which have six sliding iron hooks which can be adjusted to give the required separation between the rectangular cradles M85. When many threads were involved Adams advised the use of different colours. The velocity of the cradle was judged from the distance it was initially pulled out of the vertical by lining up the threads and observing the ruler on the backing board.

See M5 for photograph

's Gravesande 1747, Vol. 1, p. 180, Pl. 28, Fig. 2, Pl. 27, Fig. 1
Adams 1762, pp. 116–[117], Pl. 60, Fig. 194, Pl. 61, Fig. 196

Inventory No. 1927-1221/2

M85 Rectangular cradles for collision of bodies experiments

L 115mm, **W** 35mm, **T** 12mm

These brass rectangular cradles were suspended from threads from hooks on the brass rulers M84, and were swung to collide with each other or with the clay referred to in M90. The rectangular weights M86 can be held inside the cradles by the conical screw at the back while at the front the cradles carry the impact heads M83. There are slots in the upper surface for the threads. They are made of brass and together with the impact heads weigh $5\frac{1}{2}$ oz. Their weights can be doubled, trebled, or otherwise multiplied by the addition of the rectangular weights M86.

The cradles were used in a series of experiments devised by 's Gravesande to compare the force of various weights at various velocities. The velocity was estimated as described in the previous item. One cradle was lifted to a certain height and released to collide with a stationary cradle. The velocity imparted to the second cradle was observed. 's Gravesande concluded that the forces are equal when the squares of the velocities are inversely proportional to the masses.

's Gravesande 1747, Vol. 1, p. 179–86, Pl. 27, Fig. 1, Pl. 28, Fig. 4
Adams 1762, p. [117], Pl. 16, Figs 194, 197

Inventory No. 1927-1222/3

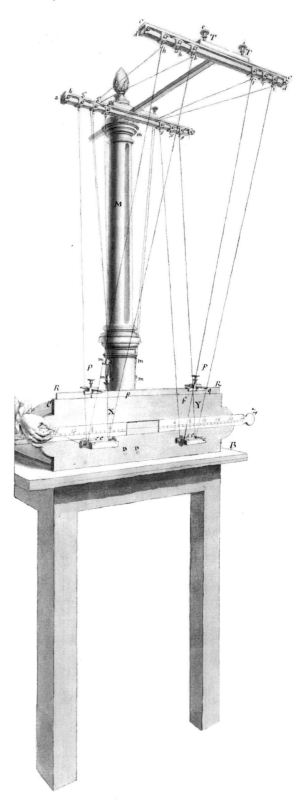

Fig. CXCIV.

Adams 1762

M86 Set of seven rectangular weights

ON SEPARATE WEIGHTS
  '2|oz|5 1/2 Tr'
  '4|oz|11 Tr'
  '5|oz|16 1/2 Tr'
  '6|oz|27 1/2 Tr'
  '8oz|38 1/2 Tr'
  '9oz|44 Tr'
  '16|oz|82 1/2 Tr'
ENGRAVED
  '2', '4', '5', '6', '8', '9', '16'

L 70mm (all weights), **W** 15mm (smallest), **W** 25mm (all others), **H** 177mm, **H** 90mm, **H** 77mm, **H** 56mm, **H** 34mm, **H** 12mm, **H** 8mm

These lead-filled brass weights were made by Adams for the experiments on 'innate' forces, or inertia and momentum, and the collision of bodies described above. They fit into the rectangular cradles by means of small screws at one end and cones at the other. The total weight of the frame could be multiplied by 2, 3, 4, 5, 6, 8, 9, or 16 by adding the appropriate weight. The corresponding values are $5\frac{1}{2}$ oz, 11 oz, $16\frac{1}{2}$ oz, $27\frac{1}{2}$ oz, $38\frac{1}{2}$ oz, 44 oz, and $82\frac{1}{2}$ oz. These values and the numbers are engraved on each. However they have been numbered wrongly. The number 3 is now missing.

Adams 1762, p. 118, Pl. 61, No. 2, Fig. 198 C 62

Inventory No. 1927-1797

M87 Five indices for collision of bodies apparatus

larger two: **L** 153mm, **W** 43mm
smaller three: **L** 112mm, **W** 37mm, **D** 7mm

It is not clear why Adams used two types of indices for the collision of bodies apparatus. 's Gravesande used one type, that corresponding to the larger pair of these. The brass indices were slotted onto the brass ruler on the upper surface of the mahogany lower back board of the Philosophical Table. They were adjusted so that they did not quite touch the thread before the cradles M85 were released, or did not quite touch the thread when they ascended after collision. They were used to give a measure of the velocities of the cradles.

Adams 1762, p. 118, Pl. 62, Figs 199, 200   **Inventory No.** 1927-1222/4, -1867

M88 Two cylinders

L 127mm (cylinder 1), **L** 66mm (cylinder 2), **D** 35mm (both)

Adams does not describe precise experiments for these cylinders but it is assumed that 's Gravesande would be followed. The cylinders are made of ivory $1\frac{1}{2}$ inches in diameter, slightly less than $2\frac{1}{2}$ and $4\frac{3}{4}$ inches long, so that they weigh $3\frac{3}{4}$ and $7\frac{1}{2}$ oz respectively, 'one exactly double the other'. One end of each is hemispherical, and the other is conical. To the conical ends are attached small brass balls with holes, through which threads can be passed. The balls were simply dropped by hand onto soft clay or wet marble from a height determined by a vertical ruler. The resulting spots on the marble or marks in the clay were compared. Adams improved on the release mechanism for the cylinders and other bodies as can be seen in the next item.

's Gravesande 1747, Vol. 1, p. 193, Pl. 30, Fig. 1
Adams 1762, p. [118a], Pl. 63, Fig. 201   **Inventory No.** 1927-1222/5

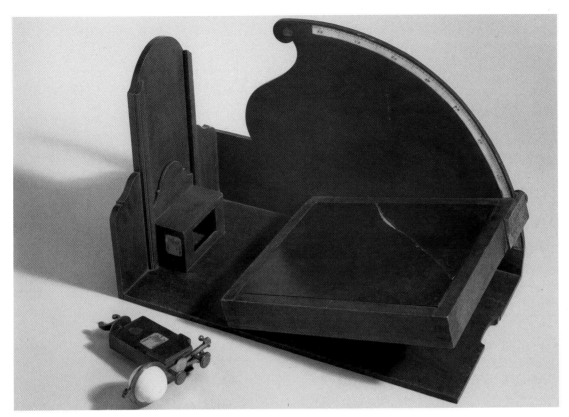

M89 Apparatus to compare the 'force of falling bodies' or rebound of falling bodies

LABEL
'203'

L 540mm (overall), **W** 340mm, **H** 350mm, **L** 156mm (pincer), **D** 51mm (ball)

Adams departs considerably from 's Gravesande for this apparatus and followed Nollet's more elegant design, though the principle is the same. The apparatus served for two experiments: brass balls were dropped onto soft clay for non-elastic collision, and ivory balls were dropped onto marble for elastic collision. Only the materials for the latter remain.

The apparatus was used on the Philosophical Table where it was placed beneath the pillar. The brass pincers could be attached to a piece of mahogany which itself could be attached to a movable arm of the pillar. The balls were released by the pincers to fall on a mahogany tray containing either wet black marble or soft clay. The tray can be inclined to the horizontal by means of a hinge and screw which acts on the backboard. An ivory scale marked in degrees up to 80° gives the angle of inclination. In the case of elastic collision the balls can be caught in a small mahogany box which slides vertically on a mahogany upright. The marble is about 1 foot by 10 inches and $1\frac{1}{2}$ inches deep; unfortunately it is now cracked. The dimensions of the hole in the box are 2 by $3\frac{1}{2}$ inches.

Originally there were six balls: three brass and three ivory of which only one ivory ball remains. The brass balls were all of the same diameter and their weights were as 3:2:1. There

were solid ivory balls with diameters of 2 inches and $1\frac{1}{2}$ inches, and this ball, which is hollow, 2 inches in diameter, composed of two hemispheres which screw together. A piece of lead can be added to give extra weight and also a helical brass spring possibly to give extra elasticity. The pincers can be used to drop two balls simultaneously: one rests on the circular hole and one is clasped by a thread in the jaws of the pincer. There were originally two pairs of pincers, one for the 2 inch balls and one for the $1\frac{1}{2}$ inch ones. The latter is M92.

The apparatus could be used to observe the relation between the forces, masses, and velocities of falling bodies in non-elastic collisions and the relation between forces, velocities, masses, and angles of inclination in elastic collisions. Adams is not specific about any experiments performed except to say that two balls were dropped simultaneously and the rebound observed. The fact that the angle of incidence equalled the angle of rebound could have been confirmed from the position of the box receiving the balls. 's Gravesande suggests wetting the marble slightly so the impact of the ball leaves a spot, the size of which can be compared with other spots. Nollet and Adams remark that the spot is elliptical when the marble is inclined.

Nollet 1754, Vol. I, p. 312
Adams 1762, p. 119, Pl. 63, Figs 202–204; Pl. 64, Figs 205, 206
C 23   **Inventory No.** 1927-1201

M90 Fixture and cavity for clay in experiments on the 'destruction' of forces

L 95mm, **W** 62mm, **H** 305mm, **L** 101mm (screws)

The solid piece of mahogany was clamped to the lower back board of the Philosophical Table by means of the brass screws. One of its sides contains a cavity 4 inches long, 2 inches wide, and $1\frac{1}{2}$ inches deep. The cavity, filled with clay, was held in a vertical position and one of the cradles carrying the impact heads M83 could be swung so that it made an impression in the clay. Although the impressions in the clay were not actually measured it was possible to compare them. The masses of the cradles could be altered as before by the addition of the rectangular weights M86, and the velocity of the impact could be altered by the amplitude of the swing. The relationship between force of impact, mass, and velocity could then be investigated. The cavity was also used in a similar manner when the impact heads were attached to the pendulum M81.

's Gravesande 1747, Vol. I, p. 196, Pl. 25, Fig. 8, p. 200, Pl. 30
Adams 1762, p. 119, Pl. 55, Fig. 183, Pl. 7, Fig. 38

Inventory No. 1927-1229

M91 Two pieces of plaster

No. 1: **L** 130mm, **W** 125mm, **T** 40mm
No. 2: **L** 140mm, **W** 120mm, **T** 8mm

These pieces of white clay, one rectangular and one oval, may possibly be connected with Adams's apparatus for falling bodies M89 or with other experiments on inelastic collision. However, 's Gravesande advised against the use of white clay 'which is almost of the nature of chalk', preferring 'that of which the commonest and meanest earthen vessels are made'. Adams may have used white clay to demonstrate its inadequacy, or the clay may have been used in another situation. One piece is broken.

Inventory No. 1927-1539

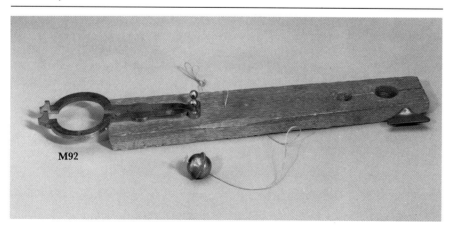

M92

M92 Release pincers

L 130mm (pincer), **L** 320mm, **W** 65mm, **DE** 35mm, **D** 40mm (hole), **D** 22mm (brass ball)

These are the brass pincers for the $1\frac{1}{2}$ inch balls mentioned under M89. They were used to drop brass and ivory balls onto clay and marble in experiments on elastic and inelastic collisions. The pincers are slightly smaller than those in M89, but are identical to the ones illustrated in the *Mechanics* manuscript. They have a circular hole approximately $1\frac{1}{2}$ inches in diameter and can be opened by moving one of the levers. The pincers have been attached to a very rough piece of mahogany which has a triangular nut soldered to one end and a hollow brass ball attached by thread to the centre.

Nollet 1754, Vol. I, p. 312
Adams 1762, p. 119, Pl. 64, Figs 205, 206

Used with M89

Inventory No. 1927-1858

M93 Six cylinders with screws and one larger cylinder with hooks

L 25mm, **D** 40mm, **L** 60mm (large cylinder), **D** 42mm (large cylinder)

Adams provided these boxwood cylinders for use with the brass cradles and impact heads M83/85 to perform the series of experiments described by 's Gravesande on the inelastic collision of bodies.

The cylinders were all used as cavities to carry clay. The six smaller ones could be screwed in turn to the end of one of the brass cradles, while an impact head was attached to the other. The momentum before and after collision could be compared and also the size of the cavity made in the clay. The larger cylinder, which has hollowed ends, could be suspended between the cradles from the small brass hooks and hit from both sides simultaneously; this was known as double collision. It is not actually mentioned in the text of the *Mechanics* manuscript but it is illustrated in the plates.

's Gravesande 1747, Vol. 1, pp. 219–33, Pl. 28, Figs 10, 11
Adams 1762, p. 123, Pl. 66, Figs 209, 210

Used with M83, M85/7

Inventory No. 1927-1859/1

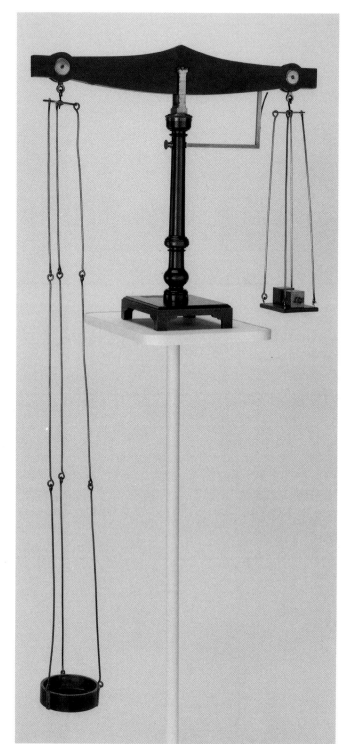

Fig: CCVII.

Adams 1762

M94 Balance beam for experiments on falling weights

L 563mm (beam), **L** 380mm (short suspension), **L** 1090mm (long suspension), **H** 90mm (beam), **T** 21mm (beam), **D** 120mm (pan)

Adams takes this apparatus from a scholium in 's Gravesande in which experiments first performed by Mersenne were repeated. Both Mersenne and 's Gravesande had had problems confirming a relationship between the height from which a body needed to be dropped into one scale-pan of a balance in order just to raise a given weight in the other.

The apparatus consists of a thick mahogany balance beam which was supported on the edge of the Philosophical Table on the small stand M20. One arm carries a circular mahogany scale-pan about an inch deep on a long linked brass suspension which hung over the edge of the table. The scale-pan would have been filled with soft clay onto which a ball would have been dropped. Adams tells us that the ball was $1\frac{1}{2}$ inches in diameter but does not give the material; it is

no longer with the apparatus. The other arm carries a shorter brass suspension which originally held an iron rectangle measuring 9 by 7 inches. The weights to be lifted would have been placed on the rectangle. The balance beam was first brought to equilibrium and then the extra weight was added. To maintain the horizontal position of the beam an L-shaped iron support was fixed from the stand to the arm with the weights. A strip of steel on the support was caught under the point of suspension so that if the arm was raised a certain amount the steel would be released, but if the raising was insufficient it would not. The ball was dropped onto the clay from various heights by cutting the thread from which it was suspended.

's Gravesande 1747, Vol. I, p. 247, Pl. 35, Figs 4, 5
Adams 1762, pp. 123–4, Pl. 65, Fig. 207

Inventory No. 1927-1122

M95 Rectangular cradle with clockwork mechanism and key for experiments on elastic collisions

LABEL
'Fig 208 pp 127–127'
L 130mm, **W** 45mm, **H** 50mm

This is a brass rectangular cradle identical in every respect to those in M85 except that there is an intricate clockwork mechanism to drive a hammer release attached to one end. This rectangle, together with one of the two in M85, could be suspended from the hooks in the brass rulers as for the previous experiments on elastic and non-elastic collisions. This provided a similar catch mechanism for the spring as in M81, but was driven by clockwork with the advantage that

the apparatus did not need to be touched. The spring from M81 was attached to one rectangular cradle and held in the catch attached to this cradle. When the clockwork ran, a hammer released the catch so that the spring separated the cradles. One wind of the clockwork could release the hammer several times. The usual experiments on the relationships between velocity and quantity of matter were performed.

's Gravesande 1747, Vol. I, p. 255
Adams 1762, pp. 125–127; Pl. 66, Fig. 208

Used with M81, M84

Inventory No. 1927-1872

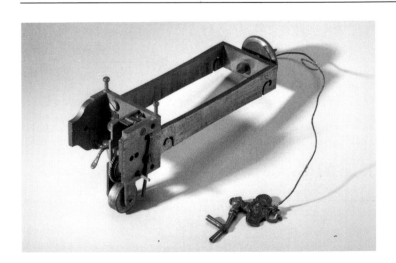

M96 Ivory sphere, 6 oz Troy, with two hooks

LABEL
'215'

D 55mm

The following three ivory bodies are all that remain of a set of eight balls and three cylinders for use with the 'iron arm of the great table' M5. The ball weighs 6 oz and the cylinders 4 oz and 6 oz respectively. The ball has two brass hooks for threads while the cylinders have four. The cylinders each have one flat and one hemispherical end. The bodies appear to be well used since the ivory is cracked where the collisions took place. They were suspended from the iron arm of the great table M5, and used in simple experiments on elastic collision following 's Gravesande. The mass and velocity of the various bodies was observed and what we would now refer to as the conservation of momentum was confirmed.

's Gravesande 1747, Vol. 1, p. 258, Pl. 26, Fig. 7
Adams 1762, p. 127, Pl. 67, Figs 212, 213, 215

Inventory No. 1927-1900

M97 Ivory cylinder 6 oz Troy, with four hooks

LABEL
'213'

L 92mm, **D** 37mm

See M96

Inventory No. 1927-1903

M98 Ivory cylinder, 4 oz Troy, with four hooks

LABEL
'212'

L 65mm, **D** 37mm

See M96

Inventory No. 1927-1902

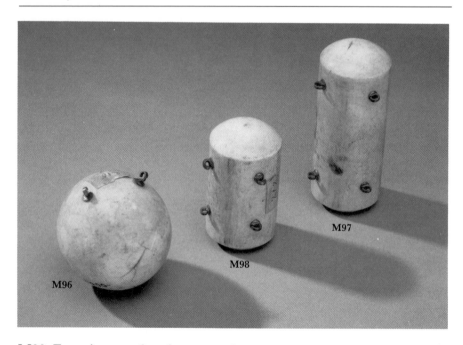

M99 Four impact heads on one base

L 76mm, **W** 36mm, **H** 25mm

These four steel impact heads on a brass frame are a variation on the single ones in M83. The frame can be screwed onto the rectangular cradles M85 and made to strike soft clay. Each cone has a vertex of 55°. Four equal cavities will be made; the purpose of having four cones rather than one is not given.

's Gravesande 1747, Vol. 1, p. 204, Pl. 28, Fig. 9
Adams 1762, p. 128, Pl. 66, Fig. 210

Used with M85

Inventory No. 1927-1859/2

M100 Apparatus for demonstrating elasticity of a spring or wire

L 1085mm, **W** 302mm, **H** 655mm

's Gravesande refers to this apparatus as a machine whereby the laws of elasticity are explored by experiments. Adams copies 's Gravesande's machine except for the design of the dial.

A large mahogany board is placed vertically on four mahogany feet. Along its upper surface a thicker piece of mahogany conceals an iron bar with a square cross-section and sides of $\frac{3}{4}$ inch, whose ends appear projecting at right angles to the board nearly 3 feet apart. The material to be investigated could be stretched and screwed between plates attached to these prominent ends. A steel spanner with a square cross-section is provided to tighten the screws. For the material Adams suggests brass wires 'such as are put into musical instruments' or two watch springs joined together. A dial engraved from 0 to 120 is set vertically in the centre of the

instrument above the board. A brass index finger rotates over the dial on an axis which also carries two small brass pulleys. A watch chain, now replaced with a thread, was attached to one pulley and led down to a brass plate through a hole through which the test material passes. A counterweight was similarly attached to the other pulley in the opposite direction to counteract the weight of the plate and to keep the watch chain taut. The original counterweight has been lost. When weights were added to the plate they move the test material downwards. This turned the pulley which moved the index finger over the dial.

Adams 1746, No. 123
's Gravesande 1747, Vol. I, p. 318, Pl. 43, Fig. 4
Adams 1762, pp. 129–31; Pl. 68, Fig. 216
C 45

Inventory No. 1927-1114

M101 Apparatus to show the elasticity of a watch spring

L 160mm, **W** 160mm, **H** 460mm (without wheel), **D** 105mm (wheel)

In making this instrument Adams may have extemporized since there are no references given to other authors. It is described rather sketchily just before the last experiment in the *Mechanics* manuscript. It consists of a brass spoked wheel about 4 inches in diameter which turns on a steel axle between two brass bearings on a mahogany base. A watch spring which is now missing was attached at one end near the circumference of the wheel and at the other to a fixed cylinder into which the axle fits. The spring was wound by turning the wheel and weights could be hung by threads from the groove in the circumference. In the illustration an index finger indicated the degree of turning by passing over marks on the wheel; however, the marks are not apparent on the instrument. Presumably the weights needed to hold the spring in various states of tension were observed.

On the same plate as the sketch of this instrument is a sketch of a helical spring which Adams described as being used to confirm the linear relationship between load and extension. If it was made it represented a very early example of a spring balance.

Adams 1762, p. [132], Pl. 69, Fig. 220

Inventory No. 1927-1860/2

M102 Apparatus for demonstrating the elasticity and compression of a spring

H 660mm, **W** 1100mm

Adams refers to this instrument as 'another machine for elasticity'. Again it is copied directly from 's Gravesande. The spring under investigation consists of one brass plate and 24 steel plates riveted end to end in a zigzag fashion which is held between two vertical brass plates and rests on another brass plate below. Threads are attached to the top of the folded spring at both sides, pass through holes in the bottom brass plate and stand, and can be attached to weights below. A weight of half a pound compresses the spring by half an inch, and additional weights a proportional amount 'till the plate can no longer be pressed together'. The whole is on a mahogany stand.

Adams 1746, No. 143
's Gravesande 1747, Vol. I, p. 326, Pl. 44, Fig. 8
Adams 1762, p. 131; Pl. 68, Fig. 217
C 46

Inventory No. 1927-1113

M103 Stand to be used in experiments on accelerating bodies

H 600mm, **L** 255mm, **W** 100mm

Adams followed Nollet in the design of an apparatus to 'shew the velocity of falling bodies is accelerated every moment', a precursor of the Attwood machine, although the body does not actually fall but is merely accelerating uniformly. Only accessories now remain. This stand held one end of a trough 12 feet long on the Philosophical Table; the other end was placed on the floor. A frame with two wheels was rolled down one edge of the trough, the inclination of which was adjusted so that the frame travelled one-ninth of the total distance as a pendulum performed one oscillation. A bell operated by the pendulum and a bell triggered by frame were heard simultaneously. The distances travelled in the time taken to perform two or more oscillations was then observed using the bell mechanisms M104 and M105. Nollet gives the relationship between the square of the time and the distance travelled.

The stand is a mahogany board with screw holes at the top for attachment to the trough. It is set on two brass levelling screws. There is a protruding piece of mahogany for the attachment of M104.

Nollet 1754, Vol. 2, p. 161, Fig. 10
Adams 1762, p. 133, Fig. 221

Used with M104, M105

Inventory No. 1927-1789

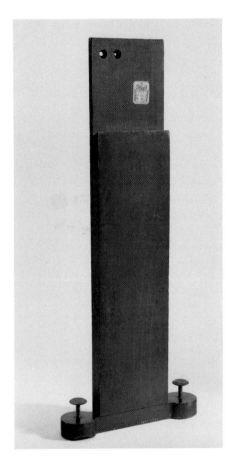

M104 Striker mechanism for use with apparatus on accelerating bodies

L 84mm, **W** 74mm, **H** 117mm

This was originally a bell and striker which was clamped to the stand M103 for experiments on acceleration described above. A pendulum was swung beneath the trough down which a frame with wheels was rolled. The pendulum struck the catch which released the striker and so the time taken for one oscillation was known. It is not clear whether the striker could reset itself for subsequent experiments when a number of pendulum oscillations were required.

Nollet 1754, Vol. 2, p. 161, Fig. 10
Adams 1762, p. [133], Fig. 221

Inventory No. 1927-1790

M105 Bell and striker for use with apparatus to demonstrate the acceleration of falling bodies

L 419mm, **W** 115mm, **H** 120mm

This steel bell and striker was clamped to the side of the trough down which a frame with pulleys was rolled in experiments on falling bodies. The striker has a tail which, when hit by the moving body, released the striker to fall on the bell. The time taken for the body to descend a known distance could then be compared with the time taken for a pendulum to make a number of oscillations.

Adams 1762, p. [133], Fig. 221   **Inventory No.** 1927-1791

M106 Set of seven hemispherical weights with loops

L 55mm (total), **D** 20mm

The loops are similar to those in M14 suggesting that the weights may be by Adams although there is no evidence. One weight has no loop but two hooks and another is slightly smaller than the rest.

Adams 1762, p. 18, Pl. 9, Fig. 41, p. 21, Pl. 11, Fig. 47
See also M14   **Inventory No.** 1927-1865

M107 Set of seven cubical weights

ENGRAVED—ON WEIGHTS
'*1*', '*1*', '*2*', '*3*', '*4*', '*5*', '*6*'

L 29mm (side, largest), **L** 15mm (side, smallest)

These cubical brass weights are similar to M37, but they have only one hook each and are not marked 'oz T'. They are not mentioned specifically in the *Mechanics* manuscript but the style of engraving is consistent with the assumption that they were made by Adams. They are marked only with a number. Their weights in ounces correspond approximately to the numbers engraved.

Adams 1762, Pl. 35, Fig. 118, No. 2
C 62   **Inventory No.** 1927-1861

M108 Two glass bubbles in box

Adams, G.?

1761–2?   London

D 15mm (bubbles), **L** 90mm (box), **W** 50mm (box)

These glass bubbles may possibly have been used by Adams to demonstrate the attraction of water to clean glass as illustrated in the *Mechanics* manuscript. They are not perfect spheres, having a small tail where the glass was blown. The balsa wood pill box is oval in shape. Two bubbles have been lost.

Adams 1762, p. 4, Pl. 2, Fig. 7

Inventory No. 1927-1573

M109 Two rods of square cross-section and apparatus on sliders

Adams, G.?

1761–2   London

L 620mm, **W** 200mm (max), **DE** 80mm

These are almost certainly by Adams since they are identical in style to the apparatus made for the mechanics course. However, they do not feature in the *Mechanics* manuscript and they have not been traced in 's Gravesande or Desaguliers.

The two pieces of mahogany can be clamped at one end, probably to an attachment of the Philosophical Table. Brass rods $4\frac{1}{2}$ inches long can be moved along the mahogany on brass sliders. The brass rods each carry two thin plates of brass about 1 inch by $\frac{3}{4}$ inch in size. These are also movable on sliders on the brass rods.

Inventory No. 1927-1809

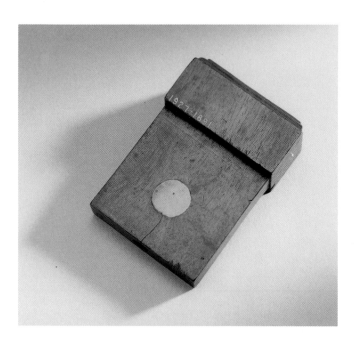

M110 Accessory for Philosophical Table

Adams, G.?

1761–2   London

L 100mm, **W** 70mm, **T** 30mm

This piece of mahogany has a brass plug at one end to receive a clamping screw. It is very similar to the piece of mahogany to which the pair of pincers in M89 is attached, and may be the corresponding piece for the second pair of pincers M92.

Used with M89, M92

Inventory No. 1927-1891

M111 Circular stand with plaster gilt decoration

c. 1762

H 430mm, **D** 250mm (max), **D** 90mm (min)

This stand has been associated with the Adams mechanics material because in the complete version of the *Mechanics* manuscript held at Windsor, a similar one is shown supporting the compound engine M50. It was probably bought for that purpose. It is made of pine and has a slotted hole in the top.

Inventory No. 1927-1666

M112 Two pulley wheels

Adams, G.?

second half eighteenth century

L 133mm (wheel 1), **H** 108mm (wheel 1), **D** 97mm (wheel 1), **D** 105mm (wheel 2)

These pulley wheels differ very slightly from one another. The smaller one has five spokes and a steel axle. The larger one has six spokes and an iron axle. It is possible that they are by Adams since they are typical of his work, but they are not mentioned specifically in the *Mechanics* manuscript

C 30

Inventory No. 1927-1883, -1884

M113 Accessory for use with the Philosophical Table

Adams, G.?

1761–2   London

L 315mm, **W** 65mm

This item was almost certainly intended for the mechanics course as it resembles other accessories made for the table. It is not featured in the manuscript.

See M110

Inventory No. 1927-2039

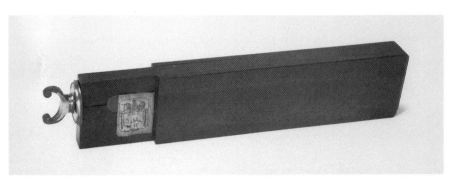

M114 Box containing accessories

Adams G.?

second half eighteenth century   London

The box contains two weighted brass rollers in frames similar to M23, M24. There are also two brass keys, four brass discs, a sliding socket, three steel hooks, two cylindrical weights, one brass and one mahogany, and other hooks. The small brass socket was possibly made for the lever M39; however, it does not quite fit at the present time. The pieces are stylistically similar to Adams's material.

Adams 1762, p. 40, Pl. 23, Fig. 90

Used with M39

Inventory No. 1927-1927

PART 3

Other seventeenth and eighteenth century material

A high proportion of material in the collection falls into this category. However, it is not a cohesive group. Some were in the possession of the royal family before George III came to the throne; a dial with the monogram of William and Mary or the Grand Orrery enlarged for George II, for example. Some other items came from the Duke of Cumberland or from George IV when he was Prince of Wales. A small number are known to have been gifts to George III or Queen Charlotte while others have no known origin.

In addition there are a few sets of instruments such as the thermometers made by James Six, or the items which may have belonged to Robert Boyle. In the latter case, there is a manuscript list of the items, dated May 1770, which states these items belonged to Robert Boyle (1627–91) and were deposited at the Observatory at Richmond (now Kew) in 1770, soon after the Observatory had been built. However, no other supporting evidence has been found to connect this material with Boyle though many of the items do appear to date from the seventeenth century.

DRAWING INSTRUMENTS

E1 Volute compass

Lyle, David

1760 London

'*Opt. max. Principi*/*GEORGIO III DEI gratia*/*mag. Brit. Fr & Hib Regi &c*/*hoc Instrumentum*/*omnium spiras geo. describendo* /*primum & fundamen*/*D.D.D. Inventor*/*suae Maj. subj. fideliss.* /*D. Lyle*/*1760*'

L 310mm (box), **W** 75mm (box), **H** 60mm (box), **L** 290mm (overall), **W** 50mm, **H** 55mm (end screw)

The inscription translates approximately as follows: To the most excellent and mighty prince George III, by grace of God, King of Great Britain, France, and Ireland, etc., this instrument, the principal and foundation of all (those used) for describing geometric curves, is given and dedicated by the inventor, his Majesty's most faithful subject D. Lyle, 1760.

Lyle's book, 'The Art of Shorthand Improved' has a dedication to Lord Bute in which he states: 'By your LORDSHIP's good offices, I was enabled to bring my new mathematical instruments to great perfection; and at your desire, I compleated a set of them for the use of his MAJESTY'. The set of instruments referred to is almost certainly that comprising the four volute compasses E1, E2, E3, and E4. The compasses are very similar to one another stylistically, even down to the decoration on the boxes.

This is the only signed instrument: it is the largest and most complex of the set. It is similar in principle to a beam compass except that the pencil in the slider is not fixed but moves with respect to the stationary point as the compass is turned. The beam is silver with a triangular cross-section. It is 7 inches long and marked in tenths of an inch. At one end of the beam and slightly to one side is the support for the fixed point. This is a silver rod with an hexagonal cross-section $3\frac{1}{4}$ inches long. In the line of the beam is a steel screw with a large dodecagonal silver head on which is the inscription. The base of the screw is broken but it will still hold one of the seven helical drums provided. A piece of catgut was wound around the drum and attached to the slider which carries the support for the pen, again slightly to one side of the beam. The pen holder is another silver rod of hexagonal cross-section which has a pair of adjustable blades at one end to hold the ink. The far end of the beam carries a second steel screw which could be used for the attachment of the helical drums in the same way as the first.

The compass is operated by turning the beam round the fixed point in such a manner that the large screw head remains in one orientation. If this is the case the slider will be moved down the beam at a rate in proportion to the rotation of the beam, and a spiral will be drawn. The various drums give various spirals. The instrument fits into a green shagreen case lined with green baize when not in use.

Little was written about the volute compass in the eighteenth century and few examples survive; they were probably not widely used owing to their limited field of application and the difficulty of manipulating them.

Lyle 1762
Adams 1791, p. 157, Fig. 6, Pl. 11
C 238
Hambly 1988, p. 97, Pl. 88
Inventory No. 1927-1924

E2 Right-handed volute compass

Lyle, David

1760   London

ENGRAVED—BOX
   '*The right handed/Volute Compasses*'
ENGRAVED—ON COMPASS
   '*No 2*'

L 120mm (box), **H** 34mm (box), **W** 63mm (box), **L** 130mm (max), **W** 30mm, **H** 220mm (max)

The main body of the instrument is made of silver while the various rods with points and ink pens are of steel. The principal arm of the compass carries a holder for one of the two hexagonal rods with points. These are 2 and 4 inches long respectively. The other arm carries the pen and is hinged to the principal arm at the lower end; it is moved with respect to the principal arm on a rack and pinion by moving the handle. There is a choice of two pens to accompany the long and short rods with points, and two tweezers for a pencil lead. The instrument fits into a shagreen box with a silver plaque on the lid when not in use.

If the handle of the instrument is held stationary with the left hand and the compass is rotated anti-clockwise on its point with the right, a right-handed spiral will be drawn. If the compass is rotated clockwise a right-handed spiral will be drawn but moving towards the centre rather than away from the centre.

Lyle 1762
Adams 1791, p. 157, Fig. 6, Pl. 11
C 236
Hambly 1988, p. 97, Pl. 88
Inventory No. 1927-1781

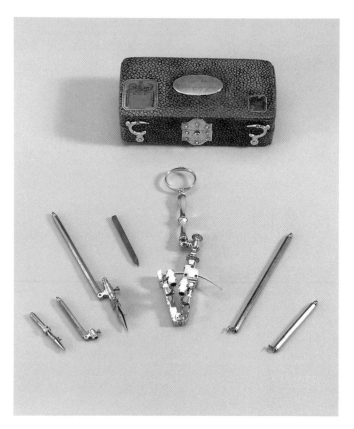

E3 Left-handed volute compass

Lyle, David

1760   London

'*The left handed/Volute Compasses*'

L 120mm (box), **H** 34mm (box), **W** 63mm (box), **L** 130mm (max), **W** 30mm, **H** 200mm (max)

The instrument is a mirror image of E2 so that the resulting spiral is left-handed rather than right-handed.

Lyle 1762
Adams 1791, p. 157, Fig. 6, Pl. 11
C 235
Hambly 1988, p. 97, Pl. 88
Inventory No. 1927-1780

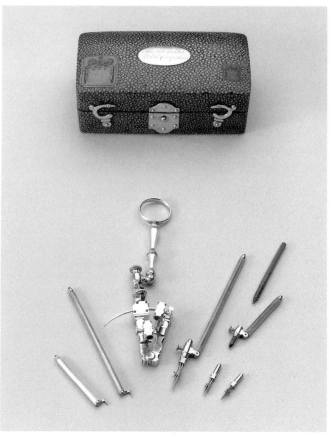

E4 Right-handed volute compass

Lyle, David

1760   London

L 180mm (box), **W** 65mm (box), **H** 55mm (box), **H** 140mm (max), **L** 90mm (max), **W** 50mm (max)

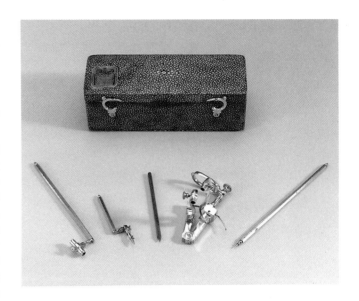

This is a slightly different design of volute compass from E2 and E3, but is based on the same principle. The main instrument is of silver while the rods for the point, pen, and tweezers are of steel. The principal arm carries both the handle, which is an oval ring, and the holder for the point. The secondary arm carries the ink pen; it is hinged to the principal arm at the lower end. The holders are further removed from the arms than in E3 and E2 and the rods for the point and pen are longer, allowing a larger spiral to be drawn. Again, turning the handle moves the secondary arm with respect to the primary arm on a rack and pinion. When not in use the instrument fits into a shagreen case.

Lyle 1762
Adams 1791, p. 157, Fig. 6, Pl. 11
C 237
Hambly 1988, p. 97, Pl. 88
Inventory No. 1927-1925

E5 Elliptical trammel

Finney, Joseph

1760–72   Liverpool

J. Finney

L 440mm, **W** 430mm, **DE** 12mm

Unlike some other Finney instruments this is not listed in the Queen's Catalogue and it is not known how it came into the collection. It consists of a brass T-shaped frame, both pieces having a central slot along which a slider can be moved or a screw fixed. A movable arm is attached to a slider which moves along the base of the T. It is fixed at one end by a screw which can be fastened at any point on the central part of the frame to provide the focus. When the arm is moved, a screw at the other extremity of the movable arm describes a semi-ellipse. The instrument can be turned through 180° and the process repeated to provide the complete ellipse. The ratio of the major and minor axes can be altered by moving the position of the screw along the central part of the frame.

Finney was a well-respected Liverpool instrument-maker working in the third quarter of the eighteenth century.

Q.C., Presents p. 20, No. 2, 'Finney's pyrometer'
Bion 1758, p. 80, Pl. 9
C 239
Goodison 1968, p. 63, Pl. 25
Fairclough 1975
Banfield 1985, Pl. 52, Fig. 194
See also D124, E21
Inventory No. 1927-1210

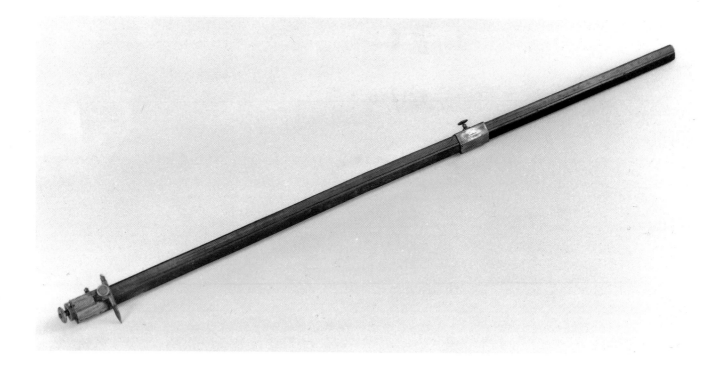

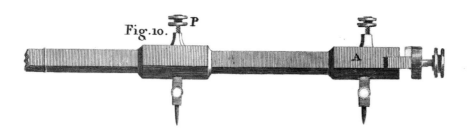

Adams 1791

E6 Beam compass

Adams, George, the Younger

last quarter eighteenth century

SCALE

$0 \left(\frac{1}{10} \right) 33$

SCALE—MICROMETER DIAL

$0 (5) 20$

L 910mm, **H** 60mm, **W** 17mm

The beam compass is almost certainly by Adams although it is not signed. It is identical to that illustrated in Adams's *Geometrical and Graphical Essays*.

A beam of mahogany has an inlaid boxwood scale running from 0 to 33 inches and marked in tenths. At one end is a fixed brass sleeve with a steel pointer attached. The pointer can be moved 'with extreme regularity and exactness' by means of

the micrometer screw on the end of the sleeve. The brass screw head is marked from 0 to 20. There is a sliding brass sleeve which can be fixed with a screw at any point on the scale. An attachment for a crayon is now broken.

Beam compasses were used to describe large arcs, bisect lines or angles, and transfer divisions from a diagonal or 'nonius' scale. They were also used with Sisson's 'scale of equal parts' which gave three or four chains to the inch enabling triangulation station lines to be laid down on a map.

Adams 1791, p. 18, Pl. 3, Fig. 10
Edinburgh Encyclopaedia 1830, under Drawing Instruments, Pl. 238, Fig. 7
C 234

Inventory No. 1927-1218

E7 Pantographer

Adams, George

c. 1765 London

ON IVORY SCALE
'*G. ADAMS Math.¹ Inst.ᵗ Maker to His MAJESTY Fleet Street London*'

L 700mm (outer arms), **W** 20mm (arms), **H** 135mm, **H** 45mm (weight), **L** 160mm (weight), **W** 66mm (weight)

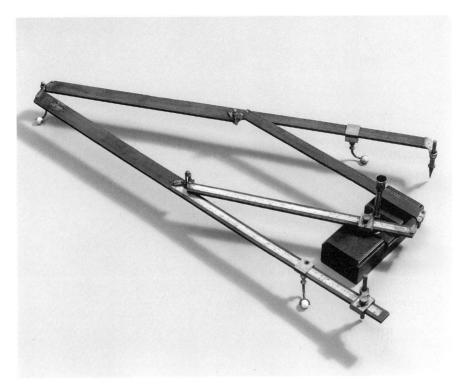

The pantographer enables drawings to be copied, enlarged, or reduced. It consists of two long arms of mahogany hinged at one end, with two shorter arms fixed to their centres and hinged together, so that the four pieces form a parallelogram. There are two sliding brass sleeves, one on the lower left-hand arm and one on the short left-hand arm, and a fixed sleeve on the lower right-hand arm. Each of these has an attachment for the fulcrum, tracer, or crayon, which are interchangeable. Both sliding sockets move over ivory inlaid scales. They are marked with the letters B and D respectively. The fixed socket is marked C. The instrument stands on three ceramic rollers.

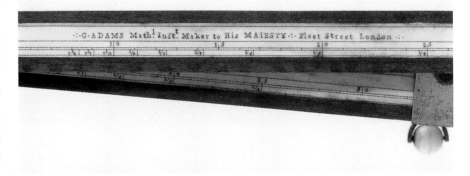

The scales are as follows: on **B** arm: 0 to 50 on whole scale, $\frac{1}{2}$ to 0 whole scale, 50 to 100 half-scale, $\frac{1}{2}$ to 1 half-scale. On **D** arm: the inscription, 0 to $\frac{1}{2}$ whole scale, 0 to 50 whole scale, $\frac{1}{2}$ to 1 short scale, 50 to 100 short scale.

Some instructions for using the pantographer signed by Adams have been found in King's College. The signature is similar to that in the *Mechanics* manuscript and is likely to be George Adams the Elder, which corresponds to the style of the instrument. The instructions are similar to those in the younger Adams's *Geometrical and Graphical Essays*. For original size copying the fulcrum is placed at D, the crayon at B, and the tracer at C, with B and D being set at the $\frac{1}{2}$ marks. For half-size copying the fulcrum is at B, the crayon at D, and the tracer at C, B, and D remaining at the $\frac{1}{2}$ marks. For less than half the original, the sliders are set at the appropriate

fraction on the long scales. For greater than half, the fulcrum is moved to D, the crayon to B, and the tracer to C, the shorter scales being used. For enlarging the fulcrum is at D, the crayon at C, and the tracer at B. The scales 0 to 100, 50 to 100, etc. are simply hundredths in the case when the fraction required is not available on the other scales.

The weight was identified as belonging to the pantographer from a plate in the *Encyclopaedie*. It consists of two rectangular brass weights either side of a short brass pillar. The pillar can be attached to an arm of the pantographer.

Since the pantographer can be moved during copying, a map or plan of any size can be enlarged or reduced. Adams the Younger claimed that his father had brought the pantographer 'to its present state of perfection' around 1750.

Adams 1746, No. 75
Diderot 1767, Vol. 2, pt 2, under Dessein p. 5 Pl. 3.
Adams 1791, pp. 374–7, Pl. 31, Fig. 19
Edinburgh Encyclopaedia 1830
C 240
Wynter and Turner 1975, Fig. 174

Inventory Nos 1927-1151, -1871

E8 Curve-bow

Wright, Thomas

ON PLAQUE
'Tho: Wright Fecit:'

L 670mm, **W** 50mm, **DE** 18mm

It is possible that this instrument came into the collection via George II since Wright was appointed Mathematical Instrument Maker in Ordinary to the King in 1721.

The instrument is made of solid rosewood. A narrow flexible piece of wood, the bow, is attached at each end to the thicker frame by means of an engraved brass fitting. The fitting supports iron rollers in such a manner that the bow can move laterally with respect to the frame. The frame has three holes bored at right-angles to the bow—one in the centre of the frame and one 6 inches either side. Each hole is surrounded by ivory and an ivory plaque on the upper surface of the frame carries the maker's name.

To operate the insrument long screws would be threaded into the holes which would push the bow away from the frame to form a curve. The degree of curvature could be varied by altering the number of turns of the screw and by the use of the side screws. The instrument was used to produce large curves simply by placing it on a piece of paper and drawing round it.

Thomas Wright is best known for his orreries, a fine example of which is given by E34.

Young 1807, pp. 101–2, Pl. 6, Fig. 85
Taylor and Wilson 1944
C 337
Taylor 1954, p. 302   **Inventory No.** 1927-1774

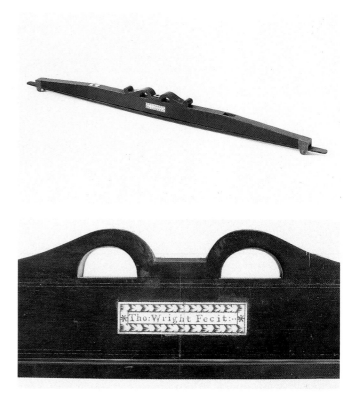

E9 Dividers

second half eighteenth century

H 105mm, **L** 40mm, **W** 23mm

The dividers consist of two parallel plates of brass, each with steel points, which can be moved with respect to each other by turning a knob in the centre of one of the plates. The knob turns on an axis on which there is a small toothed wheel which meshes with two larger ones on the other plate. The two large wheels turn screws which adjust the distance of the plates. Hence the dividers can be used with considerable precision.

The design of this instrument appears to be quite unusual; it is not featured in the standard texts on drawing instruments.

C 241   **Inventory No.** 1927-1783

E10 Drawing instrument

Adams, George (the Younger?)

last quarter eighteenth century
London

'G. Adams No 60 Fleet Street London'

L 327mm (box), **W** 90mm (box),
H 25mm (box), **L** 280mm (ruler),
W 15mm (ruler), **D** 70mm (protractor),
L 230mm (arm on protractor), **W** 11mm
(arm on protractor), **L** 265mm (other
arm), **W** 11mm (other arm)

The instrument consists of a small brass protractor marked in 2° intervals from 0° to 180°, and 1 to 10 in tenths of an inch along its base. An arm with a straight edge is rotatable from the centre of the instrument; it is 9 inches long and marked in twentieths of an inch. The angle of inclination can be read off a vernier to half a degree. A sleeve carrying a crayon holder can be slid along the arm. A second arm 10 inches long of similar width is also marked in twentieths of an inch and has a brass sleeve at right-angles at its base. At one time it would have been attachable to the first arm but has since been altered so that it can be attached only at the end of the arm. In addition to these arms, there is a ten inch ruler marked in tenths bearing the maker's name.

The precise use of the instrument is not obvious, but it appears to be one of a number of combinations devised in the late eighteenth and early nineteenth centuries. Adams invented the related 'protracting parallel rule', Bolles the rather more complex 'trigonometer', Pool the 'geometrical protractor', and Lyman the 'protracting trigonometer'. All combined a protractor, straight edge, and scale.

Bolles 1825
J.F.I. 1830
C 232
Kidwell 1988
Wing 1990

See also L9

Inventory No. 1927-1786

COUNTING MACHINE

E11 Counting machine

last quarter eighteenth century
L 230mm, **W** 126mm, **H** 50mm

The counting machine is in an oak case and has four square holes in a brass plate where the numbers appear. The counting is performed by moving a lever. The case can be removed to reveal the mechanism.

The numbers are situated round the circumference of the upper surface of four boxwood wheels. Each wheel has ten iron pillars fixed vertically around its circumference and carries an iron arm. When the lever is pulled it rotates an axle which fits an iron arm between two of the pillars on the first wheel, and moves the wheel so that the numbers move round one place. The lever then springs back into position. If this is repeated ten times the arm on the first wheel will have moved through 360° and will fit between the pillars of the second wheel, moving it on one place. This wheel will in turn move the third wheel after having been moved ten times. Hence the machine can count up to 9999. Brass knobs on the axles of the wheels allow the instrument to be reset.

C 244

Inventory No. 1927-1451

WEIGHING INSTRUMENTS

E12 Double-fulcrum precision balance

Sisson, Jeremiah

before 1789   London

ENGRAVED—BASE
'J Sisson/London'

L 500mm (case), **W** 300mm (case), **H** 380mm (case), **L** 320mm (beam), **H** 170mm (balance), **W** 60mm (balance)

The balance is almost entirely of brass and is fixed inside a glazed mahogany case. Four pillars support a platform which carries two conical steel fulcra on which the beam of the balance rests. The beam is about a foot long with box-type ends; the scale-pans are now missing. A steel pointer descends between the pillars and can move within a narrow limit set by a brass bracket. An arresting device would have lifted the beam off the fulcra but it no longer operates. The inscription is on a piece of brass running between the handle and the balance. Also on the base of the case is a spirit level; the balance can be adjusted by means of the four levelling screws at the corners. In the base of the case is a drawer which contains four pieces of brass sheet and five paper weights with the following written in ink: 1/4, 1/8, 1/16, 1/32, 1/64. The weights correspond to grains in Troy or apothecary's weights.

C 56

Inventory No. 1927-2040

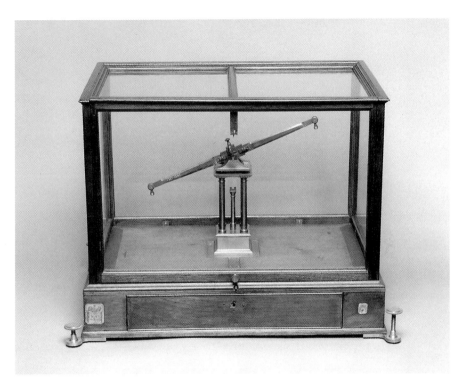

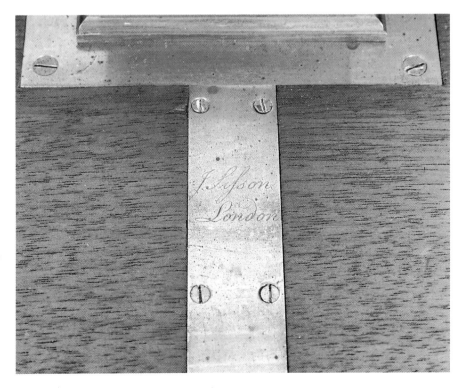

E13 Circular spring balance

1796?

SCRATCHED
'1 20 96'

SCALE
0 (10) 112

H 330mm, **D** 210mm (dial), **DE** 30mm,
D 204mm (dial)

An elliptical steel spring is attached to
the brass frame from above. When a
weight is suspended from the lower side
of the spring a rack and pinion moves a
pointer over a circular brass scale. The
scale is marked from 0 to 112 in pounds
with every ten numbered and each
division numbered up to 60 lb. An
inscription *1 20 96* at the top of the
instrument may possibly give the date.
A similar spring balance is in the Avery
Museum, the catalogue of which credits
Edme Regnier with its invention in
1790.

Benton 1940–8, Vol. 22, p. 17, Pl. 15
C 59
Kisch 1965, p. 69

Inventory No. 1927-1456

E14 Wiedeman's stilliard

second half eighteenth century

SCRATCHED
50 (14) 322
10 (10) 90

TOP—HANDWRITTEN IN INK
'Wiedeman's Stilliard'

H 300mm, **D** 165mm, **DE** 35mm, **L** 215mm (box), **W** 175mm (box), **H** 50mm (box)

This is one of the items in the collection to be uniquely identifiable from the list of instruments belonging to the Duke of Cumberland at his death in 1765.

A thick strip of steel is bent into an approximate circle. A steel pointer is attached to one end of the circle and passes through a slot in the other in such a way that if a heavy weight is suspended the ends are drawn apart and the pointer moves over a brass scale on the other side of the circle. On one side the scale is marked in pounds from 50 to 322, each stone between 56 and 322 is numbered. On the other side the scale is marked in degrees from 10 to 90. Both scales are extremely crude. Large steel hooks are used for the suspension of the instrument and to carry the weight. There is an alternative position for another weighing capacity but it is not clear how this relates to the scale. The instrument is in a fitted case which is covered in embossed leather. A paper label on the top of the case carries the inscription 'Wiedeman's Stilliard' handwritten in ink.

Crawforth states that the earliest dated 'mancur' scales, of which this is an example, are from the mid-eighteenth century.

Cumberland 1765, 'Weideman's stilliards and a case to it'
C 60
Crawforth 1984, p. 14, p. 17, Fig. 1

Inventory No. 1927-1150

E15 Steelyard

mid-eighteenth century

SCALE
$0 \left(\frac{1}{10} \right) 11$

L 338mm (arm), **L** 273mm (pointer),
L 120mm (pivot), **W** 14mm (arm),
DE 4mm (arm)

The steelyard is a lever made of box-wood with a scale on the long arm from one to 11 inches marked in tenths. The short arm has no scale but there are holes at 1 and 1½ inches from the pivot with brass wires to hold the counter-weight. The long arm carries a pointer which is a boxwood arm similar in dimensions to the steelyard. It has a point at the top and a line running along its centre. It can be moved along the steelyard on a rectangular brass sleeve. The reading is taken from an index in the centre of the sleeve.

Inventory No. 1927-1202

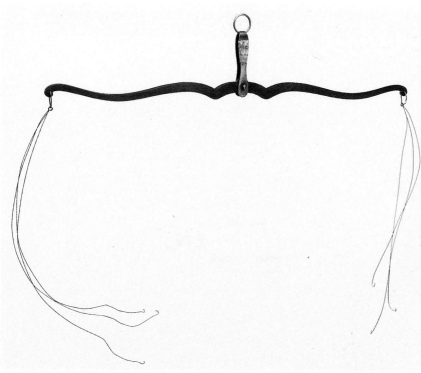

E16 Balance beam

mid-eighteenth century

L 620mm, **H** 120mm (excluding wire),
W 25mm

This is a simple boxwood balance beam 2 feet long on a brass suspension unit. It can be held by a brass ring at the top. Thick brass wires pass through holes at the end of each arm and are attached to the wires which hold the scale-pans. It is possible that it was made by Adams since it is stylistically similar to some of his apparatus, although it is not maho-gany. 'A beam of a balance' is listed in the Queen's Catalogue, possibly refer-ring to this object. The scale-pans are missing.

Q. C., Item 22; beam of a balance
Inventory No. 1927-1203

E17 Portable balance in box

eighteenth century

L 157mm (beam), **L** 170mm (box), **W** 87mm (box), **H** 25mm (box), **D** 70mm (pans), **H** 215mm

This is a standard small balance of a type made over a long period of time. The beam is made of iron 6 inches long with the usual 'swan neck' suspension for the pans. There are no markings except for a double line about half way along each arm. The pans are beaten brass about $2\frac{1}{2}$ inches in diameter. When not in use the instrument fits into an oak case, the lining of which is probably not original. The balance would have been used for weighing coins. Similar balances are described in the works below.

Sheppard and Musham 1923, pp. 45–76
C 55
Brown 1982, Cat. Nos 29–55
Houben 1982, p. 39, Fig. 114

Inventory No. 1927-1135

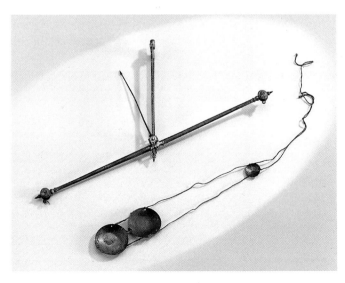

E18 Balance

mid-eighteenth century

L 277mm, **H** 145mm, **D** 35mm (two pans), **D** 15mm (one pan)

This is a brass equal-arm balance with box ends which is supported by a double steel fulcrum. There are three brass scale-pans which appear to have been linked together on one side of the beam, a smaller one above two similar sized ones, although this may not have been the case originally. The instrument may have been used for hydrostatical purposes, but since it is incomplete this is uncertain.

Inventory No. 1927-2056

E19 Gold coin balance

Anscheutz and Schlaff

last quarter eighteenth century   London

Avoird sec W, drams, Troy, Dwts, Gold Coins,
Anscheutz and/Schlaff/London/No 112

L 270mm, **W** 200mm, **H** 370mm

The balance weighs gold coin and expresses the result on three scales simultaneously. One scale gives the value of the coins directly in pounds and shillings, one in avoirdupois, and one in ounces Troy. A vertical iron dial carries the three scales, behind which is a brass lever arm with a flat disc at one end and the scale-pan at the other. The indices are in the form of a three-pointed star which is connected to the arm through slots in the dial. The instrument is supported on a mahogany stand fixed to a heavy brass base with three levelling screws. A bubble level is situated above the pillar. The balance and scale can be adjusted independently.

London Directories 1772, 1775

Inventory No. 1927-1146

MEASURING INSTRUMENTS

E20 Micrometer

Sisson, Jonathan or Jeremiah

1740–60   London

'*J Sisson London*'

L 250mm (box), **W** 160mm (box), **H** 95mm (box), **L** 220mm (inst), **W** 133mm (inst), **H** 85mm (diameter of dial)

According to Lalande, he first saw this type of micrometer in 1753 when he met Dr Bevis. Bevis had an observatory in Richmond not far from the future Royal Observatory site and so it is just possible that this is the micrometer that Lalande saw, although it is more probable that it was made in connection with the planisphere and transit telescope E44 and E188.

Smith described the type in 1738 as the 'best sort', as did Lalande. It consists of a brass rectangular plate with a rectangular hole with one side forming a concave arc. A wire extends across the hole lengthways and two sights cross it at right angles to the wire. One brass sight is fixed but the other,

which in this case is iron, is movable parallel to the first by turning a knob at the end of a long iron screw. The screw turns a pointer over a dial fixed at right-angles to the main plate. The movable sight can be attached to a brass bracket through which the screw passes. Each revolution of the screw turns the pointer through one revolution on the outer scale of the dial, and for each revolution on this scale one division is passed on the inner scale. The measurements conform to those given by Smith. The screw has 40 threads to the inch and the dial is divided into 40, giving accuracy of 1/1600 parts of an inch.

The instrument also incorporates Bradley's improvements suggested shortly before 1738 which enabled the micrometer to be turned around the point of intersection of the fixed sight and the wire without moving the telescope to which it was applied. A second brass plate, slightly shorter but of the same breadth and thickness as the first and with a slightly larger hole, is attached to the first. It is rotated by means of a long steel screw set at right-angles to the micrometer screw. The spiral thread moves the teeth of a brass arch fixed to the second plate. Since this plate incorporates the brackets for fixing to a telescope, turning the plate will turn the micrometer inside the telescope.

The micrometer could be used for measuring angles of arc since by similar triangles the angles are proportional to the distance apart of the fixed and movable sights. By observing a star in transit and finding the time taken to cross the distance between the sights, an angle could be calculated and hence angles found for all other separations of the sights.

distance apart of the fixed and movable sights. By observing a star in transit and finding the time taken to cross the distance between the sights, an angle could be calculated and hence angles found for all other separations of the sights.

Smith 1738, pp. 345–9, Pl. 50, Figs 604, 605
Diderot 1767, Tome 4, Pl. 18
Lalande 1792, Tome 2, pp. 601–4, Pl. 20, Figs 157, 158
C 147
King 1955, p. 98, Fig. 47

See also E44, E188

Inventory No. 1927-1491

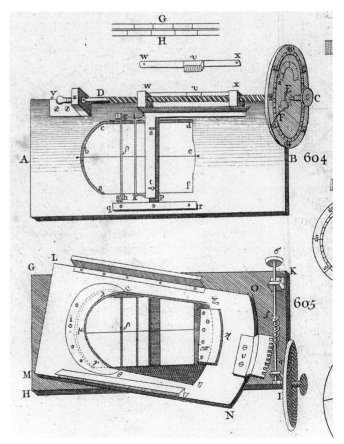

Smith 1738

E21 Dial micrometer

Finney, Joseph

1760–72   Liverpool

'Made by/Joseph Finney/LIVERPOOL'

L 970mm, **W** 340mm, **H** 185mm

One of the two micrometers by Finney originally in the collection was presented by the Earl of Warwick. The micrometer acts as a form of pyrometer, measuring the elongation of a brass bar when heated and comparing it with the elongation of another material similarly heated. A mahogany bar was originally with the instrument, probably to compare the expansion of metal with that of wood.

The instrument consists of a strong mahogany frame into which the two bars are placed. At the lower end they are fixed securely and at the upper end they press against levers which are geared to pointers which move over the dials. Each dial has a long pointer which operates on a scale marked in degrees and a short one which moves an interval of 6° each time the longer one completes a revolution. The smaller scales are marked from 1 to 60. Any expansion can be measured very precisely.

Originally there were two similar micrometers in the collection but only one has been located.

Q. C., p. 20, No. 2, Finneys pyrometer
C 83
Fairclough 1975
See also D124

Inventory No. 1927-1663

E22 Taper gauge

c. 1770

SCALE
10 (10) 100 and 3–1024
MP, MS, CP, CS, PP, PS

L 148mm, **W** 30mm, **DE** 1mm, **L** 155mm (box), **W** 35mm (box), **DE** 8mm (box)

A shot gauge is listed in the Boyle manuscript which could possibly refer to this instrument; however, there is no other evidence to link it with the Boyle material. It is a barrel gauge which would have been inserted into the bore of a gun to give the diameter and bullet size required. On one side there is a scale of lateral lines from 10 to 100 in tens, which gives the width of the gauge in hundredths of an inch. Each hundredth is marked down both edges. The other side has a cubic scale from 3 at the wide end to 1024 near the apex. At the points 11, 14, 17, 20, 28, and 34 respectively the letters MP, MS, CP, CS, PP, and PS are marked. The second P and S stand for 'powder' and 'shot'. This scale gives the number of shot per pound for each diameter. When not in use the gauge fits into a tapered fishskin case.

Richmond 1770, No. 26
C 242
Blackmore 1983, p. 513   **Inventory No.** 1927-1132

E23 Wire gauge

last quarter eighteenth century
L 150mm, **W** 8mm (max), **DE** 3mm

This is simpler than the gauges used in the early nineteenth century and is therefore probably of an earlier date. Eight holes have been made in the brass: the largest is about 3mm in diameter, and the smallest about 0.5mm.

Rees Cyclopaedia, under Wire   **Inventory No.** 1927-1898

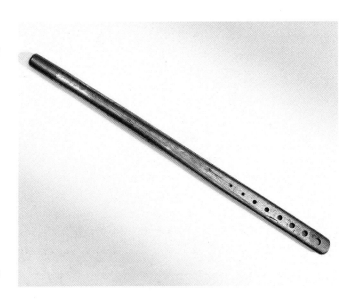

CLINOMETERS AND SURVEYING INSTRUMENTS

E24 Pocket clinometer and square in a case

Marke, J.

third quarter seventeenth century
London

'J Marke Fecit'
L 140mm (case), **L** 129mm, **W** 115mm

This item may have originated with the Boyle Collection as 'a square brass scale plate in a case'; it features on the list of 1770. It is contemporary with the Boyle material.

On the obverse is the clinometer: this consists of two pinhole sights to be used with a quadrant scale which runs along the base and up the right-hand side of the instrument. The scale is marked in $\frac{1}{2}$ degrees from 0 to 90. The left-hand side has a scale of equal parts from 0 to 10 marked in twentieths. The reverse carries an unusual form of engraved square for horary and trigonometric uses. There are two sets of parallel lines—one running vertically and one diagonally. The vertical lines are for hours, each line marking 5 minutes and every third line dotted to represent the quarters of the hours. The hours are divided as sines and graduated from 0 to 6 along the base and 6 to 12 along the top. The diagonal lines represent degrees and are also divided as sines. The left-hand side is graduated from 0 to 90 and the right-hand side from 30 through 0 to 30 so that the lines connect the same values at each side. There are two sliding pieces which run in slots down each side of the instrument. These probably carried a thread across the face of the square. This would be equivalent to having an index of variable inclination and was not unusual on instruments of this period.

A general sense of the capabilities of

the instrument can be inferred from the design and explanation of the rather similar 'geometrical square' of Samuel Foster. This square used a thread to solve problems such as 'having the latitude of the place, the declination and altitude of the Sun, to find the Hour of the day'. There are several other 'geometrical squares' of broadly similar function known on surviving instruments of this period. Perhaps the closest in appearance is a quadrant by Abraham Sharp at the Bolling Hall Museum in Bradford. This type of instrument was made to display the virtuosity of the mathematical practitioner, characteristically featuring an ingenious method of solving a standard problem in the literature of dialling and astronomical trigonometry. The square was probably made as a special commission.

Marke was a well-known London maker who was apprenticed in 1655 and still working in 1679. He was an apprentice of Henry Sutton and after Sutton's death in 1665 he took over the business.

Foster 1659   Richmond 1770, No. 35
Taylor 1954, pp. 252, 212   C 230
Abraham Sharp 1963, Item 51

Inventory No. 1927-1918

E25 Altazimuth theodolite

Cole, B.

mid-eighteenth century   London

'COLE maker at ye ORRERY in Fleet Street'
Royal coat of arms

L 325mm, **D** 200mm (dial), **H** 275mm

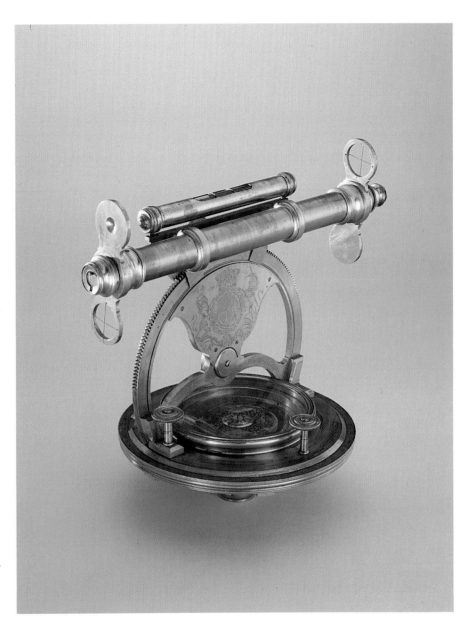

An altazimuth theodolite measures both vertical and horizontal angles, 'altazimuth' being a combination of 'altitude' and 'azimuth'. This instrument bears the coat of arms of the Duke of Cumberland and so it is fairly certain that it came into the collection at his death in 1765. It may be connected with the survey of the Highlands of Scotland in 1747.

There is a compass dial marked with eight points of the compass about 5 inches in diameter. The needle is missing. The two compass scales read 0° to 360°, and 0° to 90° each way in quadrants from north and south. The horizontal circle can be turned with respect to an outer scale and has a vernier which can be read to five minutes of arc. Another index 45° clockwise from the vernier is marked 'Building at 100 feet distance'. It has a linear scale running 100–0–100 extending 45° in both directions.

The inverted arc has the royal coat of arms engraved inside it and below this is the signature. The other side of the arc is graduated from 0° to 60° each way and from 0 to 30 which gives the number of chains to be subtracted from the surface measurement to obtain the true horizontal distance, as in E32. This scale is marked 'links of chain' while the arm carrying the sights is marked 'Depression, Elevation' and 'Diff of Hyp and Base', the last referring to the chain scale. There are backwards and forwards open sights with pinhole and cross-wire, and a telescopic sight. There are two spirit levels but no levelling screws; possibly they were attached to the tripod which is now missing. The instrument is similar to one by Heath and Wing in the Whipple Museum at Cambridge.

Cumberland 1765 note 'a small theodolite and stand in a case'
C 223
Chew 1968, Item 18
Daumas 1972, footnote 88, chapter 5
Brown 1982, cat. No. 69
Bennett 1987, p. 86, p. 149

Inventory No. 1927-1923

E26 Altazimuth theodolite

Bennett, J.

mid-eighteenth century   London

ON HORIZONTAL CIRCLE
  '*J : Bennett LONDON Fecit*'
  '*Edmond Scott Hylton Inv.*'

L 245mm, **W** 155mm (diameter of circle), **H** 145mm

The horizontal circle is divided into degrees numbered in tens, and it is signed with the maker's and inventor's names. It has two fixed slit-and-window sights. The vertical arc is marked from 0° to 50° each way in degrees and carries the telescope. A spirit level would have been situated above the telescope but is now lost. The vertical arc is mounted on a rotatable platform, the angular direction of which can be read to five minutes of arc using a vernier. Two unusual features are the addition of bevelled straight edges protruding from the rotatable platform, and a centring point consisting of a triangular index in the middle of the instrument. These suggest that the instrument was used for drawing directly onto a map or plan. There is no tripod or visible fixture for one, which may mean that part of the intersection between the instrument and stand is also missing.

C 222
Taylor 1966, pp. 197–8
Inventory No. 1927-1921

E27 Altazimuth theodolite

Hindley, Henry or Smith, John

before 1786   York

ON VERTICAL ARC
'Perp to 100s of the Base' and *'Diff of Hypoth and Base'*

L 285mm, **W** 180mm (diameter of horizontal circle), **H** 435mm

Law has attributed this instrument to Henry Hindley of York, or John Smith who was 'bred' with him and later worked for Demainbray. He cites the characteristic features as being the universal joint with two levelling screws on which the azimuth circle is mounted, and the endless screws and micrometers on the azimuth circle and altitude semicircle.

The instrument is on a brass stand on which is a universal joint with two levelling screws acting at right-angles; however, only one spirit level is provided on the azimuth or horizontal circle. This circle is graduated in degrees and numbered in tens. It is turned by means of an endless screw with a micrometer dial which can be read to one minute of arc. The compass in the centre of the circle is marked with eight points, and two scales give degrees from 0° to 360° and 0° to 90° each way from north and south in quadrants. The horizontal circle can be adjusted by means of the endless screw to a limit of about 15°, presumably to take account of variation in declination.

The altitude or vertical arc is supported on two brass pillars. It is also provided with an endless screw and micrometer. It is graduated with two scales: 'Diff of Hypoth & Base' from 0 to 30 each way, which gives the distance to be subtracted from the distance on the surface of the ground to give the true distance for that particular elevation, and 'Perp to 100s of the Base' from 0 to 100 each way which gives the height gained or lost per 100 measures on the level for each particular elevation. The other side of the vertical circle gives degrees to 70° each way numbered in 20s. There is a brass telescope with cross-wire sights in Y bearings above the circle, and another similar telescope for back sightings on a pivot in the base of the stand; both appear to be achromatic. The tripod is now missing; the wooden stand was made in the Museum. An earlier pine stand is also not original.

C 221
Law 1970–2*a*, p. 220, Fig. 16
Law 1970–2*b*
Inventory No. 1927-1920

E28 Plummet clinometer

mid-eighteenth century

'French Fathoms'

SCALES

35 (1) 0 (1) 35° and −10 (10) 90

L 153mm, **H** 112mm, **T** 3mm

This is a triangular piece of brass which has a brass plummet suspended from the apex. A circular scale gives degrees up to 35° each way, allowing the elevation of the base to be measured. Another scale along the base is entitled 'French Fathoms'. It is marked from −10 to 90 in tens with only one division split into units. The ratio of French fathoms or toises to this scale is approximately 1:1500.

The instrument may have been used for levelling mortars. A slightly similar but probably unrelated item is depicted in 's Gravesande's *Mathematical Elements of Natural Philosophy* where it is used to measure the angle of spouting mercury.

Bion 1723, p. 144, Pl. 15, Fig. R

's Gravesande 1747, p. 385, Pl. 53, Fig. 7
C 229

Inventory No. 1927-1915

E29 Plummet clinometer

Davis, J.

mid-eighteenth century   Windsor

'J: Davis Windsor'

SCALE

$0 \left(\frac{1}{4}\right) 90°$

L 210mm (overall), **H** 180mm, **T** 1mm

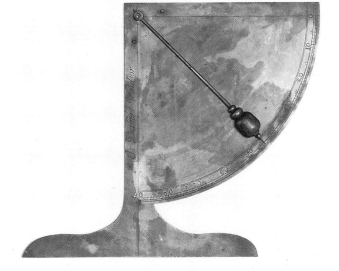

This simple instrument gives the inclination of the plane on which the base rests. A brass plumb bob with a pointer slides over the scale which is marked from 90° when the instrument is horizontal to 0° when it is vertical. The scale is marked in quarter degrees. The instrument is not free-standing, suggesting that it was attached to a larger instrument, but there is no indication of how.

There was a family of clockmakers and locksmiths by the name of Davis in Windsor from the late seventeenth century. This is probably the second John Davis who made a turret clock for Colnbrook Church, Buckinghamshire, in 1746.

C 228
Britten 1982, p. 419
Ashworth 1990   **Inventory No.** 1927-1914

E30 Gunner's quadrant

Jackson, J.

third quarter eighteenth century
London

SCALE
'*J. Jackson London*'

L 435mm, **W** 178mm, **DE** 26mm,
L 485mm (box), **W** 210mm (box),
H 60mm (box)

The instrument consists of a quadrant on a heavy brass arm of square cross-section. The brass quadrant is marked from 0° to 90° in degrees with every ten numbered. A movable index carries a spirit level so that when the index is horizontal the degree of inclination of the surface can be read. There is a vernier which allows an accuracy of five minutes of arc. When not in use the instrument fits into a mahogany case lined with green baize.

The brass arm was placed in the mouth of a cannon or mortar, the spirit level was made horizontal, and the elevation of the cannon read off the scale.

The Jackson on the signature is almost certainly Joseph Jackson who was a well-respected instrument-maker who flourished during the mid-eighteenth century; he made the first Hadley's 'quadrant'. However, the type of gunner's quadrant described above was not introduced into Adams's *Geometrical and Graphical Essays* until 1797.

Adams 1797, p. 495, Pl. 33, Fig. 8
C 226

Inventory No. 1927-1785

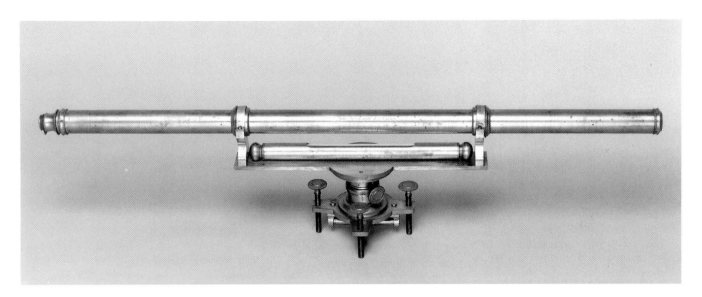

E31 Surveyor's level

Jackson, J.

mid-eighteenth century   London

'*Jos : Jackson LONDON*'

L 750mm, **W** 130mm, **H** 180mm, **L** 260mm (spirit level),
D 30mm (telescope)

The instrument is similar to the standard type of Sisson Y
level common during this period; the spirit level is slung
beneath the telescope which is reversible. However, the brass
bar carrying the telescope and level is not adjustable so the
instrument can only be set using the four levelling screws
below. There is no compass. The tripod and the objective lens
in the telescope are missing. The level is almost entirely made
of brass.

Gardiner 1737, frontispiece and p. 111
C 220
Bennett 1987, pp. 151–2
Inventory No. 1927-1916

E32 Circumferentor

Streatfield

mid-eighteenth century   London

'*Streatfield London*'

L 195mm (box), **W** 190mm (box),
H 87mm (box), **L** 305mm, **H** 215mm,
D 165mm (dial)

A circumferentor was the name given to a magnetic compass with sights. The bearing between two stations was taken with respect to the magnetic needle and a calculation made to give the bearing with respect to astronomical north. Adams states that the instrument was more useful in America than England where land was cheaper and sightlines were more likely to be obscured.

The compass is marked as follows: N, NE/NW, E/W, SE/SW, S, SW/SE, W/E, NW/NE. The complete circle is divided into 360° and also from 0° to 90° each way from north and south. A knob under the dial turns it against a vernier which can be read to a twentieth of a degree. There is a pair of detachable 'double slit and window' sights with one cross-wire missing. A spirit level in working order is situated beneath the compass. The tripod is now missing.

There is a brass cover plate for the dial which is marked with two scales. One is degrees 45–0–45 and the other 30–0–30 engraved 'Diff of Hypo and Base'. This refers to the number of links of chain needed to be subtracted from a 100-link measurement on an incline to find the distance travelled horizontally.

The instrument fits into a mahogany case.

Adams 1791, pp. 213–17, Pl. 15
C 219
Daumas 1972, footnote 88, chapter 5
Bennett 1987, p. 149

Inventory No. 1927-1919

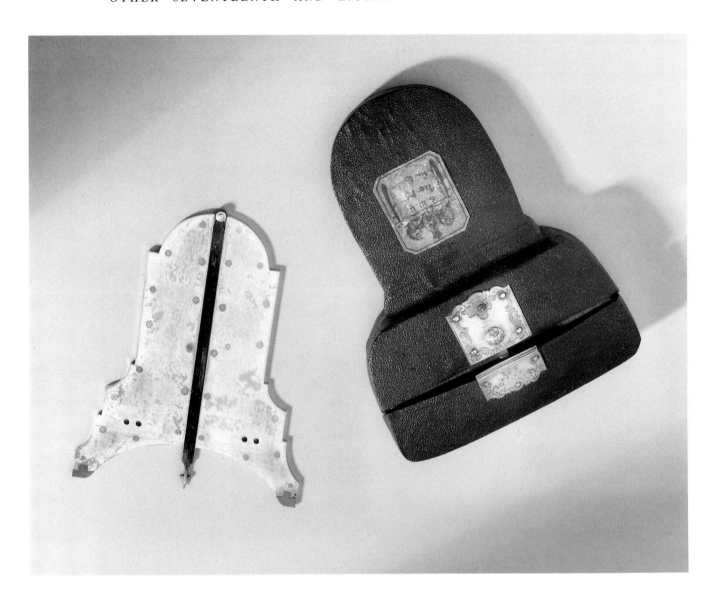

E33 Gunner's perpendicular in a case

Adams, George, the Younger

second half eighteenth century   London

H 128mm, **W** 97mm, **T** 2mm, **H** 142mm (case), **W** 107mm (case), **T** 20mm (case)

This instrument, although not signed, is almost certainly made by Adams; according to Millburn he made 152 for the Office of Ordnance between 1750 and 1772. They do not appear in print in the Adams works until the second edition of *Geometrical and Graphical Essays* in 1797. However, they do appear in Muller's *Treatise of Artillery* in 1757.

The instrument consists of a silver frame with two steel feet and a blue steel index or plunger which slides in a channel down the centre. It was used to find the centre line of a cannon or mortar, which was then used to aim the piece by sight. It was placed on the barrel so that the central plunger was vertical, and the plunger was then depressed to make a mark. A series of these marks could be joined to construct a line. In 'new pattern' perpendiculars introduced in about 1756 a spirit level was included. It is not clear whether this instrument had a spirit level originally.

Adams 1797, p. 515, Pl. 33, Fig. 9
C 227
Wynter 1976, p. 11
Millburn 1988, pp. 250–3   **Inventory No.** 1927-1913

ASTRONOMICAL INSTRUMENTS

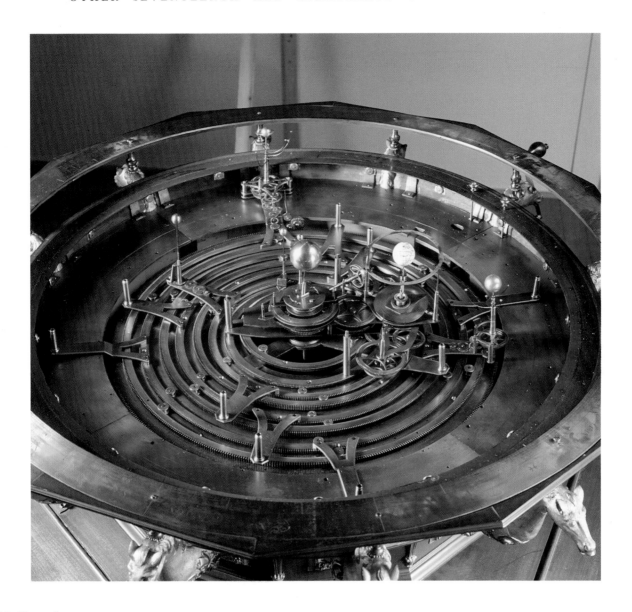

E34 Grand orrery

Wright, Thomas?

before 1733   London

'All the| Machinery of| this instrument| was made new in| 1733 | by Tho Wright| Instrument maker| to His MAJESTY- |GEORGE II'

D 1225mm, **H** 1675mm (without cover), **H** 2005mm (with cover)

This orrery has been described in some detail by Pearson in *Rees Cyclopaedia* and by King and Millburn in *Geared to the Stars*. It is not known who made the orrery in the first place, only that Wright made such extensive modifications in 1733 that it has since been referred to as Wright's grand orrery. Both in the bills for the royal household and on the instrument itself Wright is said to have made new the wheelwork, and not actually to have made the orrery. The entry in the Public Record Office LC.5/19 (115) reads 'Mr. Thomas Wright Mathematical Instrument Maker to His Majesty the sum of Three hundred Thirty Six Pounds for making new all the Machinery and Wheel-Work to perform the Motions of all the planets and Co to the Great Orrery in Her Majesty's Gallery at Kensington. 21st Dec 1733'. It seems likely that Wright was claiming it as his own in 1734 in the advertisement in Harris's 'The Description and use of the Globes, and the Orrery . . .', where he lists 'Another for his Majesty at Kensington' as an example of the large orreries he had made. The argument for it being the orrery 'for his

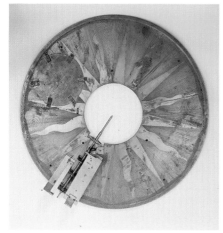

Majesty at Kensington' which was subsequently removed to Richmond, probably when the observatory was built, is very convincing. Gunther does not give sources for his assertion that the orrery was made by Wright in 1733 for about £1500.

Inspection of the internal construction of the instrument reveals that it has been built upon a 3 foot orrery, the dial plate of which has been re-used as the base plate of the new model. The old dial plate was of brass engraved with rays of the sun emanating from the centre. This is now reversed and supports the later wheelwork. A circle cut out of this dial plate, now filled with a brass plate, indicates where the Earth–Moon mechanism was located. Apart from this, little can be said about the original orrery, which Millburn suggests was probably made by Rowley. It presumably fitted into the 12-sided cabinet before it was extended by the addition of Saturn. It is possible that the Earth–Moon system is the original.

The present dial plate is in the form of concentric brass rings which have been painted blue. Each ring carries a planet and turns on rollers when the handle is turned. The armillary hemisphere rests on the calendar ring on which are engraved the signs of the zodiac, each divided into 30°. The armillary hemisphere includes half the celestial equator, the primary meridian divided into degrees from 20–0–90, the solstitial colures, a movable horizon, the celestial Tropic of Cancer, and the Arctic Circle. The calendar ring itself is supported by 12 short pillars which are mounted on gilded horseheads extending from each of the side panels of the base. The horseheads also carry the outermost ring on which the planet Saturn rotates. In this way a planet has been added and the orrery enlarged considerably.

With the exception of the Earth and Moon, which are ivory, the planets are all made of brass. Jupiter has four moons and Saturn five. Turning the handle demonstrates the revolutions of all the then known planets, the motion of the Moon, the diurnal motion of the Earth, the axial rotation of Mars, the motions of the moons of Jupiter and Saturn, and the constant orientation of Saturn's rings. The rings are activated by trains of wheel work which are operated by a '24-hour' arbor. The wheels for Saturn are particularly intricate since they translate the motion to a plane above that of the rest of the orrery. Some pinions can be disengaged by pulling the brass knobs on either side of the handle. One stops the rotation of Mars and the moons of Jupiter and Saturn. The other stops the motion of the superior planets. The whole orrery can be inclined by raising the two engraved brass handles at one side of the base and putting a steel pin through one of a series of holes in a brass arc underneath. The wood throughout is mahogany except for the internal structure which is painted pine. A large hemispherical cover with 12 panes of glass protects the mechanism when not in use.

Q. C., Item 312; orrery from Kensington
Rees Cyclopaedia 1819, under Orrery
Gunther 1920–67, Vol. 2, p. 270
Taylor and Wilson 1944, p. 15
C 140
Chew 1968, Item 16
King and Millburn 1978, pp. 161–2, Figs 9, 10
Public Record Office L.C. 3.64(84) and L.C.5.19(155)

See also D28, D118
Inventory No. 1927-1659

E35 Copernican armillary sphere and planetarium

Sisson, Jonathan

1731   London

ON BASE—BELOW COAT OF ARMS
'Made by JONATHAN SISSON Mathematical Instrument Maker To HIS ROYAL HIGHNESS FREDERICK PRINCE OF WALES. MDCCXXXI'

H 675mm, **D** 610mm (overall), **D** 455mm (sphere)

This instrument is the only one in the collection known to have a direct connection with Frederick Prince of Wales, George III's father. There must have been some interest in astronomy in this generation since Stukeley records discussing orreries with Princess Augusta, wife of Frederick, at Kew in 1754. The sphere was deposited in the observatory by Queen Charlotte sometime in the late eighteenth century.

It is a Copernican armillary sphere with the terrestrial and lunar motion provided by gearing and the other planets on fixed spindles from the central axis. The stand is of solid brass with the inscription running around the base and the royal coat of arms on one side. There are six supports for the horizon ring, which is 3 inches wide, with an outside diameter of 2 feet. The horizon ring is engraved with the calendar, the signs of the zodiac, and the points of the compass. It is also marked in degrees from 0° to 90° in each direction, starting from east and west. Owing to the changing of the calendar in 1752, the dates for the solstices, equinoxes, and boundaries between the zodiac signs are eleven days earlier than at present. The meridian circle slots into the horizon ring at north and south and into a brass ball in the stand. The latitude of the observer can be altered by rolling the circle on the ball. The meridian is marked on one side in degrees from 0° to 90° in both directions from north, and from 90° to 0° in both directions from south. At the celestial North Pole an hour ring gives the time for diurnal revolutions to five minutes' accuracy. An iron pointer which revolves with the sphere moves over the dial.

The sphere is approximately 18 inches in diameter. It consists of an equator, two tropic rings, and two polar rings linked by solstitial and equinoctial colures. The ecliptic is a brass band $2\frac{1}{2}$ inches wide; it is marked in degrees 0° to 30° celestial longitude for each sign of the zodiac and also in degrees of celestial latitude to give a lattice of squares. On the axis of the ecliptic is a brass and steel rod at whose centre is a brass ball representing the Sun. The planets Jupiter, Saturn, and Mars are held on spindles from the rod and can revolve with it or be held stationary. The Earth is situated at the end of an oval brass frame containing three gear wheels. As the frame turns about the axis, the movement of the inner wheel is communicated via the larger central wheel to the smaller outer wheel which carries the Moon around the Earth. There is no diurnal motion of the Earth. Venus, Earth, and Mercury are now missing. The sphere was in poor condition but has been conserved recently.

Jonathan Sisson, and later his son Jeremiah, worked at The Sphere, Corner of Beaufort Buildings, Strand, London.

Q. C., Item p. 20, No. 13; Sisson's armillary sphere
King and Millburn 1978, p. 151

Inventory No. 1927-1654

E36 Eight-foot mural arc

Sisson, Jeremiah

1770 London

'*J Sisson London 1770*'

H 2700mm, **L** 5050mm, **W** 150mm

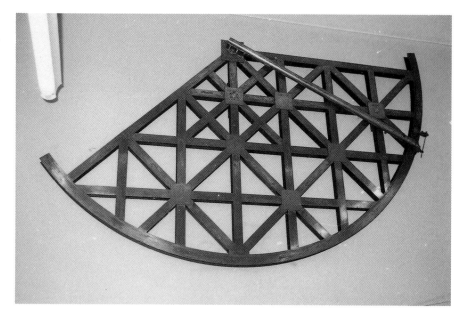

Although technically not in the King George III collection, the mural arc formed part of the King's apparatus at Richmond Observatory. It came into the Science Museum via a different route from the other items. It was made for the observatory at Richmond, now Kew, by Sisson in 1770, remaining in the observatory until 1889 when it was loaned to the Science Museum by the Office of Works on behalf of Queen Victoria. For some time during the mid-nineteenth century the telescope was used at Armagh. The arc is listed at the back of the Queen's Catalogue among some observatory instruments, mainly clocks, which appear to have been added after the original list was drawn up.

The design of the instrument is based on that of the mural quadrant made by Jonathan Sisson under Graham's direction for Greenwich Observatory in 1725. There are three significant departures; the frame is made of brass, the arc extends to 142° and there is no scale to 96 as on the earlier instrument. The frame consists of 2-foot sections of brass bolted together to form a square lattice. The telescope was also supported on a lattice frame to reduce flexure but this is now missing. A device to facilitate manipulation of the telescope, consisting of a wooden frame connected to a cranked level with a counter-poise weight, is also missing. The telescope has an object glass with an aperture of 3 inches and a focal length of 8 feet 2 inches. At the eyepiece end it is fitted with rollers to reduce friction. A clamp is provided to fix the telescope in its approximate position and a fine-threaded micrometer screw at the end of a rod bearing on the clamp gives the

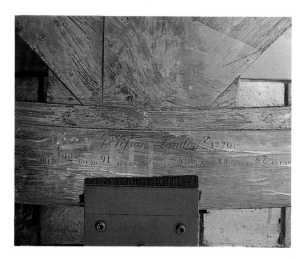

fine adjustment. The scale is graduated in divisions of two minutes of arc. However, the absence of the vernier scale and a scale on the micrometer prevents a reading from being taken.

The extension of the quadrant to an arc of 142° allows all the circumpolar stars to be viewed from London using one instrument.

Q.C., Item 311; mural arch
Smith 1738, Vol. 4, pp. 332–41, Pl. 49
Bird 1768
Le Monnier 1774
Whipple 1926, p. 9
King 1955, pp. 114–15
Howse 1975, pp. 21–4

Inventory No. 1889-39

E37 Planetary system

c. 1800

L 1120mm, **H** 345mm, **D** 325mm (base), **D** 205mm (whirler 1), **DE** 120mm (whirler 1), **D** 320mm, 205mm, 153mm (whirler 2), **DE** 120mm (whirler 2)

The instrument is in three pieces: the principal mechanism and two whirlers, which must at one time have fitted onto a board or table top and been linked to the planetary system by means of cords. The principal mechanism is on a circular black-painted mahogany base. It consists of a series of mahogany rings placed one above the other, each supporting a planet. One of the rings has a groove for a cord. With the exception of Earth the planets are on brass pillars mounted at the extremities of brass arms of various lengths. Earth is supported on a mahogany arm which has a circular groove for a cord, a circular brass plate, and a stand which tilts it at $23\frac{1}{2}°$. The Moon, which is a ceramic ball, moves on a mahogany ring by means of a rack and pinion which moves on the inside of the circular plate against the mounting of the Earth. Diurnal motion can be achieved by turning the Earth, which also moves the Moon. All the planets are made of boxwood. Earth is the largest with a diameter of 2 inches. It is delineated with the equator, tropics, ecliptic, and polar circles in a style very similar to that depicted in Adams's *Lectures on Natural and Experimental Philosophy*.

The planets represented are Mercury, Venus, Earth, Mars, Jupiter, Saturn, and Uranus. The last was then known in England as 'Georgium Sidus', so named in honour of George III by Herschel who discovered it in 1781. Jupiter has four moons and Saturn seven which correspond to the number known at the time. However, Uranus has six, whereas only two had been discovered by 1800. It is possible that the moons were added later, but more likely that they have been put onto the instrument by a user simply because they were there: it is easy to add or subtract moons and planets and there are four extra planets and three extra moons provided. There is no ring for Saturn and the Sun is now missing. One of the whirlers has three different diameters, and the other has a rosewood handle.

Adams 1794, Pl. 12, Fig. 2, Vol. 4
C 142
Inventory No. 1927-1413

E38 Terrestrial globe

Adams, George

1766   London

ON THE GLOBE IN THE PACIFIC OCEAN

'*Britann[. . .] Aug[.]stifsimo|GEORGIO TERTIO|Scienti[. . .] Cultori par-iter et Pra[.]fidio| Globum h[. .] Terrestrem,| Omnes hactenus exploratos [. . .]arum trac[.]tus, Ad| Ob[. . .] ones Navigant[. . .]nerantium, et Astro-nomo|[. . .]ntiores accuratifsime defcriptos ex[. . .]bentem| [. . .]nimi et pietatis monumentum|[. . .] D.D.Q.|Omn[. . .] et officio devin[. . .]tifsimus| G. Adams|Londini apud G.ADAMS artificem Regum| in vico Fleet-street|*'

ON COMPASS

'*Made by G.ADAMS Fleet Street LONDON*'

D 460mm (globe), **D** 605mm (horizon ring), **H** 840mm (overall), **D** 660mm (case), **H** 1360mm (case)

The inscription above translates as:

> To the most august King of the Britions, George III, equally a devotee and protector of the sciences, this terrestrial globe, showing all the places on Earth thus far discovered, from the most recent and accurately described observations of voyaging seamen and astronomers, is given and dedicated as the monument of a grateful heart and respect, with all honour and duty, by the most devoted G. Adams. London, by G. Adams instrument-maker to the King, in Fleet Street.

This globe together with the matching celestial globe were made by George Adams around 1766 for they match the globes illustrated in his book *A Treatise Describing and Explaining the Construction and Use of New Celestial and Terrestrial Globes*, first published that year. These globes must be among the first made by Adams, for though there is evidence that he planned to sell globes from the mid-1750s, he does not seem actually to have done so until a decade later.

The globe is made from two sets of 12 gores. It rotates about an axis carried by a brass meridian circle. One side of this meridian is graduated to give both latitude and polar distance. The other side of the meridian, near the North Pole, is graduated to show the declination of the Sun throughout the year. The mahogany horizon circle has a paper covering on top with printed information about winds, the signs of the zodiac, and a calendar with dates such as the birth of George III.

The globe has four features introduced by Adams. One is a semicircular brass strip which functions as an adjustable meridian with a sliding flat ring engraved with the points of the compass. Another novel feature is a semicircular brass wire whose ends are hinged to the horizon circle and which passes through a hole in the meridian circle. This wire carries two indexes and follows the equator as the globe is moved. The other two features are a brass quadrant graduated in degrees which can be set to go from the zenith point of the meridian to the horizon; a brass wire at 18° below the horizon representing the limit where twilight becomes darkness.

The route of Anson's voyage round the world between 1740 and 1744 is shown but no later voyages, such as those by Cook which appear on later globes by Adams. These details are consistent with a date of 1766 for these globes.

There is a magnetic compass in the elaborately carved mahogany base. The globe has a glazed octagonal mahogany case.

Adams 1766   C 162   van der Krogt 1984, pp. 35–44

Inventory No. 1927-1700

E39 Celestial globe

Adams, George

1766   London

ON GLOBE

'Britanniarum REGI Augustifsimo/ GEORGIO TER-TIO/ Astronomorum Patrono Munificentifsimo, Celeberrimo/ Globum hunc Coelestum,/ Novam et Emendatiorem Coeli Imaginem, Sydera apud/ Africae Promontorium Auf[. . .]ale nuperrine observata, Atq/ Stellas Catalogi Flamftediani Universas, vere exprimentem/Grati animi et pietatis monumentum/ D.D.Q./ Omni cultu et officio devinctifsimus/ G.Adams/ Londini apud G.ADAMS artificem Regum/in vico Fleet-street'.

ON COMPASS

'Made by G.ADAMS Fleet Street LONDON'

D 460mm (globe), **D** 605mm (horizon ring), **H** 830mm (overall), **H** 1350mm (case), **W** 660mm

The above inscription translates:

> To the most august King of the Britons, George III, the most generous and celebrated patron of astronomers, this celestial globe, a new and corrected image of the heavens, showing accurately the stars recently observed at the southern promontory of Africa and also all the stars in Flamsteed's Catalogue, is given and dedicated as the monument of a grateful heart and respect with all honour and duty, by the most devoted G. Adams. London, by G. Adams, instrument-maker to the King, in Fleet Street.

The celestial globe is very like the terrestrial globe in the details of its construction. The horizon circle, again of mahogany with a paper covering, carries the same information. The globe rotates about an axis through the ecliptic poles which is supported by a brass meridian circle. One side of this meridian is graduated to give both latitude and polar distance. The semicircular movable meridian carries a small flat ring engraved with the rays of a sun rather than with the points of the compass. The other features are similar to those of the terrestrial globe except that there is no adjustable quadrant to go from the zenith to the horizon.

According to Adams the globe was based on Flamsteed's star catalogue. The names of the constellations are given in English, Latin, Greek, and sometimes Arabic. Several constellations in the southern hemisphere are based on the shapes of scientific instruments and other contemporary items: Officina Sculptoris, Fornax Chemica, Horologium, Equuleus Pictorius, Machina Pneumatica, Octans Hadleianus, Sextans, Cela Praxitelis, Telescopium, Microscopium, Quadra Euclidis, Reticulus Rhomboidalis. These are Sculptor's Workshop, Chemical Furnace, Clock, Painter's Easel, Air-Pump, Hadley's Octant, Sextant, The Gravers, Telescope, Microscope, Euclid's Square, Rhomboidal Net. These were based on the constellations devised by Lacaille from observations he made at the Cape of Good Hope in 1750–2 and published in 1763. However, Adams used his own instruments such as his air pump and universal microscope as models.

There is a magnetic compass in the elaborately carved mahogany base. The globe has a glazed octagonal mahogany case.

Adams 1766
C 160
Warner 1967
van der Krogt 1984, pp. 35–44

Inventory No. 1927-1701

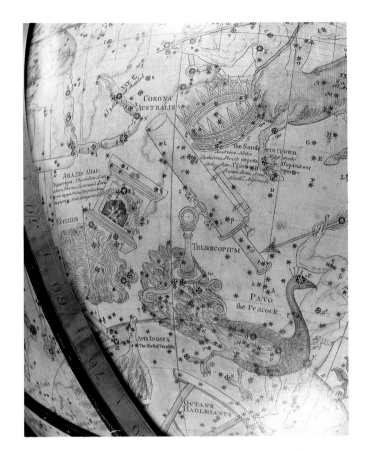

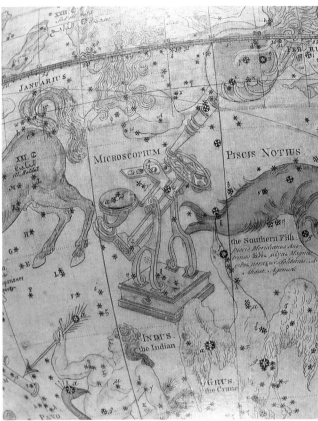

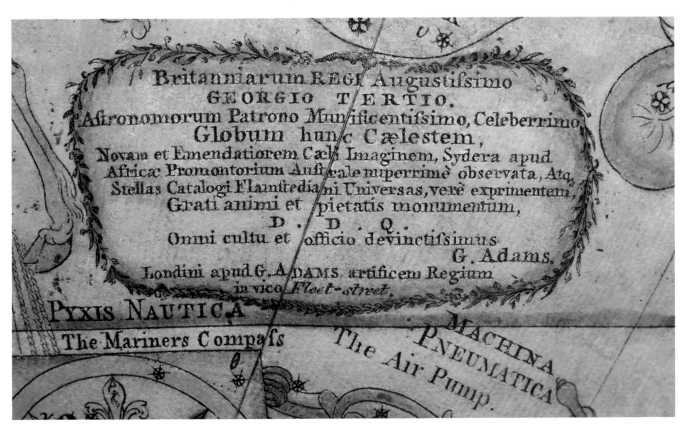

E40 Terrestrial globe

Senex, John

second quarter eighteenth century   London

'A New & Correct GLOBE of the Earth by I. Senex F.R.S.'

D 130mm (base), **D** 75mm (globe), **D** 100mm (brass disc), **H** 210mm

The globe, which is nearly 3 inches in diameter, is mounted with its axis of rotation at an angle of $22\frac{1}{2}°$ above a brass disc 4 inches in diameter; the tilt imitates the inclination of the Earth's axis with respect to its orbit. The stand consists of a brass rod on a mahogany base. The globe is marked with the equator, tropic, and polar circles and lines of longitude every 30°. The map omits Alaska and the eastern part of Australia; the western part is named New Holland. The polar regions are marked 'Incognita'. The trade winds in the equatorial regions are marked by a number of arrows. The system is similar but less complicated than that of Halley. The brass disc is numbered in hours, from 1 to 12 twice, and the quarters are given. An index at the longitude of London moves over the brass disc as the globe is turned; thus the time at various places around the globe can be found when the time at London is known. The mounting is similar but not identical to that which Senex described in 1738.

    Senex was Geographer to Queen Caroline, wife of George II.

Halley 1685	C 164
Senex 1738	Taylor 1966, p. 143

Inventory No. 1927-1464

E41 Terrestrial globe

Senex, John

second quarter eighteenth century   London

'A New & Correct GLOBE of the Earth by I. Senex F.R.S.'

D 75mm, **H** 205mm, **D** 90mm (with ring)

The globe is identical to E40 but the mounting is simpler. It is set vertically on an iron rod attached to the South Pole. A silvered brass horizon ring engraved with N, S, E, and W is movable around the globe. At four points on the ring brass indexes prevent it from becoming detached. Holes in the ring suggest it was once mounted. The wood base for the instrument was made in the Museum.

Halley 1685	C 165
Senex 1738	Taylor 1966

Inventory No. 1927-1463

E42 Terrestrial globe

Hill, Nathaniel

1754 London

'A/NEW Terrestrial/GLOBE/by/Nath Hill/1754'

D 75mm, **D** 95mm (with ring), **H** 190mm

The globe is marked with the equator, tropic, and polar circles and the ecliptic. The longitude is marked every 30° and numbered every 15°; the latitude is numbered every 10° north and south. There is no Alaska or Eastern Australia and only a small section of 'New Zeeland' is drawn. The land masses are coloured. There are the same trade wind patterns as in the two previous items and the Pacific Ocean is called The Great South Sea. The sea route taken by Anson on his voyage round the world between 1740 and 1744 is marked in red. The globe is mounted at an angle of $23\frac{1}{2}°$. Chaldecott points out that the gearing near the South Pole indicates that it once formed part of an orrery; it is now on a brass rod on a wooden base which was made in the Museum. The silvered brass horizontal ring is rather crudely engraved with the calendar and signs of the zodiac.

C 163

Taylor 1966, p. 208

Inventory No. 1927-1548

E43 Quadrant

Rowley, J.

c. 1756   London

'*Table for observing/Particular Stars with/a sight & cross wires for directing it*'
'*J. Mynde Sculp*'
PAINTED
  '*Rowley/Fecit*'
AT APEX
  '*W. Winch*'

R 300mm, **DE** 175mm (overall)

This item almost certainly came into the collection from the Duke of Cumberland since a 'printed Quadrant in Mahogany Stand' is listed in the inventory of his household furniture. The quadrant was described in detail by Thomas Woodford, writing as 'T.W.' in a treatise written in 1756. Apparently Rowley designed it when he was just out of his apprenticeship, probably around 1700. It was intended as an improvement on Collins's quadrant which was 'too full of lines, some necessary, some unnecessary'. The 1 foot quadrant was attached to the front of the book as a plate which could be removed and pasted on a smooth board.

third names it, and the fourth gives some indication of brightness. Woodford added a cross above the annuli if it had more than 12 hours right ascension. The following two annuli give the star's right ascension either on the scale 6 to 12 and 0 to 6 or 0 to 12. To find which of the two scales should be used required a celestial globe to ascertain whether the star rose after an equinoctial point or a solstitial point. Then there are four calendar rings based on the Julian calendar, which is unremarked in Woodford's work even though it had then become out of date. Weeks are marked with small dots. Then comes a 'quadrantal line of the ecliptic' or zodiac ring matching the calendar ring. Following this is the Sun's declination from 0° to 23° 30′. Beneath this is a secant scale for a radius equal to half that of the quadrant. At the right-hand end of the secant is a versed sine scale marked in hours and minutes instead of degrees, which enabled the user to find the hour from noon 'more precisely than from the other lines'. Near the circumference of the instrument is a versed sine scale of twice the radius of the quadrant with a corresponding hour scale attached. At the circumference is a line of sines which also gives degrees and hours and minutes.

Leybourn 1672, pp. 3–6
Leybourn 1682, pp. 13–15
Collins 1710, pp. 19–23
T.W. 1756
Bion 1758, p. 16, pp. 38–40, Pl. 4, Fig. 5
Cumberland 1765
C 150
Cooke, Troughton and Simms 1959
Taylor 1970, p. 294

Inventory No. 1927-1481

The quadrant has been pasted onto mahogany. One of the original brass sights remains and there is a brass butterfly nut at the back, presumably for attaching to a stand. Handwritten in ink near the butterfly nut is 'table for observing particular stars with a sight & cross-wires for directing it'. The plumb bob is now missing.

The explanation of the various lines is given to some extent by Leybourn and Bion in the texts listed below. The front surface of the quadrant is marked as follows: down the left-hand edge is a line of equal parts, or simple ruler, 10 inches long. Next to this is a line of sines for a 6 inch radius ending at 90°, and then a line of versed sines for the same radius, from 90° to 180°. Down the right-hand edge is a line of tangents of radius equal to that of the quadrant and next to it a line of sines to 90° for the same radius. The separation of the two lines was to allow for the positioning of the sights. The chart near the apex gives the minutes to be added or subtracted from the Sun's declination for each group of 10 days in the years immediately following a leap year, immediately preceding a leap year, and the leap year itself. The quadrant is set for the second year in the leap year cycle. Presumably 'W. Winch' was responsible for this rectifying table.

Moving outwards from the apex there are four quadrantal annuli containing information about the principal stars of the astrolabe. They are named as follows: Vir Spik, Aries, Arcturus, Bri No Cro (Brightest Star in the Northern Crown), Bull's Eye, Orio L Sho, Orio R Sho, Gr Dog, Lit Dog, Aquila, Lyon Har, Formahaut, and Pega Wing. One annulus is marked with an N or S to show which hemisphere the star can be found in, another gives its declination, the

E44 Disc planisphere

Sisson, Jeremiah

third quarter eighteenth century   London

'*J. Sisson London*'

D 520mm, **L** 55mm (handle), **H** 100mm

It is possible that this planisphere and the transit instrument E188 were made for the observatory at Richmond. It is one of the relatively few examples of a high precision instrument in the collection and displays fine quality workmanship. It could have been used in conjunction with a transit instrument, possibly E188, to set clocks and watches or predict the times of other transits, given the time of one.

The planisphere forms the top surface of a heavy brass cylinder about 20 inches in diameter and 4 inches deep. An ivory handle turns the wheelwork which consists of a flat toothed wheel, which meshes with the teeth on the underside of the planisphere and a wheel of similar size which is placed at an angle of about 10° diagonally inside the cylinder. The purpose of this second wheel is not clear.

The outer fixed ring of the planisphere contains the hours in roman numerals from 1 to 24 and is marked in minutes. 'Noon' and 'Night' are also engraved. Proceeding towards the centre, the next ring is the calendar ring which is marked in months and days with every 10 days numbered. This can be adjusted slightly in relation to the central disc to take account of leap years. A square key, now missing, could be used to rotate the central disc containing the celestial data while the calendar ring was stationary. The calendar ring can be removed if necessary, allowing a greater flexibility of dates, which may suggest the instrument was made around 1752 when both Gregorian and Julian calendars would have been useful. The next ring, the outermost on the central disc, is marked in hours for the right ascension. It is not marked other than with hours so the exact value cannot be found. There are then three narrow bands: the outer is empty, the central one contains the exact position of the star, and the inner is marked with a '1' or '2' for the magnitude. Then the star is named with its name if it has one, and the letter and constellation, the letter being scratched rather than engraved. The declination of each star is on the next ring, and the final ring gives the altitude of some stars with respect to London. A total of 54 stars is listed. The instrument is engraved 'J. Sisson London'. A brass index which held a wire can be moved over the planisphere to enable the information to be read accurately.

C 148

See also E188

Inventory No. 1927-1196

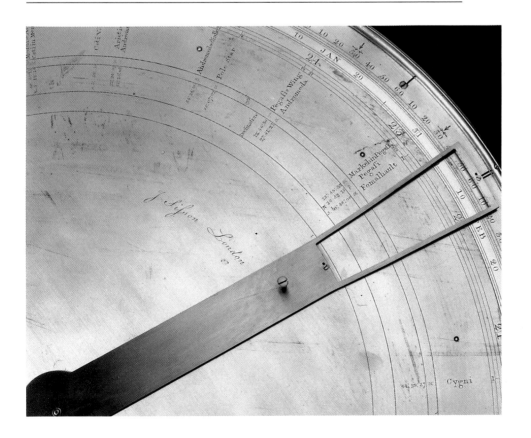

TIME MEASUREMENT

E45 Regulator clock

Vulliamy, Benjamin

c. 1780   London

'Benjamin Vulliamy/London'

L 470mm, **W** 310mm, **H** 1430mm

According to notes in the Museum's files this regulator was made for George III in about 1780 for his observatory at Richmond where it was the principal timekeeper; one note from the Patent Office stating that it was originally at Greenwich is probably a mistake. It is almost certainly the 'Vulliamy's Regulator' listed at the back of the Queen's Catalogue. Whipple identifies it as the follower but does not give reasons; this is unlikely since this is a timekeeper rather than an instrument for transferring time. It passed through many royal hands before being presented to the widow of Vulliamy's eldest son in 1845. She sold it to the Patent Office Museum from whence it came into the Science Museum in 1884.

The regulator has a 'grasshopper' escapement which is almost frictionless since the pads of the pallets merely touch the teeth of the escapement wheel and do not pass along them. There is also extensive use of friction wheels to carry the bearings of the wheelwork. When the clock is wound the great wheel is lifted from its bearings and a sun-and-planet mechanism prevents the clock stopping. The pendulum oscillates through an arc of between 8° and 10°, the extent of which can be observed on the scale beneath the bob. An almost continuous impulse is received throughout each swing. The pendulum is a gridiron comprising steel and zinc rods whose unequal expansion allows for temperature compensation; the lenticular bob is made of brass filled with lead. There is a dial for seconds and one for minutes. The clock is in a walnut and glass case.

Benjamin Vulliamy worked in London between 1775 and 1820. He is said to have given advice to King George III concerning the observatory on many occasions.

Q.C., Items 318, 320; Note: Vulliamy's Follower, Vulliamy's Regulator

Rigaud 1882, p. 282

Scott 1885, pp. 43–4

Whipple 1926, p. 10

Britton 1982, p. 633

Inventory No. 1884-79

E46 Equinoctial sundial

Hager, M. T.

before 1695   Arnstadt, Germany

ON COMPASS
'SEP', 'ME', 'OCCI', 'ORI'

ON LID
'MT Hagar a Arnstadt'

L 55mm, **D** 42mm (scale), **D** 50mm (case)

The monogram in silver studs on the outer case is that of William and Mary, making this one of the earliest instruments in the collection. The same monogram can be seen on the ceiling of the banqueting hall at the Royal Naval College at Greenwich.

The instrument has a brass face on which there is a circular dial giving latitudes between 35° and 65°; however, the index is now missing. This set the inclination of the hour ring which springs into place at the touch of a button. The hour ring is marked in Arabic numbers from 4 to 12 to 8. It is engraved on silver and attached to the inner surface of a plain brass ring. A small glass window allows the compass needle to be viewed. The compass is marked SEP, ME, OCCI, ORI for N, S, W, and E respectively. The gnomon is missing. The instrument is contained in a silver case like a watch case. The outside of the lid has engraved on it the latitude of many towns, mainly German. The silver case fits into a brass case which is studded with silver in the form of the monogram.

Bion 1723, p. 247, Pl. 24, Fig. 7   C 157
Gunther 1920–67, Vol. 2, p. 134   Ward 1981, cat. No. 257

Inventory No. 1927-1466

E47 Universal equinoctial sundial

Martin, Johann

c. 1700   Augsburg

'OC' 'OR' 'SE' 'ME'

DE 18mm (box), **D** 90mm (box), **D** 80mm (dial), **D** 40mm (compass)

Like E46 this instrument was probably in the possession of the royal family before George III's time. It is possible that the 'Octagon Silver Compass-level and Dials' in the Duke of Cumberland's inventory refers to this object.

The silver dial is in the shape of an octagon with floral engraving and a coat of arms on the upper surface. A compass is set into the plate with the markings OC, OR, SE, and ME, and a variation of 10°, the difference between true north and magnetic north. The brass hour ring reads from 3 to 12 to 9 in Roman numerals, and can be set for any latitude by the quadrant scale. The gnomon is pivoted on the hour scale and

should point to the Pole Star when the instrument is in use. The time can then be read off the hour scale. The underside of the instrument has four circular scales or volvelles. One shows the time of the southing of the Moon according to its age. Another converts hours into Babylonian hours. A third gives a perpetual calendar for the days of the week and a fourth is a circle with months giving number of days and zodiac sign, and two numbers, probably the start of each sign of the zodiac in the Julian and Gregorian calendars. A silvered disc on the inside of the lid of the box gives the latitudes of many towns, mostly German.

Bion 1723, p. 247, Pl. 24, Fig. 7, p. 233
Cumberland 1765
C 152
N.M.M. 1970, pp. 278/39
Bryden 1988, Cat. Nos 121, 178, 179

Inventory No. 1927-1467

E48 Mechanical equinoctial sundial

Sisson, Jonathan or Jeremiah

before 1753   London

ON CLOCK FACE
  'SISSON/LONDON'
ON BACK
  'J. Sisson London'

D 95mm (scale), **H** 155mm (max), **D** 50mm (clock)

Again there is a possibility that this instrument came from the Duke of Cumberland's collection. It could be the 'small brass Globe with a Wheel Dial' listed in the household inventory.

It is a sundial which gives the time by means of hands moving on a clock face. The clock face is in a brass sphere above which is the dial. According to Chaldecott it is set for a latitude of 53°, somewhat north of London. The readings given correspond to the Julian calendar. There is a circular brass dial which has an arm with a pinhole sight at either end. One end is fixed while the other moves over a circular calendar scale. The date is set, the clock face is turned approximately south, and the instrument is adjusted until the spot of sunlight from the first pinhole coincides with the second pinhole. Using the equation of time which is engraved on the back, the mean solar time can be calculated by reading the clock face and adding or subtracting the required number of minutes. An English example of this type of instrument is rather unusual.

C 151   **Inventory No.** 1927-1465

E49 Equinoctial ring dial

early eighteenth century

D 70mm, **L** 22mm (pin), **H** 85mm

The small silver dial has a meridian ring which is calibrated in degrees from 0 to 90 for the northern hemisphere only. The hour ring is numbered in Roman numerals with half-hourly markings on the inner surface. The nautical ring, which was used to measure the altitude of the Sun, is marked in degrees from 0 to 90 and a pin is provided. On one side of the bridge the calendar scale is in two halves; it is not possible to tell whether it is Julian or Gregorian. On the other side of the bridge the declination is given to $23\frac{1}{2}°$. There is no case and no other marking.

Bion 1723, p. 246, Pl. 24, Fig. 5
C 153
Bennett 1987, pp. 80–2, Fig. 76

Inventory No. 1927-1148

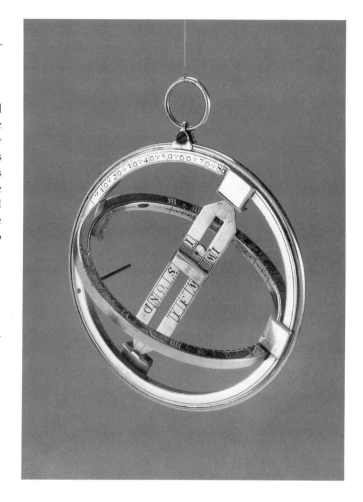

E50 Universal equinoctial ring dial

Glynne, R.

1710–29 London

'*R. Glynne Fecit*'

DE 20mm (box), **D** 130mm (box), **H** 140mm, **D** 120mm, **L** 38mm (pin)

Like the dials by Martin and Hagar, this instrument was probably in the possession of the royal family before George III's time.

The instrument is entirely of brass. The meridian ring is marked from 0° to 90° in half degrees for both northern and southern hemispheres. The 'nautical' scale, to be used in conjunction with the brass pin to give the solar altitude, is also marked in half degrees from 0° to 90°. The hour ring is engraved from I to XII twice in intervals of five minutes of time. The bridge gives the Sun's declination from $23\frac{1}{2}°$ in both directions opposite a zodiac scale, and on the other side the calendar scale in two halves which appears to be Julian. The signature is on the lower half of the meridian ring.

Like the previous item this is a common form of dial of the period with many examples in other collections. It can be used to tell the time if the latitude is first set by moving the suspension ring. The date is set by moving the index in the central bridge and the hour ring is made to lie perpendicular to the meridian ring. The spot where the point of light falls on the hour circle gives the time.

Bion 1723, p. 246, Pl. 24, Fig. 5
C 155
Ward 1981, cat. Nos 186, 187
Bennett 1987, pp. 80–2, Fig. 76
Bryden 1988, cat. No. 227

Inventory No. 1927-1147

E51 Universal equatorial sundial

Adams, George, the Younger

last quarter eighteenth century   London

ON BASE
'*G Adams London*'

H 300mm (max), **L** 140mm (sights), **D** 80mm (base plate)

The instrument is similar in principle to E48. It consists of two vertical semicircles: the lower latitude circle supports the hour ring and the upper declination circle supports the sights. The latitude circle is marked from 0° to 90° in degrees in both directions. It is moved by means of a rack and pinion against a vernier giving an accuracy of five minutes of arc. The declination circle is also marked from 0° to 90° in degrees in both directions and the sights are also moved by a rack and pinion against a vernier giving an accuracy of five minutes of arc. The sights consist of an open ring above a pinhole. The circular base of the instrument is marked in degrees and turns against a vernier giving an accuracy of five minutes of arc. There are two spirit levels and three levelling screws. Apart from the spirit levels the instrument is entirely made of brass.

To use the sundial the latitude is set on the lower scale, the Sun's declination is set on the higher one, and the instrument is turned so that the latitude circle is approximately in the meridian. The instrument is levelled and the hour circle is turned so that light from the pinhole in the first sight coincides with the second pinhole. The time can then be read off the hour scale with an accuracy to one minute.

Adams 1794, Pl. 14, Vol. 4, Fig. 2
Adams 1799*b*, Pl. 14, Vol. 4, Fig. 2, pp. 209–16
C 224
Millburn and King 1988, p. 228   **Inventory No.** 1927-1922

E52 Equinoctial ring dial

Honemann

mid-eighteenth century   Clausthal

D 68mm, **D** 70mm (case), **DE** 10mm (case), **H** 80mm, **D** 65mm

This is a silver dial which has a black fishskin case. The meridian ring is marked in degrees from 0 to 90 for the northern hemisphere only. There is no nautical ring. The hour circle is in roman numerals divided into quarter hours. On one side of the bridge is a zodiac scale and on the other a calendar scale which appears to be Gregorian. It is rather crudely marked every 10 days. Any spare surface on the instrument is utilized for the latitudes of towns, mainly in Germany.

Bion 1723, p. 246, Pl. 24, Fig. 5
C 154
Bennett 1987, pp. 80–2, Fig. 76

Inventory No. 1927-1149

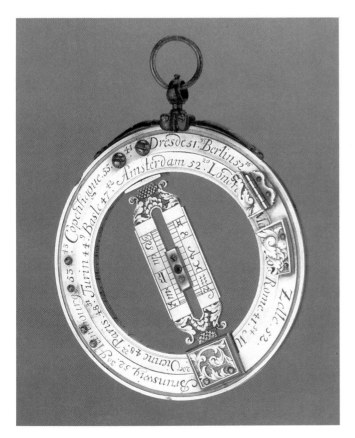

E53 Magnetic sundial

W., E.

mid-eighteenth century   Germany

'A Bedeit/die Ersten 15 Tag in allen/Monaten/B Bedeit die Letzten 15 Tag/allen Monaten/Die Sonn... durch den ausschnitt/auff die 12 scheinen alsdann/zeiget der Magnet die Stundt/E.W.'

D 50mm, **H** 16mm

A translation of the inscription reads as follows:

A covers the first 15 days in all months. B covers the last 15 days in all months. The sunrays shine through the slot onto the 12, then the magnet shows the hour. E.W.

The instrument is in a circular brass case with a lid bearing the above inscription on the inside. The compass needle occupies the centre of the instrument while the outer ring is used to carry one of six silver hour rings. Each ring is engraved on both sides, marked 'A' or 'B', with a range of hours. One side of each ring is suitable for certain days of the year, for example the first half of April and the first half of September or the last half of February and the last half of October. The rings fit on a pin in the case.

The instrument is held horizontally so that sunlight passing through a slot in the side of the case falls on the 12 or midnight mark on the hour ring. Then the magnetic needle will point at the hour.

C 270

Inventory No. 1927-1808

E54 Hour glass

second half eighteenth century or first half nineteenth century

H 153mm, **D** 63mm (max)

The hour glass is of the standard pattern and it is still in working order. It is unstable when held vertically which suggests that it could have been the 'hour glass mounted in brass' listed in the Queen's Catalogue which has since lost its mounting.

Q.C., Item 164; Hour glass mounted in brass

Inventory No. 1929-124

MECHANICS

E55 Double cone

second half eighteenth century

L 100mm, **D** 60mm

The cone may be part of the double rolling cone in wood which is listed in the Queen's Catalogue and therefore could be associated with Demainbray. It is a double cone of lignum vitae with an ivory knob at one apex. The knob originally on the other apex is missing. It would have appeared to roll up a jointed pair of rulers in the well-known mechanical paradox.

Q.C., Item 47; brass double cone with 2 brass inclined rulers
Q.C., Item 48; Do [Double cone] in wood
Leybourn 1694, mechanical recreations, pp. 12. 13
Desaguliers 1744, Vol. 1, p. 56, Pl. 4, Fig. 13
Adams 1746, No. 114.
's Gravesande 1747, Vol. 1, p. 41, Pl. 7, Fig. 1
Adams 1762, p. 22, Pl. 12, Fig. 52
C 38

See also D99, M22

Inventory No. 1927-1877

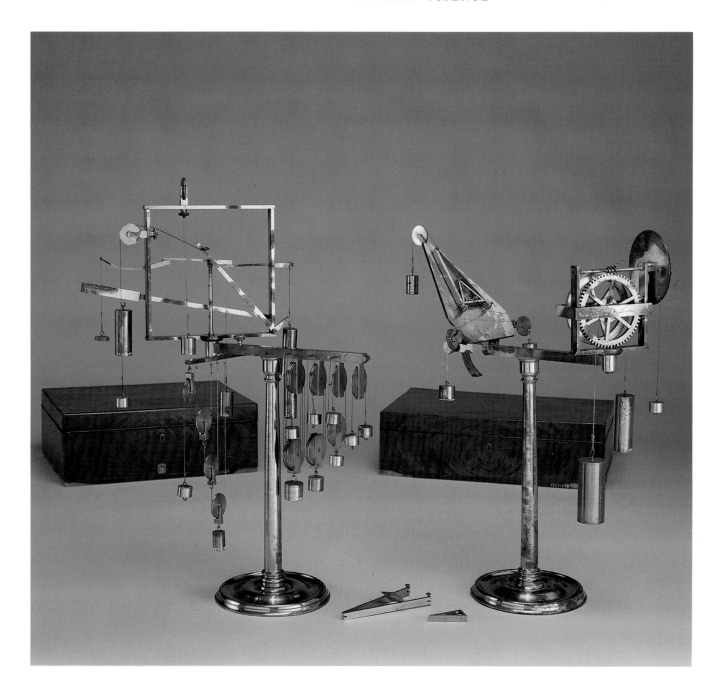

E56 Mechanical powers apparatus: pulleys and levers

Adams, George, the Younger

last quarter eighteenth century   London

L 380mm, **W** 320mm, **H** 570mm, **L** 380mm (box), **W** 290mm (box), **H** 140mm (box)

This forms a pair with the following item; together they illustrate all the mechanical powers. 'Mechanical powers in a new, commodious and elegant form' first appeared in Adams's catalogue in 1777 at the back of 'A Treatise describing the construction and explanation of the Use of the Celestial and Terrestrial Globes'. Thereafter it was described in more detail: 'The mechanical powers neatly made in brass consisting of the balance, the pulleys, levers, inclined plane, wheel and axle, screw, compound engine, compound lever, double cone, friction wheels, weights and wedge etc.' This sold for about £24 during the late eighteenth century with another 'Made on a more elegant and enlarged plan for a large auditorium' which was priced at £60.

This instrument is almost entirely made of brass and when not in use fits into a mahogany box lined with green baize. There are three sections supported on a pillar on a circular base. The first is an arm of square cross-section from which four pulley systems are suspended. The first of these is a single pulley, the second a tackle of two single pulleys, the third a tackle of two double pulleys, and the fourth a tackle of two treble pulleys. The second section consists of a bar at right-angles to the first from which is suspended a system of single pulleys. The third is the lever apparatus which is attached to a square frame. A right-angled lever is on one side while a lever with several forces in equilibrium is situated across the bottom. A compound lever consisting of three arms operates across the centre of the frame.

Adams 1777, Item 109
Adams 1794, Vol. 3, pp. 257–331, Pl. 5, Figs 3, 4, 5
Adams 1798, p. 12
C 28

Inventory No. 1927-1189

E57 Mechanical powers apparatus: inclined plane and endless screw

Adams, George, the Younger

last quarter eighteenth century   London

'*G. Adams London*'.
ON LOCK OF BOX
'*E. GASCOIGNE*'

L 290mm, **W** 255mm, **H** 520mm, **L** 430mm (box), **W** 275mm (box), **H** 120mm (box)

This forms a pair with the previous item, and illustrates the remaining mechanical powers: the inclined plane, the wedge, and the screw. A miniature rolling double cone and friction wheels are also included. It is almost entirely of brass and when not in use fits into a mahogany box with a green baize lining. Gascoigne was probably the locksmith.

It is slightly smaller than the previous instrument and consists of two sections: an inclined plane and a demonstration of the screw. The small steel friction wheel can be used alone on the inclined plane or with the wedge apparatus. The wedge is a small brass triangle which operates with its base downwards, unlike that in the wedge apparatus M45. It acts to separate two arms of a brass A-shape which are held together by weights hanging over pulleys. The endless screw is a compound engine which consists of two axes in peritrochio and an endless screw giving a total mechanical advantage of about 100.

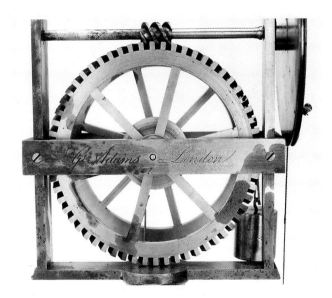

Adams 1777, Item 109
Adams 1794, Vol. 3, pp. 257–331, Pl. 5, Figs 3, 4, 5.
Adams 1798, p. 12
C 31

See also M45, M46

Inventory No. 1927-1175

E58 Hunter's screw

1781–1840

L 305mm, **W** 170mm, **H** 800mm (max)

This differential or compound screw was invented in 1781 by William Hunter, a surgeon.

The model comprises a rectangular walnut frame, the upper section containing a right-handed screw which acts vertically and is turned by two walnut handles, each 6 inches long. Within this screw is another right-handed screw, the lower end of which has a square cross-section and passes through a hole in the lower section of the frame. The pitch of the outer screw is three to an inch and the inner is four to an inch, so that when the handle is turned through one revolution, the lower screw moves 1/12 inch. In general the distance would be

$$1/a - 1/(a+c)$$

where a is the pitch of the first screw, $a+c$ is the pitch of the second screw, and c is the difference in pitches. Thus the mechanical advantage of this model is

$$2\pi \times 6 \times 3 \times 4 = \text{about } 450$$

and in general

$$2\pi ba(a+c)/c$$

where b is the length of the handle. It can be seen that the pitches of the screws should be as high as possible and the difference between them as low as possible. The mechanical advantage compares very favourably with that of a simple screw and the idea proved useful in presses of various kinds.

Hunter 1781
C 69

Inventory No. 1927-1108

MODELS

E59 Hooke's universal joint

Pease?

third quarter eighteenth century   London

L 100mm, **W** 35mm, **DE** 35mm

This could be the universal joint by Pease listed in the Queen's Catalogue but D113 is more likely to fit this description. The brass model is mounted in a spring jacket in which it can rotate. A semicircular arc carries a cross which can rotate about a second axis and the opposite arms of the cross carry another semicircle which can rotate about the third axis. A mounting for an instrument is on the second semicircle.

The joint is very similar to that illustrated in Hooke's *A Description of Helioscopes* of 1676.

Q.C., Item 71; Ditto [Model of Mr Pease's] of Universal joint
Hooke 1674 Fig. 22
Hooke 1676, Pl. 1, Fig. 11

See also D111, D112, D113   **Inventory No.** 1927-1897

E60 Model with rollers

second half eighteenth century or first half nineteenth century

L 255mm, **W** 205mm, **H** 215mm

It is not certain what this model illustrates. It appears to be some sort of machine for the manipulation of thread or fabric. It bears a slight resemblance to a patent by Nightingale in 1790 for the 'calendering, glazing and dressing' of muslin, silk, linen, cotton, wool, and various other materials. It also has features in common with illustrations in Diderot's *Encyclopaedie* under silk manufacture and trimmings.

There are two horizontal mahogany rollers and two mahogany frames for winding the material. A brass and steel handle turns one of the frames directly. A brass toothed wheel at the end of the frame meshes with another below to turn the second frame in the opposite direction. As well as turning the frames, the handle turns the uppermost roller by means of a large brass toothed wheel which turns a smaller one which carries the iron axle of the roller. The lower roller is held against the upper by iron springs on each side of the machine and is made to turn in the opposite direction. Presumably the thread was fed between the rollers and wound on the frames.

Diderot 1767
Nightingale 1790   **Inventory No.** 1927-1778

E61 Model of mechanism for a walking wheel crane

Pinchbeck, C.

after 1765   London

PAINTED—DIAL—SCALE
'1, 2, 3'
ON CERAMIC PLAQUE
'C. PINCHBECK/INVT :1766'

H 1085mm, **L** 410mm, **W** 290mm, **D** 255mm (wheel)

Like Ferguson's crane, this model was made to demonstrate an apparatus for making treadmill cranes safer. Pinchbeck received a gold medal from the Society of Arts in 1762 or 1767, both dates being recorded. The model is similar but slightly smaller than that described in the Society's collection and does not include the crane itself but only the treadmill mechanism.

The mahogany treadmill operates a calfskin bellows situated just beneath the floor of the instrument. It can be regulated by an index on a ceramic dial marked '1, 2, 3', which alters the sensitivity. If the machine accelerates to a certain critical speed the bellows lift a lifting trundle in the floor of the machine which acts on a brass latch which operates the brake system. The mechanism is described in detail by Bailey. This model is in a glass case and stands on a tall four-legged table.

The full-size crane was built at Dice's Quay or 'Key', but in 1786 it had apparently fallen into disrepair and been superseded by Bunce's centrifugal design. Keeping leather in working order out of doors was a problem.

Pinchbeck 1762
Ferguson 1764*a*
Ferguson 1764*b*, supplement pp. 1–7
Bailey 1786, pp. 146–50
C 67
Millburn and King 1988, p. 214, Fig. 49, p. 313, (ch. 14, f. 2)
Inventory No. 1927-1621

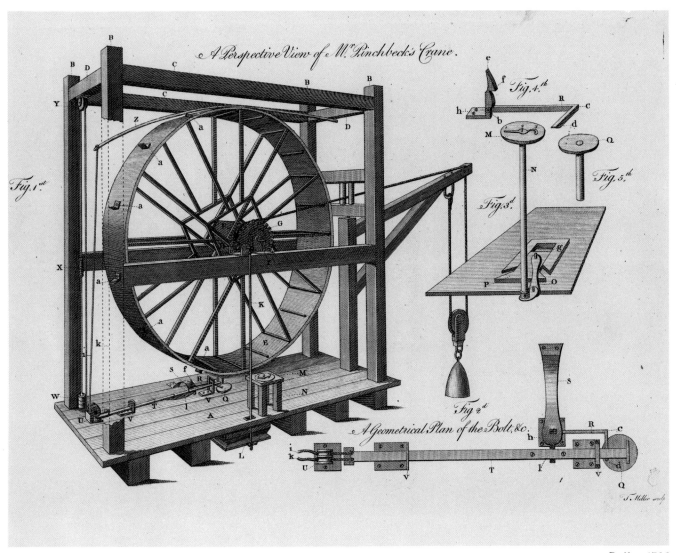

A Perspective View of Mr. Pinchbeck's Crane.

Fig. 1st.

Fig. 4th.

Fig. 3d.

Fig. 5th.

Fig. 2d.

A Geometrical Plan of the Bolt, &c.

J. Miller sculp.

Bailey 1786

E62 Model of Ferguson's crane

after 1763
L 460mm, **W** 295mm, **H** 245mm

Apparently a model of this type of crane was in the Repository of the Society of Arts in 1768. This model may have been used by Ferguson in his lectures, but since nothing else of his has come into the collection it is unlikely.

The model exactly resembles the plate in Ferguson's description of the crane in *Philosophical Transactions* in 1764. It is almost entirely made of mahogany. It has a large horizontal wheel with 60 cogs on a thick vertical axle around which turns the rope. The rope then passes between two rollers and over a boxwood pulley at the top of the arm from which the load is suspended. One of three gear wheels can be meshed with the central wheel. These wheels have 16, 8, and 4 teeth respectively; the smallest is made of boxwood. The wheels could be turned by means of winches which were twice the radius of the gear wheels, giving double the mechanical advantage. Thus in the model the mechanical advantage would be $7\frac{1}{2}$, 15, or 30 depending on which wheel was used. The winches are now lost. There is a ratchet wheel which drives a brake wheel, and the brake and ratchet are connected by a cord passing over a boxwood pulley. The brake prevents the weight from falling if the pressure on the gear wheels is reduced for any reason. The weight can be lowered gradually by pulling down the lever on the ratchet wheel which first frees the brake but then presses against a rotating wheel next to the ratchet wheel to use friction to slow the fall. Letting the lever go brings the brake back into action. The model is on a truncated circular base.

Ferguson devised the crane in 1764 to overcome two problems: the disastrous effects of letting the load drop in a common crane, and the inefficiency of raising a small load with an unnecessarily large mechanical advantage. The first problem had caused fatal accidents to workmen. Ferguson stated that the crane could be installed in a room 8 feet square.

Ferguson 1764*a*
Ferguson 1764*b*, supplement pp. 1–7
C 67
Chew 1968, item 20
Millburn and King 1988, p. 214, Fig. 49, p. 313, (ch. 14, f. 2)
Inventory No. 1927-1930

E63

E63 Model of a mechanical saw worked by a water wheel

mid-eighteenth century

LABEL—PRINTED

'1212'

L 470mm (platform), **W** 240mm (platform), **H** 75mm (platform), **D** 170mm (wheel), **L** 200mm (blade), **W** 10mm (blade)

Stylistically the model resembles those belonging to Demainbray, but the absence of a listing in the Queen's Catalogue or a mention in the surviving syllabuses makes an association unlikely.

The model is in such a poor state that it cannot be described adequately. It appears that the saw was held vertically by a thread over a pulley. It operated through a rectangular hole in a horizontal platform which was supported by two feet and the mounting of the water wheel. Eight upright pieces held the frame to which the pulley was attached. The water wheel had eight spokes and a square axle with an iron rod running through its centre. The model is made of mahogany and the blade is iron.

Inventory No. 1927-1632

E64 Manuscript describing a pendulum bucket engine

1745

L 395mm, **W** 195mm

The manuscript consists of 14 leaves bound in vellum entitled 'An explanation of the Pendulum Bucket Engine invented and executed in a Working Model, anno 1745' Nine pages describe the model engine, referring to two engravings which are now lost. Two pages describe the operation and the last three give 'particular observations with regard to the operation'.

The engine is of a general type not uncommon at the time for raising water, usually in domestic situations but sometimes on a larger scale for draining mines. The basic principle is the use of a large weight of water descending through a small distance to raise a small weight of water through a greater distance. In the model a cistern of water has two spouts with valves—one at the base and the other part of the way up one side. A large and a small bucket are placed so that they can be filled from these spouts. They are suspended from chains which pass over wheels on the same axle; the small bucket has a large wheel and the large bucket a small wheel. The chains are

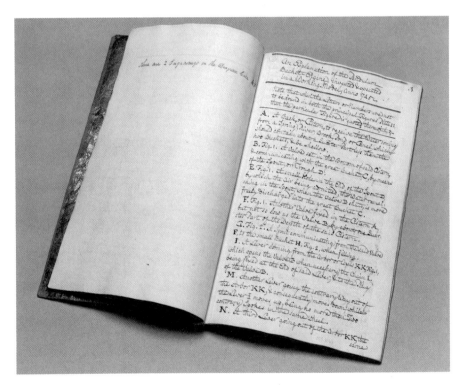

made of ivory and brass respectively. The axle is slightly more than 30 inches above the level of the cistern since this is the height that the small bucket is to be raised. This axle is geared to the 'pendulum', a weighted wheel which acts as a regulating mechanism, and thence to a fan-fly, the arrangement serving to regulate the motion of the buckets. The 'pendulum' oscillates and as the large bucket, full of water, falls 10 inches, the small one is raised 30 inches. The buckets empty by means of levers and valves and the 'pendulum' brings the buckets back to their former position where a set of levers operates the valves and the buckets refill.

Inventory No. 1927-2063

MACHINES

E65 Wheel-cutting engine

Smith, John

1770–85   Richmond, Middlesex

DE 300mm (engine), **L** 740mm (engine), **L** 750mm (glass case), **H** 530mm (glass case), **H** 715mm (table), **W** 840mm (table)

The wheel-cutting engine is thought to be by John Smith who worked for Demainbray at Kew Observatory during the 1770s. Smith grew up with Henry Hindley in York and would have been familiar with Hindley's wheel-cutting engine, on which this example is based. There is a great deal of brass in the instrument, suggesting that it was built by a clockmaker.

The engine was used to cut the teeth of wheels or pinions, and could have been used to divide the scales of instruments. It bears certain similarities to Hindley's engine which is described by Smeaton and by Law. It is set on three brass pillars on a mahogany table; the pillars are reminiscent of those in a clock movement of the period. The engine consists of a vertical arbor in a conical socket which Hindley describes as being ground in to prevent shake. The division plate, which is now missing, would have fitted on the arbor near its lower end. It was just under 10 inches in diameter and bore 360 teeth. Two mechanisms could have been used to set the plate: a Hindley endless screw and a point which was dropped into the division holes. The endless screw is shaped to accommodate the plate over an angle of about 10°. It can be turned by hand and is fitted with a micrometer which can be read to one minute of arc. The mechanism for dropping the point into the division hole appears to be incomplete. An index on this piece allows the position of the division wheel to be located.

The cutter is in a double cutting frame, the design of which is associated with Hindley. This device ensures that the traverse of the cutter is always perpendicular to the wheel being cut. When the handle is depressed to bring the cutter into contact with the wheel, the cutter is constrained by two light bar guides to move vertically. The height of the cutter can be adjusted using a screw at the top of the frame, and there is a depth stop with a screw lock for cutting contrate wheels. An index on the cutting frame allows the cutter to be aligned radially to the wheel. The cutting frame can be set at an angle in order to cut obliquely for helical worm wheels; a protractor is provided to measure the inclination. The cutting frame can be moved along the main frame of the instrument so that wheels of various diameters can be cut. Behind the cutting frame is a micrometer which also slides on the main frame and provides fine adjustment.

The above describes the principal parts of the engine. In addition to these there are various accessories. Attached to the central arbor is a device for cutting sectors of circles which provides instant stops if required. An additional mechanism which holds the piece to be cut eccentrically; it contains a silvered scale for measuring the extent of offset and a slider. Together with unidentified parts there are two alternative central mandrels, a key for the frame nuts, a screwdriver, and two brass boxes to receive the cuttings. The piece of cardboard which would have protected the division plate has also survived. The purpose of a large piece of apparatus which would have been attached by two screws has not been identified owing to the incompleteness of the engine.

Smeaton 1786
Rees Cyclopaedia 1819, under Cutting
C 345
Law 1970–2a,b

Inventory No. 1927-1942

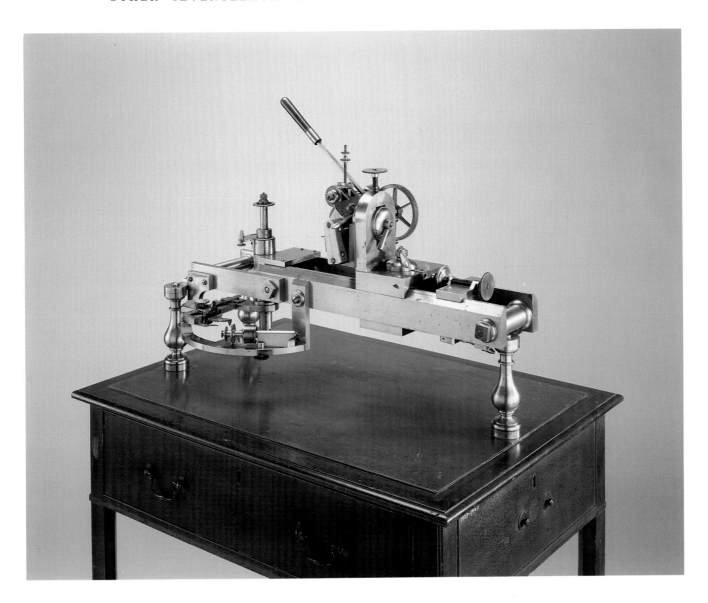

E66 Glass cutting machine used in the construction of lenses

last quarter eighteenth century or first quarter nineteenth century

L 215mm (box), **W** 215mm (box), **H** 90mm (box), **D** 205mm, **L** 130mm (side of square plate), **H** 130mm

The instrument consists of a circular brass plate above which is a smaller square plate. A piece of glass is held rigidly between a four-jaw chuck over a central circular aperture in both plates. Fine adjustment to the position of the glass can be made by means of screws applied to each of the four sides of the brass square. A long brass arm is hinged to a short pillar at one side of the circular plate. It carries a sleeve which holds a brass cylinder to which the cutting diamond could be attached. At present there is a steel cone.

When the glass is in place the position of the diamond is set depending on the lens diameter required. The square plate is then turned with respect to the circular plate using the rotating screw beneath the instrument, and a mark is made on the glass. When not in use the instrument fits into a mahogany case.

C 209

See also L32

Inventory No. 1927-1668

MAGNETISM

E67 Terrella

Adams, George?

mid-eighteenth century   London

H 180mm, **D** 95mm (base), **D** 53mm (ball)

This is an ornamental terrella, probably made by Adams in about 1765. The magnetite ball has a diameter of just over 2 inches and rests on a spiral brass stand.

'Terrella', or little Earth, was a name given to a spherical lodestone or 'loadstone' owing to the magnetic properties shared with the Earth. Gilbert defined a terrella as a 'globular loadstone'.

Gilbert 1600, preface p. 6
C 260

Inventory No. 1927-1816

E68 Compass

last quarter eighteenth century

L 150mm, **W** 150mm, **H** 20mm, **D** 90mm

The compass is set in a square mahogany box with a hinged lid. The bowl is of brass with a paper rose showing the 32 points. There are two concentric degree scales; the outer one gives 0° to 360° in both directions and the inner one gives four quadrants, each from 0° to 90°. The blued steel needle is balanced on a conical brass pivot; it is now very corroded. There is a locking device which is now broken.

C 272

Inventory No. 1927-1815

E69 Compass

last quarter eighteenth century or first quarter nineteenth century

D 70mm, **H** 18mm

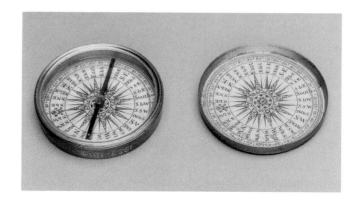

The compass is in a circular brass box with a brass cap. It has a paper rose with the 32 points and a degree scale from 0° to 90° in four quadrants around the circumference. Although each degree is marked, the width of the blued steel needle allows an accuracy of only about 2°. The needle is rectangular which is common in this period; it is balanced on a conical brass pivot. Pasted inside the brass cap is an identical paper rose to that on the compass.

C 271   May 1973, p. 72   **Inventory No.** 1927-1814

E70 Magnetic toy

Adams, George

1765   London

'LICTSGgstcil' '1765'

L 277mm (box), **W** 166mm (box), **DE** 15mm (box), **W** 47mm (metals), **H** 95mm (compass), **D** 45mm (compass)

The mahogany box originally contained six heart-shaped pieces of metal of which five remain. They are made of silver, lead, tin, copper, and iron respectively, with the missing piece having been made of gold. Each piece contains a hidden magnet at a different orientation. The ivory compass is screwed to the base of a hollow ivory cylinder which also unscrews in the centre to allow a magnifying glass to be inserted. The lens is now lost. The top of the cylinder consists of an eyepiece with a circular hole about half an inch in diameter through which the compass can be viewed. The compass card has the above letters printed clockwise around the circumference, together with what is presumably the date, 1765. The inside of the lid of the box has a label identical to that of C71.

If the lid of the box is closed and the compass is placed over it, the compass needle will point to the first letter of the metal beneath; thus the position of the

metal pieces will be revealed.

Although it does not appear in Adams's catalogues of the period, the item is not unique. When described in Adams's *Essay on Magnetism* in 1787, there are only five metals listed. Iron has been omitted possibly because of its magnetic properties.

Adams 1787a, p. 450   C 269

Inventory Nos 1927-1180, -1750

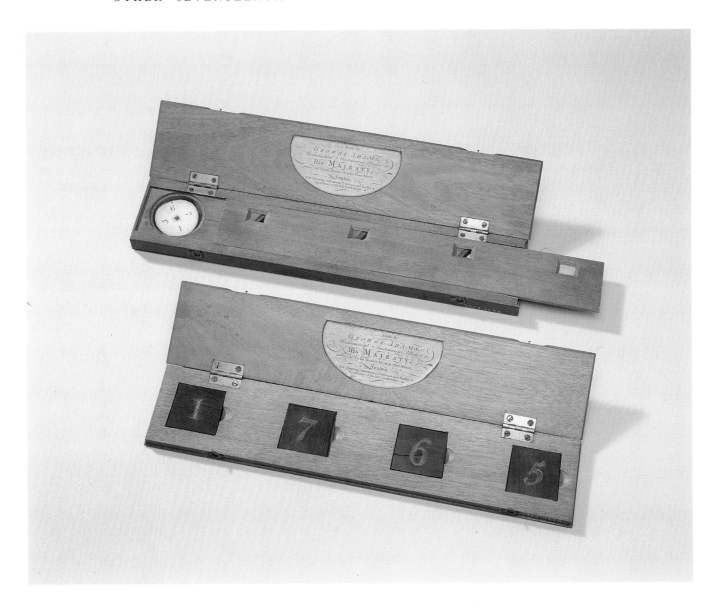

E71 Magnetic toy

Adams, George

1765   London

Made by/GEORGE ADAMS/Mathematical Instrument Maker to/His MAJESTY/at Tycho Brahe's Head in Fleet Street/London

L 350mm (boxes), **W** 73mm (boxes), **DE** 15mm (boxes)

The instrument consists of two mahogany boxes. The first contains four squares of ebony with the numbers 1765 inset in brass on the upper surfaces. Each square contains a concealed magnet. The second box has a sliding mahogany panel with four windows which can be removed to show four circular paper dials, each with a concealed magnet beneath. Each dial has the numbers 1, 5, 6, and 7 printed round the circumference. When the sliding panel is replaced and the second box is placed directly above the first, the windows reveal the numbers 1765, presumably the date. It is then found that these numbers correspond to those on the squares in the box below.

Adams 1798, p. 12
C 267

Inventory No. 1927-1278

E72 Horseshoe magnet

second half eighteenth century

H 135mm (including handle),
W 45mm, **DE** 45mm

This is probably a very early example of a horseshoe magnet. Six steel horseshoe magnets are strapped together and bound in calfskin. There is a steel ring handle at the top. The north poles are marked with a line. The magnets are badly rusted.

C 264

Inventory No. 1929-111

E73 Pillar magnet

second half eighteenth century or first half nineteenth century

H 125mm (overall), **L** 117mm, **W** 95mm

This is a steel pillar which is fairly strongly magnetized with its south pole at the top. It is set in cork on a rectangular base. It is not known to what purpose the instrument was put.

In the late seventeenth century it was known that a piece of iron kept vertical for a long period of time acquired magnetism. Desaguliers remarked on the fact in passing in 1736 and said that the top always attracted the north end of a compass needle.

C.J. 1694
Desaguliers 1738

Inventory No. 1929-100

E74 Battery of magnets

second half eighteenth century

'*N*' and '*S*'

L 90mm, **W** 19mm, **H** 27mm

Sixteen 3-inch magnets are stacked in four rows of four and bound by brass strips. Two pole pieces are attached to square steel pieces which are screwed against the ends of the bars. 'N' and 'S' are engraved on the brass binding. The battery was suspended by steel chains and was obviously used to lift weights. It is possible that it is the magnetic magazine listed in the Queen's Catalogue but this is more likely to have been the larger set D52.

Q.C., Item 120; magnetical magazine mounted in brass C 263

Inventory No. 1929-108

PNEUMATICS

E75 Nooth's apparatus

late eighteenth century

SCRATCHED—SPOUT
89; N 3027; N 3919; 36

H 665mm, **D** 270mm (base)

This apparatus was originally designed by Dr Nooth around 1775 to prepare water containing carbon dioxide. This example is made from colourless glass and consists of three main parts; the stopper is now missing. The bottom vessel contains chips of marble to which dilute sulphuric acid was added. Carbon dioxide was given off and this passed into the middle vessel which contained water in which some of the carbon dioxide dissolved. A valve arrangement allowed the gas to move upwards but did not allow the water to move downwards.

When the water was needed, it was drawn off using the tap in the middle vessel; more fresh water could be added through the topmost vessel which also acted as a mechanism for adjusting the pressure.

Drawing on earlier work by Priestley, Nooth designed his apparatus to avoid some of the problems of Priestley's method of impregnating water. Parker improved on Nooth's design and supplied many thousands of these machines from his Cut-glass Manufactory in Fleet Street, the machines themselves being made by Whitefriars Glassworks. A further improvement, suggested by Benjamin Vaughan, was to use a plano-convex lens in the valve which allowed the gas to travel up from the lowest vessel.

Later in the century other ways of preparing water were introduced as Cavallo explained:

> In Dr. Nooth's glass apparatus for impregnating water with carbonic acid gas, the quantity of gas that can be thrown into it is very moderate, yet efficacious; but the soda water which is now prepared and sold in London by a Mr. Schweppe, contains an incomparably greater proportion of carbonic acid gas, and accordingly is much more efficacious.

Nooth 1775
Priestley 1776, Vol. II, pp. 263–77, 293–303
Magellan 1783, pp. iv, 3, 7
Cavallo 1798, p. 146 fn.
Adams 1799a, Vol. 1, pp. 550–5 and Pl. 6, Fig. 5
C 139
Zuck 1978

Inventory Nos 1927-1198, -1383

E76 Suction pump

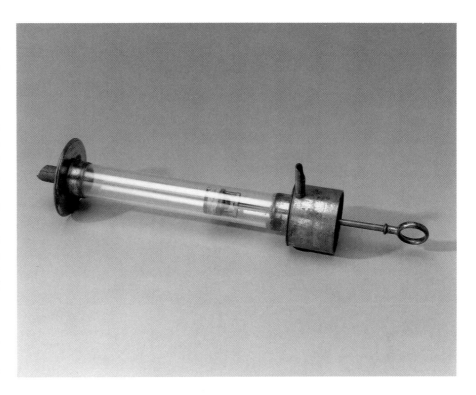

last quarter eighteenth century or
first quarter nineteenth century

H 305mm, **D** 50mm (barrel)

In this model pump, a piston is free to
move in the glass barrel. There are two
valves—one in the brass base of the
pump and another in the piston. When
the lower end of the pump is in water
and the piston is raised, water enters the
barrel through the lower valve. When
the piston is lowered, the water passes
through the valve in the piston into the
upper part of the barrel. The water
flows out of the brass spout at the top of
the pump on the next up-stroke.

Hauksbee 1714, p. 14, Pl. 3, Fig. 1
Desaguliers 1744, Vol. II, p. 164, Pl. 15,
Fig. 2
Adams 1799*b*, Vol. 1, p. 47, p. 138, Pl. 1,
Figs 21, 22
C 107

Inventory No. 1927-1287

E77 Glass globe containing a bladder

late eighteenth century

H 205mm (overall)

The bladder in this glass was made to expand or contract
either by placing the globe under a receiver on a plate on an
air pump or by using a syringe. This apparatus is similar to
the version supplied by Adams, P47.

Desaguliers 1744, Vol. 2, pp. 380–1, Pl. 25, Fig. 7
Adams 1761, p. 42, Pl. 20, Fig. 60
Adams 1799*b*, Vol. 1, pp. 171–2, Pl. 3, Fig. 9
C 114

See E78, P47

Inventory No. 1927-1295

E78 Glass globe containing a bladder

last quarter eighteenth or first quarter nineteenth century

H 170mm, **D** 95mm

This was probably used with a syringe to show the action of the lungs.

See E77, P47

Inventory No. 1927-1296

E79 Expanding bladder

last quarter eighteenth century or first quarter nineteenth century

H 200 mm (overall)

This glass bell jar has a long neck terminating in a brass collar. Into the top of this collar is screwed a nozzle and on the underside, a metal rod with bladder attached. The bell jar was placed on the plate of an air pump and the air removed. Air would then enter the bladder through the nozzle at the top of the jar and the bladder would expand. Letting air back into the bell jar would cause the bladder to collapse.

This was another version of the demonstration showing the action of the lungs.

Adams 1799a, Vol. 1, p. 169, Pl. 1, Fig. 10
C 112, Pl. 3

See E77

Inventory No. 1927-1370

E80 Two cohesion discs

last quarter eighteenth century or first quarter nineteenth century
D 55mm

These brass discs with steel handles were heated and held in the smoke of a candle before being pressed together. It then required a large force to separate them, showing the effect of cohesion. A steelyard or weights might be used to separate the plates.

C 121

See D122   **Inventory No.** 1927-1305

E81 Two cohesion discs

second half eighteenth century
L 168mm, **D** 35mm

These lead cohesion discs have brass and walnut handles.

C 121

See E80   **Inventory No.** 1927-1306

E82 Two marble slabs

second half eighteenth century
L 100mm, **W** 75mm, **H** 20mm

These marble slabs were for demonstrating cohesion. They have brass handles.

C 121

See E80

Inventory No. 1927-1307

E83 Double-barrelled air pump

Adams, George, Elder or Younger

1766–99   London

'*G. Adams No 60 Fleet Street LON-DON*'

L 600mm, **W** 330mm, **H** 520mm, **L** 245mm (cylinder), **D** 50mm (cylinder (outside))

This example of a table air pump has two brass cylinders. The pistons are operated by turning the handle in one direction. A rack and pinion mechanism raises one piston and lowers the other. The rising piston sucks air from a vessel placed on the pump plate; the one being lowered expels air from its cylinder to the open air. When the two pistons had travelled to the ends of the cylinders, the handle was turned in the opposite direction to continue the process. There are metal plates attached to the base of the pump so that it can be fixed to a table.

Adams sold table air pumps from at least 1747, though this example must date from after the numbering of Fleet Street in 1766. The mahogany baseboard has split and been repaired by adding a strip of wood to the front of the pump.

The first air pumps made in England were floor-standing but in 1737 Stephen Davenport published a description of his 'New-Invented Table Air-Pump'. His design had two barrels and valves made of 'limber' bladder.

Writing about Davenport's design in 1766, Benjamin Martin explained:

> . . . Mr. Davenport (an ingenious workman) contrived it of a Table Form, and made it in some Respects more commodious for use, as well as considerably easier in the Purchase. And for many Years they were made and sold in this Form, and under the Title of Davenport's Table AIR-PUMP.

Martin also described his own changes to this design, such as making the various connecting pieces out of solid brass drilled to provide passages for the air, an improvement over the copper pipes previously used which were prone to leak.

Martin himself had sold 'portable' air pumps from at least 1757, for then he advertised that he would put his name on his pumps in future because the design was being pirated by others.

Adams 1747*b*, p. 255, Item 193
Adams 1799*b*, Vol. 1, pp. 44–6, Pl. 1, Fig. 2
C 102

Used with E84, E85

Inventory Nos 1927-1310, -1331

E84 Flanged disc for air pump

Adams, George, Elder or Younger

1766–99 London

D 155mm (plate)

C 208

Used with E83, E85

Inventory No. 1927-1321

E85 Pipe and ring

Adams, George, Elder or Younger

1766–99 London

L 168mm (tube), **L** 45mm (ring)
This hollow brass pipe and copper ring are probably separate accessories for the air-pump by Adams, E83.

Used with E83, E84

Inventory No. 1927-2052

E86 Syringe with ball

last quarter eighteenth century or first quarter nineteenth century

L 445mm, **D** 103mm (sphere), **D** 35mm (barrel)

This seems to have been part of an apparatus for an air gun. The syringe was used to pump air into the brass ball and the tap closed. The syringe was then removed and the ball attached to an air gun which is now missing. The ball then provided the air to fire a number of lead bullets.

Adams 1799*b*, Vol. 1, pp. 132–4, Pl. 5, Figs 11, 12
C 109

See also the air gun by Adams (P54), and the one used by Demainbray (D45)

Inventory No. 1927-1327

E87 Syringe

last quarter eighteenth century or first quarter nineteenth century
L 180mm, **D** 75mm (base)

This syringe has a brass barrel with a steel and mahogany handle. A boxwood plug screws into the syringe. In turn the plug fits into the boxwood base.

Adams 1799*b*, p. 134, Pl. 3, Fig. 18
C 109

Inventory No. 1927-1348

E88 Glass receiver

last quarter eighteenth century or first quarter nineteenth century
L 300mm, **D** 215mm

This design of receiver was used to contain 'condensed' or compressed air. It is made of thick glass and well annealed, according to the younger Adams, so it may well have been supplied by him, or his successors W and S Jones.

Hauksbee 1714, p. 17, Pl. 3, Fig. 1
Desaguliers 1744, Vol. 2, pp. 394–6, Pl. 24, Figs 8, 12
Adams 1799*b*, Vol. 1, p. 129, Pl. 2, Fig. 1
C 129

Inventory No. 1927-1351

E89 Bladder with tap

last quarter eighteenth century or first quarter nineteenth century

L 300mm

This bladder is attached to a brass collar and tap. If the tap was closed and the bladder suspended in a receiver on an air pump, when the air was removed from the receiver, the bladder expanded.

Birch 1744, Vol. I, pp. 12–13
Desaguliers 1744, Vol. 2, p. 379, Pl. 25, Fig. 1
's Gravesande 1747, Vol. 11, p. 20
Adams 1761, p. 11, Pl. 4, Fig. 9
Adams 1799*b*, Vol. 1, p. 85, Pl. 2, Fig. 6

See E90

Inventory No. 1927-1357

E90 Bladder with tap

last quarter eighteenth century or first quarter nineteenth century

L 300mm

This bladder is attached to a brass collar and tap and would have been used in the ways described for the above item.

Inventory No. 1927-1400

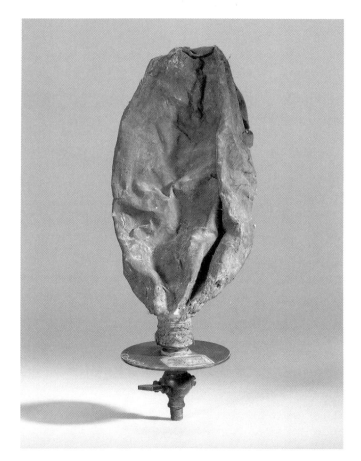

E92 (part)   **E91 (part)**

E91 Four bell jars

last quarter eighteenth century or first quarter nineteenth century

H 310mm, **D** 160mm

These glass bell jars have rims which have been folded over and ground. They also have ground necks and stoppers. Only one has been illustrated.

C 129

Inventory Nos 1927-1360, -1361, -1362, -1372

E92 Three bell jars

last quarter eighteenth century or first quarter nineteenth century

H 255mm, **D** 120mm

These glass bell jars have rims which have been folded over and ground. They also have ground necks and stoppers. Only one has been illustrated.

C 129

Inventory Nos 1927-1358, -1359, -1363

E93 Bell jar

eighteenth century

H 305mm, **D** 305mm

This glass bell jar has a rim which has been folded over. It has a glass ring attached by a stem to the top. It is of a type used by clock- and watch-makers to cover the items they were working on.

C 129

Inventory No. 1927-1352

E94 Osmotic pressure apparatus

last quarter eighteenth century or first quarter nineteenth century

H 540mm, **D** 120mm

Each of these two items consists of a long glass tube fixed into a hemispherical glass bowl by a cork sealed with wax. There are fragments of bladders which were stretched over the rim of the hemispherical bowl.

These could be used to demonstrate osmotic pressure first observed by Nollet in 1748. One vessel is filled with sugar solution and the other with water so that the levels in the tubes are the same. If these vessels are then set in water, the level of the liquid in the tube containing the sugar solution will rise as water diffuses through the membrane.

Nollet 1748
C 136

Inventory No. 1927-1374

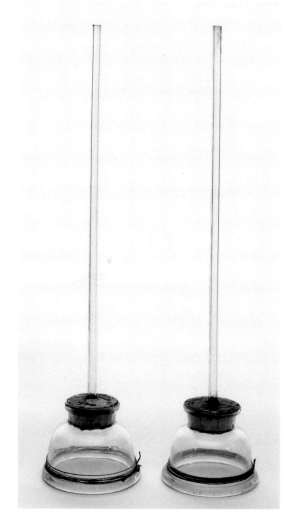

E95 Cupping glass

second half eighteenth century

H 120mm (overall), **D** 55mm

This thistle-shaped piece of colourless glass has a folded over rim and a brass collar with an outlet in its base. It is likely to be a cupping glass used with a syringe. This could be used for bleeding or cupping a patient. The cupping glass would be applied to the skin and the air removed with a syringe. It has a wooden base.

Hauksbee 1714, p. 15, Pl. 1, Fig. 9
Desaguliers 1744, Vol. 2, p. 385, Pl. 25, Fig. 17
Adams 1761, p. 42, Pl. 17, Fig. 53
C 129

Inventory No. 1927-1378

E96 Cupping glass

second half eighteenth century

H 127mm (overall glass), **D** 73mm (overall glass)

The lower end of this thistle-shaped piece of glass has a brass collar but no outlet. It is likely to be a cupping glass. It has a mahogany base.

C 129

See E95

Inventory No. 1927-1377

E97 Cupping glass

second half eighteenth century

H 127mm (overall glass), **D** 73mm (overall glass)

One end of this thistle-shaped piece of glass has a brass collar with an outlet. It is likely to be a cupping glass for use with a syringe.

C 129

See E95, E96

Inventory No. 1927-1379

E98 Aurora tube

last quarter eighteenth century or first quarter nineteenth century

L 455mm, **D** 125mm

This glass tube has brass collars at either end. At one end the collar has a pipe and tap to extract the air using an air-pump. The other end has a brass rod passing through the cap and terminating in a brass ball. An electrical discharge would be set up in the tube.

Adams 1784, pp. 136–7, Pl. 4, Figs 61, 62
C 129

See L107

Inventory No. 1927-1391

E99 Manipulator

last quarter eighteenth century

D 134mm, **L** 165mm (slip wire)

This consists of a brass plate with a 'collar of leathers' or stuffing box through which passes a brass wire with a hook at one end. It is placed on top of a receiver and used for letting down a substance or suspending an instrument in the receiver.

Desaguliers 1744, Vol. 2, p. 386, Pl. 25, Fig. 20
Adams 1799*b*, p. 176, Pl. 1, Fig. 16
C 124

Inventory No. 1927-1404

E100 Manipulator

last quarter eighteenth century or first quarter nineteenth century

L 175mm, **D** 92mm

This consists of three brass pans arranged one above the other, each pan being hinged on one side and, on the opposite side, supported by a small lug on the central wire. It was placed on top of a receiver on an air pump. Turning the central wire released each pan in turn which allowed whatever was placed in the pan to drop down in the receiver.

Inventory No. 1927-1478

E101 Porosity of wood apparatus

last quarter eighteenth century

H 220mm, **D** 78mm

A brass collar is fitted with a boxwood cup on top and a pointed painted pine stem below. It was placed on top of a receiver with some mercury in the cup. When the air was removed from the receiver, the mercury was forced through the pores of the wooden stem into the receiver.

Desaguliers 1744, Vol. 2, p. 390, Pl. 25, Fig. 30
Adams 1799*b*, Vol. 1, p. 210, Pl. 3, Fig. 3

Inventory No. 1927-1411

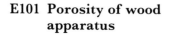

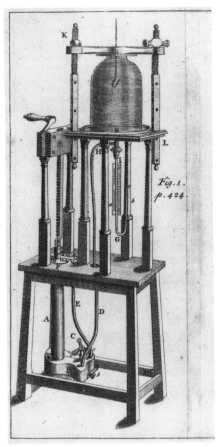

Smeaton 1751

E102 Single-barrelled air pump

Adams, George or Nairne and Blunt

1770–90   London

SCALE
10–50

H 1110mm, **L** 480mm, **W** 350mm, **L** 480mm (barrel),
D 60mm (barrel)

This pump is very similar to one designed by John Smeaton around 1750. Smeaton's design had a single barrel and the valves and pistons were carefully designed to allow as much air as possible to be exhausted. As a result, the performance of Smeaton's pump was much better than that of other contemporary pumps. Smeaton's pump could also be used to 'condense' or compress air. See P1 for a description of the valves and the operation of the pump.

Smeaton worked on the design of these pumps when he came to London in 1748. His first was constructed in 1749 and an improved version was described in a paper delivered to the Royal Society by Smeaton in 1752. According to Priestley, Smeaton made only two pumps himself and he gave the second to Priestley. That pump was destroyed when

Priestley's laboratory was attacked by a 'King and Country' mob in 1791.

Adams and, later, Nairne and Blunt made pumps to Smeaton's design. It may be that this pump was made by one of these makers. At the sale of the Earl of Bute's instruments in 1793, the catalogue mentions 'A capital standing Smeaton's air pump and condensor by Nairne and Blunt . . .'.

Smeaton had also devised a gauge to measure the pressures achieved by his pump (the example made by Adams for this collection is no longer extant). Using this gauge, Smeaton estimated the lowest pressure to be 1/1000th of an atmosphere. However, there were problems with the design of this gauge; these were described by Nairne who thought that the actual pressures were much higher than those indicated by the gauge.

Smeaton 1751
Nairne 1777*a*
Nairne 1777*b*
C 100
Turner 1987, p. 127, Pl. 115
Inventory No. 1927-1623

E103 Syringe and weight

last quarter eighteenth century or first quarter nineteenth century

H 195mm, **D** 75mm (plate)

The lead weight is screwed on the end of the brass syringe with a piece of leather (now missing) to make it air-tight. The piston of the syringe is solid. When the syringe is suspended by its piston, the weight is not enough to cause the body of the syringe to drop. However, if this trial is carried out under a receiver in an air pump, the weight drops when the air is removed, showing that it is the pressure of the air that supports the weight.

Hauksbee 1714, p. 15, Pl. 1, Fig. 11
Desaguliers 1744, Vol. 2, p. 386, Pl. 25, Fig. 20
Adams 1799*b*, Vol. I, p. 67, Pl. 1, Fig. 9

Inventory No. 1927-1628

E104 Apparatus for demonstating hydrostatic pressure

last quarter eighteenth century or first quarter nineteenth century

H 550mm, **L** 335mm (copper pipe), **D** 22mm (copper pipe)

This apparatus is used for demonstrating the effects of pressure of a liquid. It consists of a brass cylinder supported on feet and open at the bottom end. Inside is a brass piston which can move within the cylinder. A short central arm from the piston passes through the brass top of the cylinder and into the vertical copper pipe. A chain or other connection (now missing) would connect the piston to one arm of a balance. The pipe and cylinder were filled with water. To counteract the pressure of water on the upper side of the piston, weights had to be added to the other arm of the balance. The demonstration shows that it is the vertical height of the water above the piston which determines the pressure on it.

An earlier version of this demonstration used a bellows filled with water with weights placed on it.

's Gravesande 1747, Vol. I, pp. 345–50, Pl. 45, Fig. 5, Pl. 46, Figs 1, 2
Adams 1799*a*, Vol. 3, Pl. 1, Fig. 1

Inventory No. 1927-1748

E105 Valve

eighteenth century

H 75mm

This flap valve has a turned boxwood body with a leather flap to which is fixed a lead weight.

Adams 1799*b*, Vol. 1, p. 46, Pl. 1, Fig. 8

Inventory No. 1927-1629

E106 Funnel

eighteenth century

H 165mm, **D** 70mm

This glass funnel has a ground end.

Inventory No. 1927-1376

E107 Combined suction and force pump

last quarter eighteenth century or first quarter nineteenth century

L 270mm, **H** 655mm, **W** 155mm

The central glass barrel has a solid piston with a mahogany handle which enters the barrel through a brass plate and stuffing box. Raising the solid piston brings water from the lower painted metal reservoir through a valve in the brass base of the cylinder into the barrel. On the down-stroke, the water is forced through the brass pipe leading from the base of the barrel and into the glass reservoir at the top of the apparatus via a valve in its base. On that same down-stroke, water is also sucked from the lower reservoir into the barrel on the left and from there through a valve and into the other barrel above the piston. On the subsequent up-stroke, the water above the piston is forced through a valve into the upper glass reservoir. The advantage of this design was that water was pumped on both the up-stroke and the down-stroke. The pump has a mahogany stand.

A pump designed by de la Hire was similar, except that the topmost vessel was enclosed apart from a central pipe which ran almost to its base. The advantage of this version was that the jet of water provided by the pump was more continuous.

Adams 1799a, Vol. 3, pp. 500–2, Pl. 3, Fig. 2

Inventory No. 1927-1624/3

E108 Glass globe

last quarter eighteenth century or first quarter nineteenth century

H 140mm, **D** 75mm

This glass globe has a glass base and a brass collar. The collar is threaded to take a nozzle or similar piece which is now missing. This item, when complete, may have been a vacuum fountain or contained a collapsing bladder.

C 129

See E77, P47

Inventory No. 1927-1380

E109 Bell jar

last quarter eighteenth century to first quarter nine-teenth century

H 350mm, **D** 155mm

This bell jar has a long neck with a brass collar and stopper which screws on the neck. It is the same shape as the bell jars in which gunpowder could be dropped onto a hot surface when the air had been removed.

C 129

Inventory No. 1927-1371

ACOUSTICS

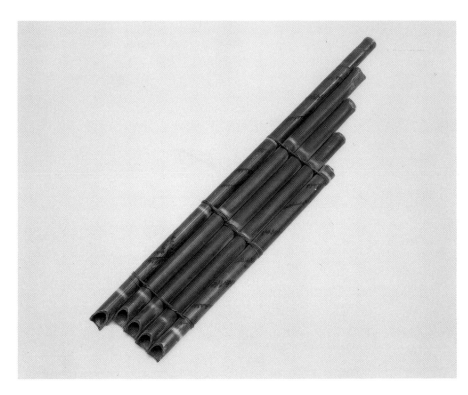

E110 Pan pipes

second half eighteenth century

W 63mm, **D** 13mm (tubes), **L**: 330mm, 290mm, 260mm, 230mm, 200mm

The instrument consists of five tubes made of reed strapped together and roughly decorated with stained cuts. The tops of the tubes are open while the bottoms are closed, so that if the tops are blown across a musical sound is produced. The lengths of the tubes give rise to a scale corresponding to fa, soh, la, te, doh.

C 257

Inventory No. 1927-1489

E111 Whistle

second half eighteenth century

L 125mm, **D** 50mm

This appears to have been a whistle. It has a mahogany cylindrical body with a boxwood mouthpiece. A projecting piece of mahogany at the end has a flap of calf-skin attached. There is no reed.

C 255

Inventory No. 1927-1484

THERMOMETERS AND METEOROLOGICAL INSTRUMENTS

E112 Double barometer and thermometer

Betali, Joseph

second quarter eighteenth century   Paris

'Thermometre/Suiv^t Mr de/Reaumur/Barometre/a 4 boulles const^t par/Joseph Betali'

H 640mm, **W** 195mm, **DE** 15mm

It is just possible that this is the 'portable' barometer listed in the Queen's Catalogue since it appears to date from around the period Demainbray was in France. However, the entry probably refers to D125. It is of the 'folded' or 'double' type designed by Amontons in 1688, constructed to enable the length of the instrument to be halved and thus making it more easily transportable. Like other examples of its type it also carries a spirit thermometer.

There are four vertical glass tubes joined together, each with a cistern placed as in the photograph. The right-hand tube is open while the others are closed. Mercury is placed in the left-hand tube and in the third from the left, and partly fills every cistern. The second tube is filled with a light oil, a slight departure from Amontons who used the division between two light oils in this tube to give a reading. This instrument, like several others, used a light oil in the right-hand tube to give a reading, but this oil has now evaporated.

Owing to the difference between the cross-section of the cisterns and the tubes a magnification of the standard mercurial scale could be given. The inscriptions are in ink on a fruitwood base. Floral patterns decorate the instrument.

At the base of the instrument is written: 'Demeure dans la Grande rue Fauberg/ Antoine vis a vis la Cour St. Louis aux/ Armes de Frances a Paris'. The scales are as follows: 15—Tres Froid, 10—Plus Froid, 5—Froid, 0—Tempere, 5—Chaud, 10—Plus Chaud, 15—Tres Chaud (Thermometer); 15—Tres Sec, 10—Beau Fixe, 5—Beau Temps, 0—Variable, 5—Pluïe ou Vent, 10—Grande Pluïe, 15—Orage (Barometer)

It is almost certain that Betali is an earlier spelling of the name Bettally. Bettally was one of the many Italian barometer makers who established themselves in England in the late eighteenth century.

Amontons 1688
de Virville 1723
Knowles Middleton 1964, p. 143
Goodison 1968, pp. 86–9
Bolle 1982, pp. 74–5, p. 226

Inventory No. 1927-1909

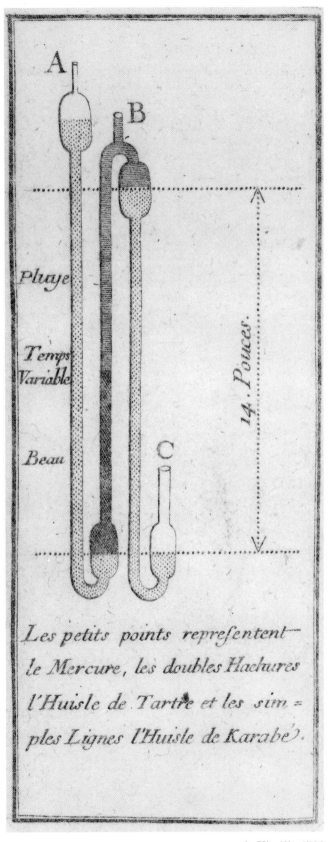

Les petits points repreſentent
le Mercure, les doubles Hachures
l'Huiſle de Tartre et les ſim=
ples Lignes l'Huiſle de Karabé.

de Virville 1723

E113 Coventry absorption hygrometer

Adams, George, the Younger

last quarter eighteenth century
London

ON ARMS HOLDING SCALE
'George Adams, Maker'
'John Coventry, Inventor'

H 350mm, **L** 350mm, **W** 105mm

The instrument is very simple. A pile of paper discs soaked in brine is threaded and suspended from the short arm of a brass lever. The long arm carries a steel pointer which can move over an ivory scale marked 0 to 100. As the humidity rises, the weight of the paper increases and the pointer indicates a higher number on the scale. Both the pivot for the lever and the scale are supported on a brass stand on a circular brass base which can be adjusted by means of a screw.

John Coventry of Southwark was a well-respected inventor and instrument-maker in the late eighteenth century. His hygrometers were 'very generally employed by the chemists and other scientific men of his day'. He was mentioned by Adams the Younger in connection with his micrometers in *Essays on the Microscope*, published in 1798.

Adams 1798, p. 61
C 74
Turner 1989, pp. 342–3
DNB Coventry, John

Inventory No. 1927-1817

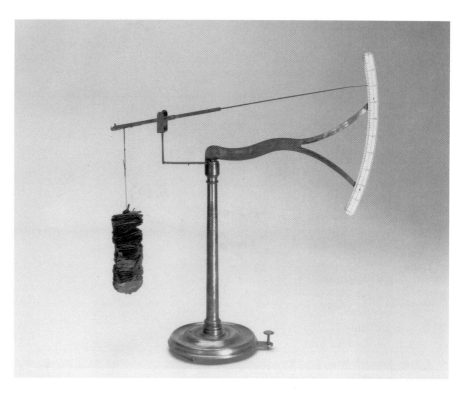

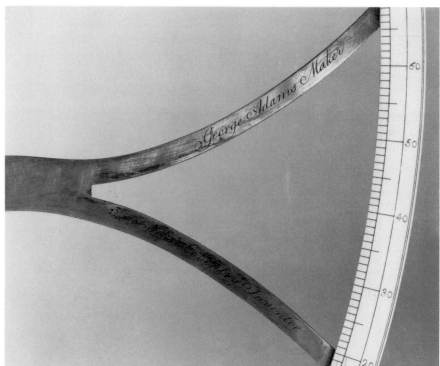

E114 Straight cistern-tube barometer

Sisson, Jeremiah

third quarter eighteenth century
London

'*J. Sisson/London*'
H 970mm, **W** 205mm, **DE** 125mm

This barometer has been described in some detail by Goodison. It is a straight barometer with an open iron-lined cistern, the top half hemisphere of which is removable so that adjustment can be made for the level of the mercury. This is done by means of an ivory pointer which can be raised or lowered to give a reading on a short scale; this reading is then added or subtracted to the reading on the main scale, an example of one of the earliest methods of compensation for cistern error. The silvered main scale is signed J. Sisson London and marked as above. A brass sliding vernier allows readings to $\frac{1}{200}$ of an inch, probably gratuitously. A small low-powered microscope is attached to the vernier. The top of the tube is covered by a brass hemisphere. The instrument is on a mahogany back. Knowles Middleton gives a date of c. 1775, whereas Goodison implies c. 1760.

C 72
Knowles Middleton 1964, p. 198, p. 222, p. 453
Goodison 1977, pp. 220–1, Pl. 135

See also D124

Inventory No. 1927-1910

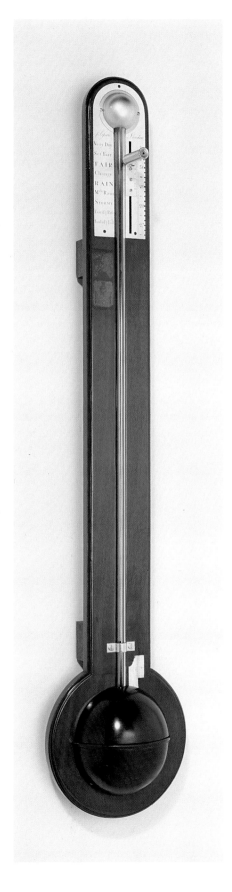

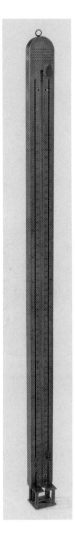

E115 Backing board for thermometer

Adams, George

c. 1758   London

'*Made by GEORGE ADAMS/in Fleet Street/LONDON/ Mathematical Instrument Maker/to his ROYAL HIGHNESS/the PRINCE OF WALES*'

SCALE

'*I. NEWTON/De Lisle/Fahrenheit/Reaumur*'

L 790mm, **W** 44mm, **DE** 16mm

Like the 'Prince of Wales' microscope (E156), this backing board can be dated to between 1756 and 1760. Adams became instrument-maker to the future George III in 1756 but was not instrument-maker to Frederick, George's father, while he was Prince of Wales.

The backing board is made of mahogany with a boxwood panel on the front and down one side. A brass cage to protect the bulb is attached to the base. There are four scales, two on each side entitled respectively Newton, De Lisle, Fahrenheit, and Reaumur. Newton's scale runs from −6° to 33°, De Lisle from 192° to 0°, Fahrenheit from −4° to 212°, and Reaumur from −16° to 84°. Down the side of the board the following are written at various intervals: Just Freeze. Winter. Spring and Autumn. Air at Midsummer. Hottest External Parts of Human Bodies. Water just tolble to the hand at rest. Water hardly tolble to the hand in motion. Melted wax grows stiff and opake. Spt of wine just boyles. Begins to boyle. Water boyles vehemently.

C 89

Inventory No. 1927-1745

E116 Six's maximum and minimum thermometer

Six, James

c. 1780   Canterbury

H 510mm (back), **H** 365mm (tube), **W** 65mm, **DE** 15mm

This is one of a collection of Six's maximum and minimum thermometers presented to King's College in about 1860 by a descendant of James Six, probably his son-in-law or grandson. It formed part of the Wheatstone Collection, until in 1926 it was transferred to the Science Museum. This was

brought to light in 1938 when the grandson of the donor attempted to see the thermometers. The significance of the items was recognized by Austin and McConnell in 1980. Biographical details are given in their book in addition to a description of the thermometers.

Six was interested in painting, astronomy, electrical machines, and thermometry, and was a regular contributor to the *Gentleman's Magazine*. In 1782 his paper describing the meteorological design of the maximum and minimum thermometer was read before the Royal Society. The innovation involved was indices inside the thermometer tubes.

The thermometers have a central reservoir filled with alcohol which acts as the thermometric fluid. The tube is then bent to go down the right-hand side of the instrument, makes a U-turn at the bottom, and continues up the left-hand side where it was stopped with a cork or a brass cap. Mercury filled the lower section of the tube and alcohol filled the remaining part. The indices in the meteorological type are short steel needles in glass cases which have long glass 'tails'. At each end is a black glass 'bugle' which makes them easier to see. When the alcohol in the central reservoir expanded, the index in the right-hand column was not moved because the glass tail held it in place. However, when the alcohol contracted, the mercury moved up the tube and dislodged the index, leaving it at the minimum temperature position. In the left-hand tube rising mercury left the index at the maximum position while falling alcohol could not move it. The indices could be re-set using a magnet and when not in use could be parked on small glass shelves at the tops of the tubes. The backing boards are pine. This design has formed the basis of domestic maximum and minimum instruments up to the present.

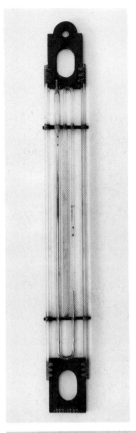

The design of the marine type was not so successful. These thermometers are slightly shorter and the glass is thicker than in the meteorological type. The indices are blued steel needles in glass cases. The principle is the opposite. The index rests on the surface of the mercury as it descends but the mercury flows past it as it rises. The right-hand tube becomes the maximum and vice versa. When not in use the indices are stored in the upper parts of the tubes away from the mercury on cylindrical glass 'shelves'. However, it was very difficult to dislodge the indices from the mercury by any means after use.

The instruments are very crude and obviously experimental in nature. It is believed that they are the prototypes for this widely used design of thermometer.

This a meteorological type of thermometer. Both indices are visible. The cork is missing and the alcohol has escaped.

Six 1782
Six 1794
C 99
Austin and McConnell 1980*a*
Austin and McConnell 1980*b*

Inventory No. 1927-1707

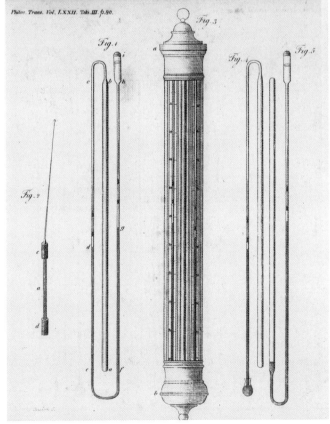

Six 1782

E117 Six's maximum and minimum therometer

Six, James

c. 1780   Canterbury

IN INK ON BACK
'37'

H 470mm (back), **H** 355mm (tubes),
W 50mm, **DE** 15mm

This is a meteorological type of thermometer. Both indices are visible.
There is no alcohol remaining despite
the cork.

As E116

Inventory No. 1927-1717

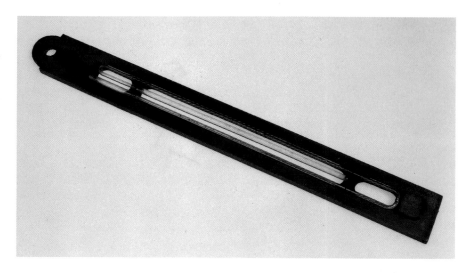

E118 Six's maximum and minimum thermometer

Six, James

c. 1780   Canterbury

IN INK ON BACK AND ON SUPPORT
'21'

H 460mm (back), **H** 360mm (tubes), **W** 50mm, **DE** 15mm

This was a meteorological type of thermometer. The central reservoir is now
missing. No indices remain. The scale is calibrated on both sides from −20 to
100°F in ink, numbered in 10s and marked in 5s.

As E116

Inventory No. 1927-1738

E119 Six's maximum and minimum thermometer

Six, James

c. 1780   Canterbury

IN INK ON SUPPORT
'3'

H 400mm (back), **H** 350mm (tube), **W** 50mm, **DE** 15mm

This was a meteorological type of thermometer. One index is visible. A scale is
calibrated from −30 to 110°F in ink on the left-hand side and −40 to 90°F on the
right-hand side. This tube is now broken.

As E116

Inventory No. 1927-1739

E120 Six's maximum and minimum thermometer

Six, James

c. 1780   Canterbury

H 630mm (back), **H** 560mm (tubes), **W** 60mm, **DE** 20mm

This is a meteorological type of thermometer with a pine back board. One index is visible. A scale is calibrated very crudely on both sides in ink from 10 to 85°F, marked in 5s. The cork is still in place and some mercury and alcohol remain in the tube.

Austin and McConnell 1980*a*, Fig. 2

As E116

Inventory No. 1927-1709

E121 Six's maximum and minimum thermometer

Six, James

c. 1780   Canterbury

IN INK ON BACK
'*35*'

H 470mm (back), **H** 360mm (tubes), **W** 50mm, **DE** 15mm

This was a meteorological type of thermometer. Written on the back in pencil is 'the Index w^h was taken out of this is in the drawer'.

As E116

Inventory No. 1927-1722

E122 Six's maximum and minimum thermometer

Six, James

c. 1780   Canterbury

H 200mm (back), **H** 180mm (tubes), **W** 50mm, **DE** 15mm

This is a meteorological type of thermometer different in a number of ways from the rest: the dimensions are more compact, there is a double cistern in the centre, and it is laterally inverted with respect to the others.

Austin and McConnell 1980*a*, Fig. 3

As E116

Inventory No. 1927-1737

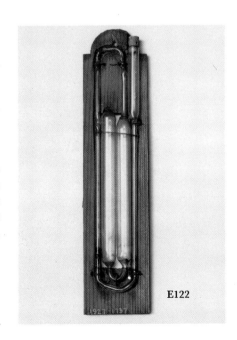

E122

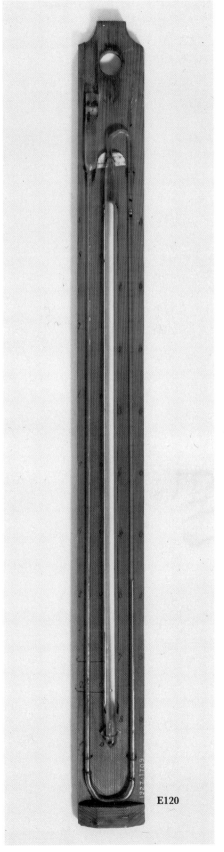

E120

E123 Six's maximum and minimum thermometer

Six, James

c. 1780   Canterbury

IN PENCIL ON BACK
'15'

H 460mm (back), **H** 370mm (tubes), **W** 55mm, **DE** 15mm

This is a meteorological type of thermometer. One index with its glass stem is visible.

As E116   **Inventory No.** 1927-1719

E124 Six's maximum and minimum thermometer

Six, James

c. 1780   Canterbury

IN INK NEAR TOP
'10'
ON SUPPORT
'9'

H 440mm (back), **H** 360mm (tubes), **W** 47mm, **DE** 15mm

This is a meteorological type of thermometer. Both indices are visible with glass tails apparent. The cork has been lost and the alcohol has escaped. The scale is calibrated in ink from −20 to 100°F on both sides, marked in 5s and numbered in 10s.

As E116   **Inventory No.** 1927-1742

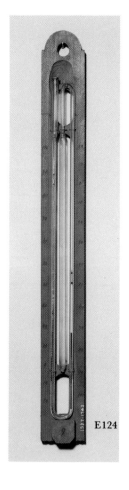

E124

E125 Six's maximum and minimum thermometer

Six, James

c. 1780   Canterbury

IN PENCIL ON SUPPORT
'2 deg too warm'

H 470mm (back), **H** 360mm (tubes), **W** 50mm, **DE** 15mm

This is a meteorological thermometer. Both indices are visible, one clearly showing the glass tail. There is a cork but the alcohol has escaped. The scale runs from −30 to 100°F marked in 5s and numbered in 10s in ink. The back board is of pine.

As E116   **Inventory No.** 1927-1714

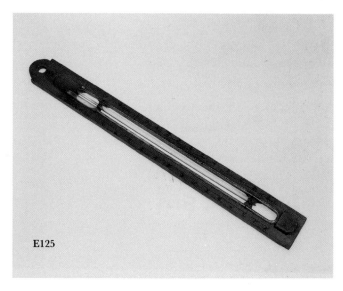

E125

E126 Six's maximum and minimum thermometer

Six, James

c. 1780   Canterbury

IN INK ON BACK
'*26*'

H 470mm (back), **H** 355mm (tubes), **W** 50mm, **DE** 15mm

This is a meteorological type thermometer. Both indices are visible and the cork, mercury, and alcohol remain. Some pencil lines serve as a scale.

As E116

Inventory No. 1927-1718

E127 Six's maximum and minimum thermometer

Six, James

c. 1780   Canterbury

IN PENCIL ON BACK
'*2*'

H 420mm (back), **H** 315mm (tubes), **W** 45mm, **DE** 15mm

This is a marine type thermometer. One blued steel index is visible in the mercury. The alcohol has escaped despite the cork.

As E116

Inventory No. 1927-1729

E128 Six's maximum and minimum thermometer

Six, James

c. 1780   Canterbury

IN PENCIL ON BACK
'*3*'

H 400mm (back), **H** 320mm (tubes), **W** 42mm, **DE** 15mm

This is a marine type of thermometer. One blued steel index is parked at the top. Some pencil lines serve as a scale.

As E116

Inventory No. 1927-1730

E129 Six's maximum and minimum thermometer

Six, James

c. 1780   Canterbury

IN INK ON BACK
'*Thick 2*'

H 430mm (back), **H** 330mm (tubes), **W** 50mm, **DE** 15mm

This is a marine type thermometer. One blued steel index is parked at the top. The cork and alcohol are now missing.

As E116

Inventory No. 1927-1743

E130 Six's maximum and minimum thermometer

Six, James

c. 1780   Canterbury

IN INK ON BACK
'*38*'

L 465mm, **H** 470mm (back), **H** 350mm (tubes), **W** 50mm, **DE** 15mm

This is a meteorological type of thermometer. The two indices are just visible in the mercury. The alcohol has evaporated.

As E116

Inventory No. 1927-1723

E131 Six's maximum and minimum thermometer

Six, James

c. 1780   Canterbury

IN PENCIL ON BACK
'*5*'

H 440mm (back), **H** 330mm (tubes), **W** 55mm, **DE** 15mm

This is a marine type thermometer but only one shelf for indices has survived. Some pencil lines serve as a scale. The cork and alcohol are missing.

As E116

Inventory No. 1927-1740

E132 Six's maximum and minimum thermometer

Six, James

c. 1780   Canterbury

IN INK ON BACK
'Thick 6'

H 410mm (back), **H** 330mm (tubes), **W** 55mm, **DE** 15mm

This is a marine thermometer. One index is visible in the mercury. Some pencil lines serve as a scale.

As E116

Inventory No. 1927-1725

E133 Six's maximum and minimum thermometer

Six, James

c. 1780   Canterbury

H 420mm (back), **H** 330mm (tubes), **W** 50mm, **DE** 15mm

This is a marine type thermometer, better finished and in better condition than the others. One blued steel needle is parked on its shelf. A brass cap covers the top end of the glass tube. The left-hand scale is marked in degrees and the numbers are stamped from 10 to 80°F. The right-hand scale is similarly marked and stamped from 20 to 90°F.

As E116

Inventory No. 1927-1741

E134 Six's maximum and minimum thermometer

Six, James

c. 1780   Canterbury

IN INK ON BACK
'Thick 3'

H 440mm (back), **H** 345mm (tubes), **W** 50mm, **DE** 15mm

This is a marine type of thermometer. One blued steel index is parked at the top. The alcohol has gone, although the cork is still present.

As E116

Inventory No. 1927-1720

E135 Six's maximum and minimum thermometer

Six, James

c. 1780   Canterbury

H 340mm (back), **H** 300mm (tubes), **W** 50mm, **DE** 15mm

This is a marine type of thermometer. One blued steel index is visible in the mercury. There are lines cut for the scale but no numbers.

As E116

Inventory No. 1927-1732

E136 Six's maximum and minimum thermometer

Six, James

c. 1780   Canterbury

H 425mm (back), **H** 345mm (tubes), **W** 55mm, **DE** 15mm

This is a marine type thermometer but no indices are visible. There is a scale cut but not numbered.

As E116

Inventory No. 1927-1731

E137 Six's maximum and minimum thermometer

Six, James

c. 1780   Canterbury

IN PENCIL ON BACK
'4'

H 400mm (back), **H** 320mm (tubes), **W** 45mm, **DE** 15mm

This was a marine type thermometer but it is now in such a poor condition that the salient features are missing. There are some pencil lines serving as a scale.

As E116

Inventory No. 1927-1728

E138 Six's maximum and minimum thermometer

Six, James

c. 1780   Canterbury

H 420mm (back), **H** 330mm (tubes), **W** 45mm, **DE** 15mm

This is a marine type thermometer. The alcohol is still present. Some pencil lines serve as a scale.

As E116

Inventory No. 1927-1726

E139 Six's maximum and minimum thermometer

Six, James

c. 1780   Canterbury

IN PENCIL ON BACK
'*1*'

H 430mm (back), **H** 340mm (tubes), **W** 55mm, **DE** 15mm

This is a marine type thermometer although only one shelf and no indices are visible.

As E116

Inventory No. 1927-1724

E140 Six's maximum and minimum thermometer

Six, James

c. 1780   Canterbury

IN INK ON BACK
'*Thick*'

H 420mm (back), **H** 330mm (tube), **W** 50mm, **DE** 15mm

This is a marine type of thermometer. No indices are visible. The alcohol has escaped despite the cork.

As E116

Inventory No. 1927-1727

E141 Six's maximum and minimum thermometer

Six, James

c. 1780   Canterbury

IN PENCIL ON BACK
'*3*'

H 430mm (back), **H** 340mm (tubes), **W** 55mm, **DE** 15mm

This is a marine type of thermometer. One blued steel index can be seen parked on its shelf. There are lines cut for a scale and pencil numbers 30 to 70 in 10s on both sides.

As E116

Inventory No. 1927-1734

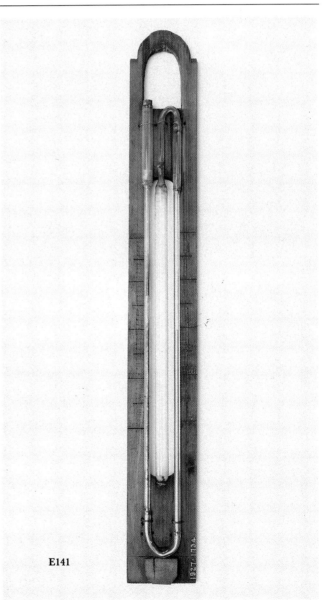

E141

E142 Mercury-in-glass thermometer

Six, James?

last quarter eighteenth century   Canterbury

H 570mm (back), **H** 490mm (tube), **W** 40mm, **DE** 20mm, **D** 17mm (bulb)

The following mercury-in-glass thermometers have been attributed to Six for the following reasons. The backing boards are pine, as are those for the maximum and minimum thermometers. The scales are similar to those on the maximum and minimum thermometers, being roughly cut in the same manner. The pencil and ink markings are also very similar. Like the maximum and minimum thermometers they appear to be trial instruments, probable tools of a man generally interested in thermometry. It is very probable that all the thermometers came from the same source and that there were not two or more sources of such similar instruments.

There is a scale cut and numbered from 30°F to 212°F in pencil on the pine back board of this instrument. It is numbered in 10s and includes 32°F. It is similar to the scale on E146.

C 90 *See also* E143–E148 **Inventory No.** 1927-1710

E143 Mercury-in-glass thermometer

Six, James?

last quarter eighteenth century   Canterbury

INSCRIPTION TOP AND BACK
'10' '10' in ink and pencil

H 530mm (back), **H** 450mm (tube), **W** 40mm, **DE** 20mm, **D** 15mm (bulb)

There is no scale on this instrument which otherwise is similar to E146.

C 91   **Inventory No.** 1927-1712

E144 Mercury-in-glass thermometer

Six, James?

last quarter eighteenth century   Canterbury

FRONT AND BACK
'9' '9'

H 520mm (back), **H** 440mm (tube), **W** 40mm, **DE** 15mm, **D** 15mm (bulb)

The instrument has a roughly marked scale. It is similar to E146.

C 91   **Inventory No.** 1927-1713

E145 Mercury-in-glass thermometer

Six, James?

last quarter eighteenth century   Canterbury

H 480mm (back), **H** 410mm (tube), **W** 40mm, **DE** 15mm, **D** 15mm (bulb)

This instrument is very similar to E146. There is a scale cut and marked in degrees and tens but not numbered. However, there is a line presumably corresponding to 32°F which enables the thermometer to be read.

C 91

Inventory No. 1927-1715

E146 Mercury-in-glass thermometer

Six, James?

last quarter eighteenth century   Canterbury

IN PENCIL ON BACK
'10 110 Equal'

H 440mm (back), **H** 370mm (tube), **W** 40mm, **DE** 15mm, **D** 15mm (bulb)

This is a mercury-in-glass thermometer with a spherical bulb on a pine back. A scale from −10° to 110°, presumably Fahrenheit, is just visible. The numbers are every 10, handwritten in pencil.

C 91

Inventory No. 1927-1744

E147 Back board for thermometer

Six, James?

last quarter eighteenth century   Canterbury

H 470mm, **W** 40mm, **DE** 15mm

Inventory No. 1927-1716

E148 Back board for thermometer

Six, James?

last quarter eighteenth century   Canterbury

H 550mm, **W** 40mm, **DE** 15mm

Inventory No. 1927-1711

E149 Wedgwood pyrometer

Wedgwood, Josiah

1786   Staffordshire

L 215mm (box), **W** 103mm (box), **H** 77mm (box), **L** 185mm (pyrometers), **W** 35mm (pyrometers), **DE** 17mm (pyrometers)

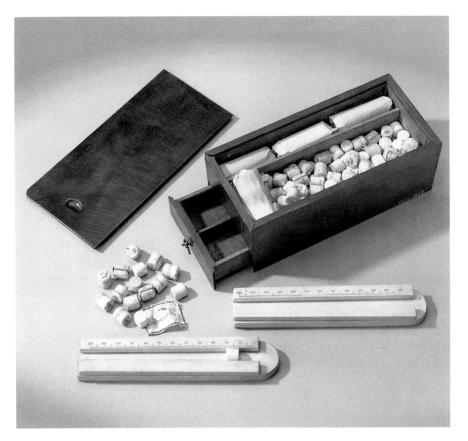

In 1765 Josiah Wedgwood became 'Potter to Her Majesty'. A private ledger in the Wedgwood Museum shows that in July 1786 this pyrometer was presented to King George III.

The pyrometer is in a mahogany box, the top part of which contains about 80 'thermometer pieces'—cylindrical pieces of porcelain or fine clay measuring half an inch in diameter. Each is individually wrapped in paper with 'Josiah Wedgwood FRS' printed on it. There are also four porcelain cases for holding the thermometer pieces to protect them from samples which are likely to melt. The drawer underneath contains two clay gauges. These consist essentially of channels of constantly diminishing width. As specified by Wedgwood they measure widths from $\frac{5}{10}$ to $\frac{3}{10}$ inch and are marked 0 to 120 and 120 to 140 respectively. They are numbered in 10s.

Wedgwood was concerned that there was no method of measuring high temperatures comparable to a mercury thermometer for lower ones. Having rejected the use of colour as being too subjective and difficult to describe, he decided on using the property of 'diminution of bulk by fire' as the measurable quantity. At first he used pieces of pure Cornish porcelain as the thermometer pieces, pared down to $\frac{5}{10}$ inch wide. The first gauge he described was made of brass 2 feet long. He divided it into 240 parts, making the wide end of the gauge $\frac{5}{10}$ inch wide and the narrow end $\frac{3}{10}$ inch wide. He also suggested using a double gauge as in L69 and shortening the gauge: as long as the ratio 5:3 and the 240 intervals were preserved, any size gauge would suffice. He assumed from observation that the diminution of the thermometer pieces was linear with temperature. He knew that continuance at a certain temperature did not affect the measurements.

A piece of ceramic under investigation was placed in the kiln together with the thermometric piece. A piece of porcelain could only be used once as it was then permanently shrunk. If a material under observation might damage the thermometer piece by melting, the piece could be placed in one of the larger porcelain cases. An indication of the temperature of the kiln could be made by retrieving the thermometer piece, cooling it, and placing it in the gauge. The shrinkage of the piece could be read on the scale.

Wedgwood attempted to link his pyrometer, which was in a 'detached state', with the Fahrenheit scale. He did this by using an intermediate measure—in this case the expansion of metals. By measuring the temperature of a piece of metal, both with a mercury thermometer and a Wedgwood pyrometer at two different values, he could give Fahrenheit equivalents for his scale. Several measurements were made on the melting points of metals at the Tower of London with Wedgwood's pyrometer.

Wedgwood 1782
Wedgwood 1784
Wedgwood 1786
C 80
Chaldecott 1975
Wedgwood 1978, Items 48, 52, 80

See also L69

Inventory No. 1927-1812

OPTICAL INSTRUMENTS

E150 Split-object glass micrometer

Dollond, P.

1760–1800   London

'Dollond London'

L 340mm, **W** 40mm, **H** 125mm

Dollond invented this improvement on the double-object glass micrometer in 1753. An objective lens of long focal length is cut in half to form two segments. To ensure that their centres are both external 'a small matter is ground away from the straight edge of each segment; and so they will more easily be brought to a coincidence'. The two halves of the lens can be moved laterally with respect to each other by means of a rack and pinion in a brass frame. The separation of the centres can be read to $\frac{1}{500}$ inch on a vernier scale. The instrument could be attached to the front of a reflecting telescope and rotated to the required angle; the angle is measured on a protractor on the reverse. When not in use it fits into a mahogany box.

The micrometer can be used to find the angular separation of two stars or the angular diameter of the Sun, Moon, or planets. The relative position of the two halves of the lens can be adjusted so that the two images produced coincide. At this position the separation between the centres of the lenses corresponds to the angular distance between the objects in view. A knowledge of the focal length of the telescope allows the angle to be calculated. If the distance of the object is known, then the actual diameter can be calculated using the geometry of similar triangles.

The split-object glass micrometer was an improvement on the double-object glass micrometer hinted at by Roemer and developed by Bouguer and Savery.

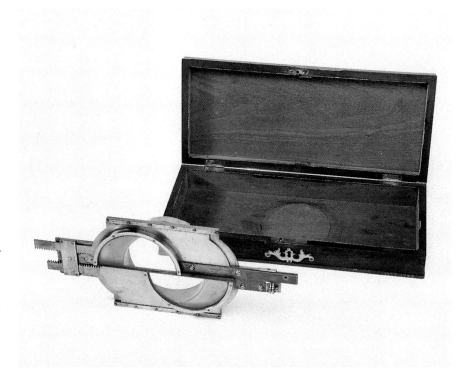

Horrebow 1740–1, Vol. 3, p. 103.
Bouguer 1748
Dollond 1753
Savery 1753
Dollond 1754
Maskelyne 1771
Lalande 1792, Tome 2, p. 641, Pl. 28,

Fig. 187
Encyclopaedia Britannica (7th edn), 1842, under Micrometer
C 146
Turner 1973, Item 258
Fauque 1983
Inventory No. 1927-1782

E151 Polyoptic telescope

second half eighteenth century
L 157mm, **D** 42mm (max)

This instrument superficially resembles a prospect glass. The objective consists of a square pyramid of glass similar to but less complicated than that in the polyoptrick pyramid D53. There was probably an eye lens but this is now missing. When used to view special pictures, a recognizable image would result from a combination of the four parts; however, these pictures are now lost. The instrument is made of walnut with a brass sliding lens cap and some painted tinplate on the body tube. Although painted tinplate suggests an association with Demainbray, the instrument is not listed in the Queen's Catalogue.

C 179

See also D53

Inventory No. 1927-1507

E152 Four scioptic balls

Adams, George, the Younger

late eighteenth century   London
L 230mm, **D** 140mm, **D** 70mm (balls)

For stylistic reasons it is thought that these scioptic balls are by Adams. Similar instruments are partially shown in the *Lectures on Natural and Experimental Philosophy* of 1794. Apart from one which has a lighter shade of wood than the others, they are all identical. The circular rims are made of mahogany and the balls are of lignum vitae. They each have an oval mirror in a brass frame, one lens, and a narrow brass tube for directing the light. Only one has been illustrated

Adams 1794, Pl. 4, Figs 9, 10, 11, 12, Vol. 2
C 192

See also D133, D134

Inventory Nos 1927-1162, -1163, -1164, -1165

E153 Scioptic ball

last quarter eighteenth century or first quarter nineteenth century

D 139mm, **D** 125mm (frame), **D** 80mm (ball)

This is probably later than the other scioptic balls in the collection. It has a knurled inner ring and a square outer frame of mahogany. The ball is made of rosewood. There are no lenses.

C192

See also D133, D134

Inventory No. 1927-1168

E154 Telescope in scioptic ball

first half eighteenth century

L 360mm (closed), **D** 125mm (max)

Three parts bearing a great resemblance to each other have been fitted together to form this object. The central section is a scioptic ball of lignum vitae in a mahogany mount with two vellum-covered pasteboard tubes having horn ends. The tooling on the tube corresponds to No. 51 in Turner's paper on the subject. One end has a draw tube also covered in green vellum. The additional pieces were an identical draw tube with a lens and lignum vitae lens mount, which was added to the other end, and a smaller lens of higher power, also in a lignum vitae mount added, to the open end. The whole forms a telescope with a magnification of about 15.

Turner 1966   **Inventory Nos** 1927-1161, -1530, -1557

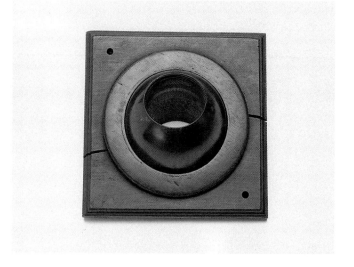

E155 Scioptic ball and mirror

second half eighteenth century

ENGRAVED—BALL
  '2'

L 300mm, **D** 95mm, **L** 175mm (mirror), **W** 55mm (mirror)

The scioptic ball is entirely of brass and is fairly heavy. The rectangular mirror is hinged to enable it to be inclined with respect to the horizontal but not rotated as is usual in a solar microscope. There is one lens of long focal length at the front of the instrument but only a circular hole at the back. Two screws are situated either side of the lens; their function is not known.

C 192   **Inventory No.** 1927-1505

E156 'Prince of Wales' microscope

Adams, George

c. 1758   London

'Invented and made by Geo Adams in Fleet Street Mathematical Instrument Maker to His Royal Highness the Prince of Wales'

L 190mm, **W** 210mm, **H** 470mm

This instrument was bought by the Museum at the auction of the Crisp collection in 1925. It is certain that it formed part of the George III collection since it carries the characteristic identification label attached to George III items at King's College. It has been known as the 'Prince of Wales' microscope because of the inscription. In a forthcoming publication Millburn suggests that Prince George formed a household in 1756 and appointed Adams as his instrument-maker at about the same time. Adams was not instrument-maker to Frederick, George III's father; hence the instrument can be dated fairly accurately.

It is a large 'Universal microscope' in brass on a box base. The body tube is covered in green fishskin. The limb of the instrument is supported on trunnions and can be inclined on an axis running through the centre of gravity. It can also be clamped in an upright position using a pin in the base. Coarse focusing is achieved by sliding the bar carrying the body, and fine adjustment by means of a fine-threaded screw. The stage has micrometer screws registering in two directions at right-angles in the plane of the specimen. These screws have 100 turns to an inch and the head is divided into 100 parts, so it could be read to 1/10 000 of an inch.

See E115

Inventory No. 1925-136

E157 Silver microscope

Adams, George

c. 1763   London

'*Made by GEORGE ADAMS in Fleet Street LONDON*'

H 740mm, **L** 380mm (base),
DE 380mm (base)

Although probably not an instrument formerly belonging to George III, this item has been included owing to its close association with the collection.

It is one of two silver microscopes known to have been made by George Adams for King George III or the Prince of Wales, later George IV, in about 1763. It was presented to the Museum in 1949 by George VI. The two instruments are almost identical except for the fact that this has a more ornate ebony base. It is very likely that this instrument belonged to the Prince of Wales since a contemporary drawing of it has the inscription 'from store room under the Clock Carlton House' across the top. The Prince of Wales paid £28 for 'cleaning repairing and burnishing the large silver microscope' in 1785. The other microscope is now at the Oxford Museum of the History of Science, having been in private hands after being sold at the auction of the Crisp collection in 1925. It is likely that the Oxford microscope was made for George III since Crisp managed to acquire George III's microscopes previously held at Richmond.

The instrument is based on the 'Universal Double Microscope' described in Adams's *Micrographia Illustrata* of 1746. However, it is so highly decorated as to be almost unrecognizable. It is made of brass and steel covered in beaten silver and it stands on an octagonal ebony base. A fluted Corinthian central pillar is supported on a horizontal cross-piece which can rotate about its axis to adjust the inclination of the instrument. The cross-piece is in turn supported on two feet which stand on the base. Two double-sided mirrors on swivel mounts are attached to the cubical base of the central pillar, and two circular spring stages can be moved vertically by means of rack and pinions. One of the stages is fitted with a revolving disc with apertures of $\frac{1}{4}$ inch, $\frac{1}{2}$ inch, and $\frac{3}{4}$ inch. The circular objective holder has sockets; into one can be fitted a simple magnifier. The holder can be rotated to bring the compound body or the magnifier over the specimen stages. The compound body fits into a collar in the centre of a wreath, the wreath being supported by two allegorical figures which stand on the objective holder. An adjustable arm with four ball-and-socket joints is mounted at the top of the pillar and carries a lens.

There are various accessories contained in the drawers in the base: an alternative eyepiece and eye cap, a stand for the jointed arm, three sets of objectives one of which has Lieberkühns, talc discs for slides, a fishplate with a glass slide attached, three magnifiers, a pair of stage forceps, a live box, a pair of tweezers, and a nosepiece.

Crisp 1925, No. 169
Clay and Court 1927
R.A. No. 29086
Illustrated London News 1949, 3 September, p. 337

Inventory No. 1949-116

E158 Solar microscope part

last quarter eighteenth century

D 153mm, **L** 97mm

This is the circular plate and tube of a solar microscope which would have been attached to a window shutter when the instrument was in use. It carries a hinge for the mirror which is now missing, and a handle which would have turned the mirror on its axis. The microscope would have been inserted into the tube.

Solar microscopes, which used sunlight from outside a room to project an image formed by a microscope inside a room, became popular in England after 1740. This all-brass design was invented by John Cuff in about 1755.

Bradbury 1967, ch. 5

Inventory No. 1927-1691

E159 Microtome in box

Cumming, A.

c. 1770   London

SCALE
5 (5) 20
500, 750, 1000
A. CUMMING LONDON

D 50mm (body), **D** 65mm (max), **L** 135mm, **L** 165mm (box), **W** 95mm (box), **H** 75mm (box)

Microtomes were used to cut very thin slices of specimens for microscopical investigation. This is probably the earliest of the three examples in the collection. It is signed by Cumming, clockmaker to George III, to whom Hill attributed the invention of the device. It is an example of the automatic type of microtome which moves the specimen upwards so that 'a hundred slices could easily be cut in a minute' at 500, 750, or 1000 slices to the inch. Hill appears to have tried this type but rejected it for the simpler manual type of which E160 and E161 are examples. If Cumming (or Cummings) made only the first 'two or three' as Hill claims, then this instrument is probably the only surviving example.

It has an ivory cylindrical body, down the length of which runs a wedge-shaped hole for the specimen. Pieces of ivory of various lengths can be fitted into the hole depending on the length of the specimen. At the top end of the body a circular

piece of bell-metal is fixed to the ivory, on to which is screwed the blade and above this a brass index. The handle, which was screwed above the index, is now missing. The body contains two brass screws which are made to press on to the wood by means of a key. At the lower end is a circular brass plate with a brass screw which turns an index over a circular engraved scale numbered in 5s from 0 to 20 and pushes the specimen upwards. Also a blued steel pointer can be set on a semicircular scale at 500, 750, or 1000 parts of an inch for the required thickness of the specimen by adjusting the throw of a small hook which pulls round the specimen advance. This mechanism is described in more detail by Bracegirdle (1986) who points out that the elasticity of the specimen did not allow for this sort of precision.

Hill 1770, pp. 3–5
C 193
Bracegirdle 1978, pp. 12–16
Bracegirdle 1986
Turner 1989, No. 312

Inventory No. 1927-1461

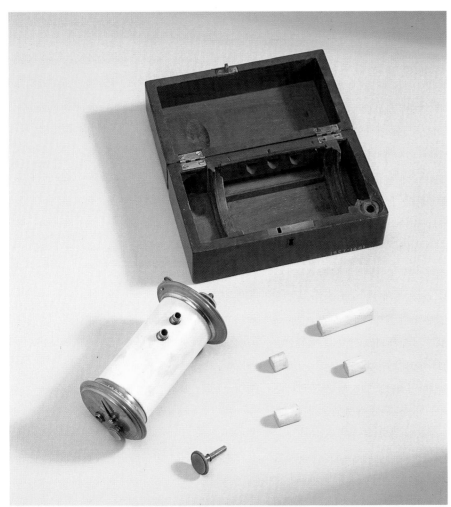

Hill 1770

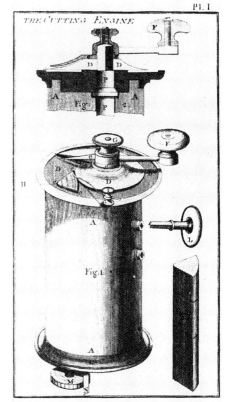

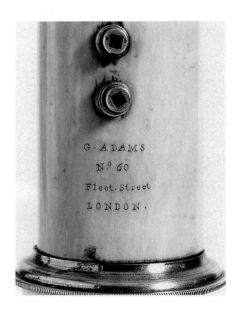

E160 Microtome in box

Adams, George, the Younger

c. 1780   London

'G. ADAMS/No.60/Fleet Street/LONDON'

L 200mm (box), **W** 110mm (box), **H** 85mm (box), **L** 165mm, **D** 85mm (max), **D** 55mm (body)

This is a manual type of Cummings microtome identical to that described by Hill. In outward appearance it resembles the previous item, but is slightly larger. It has an ivory body with a wedge-shaped cavity for the specimen running down its length. The blade, index, and handle are similar to those in E159 and E161. The lower end of the instrument has a brass plate with a screw which acts against the specimen, or piece of ivory in contact with the specimen. The raising of the screw is measured against an index which protrudes from the base plate. Since Hill tells us that the screw has a pitch of 40 threads to the inch and there are 25 divisions on the screw head, each division represents a movement of 1/1000 inch. There is a choice of four blades.

Hill 1770, pp. 3–5 plate
C 193
Bracegirdle 1978, pp. 12–16
Bracegirdle 1986
Turner 1989, No. 312

Inventory No. 1927-1462

E161 Microtome in box

c. 1780

SCRATCHED—CIRCULAR SCALE
1 (5) 25

L 165mm (box), **W** 95mm (box), **H** 75mm (box), **L** 150mm, **D** 65mm (max), **D** 50mm (body)

This is another example of a manual type of Cumming microtome and is very similar but slightly smaller than the previous item. The only significant difference is that it has a ceramic screw head instead of a brass one. The box is identical to that of E159. There is a piece of cork inside the instrument and the handle is broken, suggesting that it needed considerable force to operate with certain specimens.

C 193

As E159

Inventory No. 1927-1460

E162 Reading glass

first half eighteenth century

D 80mm, **L** 140mm, **D** 60mm, **DE** 15mm

The reading glass is in a turned ebony mount with a handle. The magnification is about 2.

C 207

Inventory No. 1927-1569

E163 Rods for adjusting lucernal microscopes

Adams, George, the Younger

1787–1800   London

L 1020mm, **D** 25mm (max)

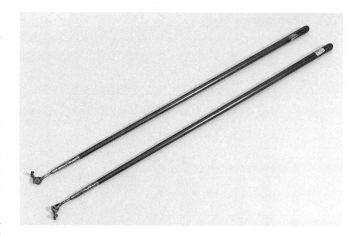

These rods are accessories for lucernal microscopes, indicating that there were at least two in the collection at one time. They are 'handles furnished with a universal joint for more conveniently turning the pinion', enabling focusing to take place without the viewer or demonstrator having to lose sight of the image.

A lucernal microscope is a solar microscope and a camera obscura combined, although an artificial light source was used rather than sunlight. Adams the Younger claimed that his father had invented the idea and part executed it, but that he himself had improved it in 1774.

Adams 1787c, p. 22, pp. 65–9, Fig. 1, Pl. 3

See also J63   **Inventory No.** 1927-1839

E164 Model of plano-convex lens

last quarter eighteenth century or first half nineteenth century

D 310mm, **H** 530mm, **W** 65mm

Together with following item, this instrument forms a pair of unusual large model lenses. The models can be attached to a table top by means of a clamp at the base. Threads can then be passed from an object through the holes in the mahogany lenses. A total of 121 threads can be used to illustrate image formation with this model.

Holes are bored as follows: one central hole, then three concentric rings of 24 holes on the convex surface. The plane surface is more complex: the outer circle consists of 24 holes but the middle and inner are duplicated to give a choice of path for the thread; one of the pair of holes is oblique. The duplication allows the paths of rays from infinity or from the focus to be traced using the threads.

C 208   **Inventory No.** 1927-1257

E165   E164

E165 Model of plano-concave lens

last quarter eighteenth century or first half nineteenth century

D 310mm, **H** 530mm, **W** 75mm

This instrument forms a pair with the above. It is slightly less complex since there is only one path for the threads to take on passing through each hole. There is one central hole and three concentric rings each with 24 holes, making a total of 73. The divergence of rays passing through a plano-concave lens can be demonstrated, but not the existence of a virtual focus.

C 208

Inventory No. 1927-1258

E166 Polished glass blank

eighteenth century

L 100mm, **W** 95mm, **H** 75mm

This is a black glass blank which has been planed and polished on one surface. On the other an oak muller, or handle, has been attached, probably with a combination of black pitch, sealing wax, and plaster. Lenses and mirrors were ground by hand in the eighteenth century, with the blank being held stationary while the tool was turned. This piece may have been intended as a Claude glass: a black glass mirror used frequently by artists in the late eighteenth century to give landscapes a quality resembling the works of Claude Lorrain.

Manwaring 1925
C 211
Bedini 1966, p. 689

Inventory No. 1927-1579

E167 Polished glass blank

eighteenth century

D 108mm, **H** 55mm

This item is very similar to the above except that the glass blank is clear rather than black. There is an oak handle attached by a mixture containing pitch to one side of the blank.

Manwaring 1925
C 211
Bedini 1966, p. 689

Inventory No. 1927-1578

E168 Lens polisher

eighteenth century

L 120mm, **W** 100mm, **H** 160mm

The instrument is a simple lens polisher of the horizontal type illustrated in Manzini's *L'Occhiale All 'occhio*, but on a much smaller scale. An ivory handle turns a brass axle on which is a boxwood pulley wheel carrying a groove for thread. The thread turns an upper smaller wheel which turns an axle carrying a boxwood disc with a protruding screw. The tool or pattern would have been attached to this disc but it is now missing; some sealing wax remains. The whole is on a walnut base which contains an ivory plaque; the maker's name is now indecipherable.

Manzini 1660, p. 158
Bedini 1966, p. 689

Inventory No. 1927-1506

E169 Glass polisher

last quarter eighteenth century or first quarter nineteenth century

IN INK ON BASE OF POLISHER
'24'

L 250mm, **W** 100mm, **H** 50mm

The instrument appears to be for demonstrating the polishing of pieces of flat glass rather than lenses. On one side a shallow circular hole contains the glass blank which is held in place by three pieces of brass. Above this is placed a brass band and a circular piece of moulded mahogany. A screw fitted down the centre around which a cord was passed, but this is now missing. The other side consists of the whirler which is another circular piece of mahogany; it has a groove cut for the cord and a small handle. The distance of the whirler from the polisher is adjustable and a groove for the cord is made between the whirler and polisher on the base of the instrument. When the handle was turned the mahogany polisher would act on the surface of the glass.

C 211

Inventory No. 1927-1661

E170 Four convex lenses

last quarter eighteenth century or nineteenth century

ETCHED ON ONE LENS
'16 feet 3 inches'

D 106mm (three lenses), **D** 110mm (one lens), **T** 5mm (one lens), **T** 4mm (three lenses)

These are high quality crown glass lenses with a green tinge and bevelled edges. They all have long focal lengths comparable with the 16 feet 3 inches inscribed on one lens.

C 207

Inventory No. 1927-1545

E171 Achromatic doublet

last quarter eighteenth century or nineteenth century

D 103mm, **T** 7mm

The doublet consists of crown and flint glass of very high quality. The diameter is 4 inches and the focal length is 52 inches.

Achromatic doublet lenses reduce the chromatic aberration or colour distortion inherent in the images of single spherical lenses. They combine two types of glass with different dispersive powers, usually crown and flint, so that the ratio of the dispersive powers is equal to the ratio of their focal lengths. If this is the case, two colours will be brought to the same focus. Although doublets had been experimented with in the early eighteenth century, the patent by Dollond in 1758 restricted their general use until the last quarter of the century.

Dollond 1758
C 207

Inventory No. 1927-1571

E172 Achromatic triplet lens

last quarter eighteenth century or nineteenth century

D 100mm, **T** 3.5mm (crown), **T** 4.5mm (flint), **T** 2mm (crown)

The triplet is made up of a concave flint glass between two convex crown glasses. The resultant focal length is very long. The triple achromat was introduced by Peter Dollond in 1765.

Dollond 1765
C 207

Inventory No. 1927-1544

E173 Two convex lenses and one concave lens

last quarter eighteenth century or nineteenth century

D 152mm, **T** 2mm (convex), **T** 5mm (concave)

The two convex lenses have a green tinge while the concave lens is colourless, suggesting that they are of crown and flint glass respectively and were used as an achromatic triplet. The convex lenses have a focal length of approximately 25 feet.

C 207

Inventory No. 1927-1585

E174

E175

E174 Convex lens in mount

last quarter eighteenth century

L 133mm (max), **D** 92mm (lens), **H** 242mm, **D** 88mm (base)

The biconvex lens has a focal length of 12 inches. It is mounted in brass on a brass swivel stand which is rather tarnished.

C 203

Inventory No. 1927-1247

E175 Convex lens in mount

last quarter eighteenth century or first quarter nineteenth century

D 155mm (lens), **H** 360mm, **D**165mm (base)

The lens is in a brass mount with steel screws, one of which is missing. The focal length is about 2 feet.

C 207

Inventory No. 1927-1428

E176 Green glass disc on stand

second quarter eighteenth century

H 350mm, **D** 155mm (glass), **D** 195mm (frame), **D** 135mm (base)

This is a green glass disc of 6 inches diameter in a walnut mount on a walnut stand. It was probably used for experiments on Newton's theory of colour.

C 207

Inventory No. 1927-1425

E177 Pair of prisms in frames

Sisson, Jeremiah

third quarter eighteenth century　　　　　　London

ON THE DIALS
'Sisson London'

H 485mm (max), **L** 195mm, **D** 75mm (dial), **L** 96mm (prism), **L** 35mm (side of prism)

These instruments are identical in every respect. Each triangular prism of crown glass is about 4 inches long with sides about $1\frac{1}{2}$ inches long. It is fixed horizontally on a brass axle which can be rotated by turning a knob on a brass dial at one end. The dial is marked from 0° to 360° in 2°, an index reads 0° when one of the vertices of the prisms is uppermost. The axle is held in a rectangular brass bracket whose lower surface carried a spirit level, now no longer operable. The height of the prism can be adjusted by a knob holding the sliding pillar in the stand. The horizontal direction of the prism can be adjusted by turning a knob in the base. The brass base stands on three levelling screws. The prisms could be used for demonstrating Newton's theory of colour, possibly with the heliostat D141.

C 214

Chew 1968, Item 6

See also D141

Inventory No. 1927-1279, -1280

E178 Prism on stand

second half eighteenth century

L 150mm (prism), **D** 105mm (stand), **H** 122mm

This is a glass prism supported on a brass stand with a ball and socket joint.

Inventory No. 1927-1281

E179 Astronomical telescope

Cock, Christopher?

last quarter seventeenth century
London

ENGRAVED—BODY
Coat of arms

L 565mm (closed), **D** 80mm (max)

It is not known how this instrument came into the collection: it may have been in the possession of the royal family for several generations before George III came to the throne. In the past it has been attributed to Cock probably because it is very similar to another instrument signed by Cock in the Science Museum. It may be the two-lens telescope listed in the Queen's Catalogue.

It is a simple reverse-taper astronomical telescope with two lenses, giving a magnification of about 30. There are eight draw tubes made of vellum-covered pasteboard. The outer tube is blotched vellum with tooling consistent with telescopes signed by Yarwell from this period, including a royal coat of arms.

Q.C., Item 209; A dioptric 2 convex lens telescope
C 171
Turner 1966

Inventory No. 1927-2055

E180 Terrestrial telescope

last quarter seventeenth century

L 325mm (telescope), **D** 36mm (telescope), **L** 360mm (case), **D** 60mm (case)

There is no indication as to how this instrument came into the collection. It is a reverse-taper terrestrial telescope with four lenses, two of which are in a removable cell forming the erector. The magnification is about 5. The body tube is covered in turquoise leather and there are four pasteboard draw tubes covered in marbled vellum. The eyepiece cover and end caps are turned walnut. The stamp consists of the letters MR with two leaves and a star above. The telescope has a case of walnut with red baize inlay.

There is a telescope with the same stamp at the National Maritime Museum, which also has a written inscription in ink on one of the draw tubes: 'Ao 1645–6'. It had been attributed in the past to Anton Maria Schyrle de Rheita. However, the characteristics of the telescope described above, in particular the removable erector, indicate a later date.

C 169
King 1955, p. 45.
NMM 1970, ref 0.71/37–74C

Inventory No. 1927-1171

E181 Terrestrial telescope

second quarter eighteenth century
L 400mm (closed), **D** 50mm

The telescope has a fishskin-covered body tube and four green vellum draw tubes. The mounts and lens caps are of ivory. The objective is non-achromatic and the two erecting lenses are in a removable cell of lignum vitae and cardboard.

C 168

Inventory No. 1927-1170

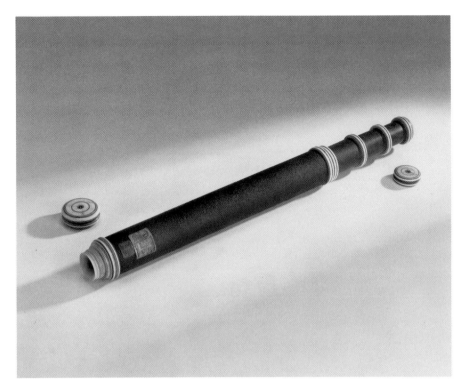

E182 Terrestrial telescope

first half eighteenth century   Italy

HANDWRITTEN—EYEPIECE
 'Ruesto cannello si allungae si Scorta Secondo le uiste'

L 510mm (partially closed), **D** 50mm

This is a fairly standard four-lens Italian telescope of the early- to mid-eighteenth century. The body tube is covered in green leather and tooled with a flower and scroll motif. The six draw tubes are covered with marbled vellum and the lens mounts are box-wood. The erectors are in a removable cell. The eye lens is loose and one draw tube is stuck so that it is difficult to ascertain the magnification or image quality. The inscription written round the draw tube nearest the eye lens translates as follows: 'This telescope is lengthened and shortened according to the views'.

In the Optics Collection at the Science Museum there is a telescope by Campani which is similar but larger.

C 167

Inventory No. 1927-1169

E183 'Day and Night' telescope

Ayscough, J.

1749–62   London

ROUND EYEPIECE
'Ayscough LONDON Inv Et Fecit'
OBJECTIVE COVER
*'Unscrew/This Small Cover/in the/Day/And the Large/
Cover in the/Night'*

L 673mm (closed), **D** 110mm

This is not quite like the slightly later 'day and night' telescopes in which the erector lenses could be removed to give additional light for night use; there is no indication that erectors were ever included. The telescope has a black fishskin-covered body with one short draw tube in green vellum. The draw tube contains a two-lens Huygens eyepiece, the second lens of which is in a boxwood and pasteboard cell which can be removed. The eye lens and objective are in brass mounts. The eyepiece lens mount is signed. The objective has two covers in brass: one extending the full diameter of the telescope for night use and one consisting of the central section only for day use. The magnification is about 10. The large aperture and fine quality lenses give a sharp image.

Ayscough was a maker of spectacles and optical instruments in London, first working for James Mann and then from about 1749 independently.

C 170
Court and Von Rohr 1929–30, p. 80
Taylor 1966, p. 168

Inventory No. 1927-1172

E184 Terrestrial telescope

Dollond, John and Peter

1752–61   London

'*DOLLOND & SON/LONDON*'

L 415mm, **L** 450mm (closed), **D** 55mm

The telescope has three green vellum draw tubes, one of which is stuck, making the erector inaccessible. The body tube is covered in green shagreen and the lens mounts are silver gilt on brass. The objective is missing and so it is not known whether it was an early example of an achromatic telescope. The eyepiece lens cap is on a separate inventory number but has been identified as belonging to this telescope.

C 172
Taylor 1966, pp. 155–6, pp. 228–9
Inventory No. 1927-1420, -1531

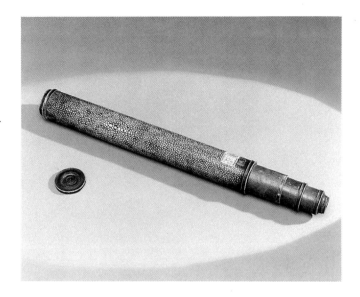

E185 Terrestrial telescope

last quarter eighteenth century or first quarter nineteenth century

HANDWRITTEN—END CAP
    '*2 pt Object Glass/Crown Glass convex/9 inches Solar Focus/Flint Glass concave/12 Inches Solar/Focus*'
HANDWRITTEN—DRAW TUBE
    '*Double Convex/2 1/2 inches*'

L 910mm (closed), **D** 65mm, **D** 75mm (max)

The mahogany-bodied telescope has one green vellum draw tube containing a three-lens eyepiece. One of the eyepiece lenses can be removed and the inscription on the cell reads 'Double Convex $2\frac{1}{2}$ inches' in ink. The inscription on the object lens cap indicates that the objective was achromatic, but the lens is now missing. There are two brass knurled screws in the body, presumably for a stand which may have been added later.

Inventory No. 1927-1421

E186 Gregorian telescope

third quarter eighteenth century

L 425mm, **D** 72mm, **H** 360mm

This is a 12 inch reflecting telescope of a type produced over a long period in the mid and late eighteenth century. The instrument is principally made of brass with a red shagreen-covered body tube of diameter $2\frac{3}{4}$ inches. Focusing was achieved by means of a screw which runs along the body. The stand is slightly unusual: it is a tripod at the lower end and a ball and socket in an oval mount at the top. The telescope screws into two slots attached to the ball and socket.

Gregorian telescopes use a primary concave mirror which reflects the light to a smaller secondary concave mirror which in turn reflects it through the eyepiece. There is a central hole in the primary mirror to allow the light to pass through. The design was invented by Gregory in 1663.

Gregory 1663, p. 94
C 175
King 1955, pp. 70–1, Fig. 33
See also D139

Inventory No. 1927-1416

E187 Gregorian telescope

Adams, George, the Younger

last quarter eighteenth century   London

ENGRAVED—BODY TUBE
  '*Adams LONDON*'

L 375mm, **D** 78mm, **H** 360mm

Only the body tube and stand of this instrument have come into the collection. It is a standard brass 12 inch reflecting telescope with a brass tripod stand. The aperture is 3 inches. Focusing is done by means of a screw thread which controls the position of the secondary mirror. Presumably the optical arrangement would have been similar to that illustrated in Adams's *Lectures on Natural and Experimental Philosophy*. The nature of the inscription indicates a relatively late date. Numerous similar instruments by Adams and other makers can be found elsewhere.

Adams 1794, Pl. 8, Vol. 2, Fig. 2
C 174
Wynter and Turner 1975, p. 204
Porter *et al.* 1985, No. 71

Inventory No. 1927-1417

E188 Transit telescope

Sisson, Jeremiah

c. 1770   London

ENGRAVED—BETWEEN TRUNNIONS
'*J. Sisson London*'
SCALES
'*130[degree]–0–50*'
'*0–90–0*'

L 135mm (brackets), **D** 630mm (semi-circle), **L** 365mm (vernier arm), **L** 1145mm (trunnions), **L** 1030mm (telescope), **D** 74mm (telescope)

It is likely that Sisson made this instrument for the new Richmond Observatory in about 1770, as he made the 8-foot mural arc, E36, for the observatory that year. It could have been used in conjunction with the planisphere E44.

A transit telescope is fixed to operate in the plane of the meridian. It is used to observe celestial bodies as they cross or make a 'transit' of this line. The telescope is mounted in a pair of conical trunnions on adjustable pivot mounts. The optics are no longer present: a rather crude clamp and slit at either end of the body tube allowed various lenses to be inserted or removed. One pivot is vertically adjustable; the other is adjustable in the azimuth. There is one suspension unit for a level which would have been slung beneath the trunnions, but both the level and the second suspension unit are now missing. An altitude arc of 1 foot radius is attached to one of the mounts. It has two scales: 0° to 90° to 0° in 20 minute divisions, and 50° to 0° to 130° in 5° divisions. The 0° on the second scale corresponds to about 51° 20′ on the first

scale which is approximately the latitude of Richmond Observatory. This enables altitude and declination to be measured as well as North Polar distance. The tangent arm has a vernier which gives an accuracy of 2 minutes of arc.

C 149

See also E36, E44

Inventory No. 1927-1692

ELECTROSTATICS

E189 Electrical machine

Adams, George

c. 1762   London

'G. Adams Mathematical/Instrument Maker to His/MAJESTY/Fleet Street, LONDON/.'

L 610mm (overall), **W** 335mm (overall), **H** 460mm (overall), **L** 590mm (machine), **W** 260mm (machine), **L** 245mm (cylinder), **D** 180mm (cylinder)

This is the only signed electrical machine in the collection. It is almost certainly the instrument Adams referred to in the *Mechanics* manuscript when he stated: 'The electrical attraction we shall here pass over although it eminently obtains at a considerable distance, and consider it when we come to explain the Electrical Machine'. Unfortunately the electrical machine was not explained, although Hauksbee's mercurial phosphor experiment was described and illustrated in the *Pneumatics* manuscript, where a small globe was rotated in an evacuated bell jar. The reference in the *Mechanics* manuscript suggests a date of around 1762.

The machine consists of a glass cylinder about 7 inches in diameter and about $9\frac{1}{2}$ inches long. Its inner surface was coated with sealing wax, as was the axis running through the centre, in order to minimize conduction. It is held by brass collars on a steel axis with a screw-threaded pivot turned by a brass butterfly nut. The axis is supported by two heavy brass pillars, one of which contains the gearing mechanism which is operated by a winch. The brass base is inscribed as above. The brass cushion backplate extends nearly the length of the cylinder and is held against it by a leaf spring which can be adjusted by means of a screw. The cushion was

covered in red leather but little of the stuffing remains. There is a 'Nooth flap' or piece of blue taffeta which may be a later addition. The machine is clamped to a stand which is not original.

If the dating is correct, then this is a fairly early example of a cylinder machine.

Adams 1761*a*, p. 27, Pl. 14, Fig. 39
Adams 1762, p. 5
C 274
Hackman 1978*b*, p. 119, Pl. 11, p. 129

Inventory No. 1927-1143

E190 Globe electrical machine

Adams, George

c. 1760   London

L 343mm, **W** 335mm, **H** 410mm, **D** 145mm (globe)

In the past this instrument has been described as a Hauksbee type electrical machine due to the gearing mechanism. However, it does not bear much resemblance to Hauksbee's electrical machine and no evidence to link him with this particular design is known to exist. Rather, it is similar to Nairne's globe machine developed later in the century. It has been attributed to Adams because the butterfly nut which clamps the machine to its stand has the same distinctive shape as those on the cylinder electric machine E189, which is signed. The design and material are consistent with Adams's instruments. A very similar instrument in the People's Palace at Glasgow is signed by him.

It consists of a glass globe which can be rotated on a vertical axis by means of gearing enclosed in a brass box. The gearing can be made visible by removing a sliding piece in the top of the box. A piece of sheepskin is held against the globe by a brass cup on a steel spring support. The device can be clamped onto a stand or the edge of a table. The stand with the instrument is not contemporary and is stamped KCL for King's College, London.

Demainbray was using a very similar machine in 1753, a sketch of which has survived in Chabrol's lecture notes. However, the instrument does not correspond in detail to the sketch, which probably depicts the 'electrical machine with apparatus' listed in the Queen's Catalogue. Another machine not to have survived was the 'Hawksbee's large electrical machine', also listed in the Queen's Catalogue.

Q.C., Item 6; Hawksbee's large electrical machine
Q.C., Item 78; Electrical machine with apparatus
Hauksbee 1709, Pl. 7, Fig. 6
Hauksbee 1714, Pl. 6 on pneumatics
Chabrol 1753, p. 3
C 273
Chew 1968, Item 12
Daumas 1972, Pl. 124
Hackman 1978a
Turner 1987, p. 153, Pl. 140

Inventory No. 1927-1186

E191 Globe electrical machine

Adams, George

c. 1770   London

'KCL 43' : '8145' : 'KCL'
LABEL—BASE
   'No 43 glass globe/Frictional Electricity'
L 515mm, **W** 280mm, **H** 310mm, **D** 220mm (globe)

This machine is very similar stylistically to the previous item and so has been attributed to Adams. The fact that it is copied from 's Gravesande's *Mathematical Elements of Natural Philosophy* adds weight to the attribution since Adams copied many instruments from this work. It is also possible that it was made for the future George IV by George Adams the Younger, as he is known to have made some electrical apparatus for the Prince of Wales in the 1780s.

The globe can be made to rotate on a horizontal axis by means of a pulley and wheel, what 's Gravesande described as a 'machine whereby glass globes are swiftly whirled about', but this mechanism has not survived. In the centre of the globe is held a small circular piece of wood to which are attached several threads. When the tube was rubbed, which was done by hand, the threads would spread out like radii towards the surface of the globe and could be affected by actions outside the globe such as moving a finger in the vicinity. As for the other globe machine, the stand is probably not contemporary and is marked KCL for King's College, London.

's Gravesande 1747, Vol. 2, p. 75, pp. 72–6, Pl. 79, Fig. 4
Prince of Wales 1783–6, 'an electrical machine'
C 279
Chew 1968, Item 12

Inventory No. 1927-1276

E192 Cylindrical electrical machine

last quarter eighteenth century or first quarter nineteenth century

INSCRIPTION—PAINTED—BASE
  '*KCL 45*'
HANDWRITTEN
  '*No 45 Globe with two rubbers and pulley*'

D 130mm (cylinder), **H** 410mm, **L** 330mm, **D** 90mm
(multiplying wheel)

This machine is difficult to date, partly because it has been
removed from its original mounting and fixed to a pine base
while at King's College. It appears to have always been
vertical, which is slightly unusual. The cylinder is mounted in

a mahogany frame on a steel axle, which also carries a
boxwood and brass multiplying wheel. The axle passes
through the centre of a circular brass plate which supports
the two insulated stands for the cushions. Apart from a
bundle of horsehair, the brass cushion supports are all that
remain of the cushion. The stands could be moved with
respect to the cylinder by means of sliders in the brass plate.
There is no collector or prime conductor.

Inventory No. 1927-1410

E193 Plate electrical machine

c. 1770

D 460mm (plate), **L** 650mm, **W** 350mm, **H** 600mm

This is a relatively early plate electrical machine with a Lane micrometer electrometer. The plate is 18 inches in diameter with four pairs of rubbers situated round the circumference. The brass cylindrical primary conductor has twelve collecting spikes in four sets of three. The Lane micrometer is attached to the mahogany base and can be raised or lowered as required. The plate is held in a cross-shaped frame and can be turned using the mahogany handle.

Lane invented this type of electrometer in 1767. It is essentially a micrometer screw which is used to adjust the distance a spark must jump. The conductor has reached a certain potential when the spark jumps a certain distance.

Lane 1767
Priestley 1767, p. 521
Prince of Wales 1783–6, 'an electrical machine'
C 277
Hackman 1978, Pl. 22
Inventory No. 1927-1245

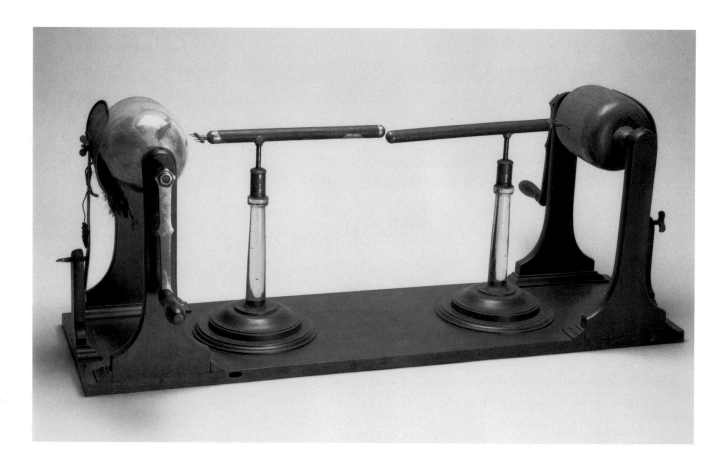

E194 Double-cylinder electrical machine

last quarter eighteenth century or first quarter nineteenth century

L 1010mm, **W** 480mm, **H** 390mm, **D** 130mm (cylinder), **L** 200mm (cylinders), **L** 325mm (conductors), **D** 25mm (conductors)

The machine stands on a rectangular mahogany base. Two glass cylinders of identical size are mounted horizontally, one at each end. They can be rotated using brass and mahogany crank handles; the axles run through their centres. One of the cylinders is painted red and both appear to have been coated internally with a resinous material. The cushions were supported by brass plates on brass mounts which were sprung against the cylinders and which could be adjusted using a thumb screw. One cushion is now missing and the other has disintegrated. Positioned towards the centre of the base are two tinfoil cylindrical prime conductors. Each carries a four-pronged comb of brass at one end and is supported on a glass insulating stand. By earthing either the cushions or the conductors positive or negative electricity could be produced.

The machine is based on Cavallo's design of 1777 which remained the standard pattern for this type of generator for over a century.

Cavallo 1777, pp. 132–41, Pl. 1, Fig. 1
C 276
Hackman 1978*b*, p. 131

Inventory No. 1927-1904

E195 Electrical chimes

Adams, George, the Younger?

last quarter eighteenth century
London

H 432mm, **D** 234mm

This apparatus was possibly made by Adams to be used with his electrical machine E189; however, it is more likely to be the 'musical bells' listed in the items made by Adams the Younger for the Prince of Wales during the 1780s. It is typical of his style and the period.

Eight bronze bells are arranged around the circumference of a circular mahogany platform which is raised on a mahogany and boxwood stand. In the centre of the platform is an insulating glass pillar on top of which is a brass 'fly' containing eight brass spokes with their ends bent at right angles. When the centre of the fly is charged it rotates and a small brass clapper attached to one of the spokes strikes the bells in turn sounding the notes of an octave. The bells have been plated with nickel silver, probably while at King's College.

The invention of the fly has been attributed to a certain Hamilton, probably Hugh, professor at Dublin; the device is sometimes known as Hamilton's mill. Apparently it was introduced around 1750 but no direct evidence has been found for this. Chew found that it was originally described by Andrew Gordon and re-invented by Hamilton.

Priestley 1767, p. 429
Prince of Wales 1783–6
Adams 1787*a*, p. 106, Fig. 34
Adams 1799*a*, pp. 162–3
C 282
Chew 1968, No. 13

See E189

Inventory No. 1927-1142

E196 Electrical chimes

last quarter eighteenth century
D 340mm, **H** 235mm

These chimes operate on the same principle as the previous example. In this case twelve bronze bells are situated round the circumference of a mahogany platform giving a range of one and a half octaves. There are four brass prongs on the fly which is on a central glass pillar. The clapper is now missing.

Prince of Wales 1783–6
Adams 1787*a*, pp. 76–7, Fig. 19
C 283
See also E195

Inventory No. 1927-1177

E197 King's electrical orrery

Adams, George?

c. 1765   London
L 390mm (max), **H** 230mm, **D** 140mm (base), **L** 365mm (overall)

This is likely to be by Adams since it is featured in his *Essays on Electricity* and is similar in style to some other electrical apparatus made around this date. Adams refers to it as 'King's electrical orrery'.

It is a tellurian with brass balls of varying sizes representing the Sun, Earth, and Moon. The Earth and Moon are balanced at either end of a brass rod on a spike, which in turn is balanced on the end of a longer brass rod which supports the Sun at the other end.

When the instrument is charged, two points, one on each rod, cause rotation and the movement of the solar system is imitated. Like the electrical chimes E195 and E196 it is based on the principle of Hamilton's mill.

The King to whom Adams refers is almost certainly Erasmus King who

was a lecturer on scientific subjects during the mid-eighteenth century. King was reputed to have had a very good collection of instruments; a globe with his name on is known to exist.

Adams 1799*a*, pp. 580–1, Pl. 4, Fig. 79
C 284
Appleby 1990

Inventory No. 1927-1182

E198 Battery of Leyden jars

last quarter eighteenth century or first quarter nineteenth century

L 1000mm, **W** 350mm, **H** 700mm, **H** 530mm (jars), **D** 280mm (jars)

It is quite likely that this item came into the collection in the 1780s when the Prince of Wales bought several pieces of electrical apparatus from George Adams the Younger. Although a battery of three jars is not mentioned specifically, money was paid for the varnishing of a large battery of jars.

There are three jars in this set which may be slightly earlier than L99. The jars are made of varnished glass coated with tinfoil. They are interconnected by a brass rod and have thick brass cages inside.

Prince of Wales 1783–6
C 288

Inventory No. 1927-1703

E199 Leyden jar

second half eighteenth century

H 180mm, **D** 165mm

This appears to be earlier than the other Leyden jars in the collection: the green glass is less perfect and the coating is lead foil. The jar is filled with iron filings.
C 286

Inventory No. 1927-1273

E200

E201

E200 Leyden jar

last quarter of eighteenth century or first quarter
nineteenth century

H 240mm, **D** 60mm

The jar is coated with black-painted tinfoil. There is a large
brass hook at the top for suspending from the prime
conductor of an electrical machine.

C 286

Inventory No. 1927-1268

E201 Leyden jar

last quarter of eighteenth century or first quarter
nineteenth century

H 220mm, **D** 60mm

This jar is similar to the one above. The two differences are
the shape of the brass hook and the fact that this jar has been
filled with iron filings.

C 286

Inventory No. 1927-1269

E202 'Thunder house'

Adams, George

last quarter eighteenth century

London

H 330mm, **L** 190mm, **W** 115mm

Again this item could have come into the collection when the Prince of Wales bought some electrical apparatus in the 1780s, but the 'powder house' listed there came with a swan and is probably not this instrument.

This is a simpler type of thunder house than is found in some other collections. The instrument stands on a mahogany rectangular base with one vertical side cut in the shape of a house with sloping roof and chimney. On the other side of the base is a vertical glass pillar which supports a brass rod in the form of an arch which terminates in a brass ball. Another brass ball is at the top of the arch; this is the point where the instrument is charged. The house side has a lightning conductor running down its length which protrudes above the chimney in a brass rod, again terminating in a brass ball. When the thunder house is charged a spark jumps between the balls above the chimney and is earthed. A small mahogany cube can be removed from the side of the house in the path of the lightning conductor. If it is replaced with the orientation of the strip of the conductor changed, it will jump out of place when the thunder house is charged. This suggests that a house may be damaged

According to Ferguson this apparatus was invented by Dr Lind of Edinburgh in order to test Franklin's theories of lightning conductors.

Prince of Wales 1783–6, 'powder house'
Cavallo 1786–95, Vol. I, pp. 302–7, Pl. 2, Fig. 1
Adams 1794, p. 383, Pl. 2, Fig. 3
Turner 1973, Nos 327, 328, 329

Inventory No. 1927-1445

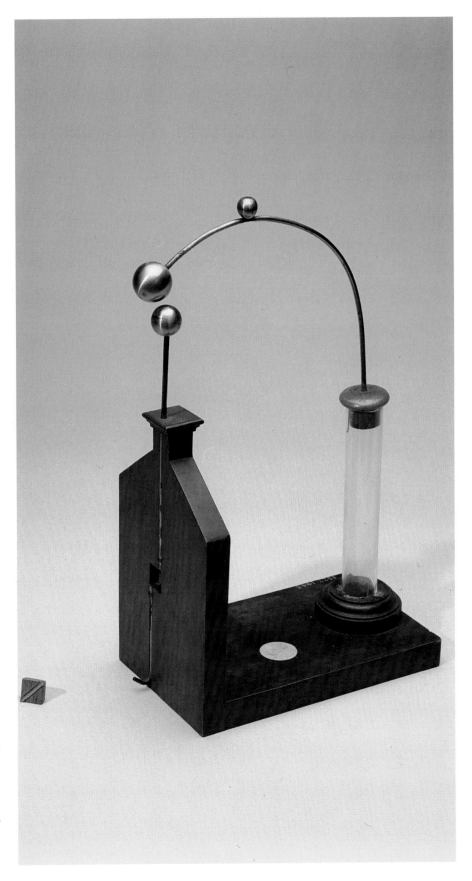

E203 Fulminating pane

last quarter eighteenth century

D 230mm, **D** 220mm (pane), **H** 110mm (stand)

This is a circular fulminating pane in a mahogany frame. It had been badly damaged but has recently been reconstructed. The pattern is star-shaped. A gilt brass ball at the top of the frame could have been charged. The pane was held vertical on a mahogany base with an iron wire set inside to carry the charge.

C 298

Inventory No. 1927-1520

E204 Fulminating tubes

Adams, George, the Younger

last quarter eighteenth century   London

H 415mm, **D** 180mm, **H** 295mm (tubes)

This could be the 'set of spirals' listed in the apparatus that the Prince of Wales bought from George Adams the Younger during the 1780s. Although it is not possible to identify the instrument precisely from the list it is very likely that the item came into the collection by this means at about this time.

The instrument has been attributed to Adams because it is similar to one illustrated in *Lectures on Natural and Experimental Philosophy* and it is stylistically comparable with other instruments of the period by him. Five fulminating rods with metal 'spangles' in a spiral pattern are arranged round the circumference of a circular mahogany platform on a mahogany and boxwood stand. The rods have brass knobs at the top and brass bases through which they are earthed by copper wire. A central glass rod holds a pair of brass balls on a brass wire which can rotate so that the balls and the spherical knobs nearly touch. When the top of the central rod is charged, a spark will jump to one of the fulminating rods which will be passed down through the tinfoil 'spangles'. If the two balls are rotated and kept charged, each fulminating tube will light up when one of the balls passes by. According to Adams this was a 'most beautiful species of illumination'.

Prince of Wales 1783–6

Adams 1794, Vol. 4, p. 333, Pl. 2 (Vol. 4), Fig. 4

C 301   **Inventory No.** 1927-1193

E205 Spiral fulminating tube

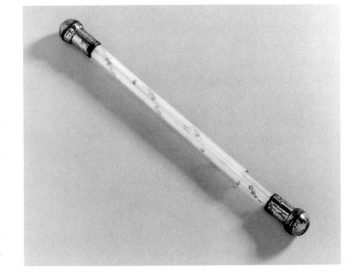

last quarter eighteenth century

D 15mm (outer tube), **D** 11mm (inner tube), **L** 340mm, **D** 24mm (brass end cap)

The fulminating tube has an inner and outer section: the inner has a spiral pattern of tinfoil circles attached; the outer is plain. The ends consist of the standard brass knob of the period. Apparently the tube was 13½ inches long before it was broken in 1960.

Prince of Wales 1783–6, 'a set of spirals'
Sturgeon 1842, pp. 84–5
C 295

Inventory No. 1927-1178/1

E206 Spiral fulminating tube

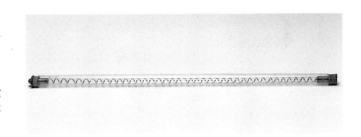

last quarter eighteenth century or first half nineteenth century

L 895mm, **D** 35mm (jacket), **D** 20mm (tube)

The fulminating tube has a continuous tinfoil spiral running around it and brass end caps. It fits loosely into a glass jacket which is sealed with cork and wax at both ends.

Prince of Wales 1783–6
Adams 1787a, p. 91, Fig. 31
C 296

Inventory No. 1927-1262

E207 Aurora tube

last quarter eighteenth century or first quarter nineteenth century

L 610mm, **H** 340mm, **D** 45mm (tube), **D** 165mm (bases)

Like a number of other electrical items in the collection this could have been acquired by the Prince of Wales from George Adams the Younger during the 1780s.

It is a horizontal aurora tube supported on two glass insulating stands. The stands have painted iron cradles at the top and mahogany bases; one stand is varnished. The aurora tube has spherical brass knobs at its ends, each having a small copper rod with a brass ball screwed into it. Inside the tube one of the copper rods terminates in a brass ball and the other in a steel point. The tube is partially evacuated.

When one of the ends is charged, the inside of the tube becomes luminous. The effect was first reported by Watson in 1752 and became a popular demonstration piece soon afterwards.

Watson 1752
Prince of Wales 1783–6, 'luminous conductor and two auroras'
Adams 1787a, p. 119, Fig. 49
C 294

See also L107   **Inventory No.** 1927-1176

E208 Aurora flask

last quarter eighteenth century or first quarter nineteenth century

L 330mm, **D** 90mm (max)

The flask is partially exhausted of air and has a brass conductor at the neck. A brass spike protrudes about 2 inches into the interior. When the conductor was charged a luminous discharge appeared from the point and filled the flask. The name 'aurora' was given because the flashing resembled the aurora borealis. Adams also described a 'Leyden flask' which was similar but the glass surfaces were coated with tinfoil as in a Leyden jar.

Adams 1787*a*, p. 119, Figs 59, 60, 49
Adams 1799*a*, pp. 297–8
Cuthbertson 1807, pp. 94–96
C 292

Inventory No. 1927-1439

E209 Conductor

second half eighteenth century

L 1290mm, **D** 220mm

This conductor could be relatively early since it was designed to be suspended rather than supported. It could be the large tin conductor listed in the apparatus bought by the Prince of Wales in 1785 from George Adams the Younger, although the use of beech suggests an association with Demainbray. It consists of a large beech cylinder with rounded ends which was coated in tinfoil. There are three small brass rings, which are now badly corroded, along one side, and two larger ones on the other which could be later.

Prince of Wales 1786, 'a large tin conductor'
Hackman 1978, p. 117

Inventory No. 1927-1617

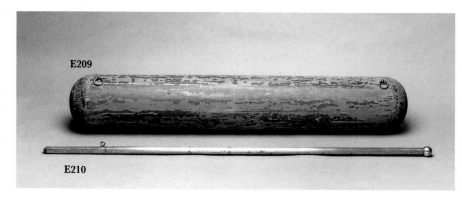

E210 Conductor

last quarter eighteenth century or first quarter nineteenth century

L 1430mm, **D** 22mm, **D** 42mm (end cap)

This is a long steel conductor with one remaining spherical brass end cap and one suspension ring. A second suspension ring has been broken off.

Inventory No. 1927-1679

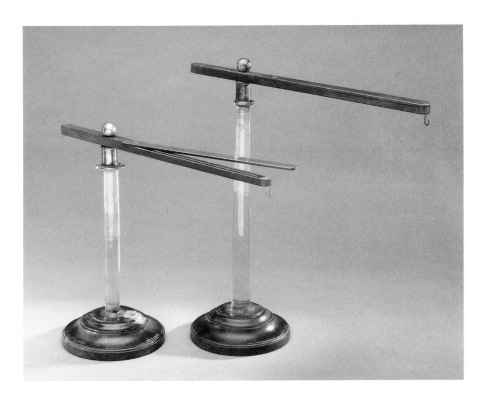

E211 Canton electrometer

last quarter eighteenth century or first quarter nineteenth century

H 330mm, **L** 410mm, **D** 170mm (base), **W** 20mm (box), **L** 260mm (box), **D** 15mm (box)

These instruments, E211 and E212, have been identified as Canton electrometers. They consist of a long rectangular piece of mahogany fixed horizontally by a brass screw near one end to a glass stand on a mahogany base. The long arm of the piece of mahogany contains two channels. At the end of the long arm is a brass hook. The channels have a lid which swivels on a screw. Across the top of the channels at the end nearest the support are two brass wires.

This was the first portable electroscope. It was used with pith balls which would have been suspended from the brass wires or alternatively from the hook. When not in use the balls were stored in the mahogany channels. It was thought to indicate the quantity of electricity in the air near the rubbed electric.

Canton 1753–4
Hackman 1978, p. 17, Pl. 6
Inventory No. 1927-1441

E212 Canton electrometer

last quarter eighteenth century or first quarter nineteenth century

LABEL—PRINTED—BASE
'8152'

H 415mm, **L** 390mm, **L** 260mm (box), **D** 150mm (base), **W** 20mm (box), **DE** 15mm (box)

This instrument is almost identical to the above.

As E211

Inventory No. 1927-1442

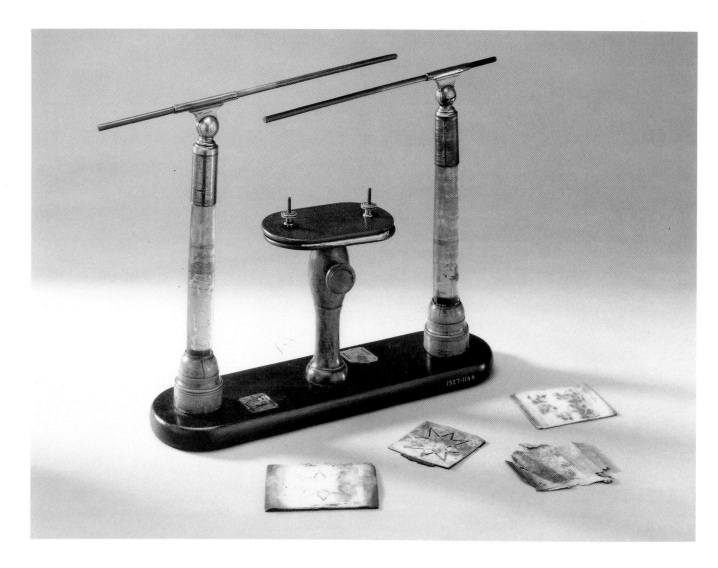

E213 Henly's universal discharger

1785–1820

L 350mm, **W** 100mm, **H** 310mm, 305mm, 355mm (pillars)

This instrument is very likely to have entered the collection in 1785 when the Prince of Wales bought two universal dischargers from George Adams the Younger.

This example is fitted with the 'press' illustrated by Cavallo and others. It consists of two glass pillars on boxwood bases placed at each end of an oval mahogany board. On the top of each pillar is a brass rod which can be turned vertically or horizontally on a brass ball and socket joint. In the centre of the instrument is a boxwood stand which can be raised or lowered. On the stand is an oval mahogany platform with another similar platform held about half an inch above on

two brass screws, which constitutes the 'press'. With this arrangement the instrument was used to pierce holes in pieces of card: several cards with burnt patterns remain. Henly reported that a Mr Lullin of Geneva had performed this experiment prior to 1778. In general the instrument was used to pass shocks through wood or glass and to melt wires.

Henly 1778
Cavallo 1782, pp. 169–70
Prince of Wales 1783–6, 'universal discharger'
Adams 1787*a*, p. 47, Pl. 2, Fig. 3
C 304
Turner 1973, No. 306

Inventory No. 1927-1144

E215 Model bucket

last quarter eighteenth century or first quarter nineteenth century

H 60mm, **D** 50mm

The bucket is similar in style to Adams's work, but is not referred to in either of his manuscripts. It has a small chain and a brass wire which protrudes through a hole in the base. It could have been used for electrical experiments since it resembles the model buckets on chains used with prime conductors described by Noad and others. The bucket was filled with water which dripped through the hole in the base. However, when the bucket was hung on a charged conductor the water came out 'like a stream of fire'.

Noad 1859, p. 85

Inventory No. 1929-130

E214 Box with adjustable glass columns for electrical experiments

last quarter eighteenth century

H 240mm (one pillar), **H** 200mm (one pillar), **L** 1260mm, **W** 120mm

This is a long mahogany box containing four bundles of shredded paper attached to cardboard cones and two bundles of hair for electrical experiments. There are two glass pillars, one at each end. Although they resemble legs, they do not support the box and must have been used to carry apparatus for various experiments. One of the pillars is slightly larger than the other and can be adjusted with respect to the other by means of a mahogany screw. It is possible that the cardboard cones were placed on the glass pillars and a charged conductor was made to act on the paper or hair.

Inventory No. 1927-1448

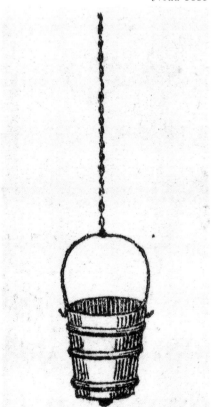

Noad 1859

E216 Pair of insulated stools

last quarter eighteenth century or first half nineteenth century

W 300mmm **L** 395mm, **H** 195mm.

The stools have varnished glass legs. There is evidence of considerable use.

Prince of Wales 1783–6, 'four feet for a stool'
C 280

Inventory No. 1927-1145

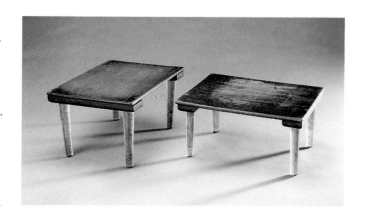

WEAPONS

E217 Pair of revolving double-barrelled flintlock pistols

Deveux

c. 1685   Liège

ON LOCK PLATES
'I. DEVEUX'

L 515mm, **L** 326mm (barrel), **D** 75mm

The pistols probably came into the possession of the royal family in the late seventeenth century. They are made of silver with mounts of chiselled engraved steel. The barrels can be revolved by operating levers on the trigger guards, which suggests the pistols were used for demonstration purposes. Deveux worked in Liège in the last quarter of the seventeenth century.

C 343

Inventory Nos 1927-1299, -1300

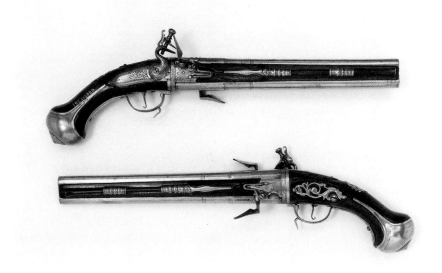

E218 Model bow and arrow

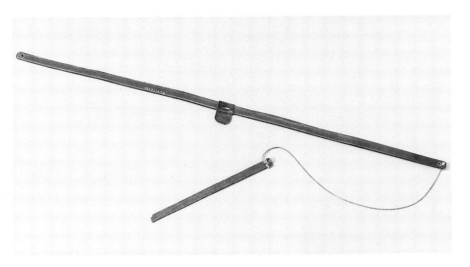

second half eighteenth century or first quarter nineteenth century

L 790mm (bow), **W** 18mm (bow), **D** 5mm (bow)

This may be a device to test the elasticity of the material in the bow which is horn. It has a brass slot in the centre through which the oak arrow would have passed. The arrow is now broken.

C 336

Inventory No. 1927-1474

MISCELLANEOUS

E219 Lamp

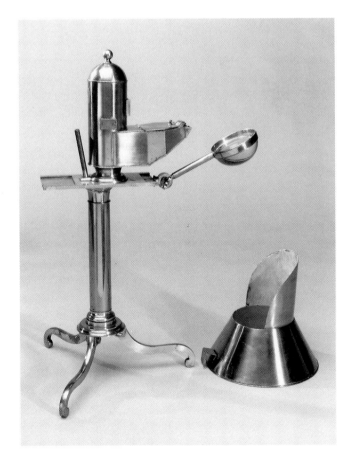

second half eighteenth century

L 290mm, **W** 235mm, **H** 490mm

The lamp is designed for the use of vegetable oil as fuel and has three wicks which are aligned horizontally. The oil is stored in a reservoir in the brass dome which contains a drip feed, allowing the wicks to be supplied adequately. The 3 candle-power light could be focused by the hemispherical lens which is adjustable on a hinge attached below the stand. A brass hood protects the user's eyes. The lamp, reservoir, and hood can be moved horizontally on a brass slider which moves in a frame set on a heavy brass pillar on a tripod base. A pair of tweezers is provided to pull up the wicks.

Since this is not an Argand lamp, it is unlikely to be later than 1800. On arrival at the Museum it was described as a lamp for microscopical work but there is no evidence for this.

C 191
O'Dea 1958

Inventory No. 1927-1157

E220 Lamp housing

last quarter eighteenth century or first quarter nineteenth century

H 132mm, **L** 105mm (bracket), **D** 73mm

The spherical brass lamp would have been used with a candle or similar source of illumination. It has a convex lens and a concave reflector. The lamp is surmounted with a rotating cylinder, on which were mounted three brackets. Only one of these brackets remains. The top of the lamp has a ring of ventilation holes. It may be the 'lanthorn' listed in the Queen's Catalogue.

Q. C., Item 129; smoak lanthorn and apparatus

Inventory No. 1927-1504

E221 Lamp housing

second half eighteenth century

H 135mm, **L** 75mm, **W** 62mm

The object consists of a brass spherical chamber with removable plane glass at the front, a chimney above, and a tube below. It appears to have been used with a candle which would have fitted inside the tube. The polished brass surface behind the flame would have enhanced the light a little. There are four holes near the base of the chamber to let in air. The purpose of the lamp is not obvious: it is possible that it was used in conjunction with microscopical work, but the method of support is not apparent. It could be part of the 'smoak lanthorn and apparatus' listed in the Queen's Catalogue.

Q.C., Item 129; smoak lanthorn and apparatus
C 192

Inventory No. 1927-1503

E222 Paper knife

last quarter eighteenth century or first quarter nineteenth century

L 520mm, **DE** 50mm, **W** 25mm

The knife is made of stained mahogany.

Inventory No. 1927-1471

E223 Pounce sprinkler

second half eighteenth century

L 150mm, **D** 70mm (max), **D** 45mm (min)

The pounce sprinkler is in the form of a concertina made of green leather with boxwood ends and caps. There is a linen sieve near the narrow end. Presumably the pounce was put into the wide end of the sprinkler and shaken out through the sieve in the narrow end.

Pounce was a powder which was used to dry ink on unsized paper. A similar device could also have been used to sprinkle coal dust over perforated paper to transfer patterns. However, there is no indication of black powder having been inside the object and so it must have been used in the former manner.

C 347

Inventory No. 1927-1173

E224 Stoppered glass jar containing 'smalt'

second half eighteenth century

ON LABEL
'Smalt'

H 170mm, **D** 65mm

The glass jar contains a fine powder of a deep blue colour. It has a label bearing the word 'Smalt' written in ink. Smalt is an artist's pigment which was used extensively in the seventeenth and eighteenth centuries before it was superseded by cobalt blue. It is a potassium cobalt silicate, sometimes with aluminium and other impurities such as arsenic. The smalt industry was described by Kunckel in 1679 and by Baron d'Holbach in 1752. In the mid-eighteenth century there was a debate as to the source of the blue pigment.

Kunckel 1679
d'Holbach 1752
Partington 1962, pp. 168–71
Harley 1970, pp. 51–3

Inventory No. 1927-1532

THE BOYLE COLLECTION

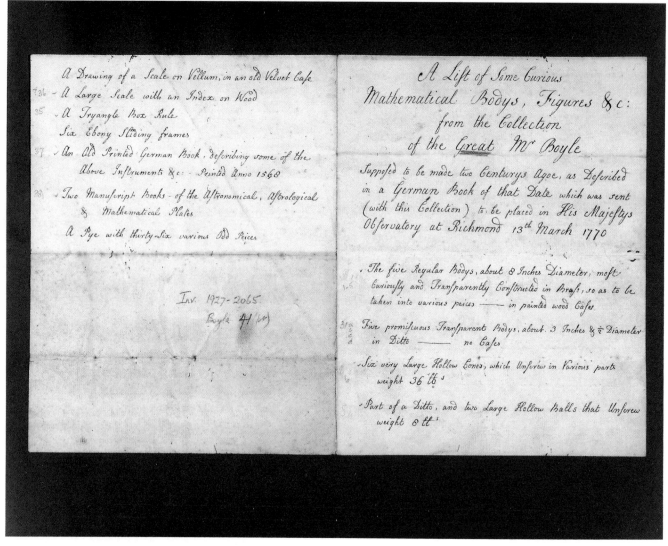

Richmond 1770. A list of the items then in the Boyle Collection (pages 1 and 4).

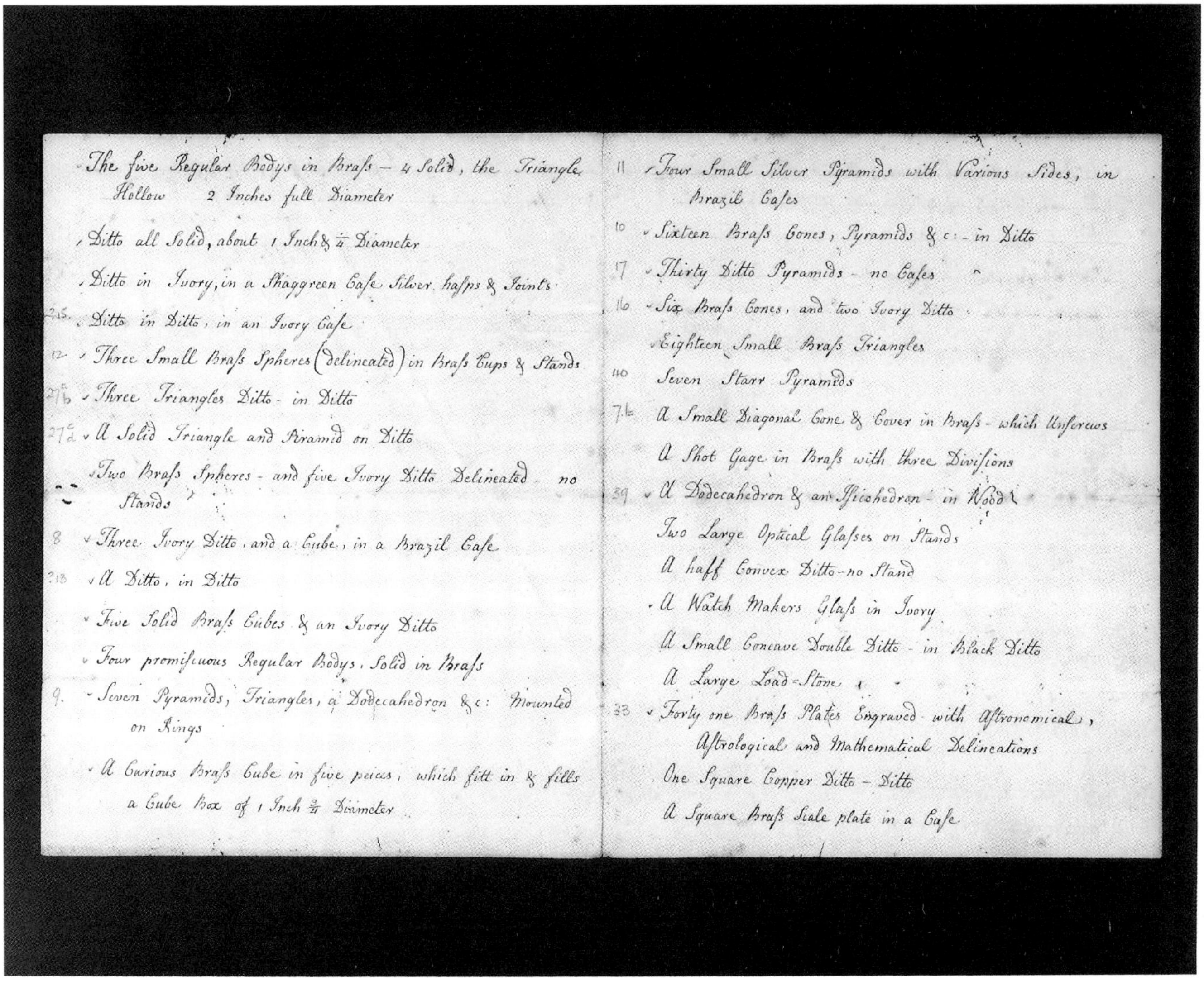

The five Regular Bodys in Brass — 4 Solid, the Triangle Hollow   2 Inches full Diameter

Ditto all Solid, about 1 Inch & ¾ Diameter

Ditto in Ivory, in a Shaggreen Case Silver hasps & Joints

Ditto in Ditto, in an Ivory Case

Three Small Brass Spheres (delineated) in Brass Cups & Stands

Three Triangles Ditto - in Ditto

A Solid Triangle and Piramid on Ditto

Two Brass Spheres - and five Ivory Ditto Delineated - no Stands

Three Ivory Ditto, and a Cube, in a Brazil Case

A Ditto, in Ditto

Five Solid Brass Cubes & an Ivory Ditto

Four promiscuous Regular Bodys, Solid in Brass

Seven Pyramids, Triangles, a Dodecahedron &c: Mounted on Rings

A Curious Brass Cube in five pieces, which fitt in & fills a Cube box of 1 Inch ¾ Diameter

Four Small Silver Pyramids with Various Sides, in Brazil Cases

Sixteen Brass Cones, Pyramids &c: - in Ditto

Thirty Ditto Pyramids - no Cases

Six Brass Cones, and two Ivory Ditto

Eighteen Small Brass Triangles

Seven Starr Pyramids

A Small Diagonal Cone & Cover in Brass - which Unscrews

A Shot Gage in Brass with three Divisions

A Dodecahedron & an Icosahedron - in Wood

Two Large Optical Glasses on Stands

A half Convex Ditto - no Stand

A Watch Makers Glass in Ivory

A Small Concave Double Ditto - in Black Ditto

A Large Load-Stone

Forty one Brass Plates Engraved - with Astronomical, Astrological and Mathematical Delineations

One Square Copper Ditto - Ditto

A Square Brass Scale plate in a Case

Richmond 1770. A list of the items then in the Boyle Collection (pages 2 and 3).

E225 The five regular polyhedra in skeletal form with inscribed stars, and their boxes

last quarter seventeenth century

L 180mm (side tetrahedron), **L** 100mm (side cube), **L** 150mm (side octahedron), **L** 80mm (side dodecahedron), **L** 120mm (side icosahedron)

These five brass polyhedra correspond to the first item on the list of the Boyle Collection. They have been extensively restored. All the oak cases have survived. They probably date from Boyle's lifetime and do not resemble figures from Jamnitzer very precisely.

Jamnitzer 1568
Richmond 1770, No. 1
Inventory No. 1927-2065/1

E226 Four skeletal figures: cube, octahedron, icosahedron and hexagonal prism

last quarter seventeenth century

L 60mm (side cube), **L** 80mm (side octahedron), **L** 50mm (side icosahedron), **H** 40mm (side hexagonal prism), **H** 70mm (side hexagonal prism)

These brass figures correspond to the second item on the list of the Boyle Collection; one of the bodies is missing. They were probably made from clock frames in the late seventeenth century.

Jamnitzer 1568
Richmond 1770, No. 2
Inventory No. 1927-2065/2

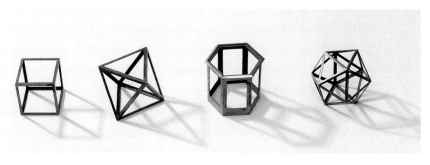

E227 Six hollow cones and one truncated cone

last quarter seventeenth century or first quarter eighteenth century

H 190mm (max), **D** 200mm (base (smallest)), **D** 290mm (base (largest))

These heavy brass cones correspond to the third item on the list of the Boyle Collection. Three have threaded mandrels at the apex. The cones fit together in sections, possibly indicating a use as nested weights. The three with mandrels have unnumbered scales around the inside of their rims. There are 360 divisions, probably referring to degrees. Again they probably date from the late seventeenth century.

Jamnitzer 1568
Richmond 1770, No. 3

Inventory No. 1927-2065/3

E228 One hollow ball and one conic section

last quarter seventeenth century or first quarter eighteenth century

D 55mm (ball), **H** 40mm (cone), **D** 90mm (base, cone, max)

One of the large hollow balls that unscrew remains, together with part of a skewed cone that together may correspond to the fourth item on the list of the Boyle Collection. Two other brass balls on stands have been listed with E233.

Jamnitzer 1568
Richmond 1770, No. 4

Inventory No. 1927-2065/4

E229 Solid cube, dodecahedron, and icosahedron

last quarter seventeenth century or first quarter eighteenth century

D 50mm

Only three of the five brass regular bodies of 2 inches full diameter which are the fifth item on the list of the Boyle Collection appear to have survived.

Jamnitzer 1568
Richmond 1770, No. 5

Inventory No. 1927-2065/5

E230 The five regular solid bodies

last quarter seventeenth century or first quarter eighteenth century

D 35mm

All the regular brass bodies are present from the sixth item on the list of the Boyle Collection.

Jamnitzer 1568
Richmond 1770, No. 6

Inventory No. 1927-2065/6

E231 Set of five regular solids in box

last quarter seventeenth century or first half eighteenth century

D 25mm (bodies), **L** 130mm (box), **W** 30mm (box), **H** 30mm (box)

The five polyhedra are made of ivory and are in a rectangular shagreen-covered box with silver clasps.

Jamnitzer 1568
Richmond 1770, No. 7

Inventory No. 1927-2065/7

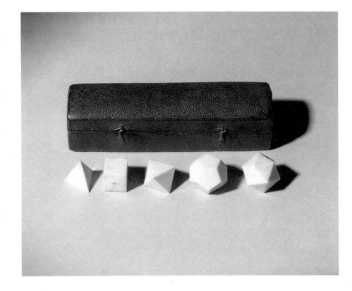

E232 Four polyhedra in case

last quarter seventeenth century or first quarter eighteenth century

D 20mm (bodies), **L** 85mm (box), **D** 30mm (box)

This item does not exactly correspond to the eighth item on the list of the Boyle Collection since the bodies are not the five regular solids; however, it is very similar. The ivory bodies are an icosahedron, an octahedron, a cuboctahedron, and another polyhedron. The case is also of ivory in the form of a cylinder with a screw cap.

Jamnitzer 1568
Richmond 1770, No. 8

Inventory No. 1927-2065/8

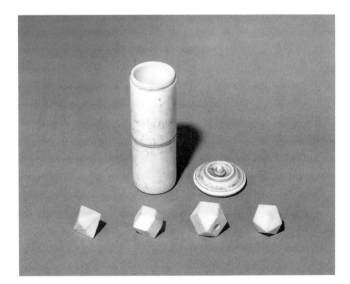

E233 Three spheres in cups and stands

last quarter seventeenth century or first quarter eighteenth century

H 55mm (small sphere), **D** 25mm (small sphere), **H** 100mm (large sphere), **D** 45mm (large sphere)

Again, the objects present do not exactly correspond to the list of the Boyle Collection. There are three brass spheres on pedestal stands, one smaller than the other two. The small one and one of the larger ones are delineated. The other larger one unscrews and a smaller brass ball is contained inside. The style of the stands indicates a late seventeenth century date.

Jamnitzer 1568
Richmond 1770, No. 9

Inventory No. 1927-2065/9

E234 Two hollow inverted tetrahedrons on pedestals, and one without

last quarter seventeenth century or first quarter eighteenth century

H 125mm, **D** 45mm (base)

These brass items may correspond to item 10 on the list of the Boyle Collection.

Jamnitzer 1568
Richmond 1770, No. 10

Inventory No. 1927-2065/10

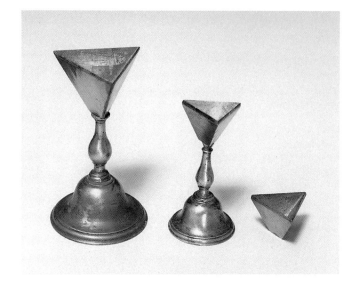

E235 Solid tetrahedron and pyramid on stands

last quarter seventeenth century or first quarter eighteenth century

H 115mm (pyramid), **D** 35mm (pyramid), **H** 70mm (tetrahedron and stand), **D** 25mm (tetrahedron)

The pyramid certainly corresponds to item 11 on the list of the Boyle Collection; it is on a pedestal stand. There is a tetrahedron on a cubic base which may correspond to the 'solid triangle'. The pedestal stand indicates a late seventeenth century date.

Jamnitzer 1568
Richmond 1770, No. 11

Inventory No. 1927-206/11

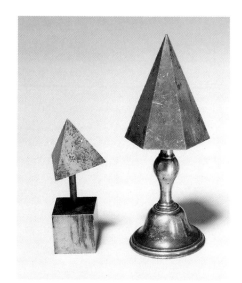

E236 Two brass and four ivory spheres

last quarter seventeenth century or first quarter eighteenth century

D 35mm (brass balls), **D** 35mm (ivory undelineated), **D** 40mm (two ivory balls), **D** 50mm (one ivory ball)

The two brass spheres and three of the delineated ivory ones are present. An undelineated ivory sphere in addition to those in item 12 on the list of the Boyle Collection has also been included.

Jamnitzer 1568
Richmond 1770, No. 12

Inventory No. 1927-2065/12

E237 Three ivory spheres and a cube in a case

last quarter seventeenth century or first quarter eighteenth century

H 137mm (case), **D** 50mm (case), **L** 18mm (side of cube), **D** 25mm (spheres)

This item is complete and consists of three ivory spheres and a cube in a cylindrical Brazilian ebony case with a screw top.

Jamnitzer 1568
Richmond 1770, No. 13

Inventory No. 1927-2065/13

E238 Ivory sphere in case

last quarter seventeenth century or first quarter eighteenth century

H 35mm (case), **D** 40mm (case), **D** 25mm (sphere)

The small ivory sphere is in a cylindrical Brazilian ebony case which unscrews into two halves. It corresponds to item 14 on the list of the Boyle Collection.

Jamnitzer 1568
Richmond 1770, No. 14

Inventory No. 1927-2065/14

E239 Four solid brass cubes

last quarter seventeenth century or first quarter eighteenth century

L 45mm (largest side), **L** 30mm (two cubes), **L** 13mm (two cubes smallest)

Only four of the brass cubes remain of item 15 on the list of the Boyle Collection, unless the cubes in E229 and E230 should more properly belong here. There is no ivory cube.

Jamnitzer 1568
Richmond 1770, No. 15

Inventory No. 1927-2065/15

E240 Seven solid regular bodies

last quarter seventeenth century or first quarter eighteenth century

D 35mm (rhombic dodecahedron), **D** 50mm (cuboctahedron), **D** 25mm (dodecahedron), **D** 50mm (polyhedron), **D** 18mm (octahedron), **D** 30mm (octahedra)

Four of these seven brass bodies probably correspond to item 16 on the list of the Boyle Collection; however, it is difficult to be more specific or to place the other three.

Jamnitzer 1568
Richmond 1770, No. 16

Inventory No. 1927-2065/16

E241 Six pyramids mounted on rings and one ring

last quarter seventeenth century or first quarter eighteenth century

H 40mm, **D** 40mm

The six brass pyramids have three, four, five, six, seven, and eight sides respectively. They correspond to item 17 on the list of the Boyle Collection. One ring base also remains.

Jamnitzer 1568
Richmond 1770, No. 17

Inventory No. 1927-2065/17

E242 Cubical box with brass piece

last quarter seventeenth century or first quarter eighteenth century

L 35mm (side of box)

The brass box is the cubic box of $1\frac{3}{4}$ inches diameter corresponding to item 18 on the list of the Boyle Collection. There is no indication that the irregular piece of brass is one of the five pieces that made up the cube, but it is possible.

Jamnitzer 1568
Richmond 1770, No. 18

Inventory No. 1927-2065/18

E243 Three pyramids on star bases

last quarter seventeenth century or first quarter eighteenth century

D 35mm, **H** 20mm (smallest), **H** 35mm (largest)

Three small silver pyramids without cases are all that remain of item 19 on the list of the Boyle Collection.

Jamnitzer 1568
Richmond 1770, No. 19

Inventory No. 1927-2065/19

E244 Seventeen brass cones and pyramids in cases and five ivory cones in cases

last quarter seventeenth century

H 90mm (max), **D** 80mm (max base)

In addition to the 16 brass pyramids and cones in ebony cases corresponding to item 20 on the list of the Boyle Collection, we have another brass cone and five ivory cones in similar cases. These objects are compatible with a date in the late seventeenth century.

Jamnitzer 1568
Richmond 1770, No. 20

Inventory No. 1927-2065/20

E245 Thirty pyramids

last quarter seventeenth century or first quarter eighteenth century

H 50mm (max), **D** 40mm (max base)

These small brass pyramids constitute item 21 on the list of the Boyle Collection.

Jamnitzer 1568
Richmond 1770, No. 21

Inventory No. 1927-2065/21

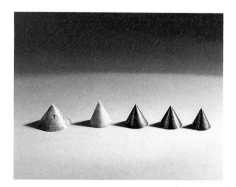

E246 Three brass and two ivory cones

last quarter seventeenth century or first quarter eighteenth century

D 35mm (max), **H** 25mm (max)

Only three of the small brass cones remain of the six corresponding to item 22 on the list of the Boyle Collection.

Jamnitzer 1568
Richmond 1770, No. 22

Inventory No. 1927-2065/22

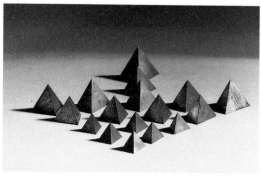

E247 Eighteen tetrahedra

last quarter seventeenth century or first quarter eighteenth century

D 25mm (largest), **D** 12mm (smallest)

These brass figures correspond to item 23 on the list of the Boyle Collection.

Jamnitzer 1568
Richmond 1770, No. 23

Inventory No. 1927-2065/23

E248 Star pyramid

last quarter seventeenth century or first quarter eighteenth century

H 80mm, **D** 70mm

This brass figure is possibly one of the 'seven starr pyramids' of item 24 on the list of the Boyle Collection.

Jamnitzer 1568
Richmond 1770, No. 24

Inventory No. 1927-2065/24

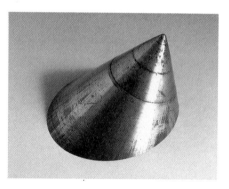

E249 'Diagonal' cone

last quarter seventeenth century or first quarter eighteenth century

H 60mm, **D** 90mm (max)

Although this brass skew cone appears to correspond to item 25 on the list of the Boyle Collection, the 'cover' is not in evidence.

Jamnitzer 1568
Richmond 1770, No. 25

Inventory No. 1927-2065/25

E250 Icosahedron

last quarter seventeenth century or first quarter eighteenth century

L 65mm (side)

Only the icosahedron remains of the original pair corresponding to item 27 on the list of the Boyle Collection. It is probably made of stained boxwood.

Jamnitzer 1568
Richmond 1770, No. 27

Inventory No. 1927-2065/26

E251 Watchmaker's glass

last quarter seventeenth century or first quarter eighteenth century

D 30mm

This small magnifying glass set in ivory is easily recognizable as the item 30 on the list of the Boyle Collection.

Richmond 1770, No. 30

Inventory No. 1927-2065/27

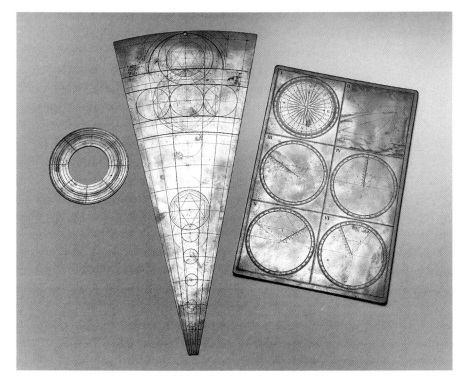

E252 Forty-six brass plates for book illustrations

last quarter seventeenth century
D 330mm (largest), **D** 40mm (smallest)

There appear to be five more plates than in item 33 on the list of the Boyle Collection. They are in many sizes and shapes: octagon, pentagon, hexagon, squares, circles, and rectangles. All are engraved with various geometrical patterns and some have compass cards marked. Several have been used to make the plates in the manuscript book E256.

Jamnitzer 1568
Richmond 1770, No. 33
Inventory No. 1927-2065/28

E253 Delamain type horizontal quadrant

Brown, John?

mid seventeenth century   London
L 555mm, **W** 345mm, **T** 12mm

This is probably the 'large scale with an index on wood' corresponding to item 37 on the list of the Boyle Collection. The horizontal quadrant is made of sycamore. On the obverse is a form of projection first published by Delamain in 1632. It folds in half Oughtred's projection, which was also published in 1632 although devised about 1600. It is half a stereographic projection of the celestial sphere onto the plane of the horizon. The ecliptic, or apparent path of the Sun, appears as two arcs. A

variety of horary and astronomical questions, such as the length of the day or the Sun's declination, can be answered without calculation. The outer limb is divided into degrees (90–0–90) and marked at 10′ intervals. The equinox is at 10 March and the declination lines are dotted at 10° intervals. The hour lines extend beyond the tropics; each hour line is dotted and there are subdividing lines at 15′ intervals. There is a small central hole for the index rule which is now missing.

The reverse of the instrument has three features: 'An Almanack for ever', a table of fixed stars, and a series of concentric circular scales. The almanac

is similar to those found on other instruments by Brown, giving an indication of the maker. It gives days, months, dominical letters, epacts, golden numbers, and leap years. The table of stars gives the right ascension, declination, and distance from the pole. A comparison with Tycho's values for 1600 and 1700 indicates a date of 1629 for the table. However, this does not date the instrument as the table was probably copied at a later date. The circular scales have a central hole for a rotating index which again is missing.

The concentric circular scales are in three parts.

(a) The outer scale is a circle of 360°

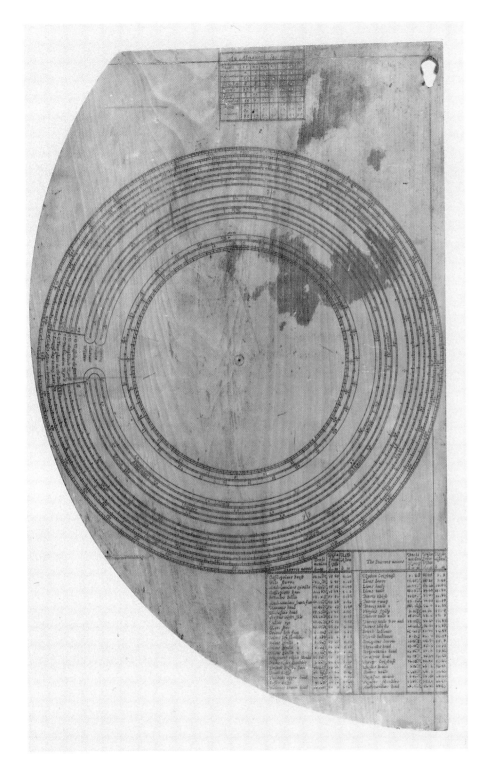

divided to half degrees.

(b) The inner scale is a double circular scale which can be identified as correlating the time of the southing of the Moon with its age. It consists of two circles of equal parts, the inner one divided 0–12, 0–12 to give the hours, and the outer from 0 to approximately 29.7 for the days. These scales could be used to find the time from the shadow of the Moon on a sundial. There is an example of this circular type of lunar scale on a Sutton quadrant of 1656 and on some nocturnals, notably one by John Brown stamped 'JB'.

(c) Between the above scales is a calendar and a set of six solar and horary scales. The calendar is a continuous 'spiral' of the months and days of the year in four segments labelled 'Spring' 'Sumer' 'Autume' and 'Winter'. It is designed to be used with the adjacent solar and horary scales. These scales give, for each day of the year, the time of sunrise and sunset, and the Sun's amplitude, declination, and true place in the zodiac. A latitude scale also allows the longest day in any latitude to be found. This group of scales is also to be found on a number of ivory or wooden quadrants signed by or attributed to Brown.

This is only the third known example of a Delamain type projection, the other two being by John Allen and Pigot. Both these others conform to Delamain's specification of 'a small portable instrument'. The large size of this example suggests that it was made independently of Delamain.

Delamain 1632
Wing and Leybourn 1649, p. 127
Richmond 1770, No. 37
Michel 1947, p. 129, Pl. 24
Turner 1981
Ward 1981, Item 275, Pl. 42, Item 276
Turner 1985, pp. 223–5

See also 1889-49 Science Museum Astronomy Collection, N.20/37-132 National Maritime Museum

Inventory No. 1927-2065/29

E254 Triangular quadrant

Brown, John?

last quarter seventeenth century   London

L 310mm (sides), **DE** 12mm

This is the 'tryangle box rule' corresponding to item 38 on the list of the Boyle Collection. Although it is not signed it is a form of 'Trianguler quadrant' invented and described by John Brown in 1671. Brown was attempting to improve on his 'joynt rule' described in 1661 and on Twysden's 'semi-circle on a sector' described in 1667. However, the scales do not conform very closely to his specifications. Whereas Brown spreads the astronomical material round the 'quadrantal' side and compresses the sectorial material onto the other, this has all the astronomical material on the 'loose piece' across the bottom; with this removed it is a fairly standard sector. It is possible that it is an earlier version of the published triangular quadrant since the instrument was evolving from 1660.

Brown devised the instrument as a combination of the sector, the quadrant, and Gunter's rule. He states that 'A better Contrivance and more general hath not yet to my knowledge been produced'. Examples could be made from 6 to 36 inches in radius—the smaller ones for land use, the larger ones at sea and for astronomical observations. Brown stressed their convenience and compactness. The number and type of lines could be varied according to the price but some were fundamental to the instrument. It is not intended to explain the use of all the lines; this is done to some extent by Leybourn and Bion in the texts below.

The instrument is in the form of a sector in boxwood with a third arm or 'loose piece' which fits into the other two to form an equilateral triangle. Each arm of the triangle measures 1 foot. Brown defines a 'sector' side and a 'quadrantal' side, but since a quadrant is present on both sides of this example it is not such a clear distinction. The side with the astronomical features, which Brown considered to be largely innovatory, will be called the quadrantal side.

On the 'sector' side the scales are as follows: placing the 'head' at the top, on the left-hand arm are a line of chords for a radius just over 8 inches, a line of chords for a 5 inch radius, a line of sines for a 3 inch radius followed by a secant scale for a 3 inch radius, a line of sines for a 12 inch radius, a line of secants for a 6 inch radius, and a line of chords to 60° for a 12 inch radius. The last three mentioned have corresponding scales on the right-hand arm. There is a hole for the attachment of a 'thred' or plumb bob and a scale for use with this running in a semicircle 90–0–90. It starts at the head, and continues down the right arm and along the loose piece to the base of this left-hand arm. It is marked concurrently 30–0–60. Brown describes this scale as being on the other side of his instrument. The 'loose piece' is empty apart from the degree scale just mentioned. On the right-hand arm are the following: a line of chords for a 5 inch radius, a line of chords for a 3 inch radius, and the paired sectoral lines listed above. All the scales are based on a principal radius of rather less than 12 inches (approximately $11\frac{7}{10}$ inches). The values have been rounded up in the description.

On the quadrantal side on the left-hand arm are the following: a ruler numbered both ways in inches and marked in tenths from 0 to 24 covering both arms on the outer edge, a line of lines 3 inches long numbered 1–10 and marked in tenths which is effectively continued by two brass points further down the line, a line of lines 12 inches long similarly marked and numbered, a line of tangents for radius 3 inches from 45° to about 76°, and a line of tangents for radius 12 inches from 0° to 45°. The last three are paired with lines on the right-hand arm. In addition there is a

continuation of a scale on the loose piece; this section runs from from 55° to 62° on the inside edge of the arm and relates to a plumb bob attached to a point near the inside of the right-hand arm of the instrument. The right-hand arm contains the matching sectoral lines mentioned above and a scale which starts at one of the thread holes running from 0 to 90 which may be the 'general scale' from Brown's quadrantal side.

The 'loose arm' on this side has a variety of scales for astronomical use. In Brown's description these are placed on the right-hand arm. Starting at the inner edge there is a scale for the Sun's right ascension 6–0–6 and 6–12–6 with the 0 being marked as 24. The midpoint is vertically below the point for the plumb bob mentioned above. Directly below this is a calendar scale for the Julian calendar. Then come two scales which find the hours and the azimuth respectively for a particular latitude; their centres correspond at 6° and 90° and they extend from 12 to 3 and 0 to 9 hours and from 0° to 130°. Then there is a line of versed sines also with an hour scale. These scales have 60° and four/eight respectively on the line of the mid-point of the other scales. The versed sine scale runs 0–90 and the hour scale both ways 0–6 and 12–6, with the 6 and the 90 figures being off the scale. Beneath this is an unidentified scale which begins at the centre line of the other scales and extends rightwards 0–60; again the 60 figure is not on the scale. Lastly, along the bottom edge is a scale of degrees relating to the plumb bob. This is numbered both ways: 30–0–62 one way and 0–90 the other.

On the outside surface of the two sector arms are four scales, namely a scale of artificial numbers or logarithms, artificial or logarithmic tangents, artificial sines, and artificial versed sines. The inside edges are unmarked despite Brown's recommending a scale of inches, a foot measure, a line of 112 or a meridian line, 'or any other more useful line'.

Another example, more closely following Brown, but now incomplete, is described by Wynter and Turner. Twysden's work, mentioned above, has not been traced.

Brown 1661
Brown 1671
Leybourn 1672, pp. 3–6
Leybourn 1682, pp. 13–15

Bion 1758, p. 16, pp. 38–40, Pl. 4, Fig. 5
Richmond 1770, No. 38
Wynter and Turner 1975, p. 64, Fig. 72

Inventory No. 1927-2065/30

E255 Book entitled 'Perspectiva Corporum Regularium'

Jamnitzer, Wenzel

1568   Nuremberg

L 330mm, **W** 250mm

This is the book referred to corresponding to item 40 on the list of the Boyle
Collection. The book consists of 49 leaves of illustrations and three of text. The
illustrations are as follows: a title page, a page introducing each of the five regular
solids, four pages depicting solids based on each of the five, and 23 other
illustrations of more complex geometrical solids.

Richmond 1770, No. 40   **Inventory No.** 1927-2065/31

E256 Manuscript book

last quarter seventeenth century

L 180mm, **W** 110mm

The manuscript is one of the two corresponding to item 41 on the list of the Boyle Collection; the other appears to have been lost. It is bound in calfskin and contains 12 plates of geometric drawings in ink and crayon; some have been filled in with water colour. Christ on a Cross is superimposed on one drawing. Joy Hancox's forthcoming publication explores the connection with the Byrom Collection.

Richmond 1770, No. 41

Inventory No. 1927-2065/32

E257 Thirty-three pieces of 'pye'

last quarter seventeenth century or first quarter eighteenth century

L 50mm (side max), **L** 5mm (side min)

These are probably 33 of the original 36 brass pieces of 'pye' corresponding to item 42 on the list of the Boyle Collection. They are triangular pyramids on irregular bases with various lengths of side. Presumably if they were assembled correctly they would form a flat cylinder.

Jamnitzer 1568
Richmond 1770, No. 42

Inventory No. 1927-2065/33

E258 Sixteen pyramids on equilateral triangle bases

last quarter seventeenth century or first quarter eighteenth century

H 50mm (max), **H** 25mm (min), **L** 20mm (base max), **L** 10mm (base min)

Although these brass pyramids do not appear to be listed, they form part of the Boyle Collection.

Jamnitzer 1568

Inventory No. 1927-2065/34

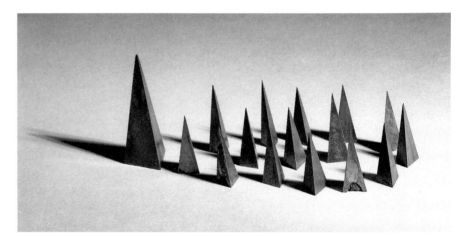

E259 Two sets of pyramids on equilateral triangle bases

last quarter seventeenth century or first quarter eighteenth century

L 45mm (set one (side)), **H** 22mm (set one), **L** 22mm (set two (side)), **H** 15mm (set two)

There are five in each set of brass pyramids. They do not appear on the list of the Boyle Collection.

Inventory No. 1927-2065/35

E260 Icosahedron on pedestal and dodecahedron

last quarter seventeenth century or first quarter eighteenth century

D 32mm (both), **H** 80mm (icosahedron)

The solid brass icosahedron and dodecahedron form a pair. The dodecahedron has a hole where the stand would have fitted. It is possible that the tetrahedron listed in E235 is linked with these items which do not appear in the list of the Boyle Collection. Only the icosahedron is illustrated.

Inventory No. 1927-2065/36

E261 Miscellaneous pieces

last quarter seventeenth century or first quarter eighteenth century

H 30mm (square pyramid), **L** 40mm (square pyramid (side)), **H** 20mm (irregular pyramids), **L** 20mm (irregular pyramids (bases)), **L** 10mm (hollow pyramids (sides)), **L** 50mm (triangular base (side)), **L** 75mm (tetrahedron (sides)), **H** 55mm (hexagonal pyramid), **D** 40mm (base)

These brass pieces do not relate to any mentioned in the list but are part of the Boyle Collection. They are a delineated hollow tetrahedron, a hexagonal pyramid, a delineated square pyramid, three irregular pyramids which may have fitted on stands, three small hollow triangular pyramids, and a triangular base which may be part of another item.

Inventory No. 1927-2065/37

E262 Three pyramids

last quarter seventeenth century or
first quarter eighteenth century

H 42mm (1), **H** 57mm (2), **H** 70mm
(3), **D** 85mm (all bases)

These brass pyramids are on hexagonal
bases. Again, they do not appear to be
listed.

Inventory No. 1927-2065/39

E263 Portrait of Robert Boyle

Holl, W.

1829

'The Honourable Robert Boyle'

H 320mm, **W** 280mm, **H** 130mm
(print), **W** 105mm (print)

The engraving was added some time
after 1829 but has been kept in this
section to maintain the integrity of the
Boyle Collection.

It has been identified from Maddison
as being a line and stipple steel engraving by William Holl, possibly from a
drawing made by Derby in 1828. It
originally appeared in Lodge's *Portraits
of Illustrious Personages of Great Britain* in
1829. Boyle is shown sitting on a high-
backed chair facing slightly to the left
holding a book in his left hand. He is
very young and has long fair curly hair.
The inscription runs underneath.
Apparently the engraving in Lodge also
included the following: 'From the original collection of The Right Honourable
Earl of Liverpool/Drawn by W. Derby
and engraved (with permission) by W.
Holl/ London. published Jan 1st 1829
by Harding and Lepard, Pall Mall
East'. However, cardboard in the frame
in this item excludes any inscription,
other than that given above, which may
be present.

The attribution to the Earl of Liverpool poses a problem since no records
have been found in his collection referring to this engraving. Two portraits of
Boyle reputedly by Dahl are listed, and
Maddison believes that one of these was
engraved by Holl. Dahl could not have
painted the portraits from life as he did
not come to England until 1681 when
Boyle would have been very much
older.

Maddison 1959

Inventory No. 1927-2065/38

PART 4
Nineteenth century material

There is no distinct break in the collection, but the majority of the later items have a different purpose, having been acquired by King's College as teaching material. Quite a number of the instruments are by Newman and date from the second quarter of the nineteenth century. Other apparatus is unsigned but it is equally clear that it was used with students and is not of a very high quality. A few further items, notably a model steam engine, were donated to the King George III museum, but there is little evidence of many such gifts.

The items within this group reflect developments in photography, electromagnetism, polarization and spectroscopy. One group which seems anomalous are the weapons which date from the first decade of the nineteenth century.

GEOMETRICAL FIGURES

L1 Parabola

mid-nineteenth century

'lats rectin 13 3/4 ins/Paraboloid'

H 150mm, **L** 242mm, **W** 76mm

This figure forms a set with the following five items. The set probably dates from the mid-nineteenth century. The pieces are made of mahogany, and most have a symbol at the centre of gravity and their names and principal dimensions painted on their surfaces. Although Adams did show a triangle and parabola in his *Mechanics* manuscript, it is doubtful whether these were made by him. This figure has one brass hook.

C 246

See also J148

Inventory No. 1927-1643

L2 Parabola

mid-nineteenth century

PAINTED
    'Lat Rect Ln 13 3/4 ins'

L 400mm, **H** 100mm, **T** 8mm

C 246

Inventory No. 1927-1847

L3 Paraboloid frustum

mid-nineteenth century

'Paraboloid/lat: rect: 13 3/4 In'

L 170mm, **D** 240mm

This item has three iron feet and appears to have been well used.

Inventory No. 1927-1843

L4 Cycloidal sector

mid-nineteenth century
'Cyloid/Rad: Gen [circle with dot in middle] 4 1/2In'
D 430mm, **H** 70mm
Inventory No. 1927-1842

L5 Cycloid

mid-nineteenth century
'Cycloid Rad: Gens 4 1/2 In:'
L 710mm, **H** 225mm, **W** 8mm
C 246
Inventory No. 1927-1848

L6 Triangle

mid-nineteenth century
L 500mm (side), **L** 430mm (side), **L** 345mm (side), **DE** 8mm
C 246
Inventory No. 1927-1846

L7 Hollow square pyramid

Parker, John W.

1833–51   London

'*John W. Parker/West Strand/London*'

L 275mm (base), **H** 140mm (sides)

At a first glance this item appears to form part of the set of geometric figures listed in Demainbray's collection. However, the colour is not the same, the wood is not apple, and there are no distinctive markings. In the base of the figure is a small rectangular hole, beneath which is stamped 'John W. Parker West Strand London'. Parker was a publisher who was active in West Strand from 1833 to 1851. It is probable that the figure was purchased by King's College and used with the earlier set.

C 247

The Post Office London Directory

See also D77 etc.

Inventory No. 1927-1644

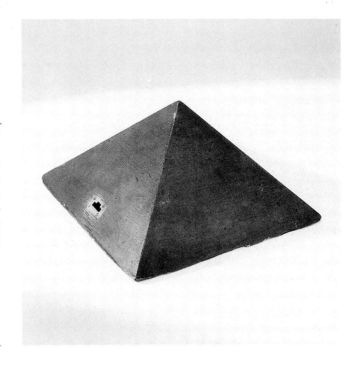

L8 Two honeycomb models

mid-nineteenth century

ON 1

'*Contributed by King's College/Honeycomb*'

L 236mm (both), **W** 140mm (both), **H** 220mm (1), **H** 180mm (2)

The slightly larger model has a piece of card pinned to its base with the above inscription in ink. There are five hexagonal cylinders of plaster with circular interiors, one end open and one closed and truncated to form three parallelograms. They are stacked two above three. The base is painted with a grey/green wash.

The smaller model is identical except that there are six plaster cylinders, two of them broken, and the base is slightly shorter.

Inventory No. 1927-1672, -1673

DRAWING INSTRUMENT

L9 Protractor and coordinatograph

early nineteenth century

L 225mm (instrument), **W** 137mm (instrument), **DE** 15mm (instrument), **L** 270mm (box), **W** 185mm (box), **H** 55mm (box)

The instrument consists of a brass semicircular protractor with a scale from 0° to 180° in half degrees. On the left-hand side of the base is a fixed piece of brass with a sine scale from 0° to 90° starting at the centre. A brass sliding arm on a sleeve can move along this scale at right-angles to the base; this arm has a linear scale marked from 0 to 70. Rotatable around the centre of the protractor is a piece of brass with an identical sine scale to the one described above. It also has a sliding arm on a sleeve fixed at right-angles, but this arm is simply a straight edge and has no scale. The function of the instrument appears to be similar to that of E10 although their proportions are different. The instrument is raised from the operating surface by three small brass feet. When not in use it fits into a mahogany case lined with green baize. Empty compartments in the case suggest that it is not complete. The precise purpose of the instrument is not known.

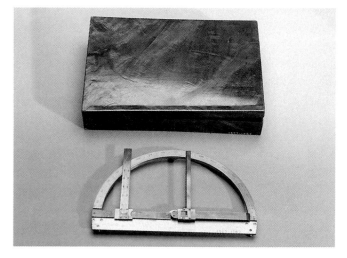

Bolles 1925
J.F.I. 1830
C 232, 233
Kidwell 1988
Wing 1990

See also E10

Inventory No. 1927-1787

CALCULATING INSTRUMENTS

L10 Calculator for treadwheel labour

Bate, Robert

1822–51   London

ALONG TOP
'*SLIDING SCALE/for Tread-Wheel Labour*'
'*Constructed & Published by R.B. Bate 17 Poultry, London March 31st 1823/(For the Prison Discipline Society)*'

L 375mm, **W** 105mm, **T** 10mm

The instrument is constructed on the same principle as a slide rule. Paper scales are pasted on a mahogany board, one section of which slides laterally with respect to the rest. The left-hand part, which is fixed, contains the 'Feet Ascent' and 'Steps per Minute' scales, the former between 2000 and 30 000 and the latter between 20 and 80. The sliding scale has time in hours down both edges from 2 to 14, and the left-hand scale is compressed onto the lower part of the slider with an index above for use with the 'Feet Ascent' scale. A short scale with height of step in inches is included with this index. If the height of the steps was outside the given range of 7 to 8 inches the instrument had to be individually calibrated. The right-hand side of the instrument gives the proportion of prisoners working on the wheel at one time for numbers of prisoners up to 32. The area is divided into nine vertical scales. The first gives the simplest fractions between $\frac{1}{4}$ and $\frac{4}{5}$ with a diagram to show in which columns the other fractions can be found. The other scales give various other fractions covering any possible number of prisoners working out of any number from 6 to 32. Below the scales is the index for use with the sliding scale to set the time to be worked.

The instrument can be used in various ways. The following example is given by Hase and Bate: the wheel runs for 11 hours, two-thirds of the prisoners are on it at any one time, the steps are 8 inches high, and the velocity is 41 steps per minute. What is the feet ascent per prisoner? The index on the right-hand side is set to 11 and the time is read off the sliding scale for the fraction $\frac{2}{3}$, giving 7 hours 20 minutes as the time that any one prisoner is on the wheel. The sliding scale is now moved until 7 hours 20 minutes on the left-hand edge coincides with 41 on the steps per minute scale. The index at the top left-hand of the sliding scale which is marked at 8 inches on this instrument now gives the Feet Ascent, which is slightly above 11 000.

Any one unknown could be found given the other information. Prison governors would probably have a set 'feet ascent' per prisoner and time of running the wheel per day. They would need to know the proportion of prisoners to be put on at any time or the number of steps per minute required. With this calculator 'The actual labour . . . be rendered nearly uniform at every prison—an object of much importance to the prison discipline of the country.'

Hase and Bate 1824, pp. 18–24
C 245

Inventory No. 1927-1526

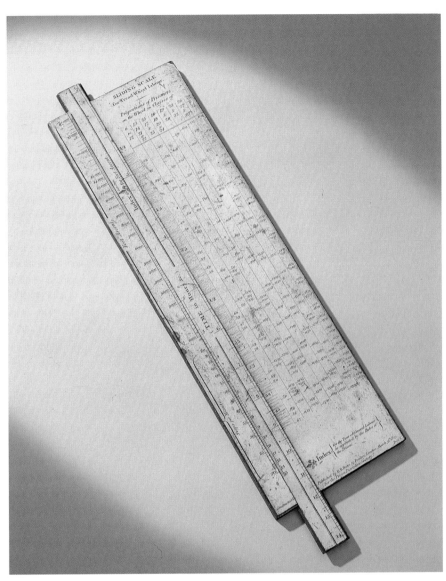

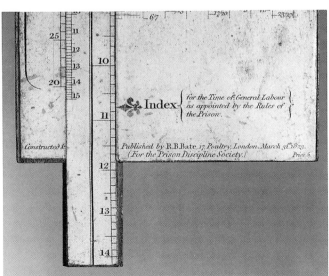

L11 Counting machine

mid-nineteenth century

'8134'

L 490mm, **H** 390mm, **W** 160mm

This is the cruder of the two counting machines in the collection; it was probably used at King's College. Three mahogany wheels have paper pasted to them with the numbers 0 to 9 round the circumference. The wheels have brass pins between the numbers. A lever at the right-hand side operates a boxwood arm which catches the pins and rotates the wheels. Three boxwood discs on brass springs at the base of the instrument fit between the pins and allow the wheels to be moved round one number at a time. A piece of brass on the first wheel moves the second after ten operations of the lever, and similarly for the second wheel acting on the third. It is not obvious which number is to be read at any time.

C 244

See also E11

Inventory No. 1929-121

WEIGHING INSTRUMENTS

L12 Balance

mid-nineteenth century

L 610mm, **H** 565mm, **W** 220mm, **D** 40mm (weight),
H 125mm (weight)

The balance stands on a mahogany base with three brass levelling screws. It consists of a brass pillar with a brass beam and boss on two steel fulcra. A vertical steel pointer gives an indication of equilibrium. It has a scale from 0 to $13\frac{1}{2}$ inches each way, numbered in inches and marked in tenths. The beam ends are simply truncated and the pans would have been suspended from crude brass hooks. One weight is with the instrument but it may not belong with it; it is a rough iron cylinder on a rod. There is an arrestment device which is operated by a mahogany lever under the base. It lifts a section of the mahogany pillar behind the balance which raises two arms to support the beam.

Inventory No. 1927-1224

L13 Balance

Watkins and Hill

1820–56   London

ON BEAM
  '*Watkins & Hill/Charing Cross London*'
STAMPED ON PANS
  '*By His Majesty's/Royal/Letters Patent*'
SCALE
  $9\left(\frac{1}{10}\right) 0 \left(\frac{1}{10}\right) 9$

D 140mm (pans), **L** 510mm (beam), **H** 405mm, **D** 125mm (base), **DE** 27mm (beam), **W** 7mm (beam)

This is a brass beam balance with a beam 18 inches long and 1 inch deep, divided into inches and tenths from the centre both on one side of the beam and along its upper surface. The beam has an adjustable steel knife-edge which rests on a steel fulcrum at the top of a brass pillar. In addition to the knife edge being vertically adjustable, the hooks which carry the cords for the pans are also vertically adjustable; a screw of high pitch situated at the ends of the beam allows the hooks to be raised or lowered. A scale reading from 0 to 40 gives the distance in $\frac{1}{40}$ inch from the upper surface of the beam. The pans are of brass, 5 inches in diameter and each stamped with 'By His Majesty's/Royal/Letters Patent'.

Inventory No. 1927-1253

L14 Assay balance

early nineteenth century

L 235mm, **W** 105mm, **H** 220mm

Presumably this small assay balance was in a glazed case but none of the glass remains. The frame and drawer beneath are mahogany with the drawer having an ivory knob. The beam is of steel, $5\frac{1}{2}$ inches long with swan-neck ends. The pointer is now broken. The scale-pans are of brass, $1\frac{1}{4}$ inches in diameter. There are two spirit levels at right-angles on the base. It is possible that there was a relieving mechanism but this is now incomplete. There are 18 weights consisting of flat pieces of brass with stamped numbers on them which correspond to grains in pennyweights Troy.

Crawforth 1984, p. 3, Fig. 1

Inventory No. 1927-1138

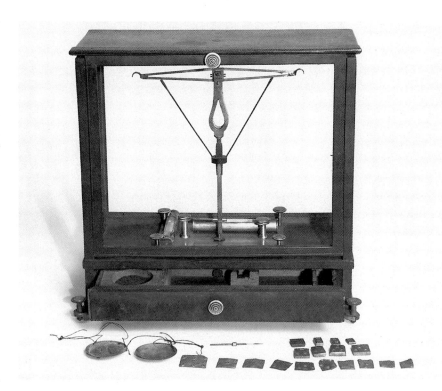

L15 Steelyard

first half nineteenth century

SCALE
1–0–24

L 1300mm, **H** 180mm, **W** 75mm

This is a mahogany T-section steelyard with a brass fulcrum and brass ring for the load. It is stamped from 1 at the load and 0 at the fulcrum to 24 along the arm. Each division is 2 inches long.

C 48

Inventory No. 1927-1228

MEASURING INSTRUMENTS

L16 Micrometer

first half nineteenth century

L 90mm (max), **W** 20mm, **H** 80mm (max), **D** 24mm (screw head)

This instrument is more properly called a comparator since it does not measure absolute length. Although the hand-held micrometer is usually thought to date from the 1840s, there are earlier similar instruments for specific purposes and this could be one of those. It consists of a brass base plate and two uprights: one fixed and one on a slider. The fixed upright has a steel pin and the movable one has the steel micrometer screw running through it. The screw thread is turned by moving the circular dial on the screw head against an index. The circle is divided into 18, but there are no numbers given. There are no markings on the index. The base plate has a scale half an inch long divided into fortieths of an inch. Again, there are no numbers. The thread of the micrometer screw is about 55 to the inch. There are two screws in the base plate: one locks the slider and the other was probably used to support the specimen. The instrument appears to be unfinished.

C 243

Hume 1980, p. 146

Inventory No. 1927-1784

L17 Two gauges

first quarter nineteenth century

L 425mm, **H** 37mm, **W** 20mm (max), **W** 3mm (rulers)

The gauges are probably for use with a pyrometer. One is of steel and the other of brass. They consist of rulers with no markings which are clamped between end-pieces holding brackets with circular holes above. The holes are probably for rods of various materials which have been heated.

Inventory No. 1927-1696

TIME MEASUREMENT

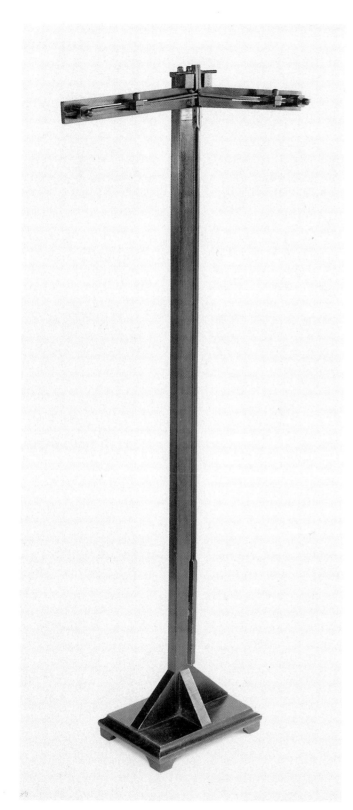

L18 Pendulum

Newman, J.

1828–57   London

'*I. NEWMAN/122/REGENT STREET/LONDON*'

D 15mm (pendulum), **L** 1150mm (pendulum), **H** 1360mm, **L** 630mm (base), **DE** 220mm (base), **L** 110mm (steel piece), **W** 10mm (steel piece)

This item could have come into the collection via King's College. It is a mahogany pendulum 3 feet 9 inches long, which is suspended from a short steel spring. The upper end of the spring is held stationary between two mahogany pieces with handles which rest on slots in the frame. The lower end is screwed into a slit in the top of the pendulum. The open part of the spring, which is about 1 inch long, swings between two pieces of mahogany which can be adjusted to regulate the amplitude of the oscillation. The mahogany pieces are screwed onto the two horizontal arms of the instrument. The pendulum has a diameter of about $\frac{1}{2}$ inch and the bottom part, which is 9 inches long, has a square cross-section. The maker's name and address are on an ivory plaque behind the spring.

Inventory No. 1927-1225

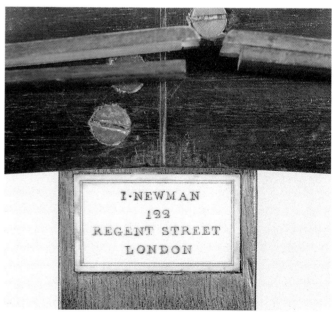

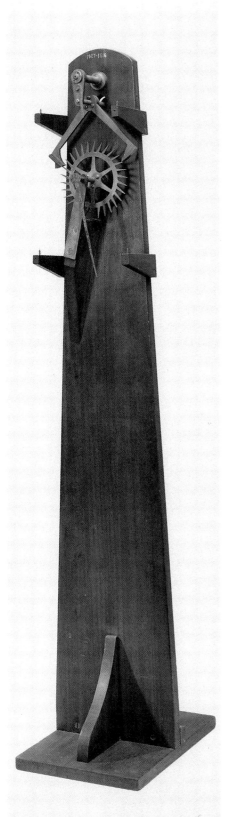

L19 Model of an escapement on a stand

mid-nineteenth century

H 1250mm, **W** 270mm, **DE** 380mm

Again, this is probably one of the items introduced into the collection at King's College. It is a model of a weight-driven dead-beat escapement with a seconds pendulum which is now missing. The brass escape wheel has a steel axle on which hangs the weight and to which an iron pointer is attached. The pointer originally moved over a dial which is now missing. The brass anchor has a steel axle running through the mahogany stand and oscillates on another steel axis above. A piece of iron hangs down the back of the stand; it has two prongs to constrain the movement of the pendulum.

Inventory No. 1927-1616

L20 Compensated pendulum

mid-nineteenth century

H 750mm, **L** 295mm (side of base), **L** 370mm (compensation bars)

This is a simple compensated pendulum on the gridiron principle. It is fixed to the mahogany stand at the base of the outer brass rods. The central steel rod is attached to the top of the brass rods so that if they expand upwards the steel rod will expand downwards to compensate. The copper pendulum bob is attached to the steel rod.

A Graham compensated pendulum and one by Julien le Roy were listed in the Queen's Catalogue but have not survived. Another nineteenth century compensated pendulum from the King George III Collection was sold at Christie's in 1987.

Q. C., Item 115, Graham's compound pendulum: Item 116, Julien le Roy's compound pendulum

Thiout 1741, Pl. 5, Fig. 6

C 85   **Inventory No.** 1927-1615

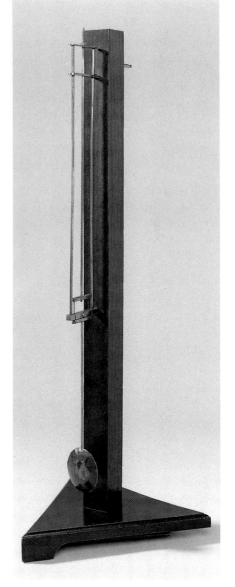

MECHANICS

L21 Pair of pulley blocks

Watkins and Hill

1820–56 London

ON BOTH BLOCKS
'WATKINS & HILL CHARING CROSS'

D 155mm (1), **D** 145mm (2),
L 210mm (1) **L** 190mm (2),
DE 48mm (1), **DE** 45mm (2)

The blocks are stepped, one block having six wheels with a maximum diameter of 6 inches and the other having five wheels with a maximum diameter of $5\frac{1}{2}$ inches. Each block has a steel hook and a brass separator for the threads. The inscription is on the frames.

C 29

Inventory No. 1927-1822

L22 Pair of pulley blocks

Watkins and Hill

1820–56 London

'WATKINS & HILL/Charing Cross-/LONDON'

L 136mm, **DE** 25mm, **D** 56mm (max), **D** 48mm (min)

The two brass pulley blocks are identical, each having six wheels in two coaxial sets of three.

C 30

Inventory No. 1927-1208

L23 Apparatus to demonstrate the mechanical advantage of a screw

Bate, Robert

1808–51   London

ON BRASS CROSS-PIECE
   ‘*Bate London*’

L 200mm (screws), **D** 130mm (two large screws), **D** 90mm (small screw), **H** 610mm, **L** 355mm, **W** 200mm, **L** 90mm (counterweight), **D** 40mm (counterweight)

One of three large boxwood screws is positioned vertically so that it can rotate about its axis. It is held in a brass frame which in turn is attached to a heavy mahogany frame consisting of two pillars and two cross-pieces on a mahogany base. A rectangular brass frame with two wheels at the lower corners can move vertically in front of the screw. The lower wheels move up a brass rectangle and the upper part of the

frame moves between two brass wheels attached to the mahogany frame. A cross-piece has a protrusion which fits into the screw, and so turning the screw moves the frame. At each side of the instrument there is a pulley wheel.

A cord is attached to the base of the brass frame, passes over the pulley wheel, and carries a brass counterweight. The movement of the screw would act against the counterweight and a comparison would be made. However, it is not clear how it would be measured since the instrument is incomplete. There are three screws, two of a larger diameter than the third. One of the larger ones and the smaller one have a pitch of one turn per inch. The other has one turn in $1\frac{1}{2}$ inches.

Inventory No. 1927-1705

L24 Pair of wheels with bracket on friction rollers

Newman, J.

1828–56   London

'*I. NEWMAN/122 Regent St London*'

D 155mm, **W** 130mm, **W** 10mm (axle)

This model demonstrates wheel bearings. The large spoked brass wheels can be turned independently of the steel axle, the central part of which has a square cross-section of side length nearly $\frac{1}{2}$ inch. A brass. bracket is slung on the axle at its extremities which have a circular cross-section; two friction rollers form part of the bearing at each side. Thus the brass bracket can be kept upright while the axle rotates. Alternatively a screw in the middle of the bracket can be used to secure it rigidly to the axle, in which case it turns with the axle and the bearings are overridden. The performance with and without the use of the friction bearings would be compared. The bracket has a ring and two hooks for attachment to a load or weight to pull the machine. The inscription is on each side of the bracket above the bearings.

Inventory No. 1927-1876

L25 Compound axis in peritrochio

first quarter nineteenth century

ON MAHOGANY WHEEL
  '2 9/10'
ON MIDDLE WHEEL
  '1 1/10'
ON AXLE OF LARGE WHEEL
  '3/4'
ON LARGE WHEEL
  '4 4=/10'

L 510mm, **H** 280mm, **W** 90mm

The model has been dated as being from the early nineteenth century for stylistic reasons. It consists of three axes in peritrochio, linked in an analogous manner to the three levers in M48 and D2. The first wheel is solid mahogany with a groove to carry the thread. On the same axis attached to this wheel is a small brass toothed wheel which meshes with the second wheel, which is also a brass toothed wheel. The small wheel makes four revolutions as the larger one makes one. The second wheel also has a small toothed wheel attached to the same axis which in turn meshes with the third wheel and makes four revolutions as the third wheel makes one. The weight could be attached to the smooth brass axle of the third wheel. The combined theoretical mechanical advantage will be the mechanical advantages of each axis in peritrochio multiplied together, which is about 50. The three wheels are mounted in a brass frame.

See also D105, M50

Inventory No. 1927-1874

L26 Apparatus to show the parallelogram of forces

first half nineteenth century
L 510mm, **W** 210mm, **H** 360mm

Together with the following two items, this piece of demonstration apparatus probably came into the collection via King's College. It is remarkably simple. A blackened mahogany board is fixed in a vertical plane on a horizontal base. Pulleys whose heights can be adjusted are clamped on each of the vertical sides. Weights hung from the pulleys sustain a thread in a horizontal position on which is then suspended a third weight. If the forces are represented by lines on the board in the directions in which they apply and in lengths proportional to the weights, it can be seen that the diagonal of the parallelogram corresponds to the force of the weight on the thread.

C 25

Inventory No. 1927-1657

L27 Apparatus to show the parallelogram of forces

first half nineteenth century
L 395mm, **W** 230mm, **H** 630mm

The apparatus is similar in most respects to the previous item. However, the lower left-hand side of the vertical board contains a horizontal slot in which are placed three pulleys. Presumably experiments similar to the parallelogram of forces demonstration were performed.

C 26

Inventory No. 1927-1658

L28 Apparatus to show the parallelogram of forces

first half nineteenth century
L 420mm, **W** 215mm, **H** 630mm

The apparatus is similar to the previous item. A blackened mahogany board is fixed in a vertical plane on a horizontal base. The two vertical sides have slots for the attachment of pulleys, on which can be hung weights. There are signs that pulleys were also attached to the lower centre of the board. Presumably, variations on the parallelogram of forces demonstration were performed.

C 26

Inventory No. 1927-1656

L29 Conical and cylindrical axle

first half nineteenth century
L 243mm, **W** 78mm, **H** 100mm

It is possible that this machine demonstrated the heat produced by friction. The mahogany axis is half cylindrical and half conical, with the conical part coming to a point against one side of the mahogany stand where there is a small hole. Turning the handle at the other end could produce the desired effect. The machine appears to have been bolted to a surface since there are two large screw holes in the base. It is difficult to see how the axis could have been used for lifting weights since only one end is firmly attached. The handle is iron with a rosewood knob; it is now bent and immovable.

Inventory No. 1927-1878

L30 Stand for an Atwood's machine

nineteenth century
H 2680mm, **L** 450mm (stand) **W** 360mm (stand)

The object consists of a hollow brass tube with a boxwood scale marked in feet and inches from the base to a height of 8 feet. It stands on a mahogany base which is supported by three brass levelling screws. It is probably the stand of an Atwood's machine, a device to demonstrate acceleration due to gravity, which has not survived.

Inventory No. 1927-1618

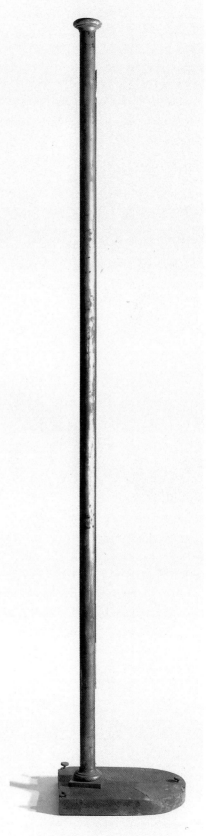

L31 Model of a feathered paddle wheel

c. 1830

ON LABEL
'1204'

L 505mm, **W** 237mm, **H** 482mm

This is a model of a feathered paddle wheel of a type used on steam ships in the second quarter of the nineteenth century. The steel paddles are held in pairs on an iron bracket attached to a central iron wheel. The inclination of each paddle is determined by the position of a pinion which connects it with one of two wheels which are set obliquely to the axis of rotation. At the point where the oblique wheels are nearest the central wheel, the paddles are feathered. When they are at their furthest point, the paddles each make an angle of 45° to the flow, with the edges of each pair nearly touching. The model is set on a mahogany base.

Turning the paddles in the plane of the direction of flow was an idea probably derived from Robertson who patented a paddle wheel in which the paddles were fixed in a V-shape in 1829. The feathering is an improvement on this design. Many patents for paddle wheels were filed during this period.

Robertson 1829
C 346
Reynolds 1983, pp. 293, 259
Evans 1988

Inventory No. 1927-1631

MACHINES

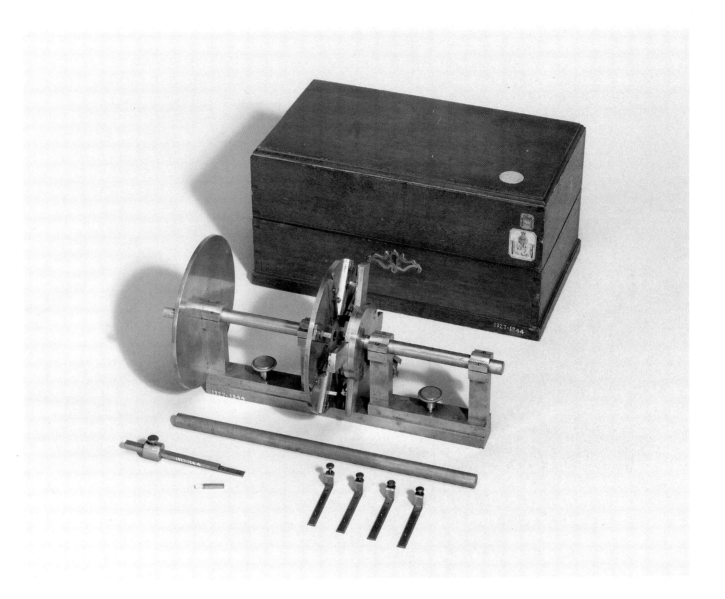

L32 Apparatus for centring lenses, in fitted box

first half nineteenth century

H 305mm, **D** 155mm, **L** 320mm (box), **W** 180mm (box), **H** 160mm (box)

The machine consists of a central pillar on a circular base which is itself supported on three feet. The pillar holds a circular platform on which are eight stops on sliders placed radially to keep the lens in place. There is a side pillar which carries the upper mechanism; this also consists of a circular plate with four stops on sliders. The upper plate is rotatable and can be raised or lowered. Loose parts include a solid brass rod 1 foot long, a weighted arm, and four sliders. The machine is made entirely of brass and fits into a mahogany box. It is not clear how it would function and it may not be complete.

C 210

Inventory No. 1927-1344

HYDROSTATICS

L33 Hydrostatical steelyard

mid-nineteenth century

ENGRAVED—SHORT ARM
'7.75' '1.5', '2'

H 420mm , **L** 293mm (beam), **L** 325mm (box), **W** 110mm (box)

The steelyard consists of a brass beam with the short arm about an inch deep and the long arm tapering to a point. The steel fulcrum can be raised to rest in slots by turning a lever in the brass pillar. The pillar screws into the lid of the mahogany box which alternatively can be removed so that the steelyard stands on three brass feet. One of these can be raised or lowered to provide levelling. When the instrument is not in use it can be fitted into the box. '7.75' is engraved on the short arm and the scale of the long arm is marked from 1 at the fulcrum to about 10 at the end. Only '1', '1.5', and '2' are numbered, and the scale is exponential. The engraving of the scale is particularly crude.

The steelyard is in equilibrium with no weight on the long arm. A vessel, probably glass, containing lead shot would have been suspended from the notches on the long arm into a container of the liquid under observation. A liquid with a high specific gravity would reduce the apparent weight of the vessel more than a liquid with a lower one, and therefore the former vessel would have to be suspended further from the fulcrum. The specific gravity could then be read off the scale. A counterweight of some description would have to be added to the short arm.

Inventory No. 1927-1454

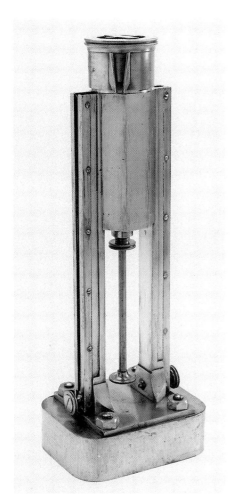

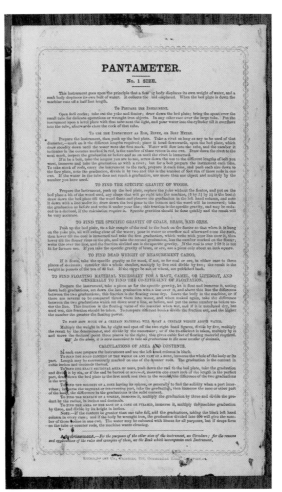

L34 Pantameter

mid-nineteenth century

L 510mm (box), **W** 210mm (box), **H** 150mm (box),
H 470mm, **L** 170mm, **W** 130mm, **D** 80mm (cylinder)

The pantameter probably came into the collection via King's College. It measures specific gravity by using Archimedes' principle that floating bodies displace their weight and submerged objects displace their volume.

A central octagonal brass vessel is filled with water which may be coloured for ease of reading the measurements. A large steel rod screws into the base of the vessel and acts as a plug. An overflow spout at the top can be directed into one of two measuring cylinders which support the central bath. The measuring cylinders have a square cross-section and are made of brass with glass fronts. The scales are as follows: 'Inches' numbered 0 to 3 upwards, each inch being actually 4 inches. Each mark on the scale corresponds to $\frac{1}{100}$ of an 'inch'. On the same cylinder is another scale engraved '100 w. Iron in one cut'. This is a logarithmic scale running downwards

from 1 to 5. On the other cylinder, which is slightly larger, there is another 'inches' scale which is numbered 0 to 7 upwards. The 'inches' are actually 2 inches long with every $\frac{1}{50}$ 'inch' marked. Next to this is a scale entitled '100 in one cubic foot immersed bodies' which runs from $2\frac{1}{2}$ to 10 downwards logarithmically. Another scale for 'grains' runs from 0 to 17 upwards and finally a scale entitled '100 in one cut floating bodies' runs downward logarithmically. The instrument stands on a heavy copper base.

A yoke is used to submerge the body in the water and the overflow spout is directed into one of the cylinders. The excess water is collected and the appropriate reading taken. A large sheet of instructions is pasted on to the inside of the lid. There is no maker's name but the printer was Reynolds & Co., 293 Commercial Road, Lindport.

C 132

Inventory No. 1927-1284

L35 Stand for Pascal's vases apparatus

Jones, W. and S.

1808–60   London

'W & S JONES/No 30 HOLBORN/LONDON'

H 540mm, **L** 495mm, **W** 270mm

The glass vases for this apparatus had been broken. In 1991 they were reassembled using the original glass wherever possible. However the original stand had warped and shrunk so that the vases no longer fitted. A new stand was made to take the vases and the original is shown beside it in the photograph.

The original stand consists of a mahogany frame with a painted tinplate trough in the base. There are three vases: conical, inverted conical, and cylindrical. All have a brass plate of equal area in the base.

Q. C., Note: 174 Hydrostatic paradox machine, pint cylinder, quart vessel
Pascal 1663, Figs 1, 2, 3, 4, 5
C 133

See also D29

Inventory No. 1927-1746

PNEUMATICS

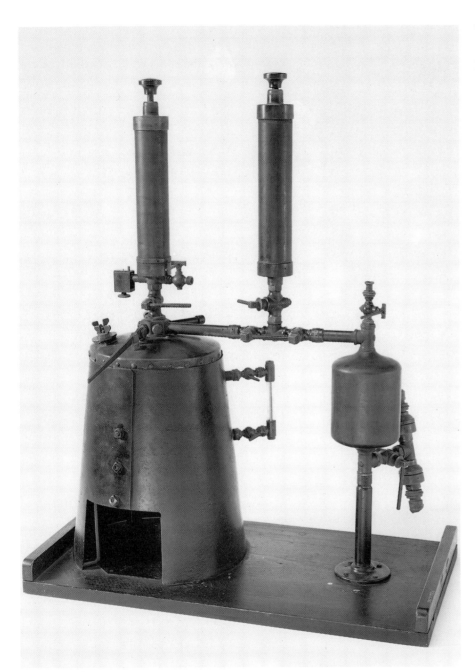

L36 Demonstration steam engines

mid-nineteenth century
L 685mm, **W** 380mm, **H** 840mm

This item probably came into the collection via King's College. It is a combination of demonstration steam engines made from pieces which resemble gas fittings. A single boiler can be made to operate a Savery engine, an atmospheric engine, or a Watt's steam engine.

The copper boiler is encased in a sheet iron cover which was probably a later addition. It has no uptake for smoke from the fire. The boiler is equipped with a water gauge on the side and a lever safety valve and filling plug on the crown. A steam pipe with a cock rises from the centre to a cylinder and piston which demonstrates the functioning of the atmospheric engine. The cylinder is fitted with an injection cock for water and a snifting valve. A horizontal pipe from the boiler crown leads through a cock to a T-piece. Above this is another cock and then a second cylinder and piston having no other connections. The horizontal pipe continues through another cock to the 'receiver' or steam vessel of a Savery engine. There is a venting cock at the top of the vessel and a cock at the bottom leading to two non-return valves. The delivery pipe leading from the non-return valves is missing. The arrangement is such that the receiver of the Savery engine can be used as a separate condenser in conjunction with the second cylinder and piston to demonstrate the functioning of the Watt engine. The whole apparatus is mounted on a softwood board.

Law 1965, pp. 6–11

See also D50, D51

Inventory No. 1927-1760

L37

L38

L37 Model forcing pump

mid-nineteenth century
D 210mm, **L** 125mm, **H** 570mm

This item is in the same style as the previous item, being made of similar parts. It consists of a brass cylinder on a cast iron stand and base, connected at the lower end to a pipe leading to two non-return valves. The piston is missing.

C 109

See also D36

Inventory No. 1927-1761

L38 Pipe with brass union and T-piece with tap

mid-nineteenth century
L 350mm, **W** 125mm, **D** 42mm (max)

The pipe is copper with brass unions and a soldered tap. It was used with the model steam engines above.

Inventory No. 1927-1762

L39 Model of beam engine in case

Peel and Williams

1821   Manchester

'George 4th/Made by Peel and Williams Manchester/July 19 1821'
ON BRASS PLAQUE IN CASE
  'Presented to/THE MUSEUM OF GEORGE III/IN KING'S COL-LEGE, LONDON/BY/ Joseph and George Peel/ENGINEERS SOHO, IRON WORKS/MANCHESTER/ 1843'

L 510mm, **H** 410mm, **W** 255mm,
D 300mm, **L** 245mm, **L** 600mm (case),
H 565mm (case), **W** 415mm (case)

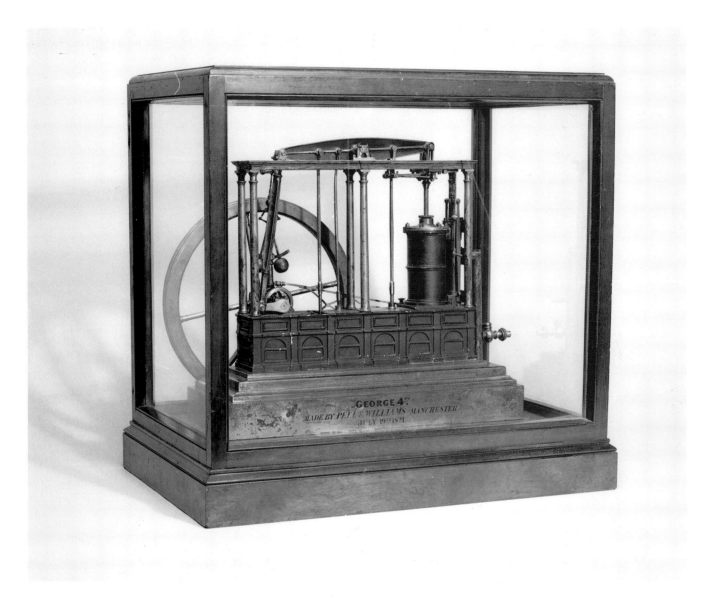

The model engine was probably made by Peel and Williams for the ascension to the throne of George IV, and presented in 1843 when the George III Museum had become established in King's College.

In most respects it is typical of its time and shows no unusual features, apart from the shape of the piston valve which is circular rather than semicircular or flat. It is an independent condensing engine with air pump, boiler feed pump, and house pump. The cistern is in the base and the boiler feed pump is above the hot well. It has a Watt governor and unequal bevel wheels from the shaft which act on the throttle. There is a gab gear for starting the machine but no hand lever to the weigh shaft. The splasher case for the crank and correcting shaft is slightly unusual but probably reflects the lack of space. Apart from the width of the groove and the low number of arms on the fly wheel, the model follows full-size practice even down to the architectural detail around the base.

Similar engines can be seen illustrated in works by Farey and by Bourne, in particular those engines made by Jukes Coulson and Company for a sugar mill, which are described by Farey.

Farey 1827, Pls 22, 8

Bourne 1846

C 327

See also D50, D51, L36

Inventory No. 1927-1941

L40 Blowpipe

Dymond

early nineteenth century

Dymond/Holborn Bars

H 240mm, **D** 20mm

This blowpipe consists of a brass tube with a nozzle at one end. The other end is threaded, probably to allow it to be attached to the reservoir of an air pump. There is a painted wooden base to support it when it is not in use.

The blowpipe was commonly used in chemical assays.

C 77

Inventory No. 1927-1480

L41 Glass syringe

nineteenth century

H 440mm (piston near bottom of barrel), **D** 37mm (glass barrel)

This syringe consists of a glass tube with a leather piston on a metal shaft. The ends of the tube have brass caps. The base and the handle of the piston are boxwood.

The lower end of this tube is sealed so that pushing the piston down compresses the air in the glass tube. The syringe can be used to demonstrate the resistance of the air to compression.

C 110

Inventory No. 1927-1209

L42 Wollaston's cryophorus

mid-nineteenth century

H 280mm, **D** 215mm (jar)

Two glass bulbs are connected by a glass tube. The glass tube passes through the wall of a bell jar so that one bulb is outside and one inside the bell jar. The glass bulbs held some water and little air; the bulb inside the bell jar also contains a mercury-in-glass thermometer, graduated from 20 to 90°F.

This form of cryophorus was invented by W. H. Wollaston in 1812. First the water was collected in the bulb inside the bell jar, and the empty bulb was surrounded by a freezing mixture. Water vapour then condensed in this latter bulb. If the bell jar was placed on the plate of an air pump and the air removed, the temperature of the water in the bulb inside the bell jar was seen to fall.

Wollaston 1813, p. 73
C 75

Inventory No. 1927-1388

L43 Magdeburg hemispheres

first half nineteenth century

D 110mm, **L** 465mm (assembled)

These hemispheres are made of brass. See P10 for a history and explanation.

Adams 1799a, p. 175, Pl. 1, Figs 19, 20
C 123

Inventory No. 1927-1312

L44 Air pump

Watkins and Hill

1820–56 London

'WATKINS & HILL/5 CHARING CROSS/LONDON'

L 275mm, **W** 215mm, **H** 190mm, **L** 260mm (barrel), **D** 180mm (pump plate)

This pump has a horizontal brass barrel arranged under the brass pump plate which in turn sits on four brass supports on a mahogany base. The base can be fixed to a table using four brass brackets fitted to its underside. The piston is moved by a mahogany handle attached to the piston rod. The valves are made of strips of oiled silk; one of these, the outlet valve, can be seen because its cover is missing.

C 103, Pl. 3
Turner 1987, p. 129, Pl. 117

Inventory No. 1927-1290

L45 Bell jar

first quarter nineteenth century

H 265mm, **D** 125mm

This glass bell jar has a brass collar and ground top.

C 129

Inventory No. 1927-1368

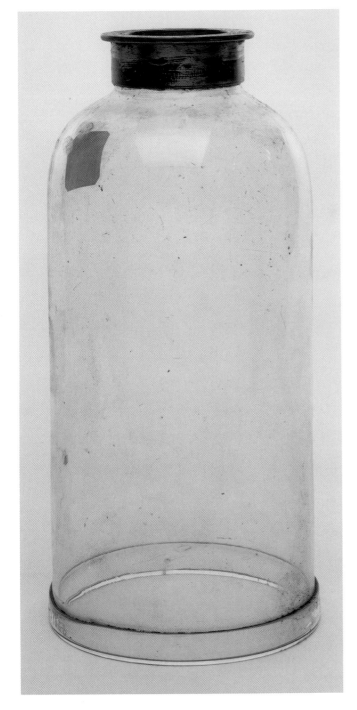

L46 Bell jar

first quarter nineteenth century

H 228mm, **D** 102mm

This glass bell jar has a threaded brass collar into which is screwed a brass tap.

C 129

Inventory No. 1927-1366

L47 Bell jar

Newman

1816–57   London

STAMPED ON TAP
'NEWMAN'

H 250mm, **D** 165mm

This glass bell jar has a threaded brass collar into which is screwed a brass tap.

C 129

Inventory No. 1927-1365

L48 Glass jar with ground base

first quarter nineteenth century
H 215mm (overall), **D** 100mm, **T** 7mm (approx)

This may be part of an apparatus to demonstrate adhesion. The brass collar is threaded and a ring may have been attached to this end. The other end of the jar has been ground flat to fit a metal, glass, or other similar plate.

C 129
Turner 1973, Item 165
Inventory No. 1927-1382

L49 Glass jar

first quarter nineteenth century
H 240mm (overall), **D** 125mm

This glass jar has a threaded brass collar.

C 129
Inventory No. 1927-1364

L50 Pressure fountain

Newman, J.

1816–57

STAMPED ON TAP
'I. NEWMAN'

H 300mm (overall), **D** 150mm

This apparatus is for demonstrating a fountain and is larger but otherwise similar to P45 and P46. The glass globe is partly filled with water so that the level comes above the bottom of the metal pipe attached to the brass collar of the globe. Using a syringe or an air pump, air was forced into the globe and the tap closed. The syringe was removed and the tap opened. The water in the globe was then forced out in a fountain.

Desaguliers 1744, Vol. 2, pp. 118–19, 202, Pl. 19, Fig. 6
Adams 1799*a*, Vol. 1, pp. 89, 176, Pl. 2, Fig. 8

Inventory No. 1927-1297

ACOUSTICS

L51 Musical drum

Ward, Cornelius

after 1836   London

PAINTED ON ONE SIDE
'C. WARD INVENTOR & /Patentee—36 Gt Titchfield St London/

L 265mm, **W** 265mm, **H** 145mm

This is a square drum consisting of a parchment membrane from an old deed stretched on a mahogany frame above a square pine box. The arrangement for tightening the membrane is as follows. Two small brass pulleys are situated on each side of the frame. A cord connects these via a pulley at the base of the midpoint of each side of the box. The cord also passes over pulleys towards the corners of the box which take the cord inside the box in four loops. Each loop passes over a pulley attached to the end of one of two fruitwood rods. The rods are threaded on a thick beech screw which is in two parts and runs across the centre of the instrument. One half of the screw is right-handed and the other is left-handed. When a knob is turned the screws move the rods apart or together to loosen or tighten the membrane; thus the whole drum can be adjusted by one operation and a more uniform effect is achieved than by adjusting many screws.

Cornelius Ward patented the drum in 1837. His other improvements were to use parts of a skin of similar thickness, to provide air holes under the drum head, and to dispense with cords for drums with two heads.

Ward 1837
C 259

Inventory No. 1927-1289

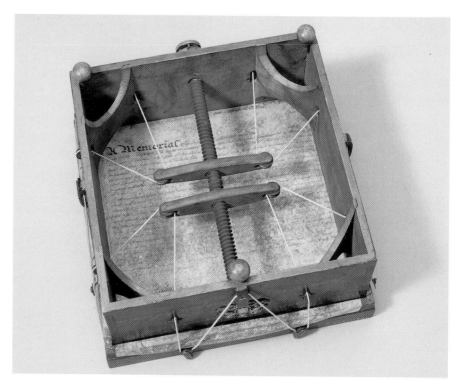

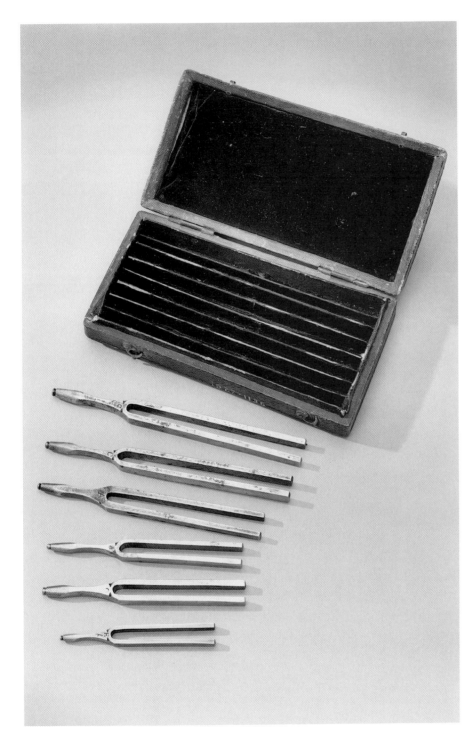

L52 Box of six tuning forks

first half nineteenth century

ON EACH FORK RESPECTIVELY
C, C, A, A, C′, E♭′

L 146mm (box), **W** 85mm (box),
H 33mm (box)

The box was made for eight forks but only six are now present, probably from two separate sets. Two are at middle C pitched at 250Hz and 256Hz respectively, two at A pitched at 420Hz and 444Hz respectively, one at C′ at 520Hz and one at E♭′ at 600Hz. The forks are made of steel and the letters are engraved on each fork. The red leather box suggests a relatively late date.

These instruments have been associated with Charles Wheatstone because he performed many experiments with tuning forks which he described in his papers on sound between 1823 and 1832. In particular, he gives a detailed description of a tuning fork in 'Experiments on audition' in 1827. Popular myth has it that the tuning fork was invented over a century before this by a trumpeter called John Shore, but the need for the description suggests that they were not widely used.

Wheatstone 1823
Wheatstone 1827
Wheatstone 1828
Grove 1954, under Tuning fork
C 258
Bowers 1975, p. 26

Inventory No. 1927-1136

L53 Wooden framed organ pipe, parchment sides

first half nineteenth century

'*B [♭]*'

W 100mm, **DE** 100mm, **H** 460mm

It is quite likely that this item came into the collection when it was at King's College and was used by Charles Wheatstone in his experiments on sound. In his paper on the vibrations of columns of air he mentions the use of organ pipes.

The organ pipe has a pine frame of square cross-section over which parchment is stretched. The mouth is also made of pine. The note B♭ is written in ink on one side. According to Chaldecott the pitch is about 225Hz.

Wheatstone 1828
C 256

Inventory No. 1927-1488

L54 Triangular framework

nineteenth century

L 275mm, **W** 240mm, **H** 47mm

The instrument was probably used for acoustic experiments since a membrane, fragments of which still remain, appears to have been stretched over the framework. The framework is made of stained pine.

Inventory No. 1927-1486

L55 Instrument to show the mode of vibration of a glass bell

first half nineteenth century

H 460mm, **L** 270mm, **W** 175mm, **D** 180mm (glass)

An upturned glass bell resembling a large wine glass stands on a walnut base. A vertical brass rod supports a mahogany eight-pointed star, from each point of which is suspended a pith ball. The pith balls touch the inside of the rim of the glass bell. If the bell is held at one point of the rim and bowed at another, the motion of the pith balls will indicate the vibration pattern of the bell. In this case alternate pith balls vibrate while the others remain motionless. The instrument illustrates Chladni's work on the vibration of bells.

C 252A
Chew 1968, No. 10

Inventory No. 1929-97

L56 Pair of ear trumpets

nineteenth century

H 168mm, **W** 120mm, **L** 150mm (approx)

There is only one auricular tube listed in the Queen's Catalogue, which is probably D123, so there is no indication as to how this instrument entered the collection. It is of relatively late date.

The japanned copper trumpets have ivory earpieces and are linked by a sprung steel headband. Ear trumpets of a similar design were available throughout the nineteenth century.

Rees Cyclopaedia 1819, under Trumpet, plates Vol. 3, Pl. 25, Fig. 6
C 248

Inventory No. 1927-2020

L57 Hopkins's fork

after 1834

L 450mm, **W** 100mm (max), **D** 30mm

The instrument is in the shape of a two-pronged fork with a hollow square cross-section. The 'handle' of the fork is slightly wider than the two prongs and had tissue paper stretched across its end. It is made of mahogany.

The fork was used in Hopkins's acoustic experiments to demonstrate the interference of sound. A glass plate was set vibrating, sand was sprinkled on the surface to show the position of the nodes, and the two prongs of the fork were placed close to the plate. If the apertures were close to sections of the plate that were vibrating in phase, the sound interfered constructively and the membrane could be observed vibrating. If the apertures were close to sections of the plate that were vibrating out of phase, then the sound would interfere destructively and no motion of the membrane would be seen.

Hopkins 1835, Pl. 8, Fig. 3
C 254

Inventory No. 1927-1483

L58 Loop-shaped hollow framework with two movable shutters

after 1834

L 302mm, **W** 32mm, **H** 178mm

This instrument is similar in style to the previous item, Hopkins's fork. It was almost certainly used for acoustic experiments.

The hollow framework has two square apertures on each side. The square-shaped tube splits into two rectangular passages which rejoin, forming a loop. The shorter passage has two movable shutters with knobs. One of the two fixed apertures appears to have been covered by a membrane, some of which is still visible.

Hopkins 1835, Pl. 8, Fig. 3

Inventory No. 1927-1482

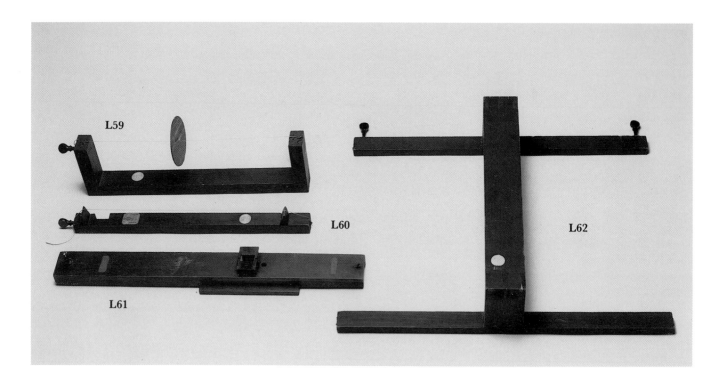

L59 Monochord

mid-nineteenth century

L 510mm, **W** 125mm, **H** 135mm

This group of monochords probably came into the collection via King's College in the mid-nineteenth century. They are likely to have been used by Sir Charles Wheatstone.

   This example has a pine framework which has been stained dark brown. The string is held well above the soundboard and fastened with a rosewood peg, probably bought from a violin-maker. There is a circular pine mute.

C 250

See also D46

Inventory No. 1927-1509

L60 Monochord with notched base

mid-nineteenth century

H 49mm, **L** 520mm, **W** 31mm

The pine base and two hardwood bridges of the monochord have been stained. Again, the rosewood peg was probably supplied by a maker of musical instruments.

C 250

Inventory No. 1927-1510

L61 Monochord with adjustable resonance box

mid-nineteenth century

L 615mm, **H** 80mm, **W** 95mm

The top of the pine resonance box has an adjustable aperture, presumably in order to observe variations in the sound quality.

C 251

Inventory No. 1927-1513

L62 Double monochord mounted on H-shaped framework

mid-nineteenth century

H 128mm, **L** 483mm, **W** 608mm

The instrument was probably used to compare strings of varying gauges and tensions. Again it has a stained pine base and rosewood pegs.

Inventory No. 1927-1511

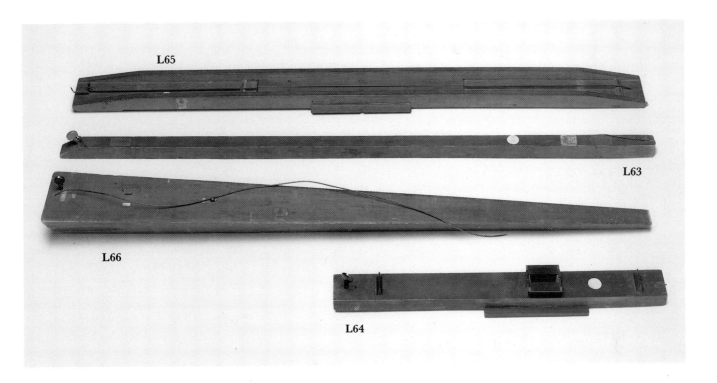

L63 Monochord with tapering end

mid-nineteenth century

L 1280mm, **W** 30mm, **H** 65mm

The bridges at each end of the monochord are missing; recesses in the base show their original position.

C 251

Inventory No. 1927-1512

L64 Monochord with adjustable resonance box

mid-nineteenth century

L 605mm, **W** 92mm, **H** 53mm

The top of the resonance box has an adjustable shutter to be used as L61. At one end there is an ebony knob with mother-of-pearl decoration.

C 251

Inventory No. 1927-1514

L65 Monochord with resonance box in channel-shaped base board

mid-nineteenth century

L 1240mm, **H** 53mm, **W** 85mm

The monochord has the usual pine frame but with a mahogany soundbox. It has been strung with a piece of copper piano wire.

C 251

Inventory No. 1927-1515

L66 Monochord with triangular tapering hollow base

mid-nineteenth century

L 1160mm, **W** 115mm (max), **W** 34mm (min), **H** 110mm (max), **H** 32mm (min)

The monochord bridges and frets are missing. Again, there is an ebony knob with mother-of-pearl decoration at one end.

C 251

Inventory No. 1927-1516

THERMOMETERS AND METEOROLOGICAL INSTRUMENTS

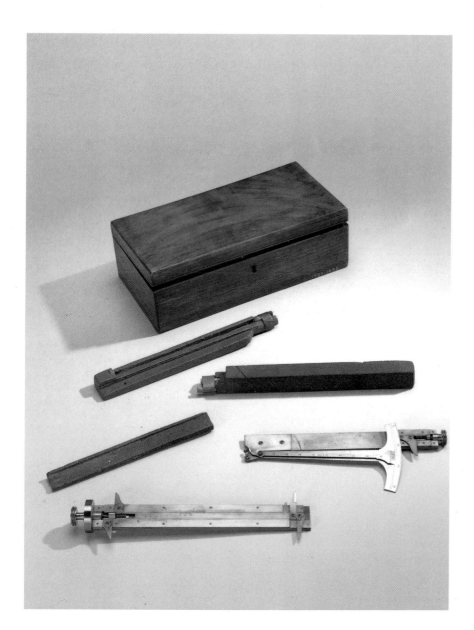

L67 Register 'thermometer'

after 1829
'4, 992'
SCALE
0 (1) 30 seconds
H 74mm (box), **W** 120mm (box),
L 235mm (box), **L** 220mm (register),
L 130mm (scale), **L** 210mm (callipers)

The instrument is an example of Daniell's register 'thermometer' or pyrometer, a device we would now call a dilatometer. It consists of a 'register', a black lead earthenware bar about 8 inches long with a cylindrical hole drilled through most of its length. Into the hole is placed the test piece, a cylinder of metal about $6\frac{1}{2}$ inches long. Above the test piece in the hole is placed the 'index', a piece of porcelain with a cross-section similar to the test piece and about $1\frac{1}{2}$ inches long. The index protrudes slightly above the top of the register. The brass scale described by Daniell was designed to fit against the register so that the short brass lever would have its short arm pressed against the top of the index in order to record any movement on the circular scale. Although very similar to Daniell's device, the scale of this instrument does not fit against the register in precisely the same way. Presumably the principle is the same.

On heating, the test piece is elongated, the index is thrust forward, the short arm of the lever is moved, and the long arm, which is ten times as long as the short arm, moves over the scale. The scale is marked in degrees from 0° to 30° and divided into minutes. The vernier on the lever allows the scale to be read to $\frac{1}{10}$ second. The lever is marked from 1 to 9 along its length and bears the number 4992. In addition to this scale there is another brass accessory not

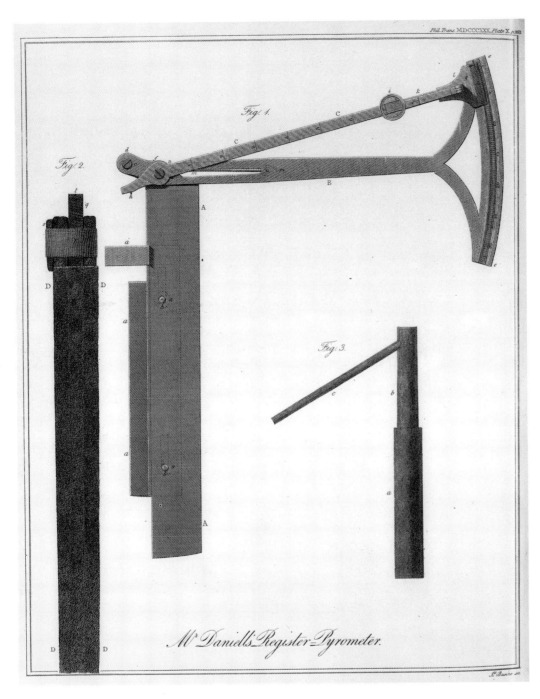

Phil. Trans. MDCCCXXX. Plate X. p.263.

Fig. 1.

Fig. 2.

Fig. 3.

Mr Daniell's Register=Pyrometer.

Daniell 1830

not mentioned by Daniell. This is a pair of callipers with a brass scale which can measure internal and external dimensions. Since the side of the scale is marked from 6 to 7 inches, it is obviously intended to measure the length of the test piece, hot or cold. The other side of the scale is marked from zero to one. Both scales have verniers giving an accuracy of $\frac{1}{200}$ inch. The instrument is in a fitted mahogany box lined with faded blue baize.

In 1830 Daniell wrote that (after 1821) 'The great desideratum still remained of a pyrometer which might universally be applied to the high degrees of heat as the thermometer has been to the lower. The indications of the Wedgwood pyrometer needed connecting to the mercurial scale.' Daniell credited Guyton de Morveau with the basic idea but had brought the instrument to fruition himself.

Daniell 1830
C 82

Inventory No. 1927-1495

L68 Apparatus to demonstrate thermal expansion of liquids

mid-nineteenth century

ON PAPER LABEL
'EXPANSION OF LIQUIDS MERCURY SPIRIT OIL WATER'

H 570mm, **L** 290mm, **W** 128mm

This is probably a teaching aid which has come into the collection via King's College. Five thermometers with large spherical bulbs are mounted in a row in a mahogany stand which has a lined trough in the base. The liquids in four of the thermometers are named by stamped letters in a boxwood strip across the top of the stand. They are labelled mercury, spirit, oil, and water; the mercury thermometer is now broken. The fifth thermometer is not named and appears to contain only air. The first four have scales in degrees Fahrenheit, starting at 32 and going up to 610, 160, 180, and 212 respectively. The scales for oil and water are not linear. The fifth thermometer has no scale. Presumably hot water was poured into the trough and the response of the thermometric fluids was noted.

C 87

See also J32, J33

Inventory No. 1927-1626

L69 Wedgwood pyrometer

Newman, J.

1828–57   London

'*J. Newman 122 Regent St London*'

L 180mm (box), **W** 50mm (box), **H** 63mm (box), **L** 170mm (gauge), **W** 43mm (gauge), **D** 8mm (gauge)

Wedgwood pyrometers and their operation are described in detail in E149. This is a later example which was possibly used for teaching. It is similar to E149 but, like the first pyrometer Wedgwood described, it is made of brass. The two scales 0 to 120 and 120 to 240 are combined on one gauge. The gauge fits into the top of the mahogany box while the thermometric pieces are in the lower part. There is one ceramic case to protect the pieces from the sample if necessary.

After a short time Wedgwood found that he could not repeat his results because the available clay samples were not identical. He managed to overcome this problem by making an artificial mixture of clay and alumina which was used thereafter.

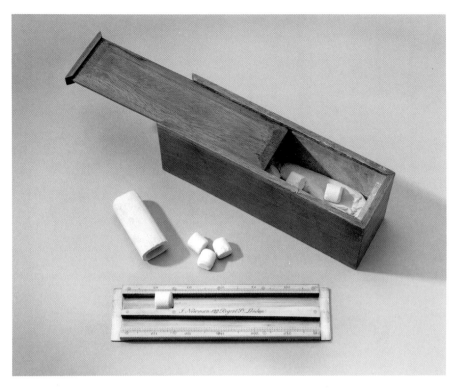

Wedgwood 1782
Wedgwood 1784
Wedgwood 1786
C 81
See also E149

Inventory No. 1927-1813

L70 Six's maximum and minimum thermometer

Newman, J.

1828–57   London

STAMPED NEAR THE TOP
'*J. NEWMAN/122 REGENT STRT/LONDON*'
SCALE
−10 to 120°F

L 380mm (back), **H** 330mm (tubes), **W** 65mm, **DE** 20mm

This is the meteorological type of Six thermometer which was developed in preference to the marine type. It has a sealed glass tube and the lower end of the instrument is protected by a brass bracket. The minimum scale, which is on the left-hand side, unlike the original thermometers, is calibrated from −20 to 120°F. The maximum scale is calibrated from 0 to 120°F. Both scales have the word 'Freezing' stamped at 32°F. There is a brass ring at the top by which to hang the instrument. A full description of the operation of the thermometer can be found under E116.

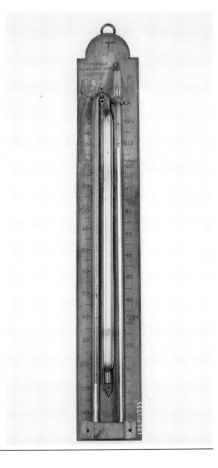

Six 1782
Six 1794   Austin and McConnell 1980*a*
C 97   Austin and McConnell 1980*b*   **Inventory No.** 1927-1733

L71 Back plate from thermometer, with Fahrenheit scale

Newman, J.

1820–8 London

'*J. NEWMAN/LISLE STRT/LONDON*'
'*Freezing*'
SCALE
0 (10) 197

L 256mm, **W** 21mm, **T** 2.5mm

The thermometer bulb and tube are missing. The back plate is made of ivory. The scale runs from 0°F up to 197°F; the freezing point of water at 32°F is also marked. Newman was at Lisle Street, Covent Garden, before he moved to Regent Street.

C 88

Inventory No. 1927-1706

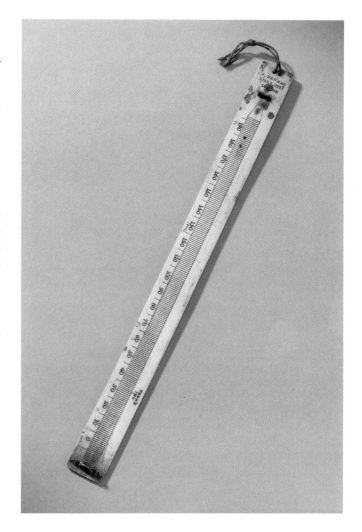

L72 Rod and gauge

first half nineteenth century

L 187mm (rod), **D** 19mm (rod), **L** 270mm (handle), **L** 280mm (gauge), **W** 55mm (gauge), **DE** 2mm (gauge)

This apparatus is usually known as 's Gravesande's rod and gauge since he devised the simple demonstration for teaching purposes. In this version a cylindrical brass rod fixed by a brass shaft to a rosewood handle can be fitted lengthwise into a brass gauge, and its cross-section can be fitted into a circular hole in the side of the gauge. When heated the rod will no longer fit, expansion being shown to have taken place in both length and in cross-section.

's Gravesande 1747, Vol. 2, p. 67, Pl. 77, Fig. 1
C 78

Inventory No. 1927-1117

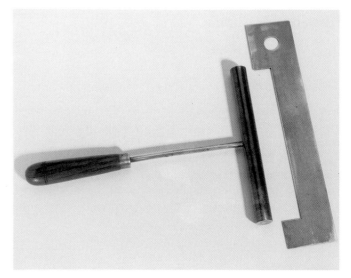

OPTICAL INSTRUMENTS

L73 Melloni's apparatus

Locatel et Cie (oil lamp)

second half nineteenth century   Paris

ON LAMP

   *'BREVET D'INVENTION/LOCATEL ET C*ie*/A PARIS'*

L 985mm (base), **W** 260mm (base), **L** 195mm (copper rectangles), **W** 130mm (copper rectangles), **H** 460mm, **D** 110mm (tin circles), **L** 70mm (sides of cube), **D** 125mm (mirror)

'Melloni's apparatus' can be used to perform various experiments on 'radiant heat' or infrared radiation. The apparatus is quite crude and was probably used for teaching purposes at King's College. It consists of a mahogany base with a central brass slot down its length, along which one can slide various stands. The stands can support one of two double copper rectangles, a linen circle, one of three tin circles, a copper rectangle, or one of three cardboard rectangles with circular or rectangular holes. There is a circular brass platform to hold the source of radiation which is an oil lamp with an oval plaque on which the inscription appears. There is also a square brass platform which carries a Leslie cube with various surfaces. The cube could be filled with liquid. In addition there is a copper reflector, a plane mirror in a swivel mount, and a rectangular mahogany funnel. It is not possible to reconstruct particular experiments with this diverse collection of accessories, but various combinations of them would have been used to demonstrate a variety of properties of infrared radiation.

From 1831 Melloni performed many experiments on the emission, absorption, reflection, refraction, and transmission of infrared radiation. He used a source which was usually an oil lamp and a detector which consisted of a thermopile and galvanometer. His many papers on the subject are listed in the *Dictionary of Scientific Biography*.

D.S.B. 1970–80, under Melloni

Used with J158

Inventory No. 1927-1926

L74 Cell for liquid for optical apparatus

second half nineteenth century
L 120mm, **H** 100mm, **D** 40mm

A cylindrical tube of brass has a plate of glass at each end. The glass is mounted in leather washers in octagonal end-caps. A side tube or handle is not connected internally with the cylinder. Presumably the cylinder was filled with an optically active liquid and viewed through an analyser in polarized light.

Inventory No. 1927-1528

L75 Convex lens in mount

first half nineteenth century
H 610mm, **L** 340mm, **D** 230mm (base), **D** 225mm (lens)

This is a standard biconvex lens with a diameter of 9 inches and a focal length of about 18 inches. It is in a walnut mount in a brass swivel holder on a walnut stand. The knurled knobs indicate a relatively late date.

C 204

Inventory No. 1927-1248

L76 Claude Lorrain mirror

first quarter nineteenth century

D 3mm (glass), **L** 125mm (glass), **W** 90mm (glass), **L** 180mm (box), **W** 105mm (box)

The object is simply a rectangular piece of black glass in a red leather case, both now in very poor condition. The shape of the glass and style of the case have been used to date it.

These pieces of black glass, called Claude glasses or Claude Lorrain mirrors, were very popular in the late eighteenth and early nineteenth centuries for sketching landscapes. Viewing the scene in the glass drained it of colour and enabled the artist to imitate the style of Claude of Lorrain who was greatly admired in England at the time.

Manwaring 1925
C 218

Inventory No. 1927-1430

L77 Newton's disc

mid-nineteenth century

D 300mm, **DE** 27mm

This item probably came into the collection via King's College. The mahogany disc has a brass mount for attachment to a whirling mechanism. There are seven colours on paper which is pasted onto the disc. The 'blue' and 'violet' are very faded.

Inventory No. 1927-1640

L78 Stereoscope

L 190mm, **W** 105mm, **H** 235mm

This instrument must have come into the collection via King's College since this type of stereoscope was not invented until 1849, and only became popular after the Great Exhibition.

It is a folding Brewster stereoscope made of mahogany which has been painted black. It is supported on a base which could be placed on a table top and which held the stereoscopic photographs. The viewer is hinged on a piece of mahogany which can be extended for focusing; when not in use it can be folded onto the base. The eyepieces are prismatic which is common in the 'lenticular' form of the instrument. Optional lenses on a folding piece of mahogany can be added to the eyepieces to give magnification if required.

Stereoscopes were invented as a result of Wheatstone's work on vision in the 1830s. By presenting each eye with a slightly different view, a two-dimensional object can appear to be three-dimensional. Brewster's form of stereoscope uses two photographs side by side taken from slightly different viewpoints. Using either two half-lenses or prisms, the eyes see the two images superimposed between the photographs and this gives the stereoscopic effect. This instrument was probably for use by students since it is not as ornate as the type sold for domestic use.

Wheatstone 1838
Brewster 1849
Brewster 1851
Brewster 1856
C 197

Inventory No. 1927-1256

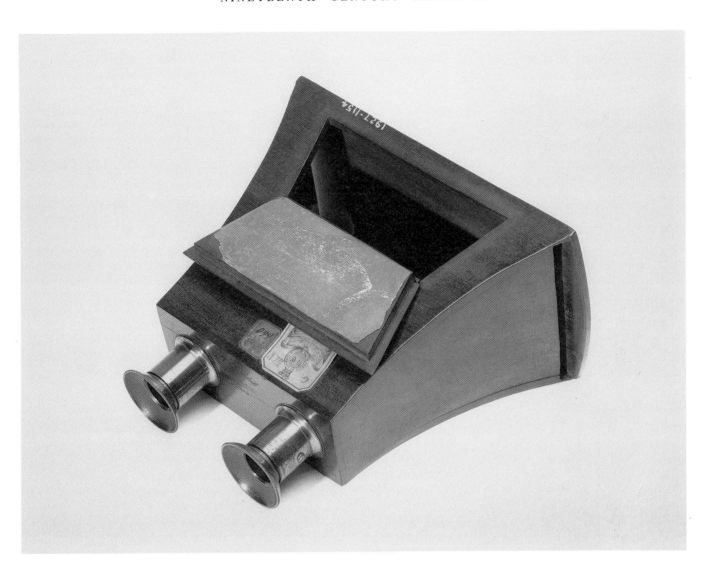

L79 Stereoscope

Duboscq Soleil

c. 1852   Paris

STAMPED ON WOOD
'*Duboscq Soleil/RUE DE L'ODEON/a Paris*'

L 170mm, **W** 160mm, **H** 95mm

This is an early stereoscope of the type invented by Brewster in 1849. Duboscq was the first maker to produce these instruments and the name Duboscq Soleil also suggests a date soon after the invention.

The instrument is made of mahogany with prismatic eyepieces mounted in brass. Focusing is provided by sliding the brass mounts. There is a flap to let in light in the upper surface and a slot to insert the cards. Unlike the previous item the instrument has all the elegance of the domestic varieties.

Wheatstone 1838
Brewster 1849
Brewster 1851
Brewster 1856
C198

Inventory No. 1927-1154

L80 Stereoscopic photographs and designs in box

c. 1852

L 180mm, **W** 80mm

There are eleven photographic cards depicting architecture, sculpture, and scenes from European cities; these are likely to have accompanied the stereoscope described in the previous entry. There are also eight black and white geometric cards of the type used by Wheatstone in his researches on vision; these may belong with the Wheatstone material from King's College. In addition there are five blank cards.

Wheatstone !838
Brewster 1849
Brewster 1851
Brewster 1856
C 198

Used with L79, 1876-521 (Optics Collection)

Inventory No. 1927-1155

L81 Model of a stained glass window to show effects of polarization

Newman, J.

1828–57   London

LABEL—PRINTED
'*J NEWMAN/Optician/122 Regent Street/LONDON*'

W 110mm, **H** 140mm, **L** 160mm

The model window is made of selenite cut in different thicknesses. It has a cardboard surround and is set at an angle of approximately 60° to the horizontal corresponding to the Brewster angle of maximum polarization for glass. The base of the frame would have had a black glass mirror set inside it to polarize the incident light and reflect it upwards through the window to the viewer. When viewed through a Nicol prism a coloured 'stained glass' window would have resulted.

The colour effects of viewing thin sheets of some crystals in polarized light were first studied by Arago in 1811. Many optical toys were based on the discovery.

Arago 1854–62, pp. 349–69
Encyclopaedia Britannica (7th edn) 1842, under Polarization of light

Inventory No. 1927-1475

L82 Ritchie's photometer

Newman, J.

1828–57   London

*'I.NEWMAN/122/REGENT
STREET/LONDON'*

L 205mm, **W** 115mm, **H** 275mm

This item probably came into the
collection via King's College, as did the
other material by Newman.

The instrument is very simple. A
sliding viewer is situated above a mahogany box which has two open sides to
the right and left. Matt white surfaces
inclined at 45° to the vertical reflect
incident light from outside upwards
onto a piece of white linen paper which
acts as a screen. Light from the left
arrives on the left-hand side of the paper
and vice versa. The intensities of the
light for each side can be compared very
easily by adjusting the distance of the
sources until the whole screen shows the
same shade. Then, by Bouguer's principle of the inverse square law of light
intensity, the two sources can be compared. This type of photometer was
invented by Ritchie in 1826; it was
usually used for testing lighting by oil
and coal gas.

Ritchie 1826
C 196

Inventory No. 1927-1140

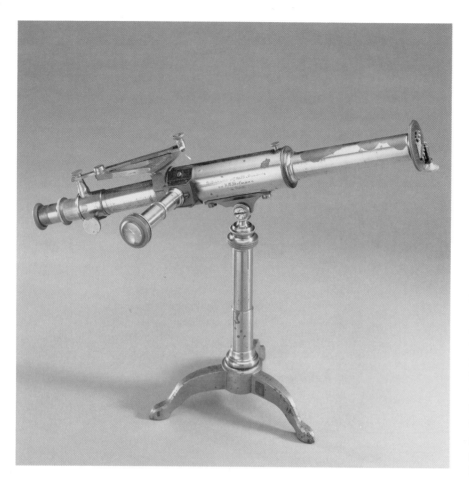

L83 Direct vision spectroscope

Hofmann, J. G.

last quarter nineteenth century
Paris

ON TUBE
'SPECTROSCOPE-HOFMANN
à Vision directe/par J.G. Hofmann/à
Paris'
ON BASE
'KCL'

H 30mm, **L** 500mm, **W** 190mm

This item came into the collection via King's College and is one of the latest instruments present. It is almost entirely made of brass and consists of an adjustable slit in a collimator tube, a train of prisms, a lateral tube containing a photographic scale, which would have been illuminated separately, and the eyepiece. The eyepiece can be rotated horizontally 35° in one direction and 15° in the other by moving a screw operating a rack and pinion over a circular scale above the tube. There is both fine and coarse focusing.

Bennett 1984, cat. No. 4

Inventory No. 1929-118

ELECTRICAL APPARATUS

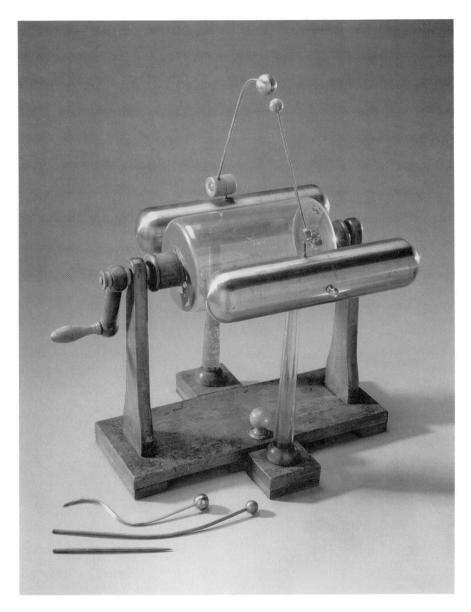

L84 Cylinder electrical machine

Newman, J.

1828–57   London

STAMPED ON BASE
'J. NEWMAN/
REGENT STREET/LONDON'

L 560mm, **W** 440mm, **H** 750mm (with rods)

This electrical machine probably came into the collection via King's College; it appears extremely well used. It is very similar to Nairne's medical electrical machine introduced in 1782. The cylinder is about 7 inches in diameter and coated with sealing wax at each end. It can be turned using a winch with a boxwood handle. It is set on a crude mahogany stand on a mahogany base. There are two hollow brass cylindrical conductors: one acts as the prime conductor with a 13-pronged comb, and the other carries the cushion. Both are on glass supports. The conductor carrying the cushion can be moved with respect to the cylinder by means of a channel in the base, but the other conductor is fixed. Five brass rods can each be attached to either conductor.

Both positive and negative charges of equal magnitude could be obtained from this machine.

C 275
Turner 1973, No. 282
Hackman 1978*b*, p. 135

Inventory No. 1927-1246

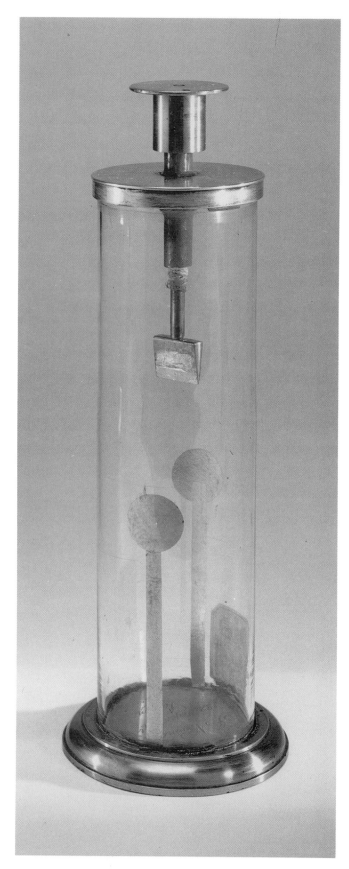

L85 Gold leaf electroscope

Newman, J.

1828–57   London

'*J Newman/122 Regent St^t./London*'

H 250mm, **D** 83mm

This is a standard form of gold leaf electroscope in a glass cylinder with a brass base and cap. The pieces of gold leaf were attached to the sides of a thin triangular brass wedge, but are now separate. Two circular pieces of tinfoil on the internal surface of the glass cover the region the leaf may touch or approach in order that there is no interference from charge on the glass. The instrument exactly resembles that described by Bennett who invented the gold leaf electroscope in 1787.

Bennett 1787
C 309
Hackman 1978*a*, p. 23
Inventory No. 1927-1191

L86 Gold leaf electroscope

Newman, J.

1828–57   London

'*I. NEWMAN/122/REGENT STREET/LONDON*'

H 480mm, **D** 170mm

The following three instruments form a set. This one is complete except that the gold leaf has become detached. It consists of a glass jar with a brass cover on a circular mahogany platform on a mahogany base. The top section is a flat brass cap which is connected by a rod to a triangular brass wedge inside the jar to which the gold leaf was fixed. Two tinfoil strips are attached to the inside surface of the glass to prevent interference. Like L88 there is a rectangular hollow where the plaque containing the signature was situated.

The brass cap would have been charged and the divergence of the gold leaves observed.

Bennett 1787

Inventory No. 1929-94

L87 Gold leaf electroscope

Newman, J.

1828–57   London

'*I. NEWMAN/122/REGENT STREET/LONDON*'

D 170mm, **H** 430mm

This electroscope appears to have been more elaborate than L86 and L88. However, the top section is now missing. The base contains an ivory plaque with the maker's name. The glass jar is identical to that on the previous item but without the tinfoil strips on the inner surface.

Bennett 1787
C 310

Inventory No. 1929-93

L86

L88 Gold leaf electroscope

Newman, J.

1828–57   London

'I. NEWMAN/122/
REGENT STREET/LONDON'

D 170mm (stand), **H** 170mm (stand),
D 108mm (top), **H** 140mm (top)

The base and top part are identical to
those of L86 but the glass jar has not
survived. A rectangular hollow indicates where the plaque with the signature was situated.

Bennett 1787

Inventory No. 1929-95

L89 Parallel plate condenser

Newman, J.

1828–57   London

'I. NEWMAN/122/
REGENT STREET/LONDON'

D 235mm (discs), **L** 235mm (base),
W 148mm (base), **H** 380mm

The 'condenser', now known as a
capacitor, consists of two parallel circular brass plates. One of the plates is on a
brass stand which can be moved with
respect to the other on a slider in the
base. The other is fixed on an insulated
glass stand. Both have brass balls for
charging. The maker's name is on an
ivory plaque in the mahogany base.

This is a standard condenser following the principles of Volta's original
design which was introduced in 1782.

Volta 1782
Noad 1859, pp. 62–3
C 290

Inventory No. 1927-1183

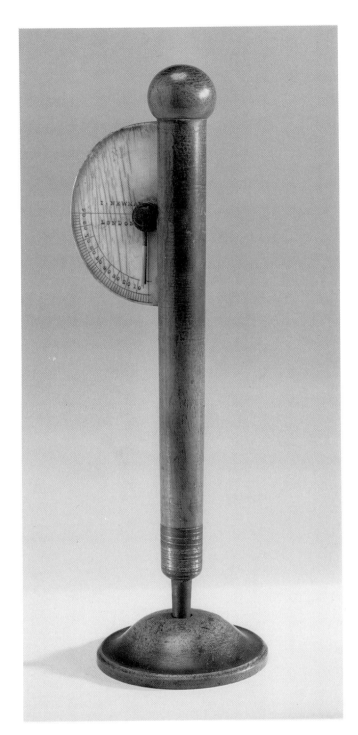

L90 Quadrant electrometer

Newman, J.

1828–57   London

'*I. NEWMAN/LONDON*'

H 180mm, **D** 55mm (base), **H** 30mm (radius of quadrant)

The quadrant electrometer consists of a boxwood pillar with a brass ring at the lower end on a boxwood base. A semicircular ivory quadrant scale is attached to the side of the pillar and gives degrees from 0° to 90° numbered in tens. At one time the pith ball hung on a piece of straw from the centre of the scale so that it touched the brass ring; the pith ball is now lost.

Henly designed this type of electrometer shortly before 1772. The pith ball is repelled in proportion to the quantity of charge on the brass ring and its deflection can be read on the scale. According to Adams it was 'the most useful instrument yet'. It was removed from its base and attached to the prime conductor where it could show the variation of the charge being generated by the machine. It could be left in place during experiments.

Priestley 1772
Cavallo 1777, pp. 161–3, Pl. 1, Fig. 7
Adams 1787*a*, p. 49, Fig. 6
Adams 1799*a*, p. 344, Pl. 1, Fig. 17
C 307

Inventory No. 1927-1905

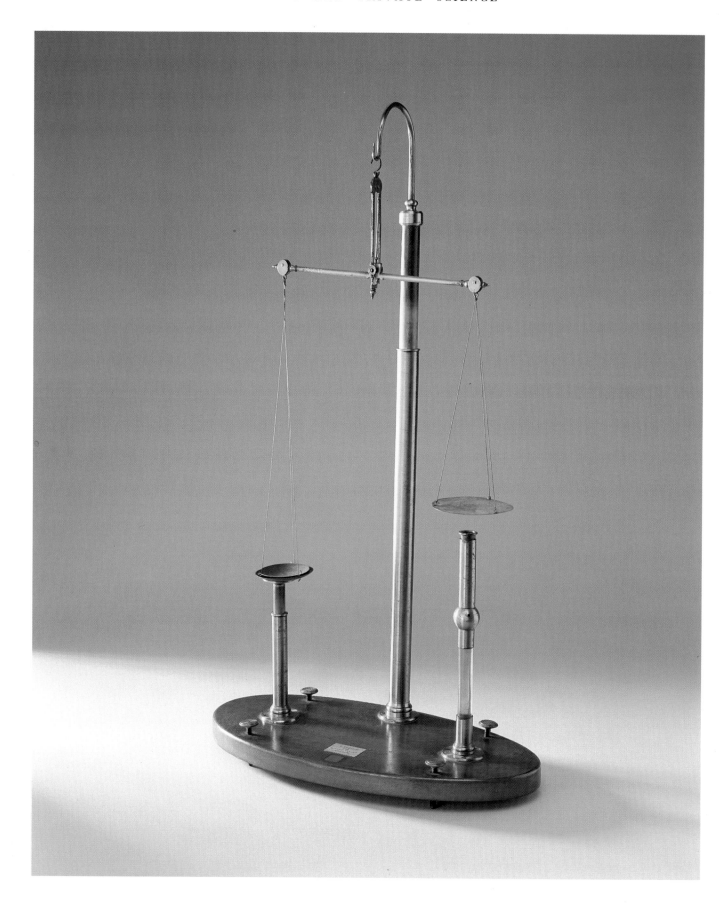

L91 Harris's balance electrometer

Newman, J.

1834–57   London

ON IVORY PLAQUE IN BASE
'*I. NEWMAN/122/
REGENT STREET/LONDON*'

H 510mm, **L** 345mm, **W** 180mm,
L 235mm (beam)

This is one of two Harris balance electrometers in the collection; it is more complete and probably earlier than the following item.

The instrument stands on an oval mahogany base on four brass levelling screws. The balance is supported on a tall brass pillar which is positioned towards the back of the base. It can be raised or lowered as required. The balance beam is made of brass about 9 inches long with box-type ends. A vertical pointer enables equilibrium to be found. The left-hand arm carries a double scale-pan of black painted brass which held the weights. The right arm carries a copper plane which would have been attracted to the plane below by electrostatic force. Under the left-hand scale-pan there is an adjustable brass pillar on which the weights could be supported when the experiments were set up. The pillar under the right-hand scale-pan is insulated. The upper part is brass which can be raised or lowered, and is marked with an unnumbered scale to give an indication of distance from the scale-pan. The top of the pillar consists of a gilt brass-covered plane of wood.

Snow Harris designed this instrument in 1834 to compare the quantity of electricity imparted to a substance with its attractive power. In an experiment similar to van Musschenbroek's concerning magnetism a century before, he weighed the force of the electric attraction on a balance. He was also able to carry out experiments involving the effect of the distance of the planes and challenged Coulomb's $1/r^2$ law. The

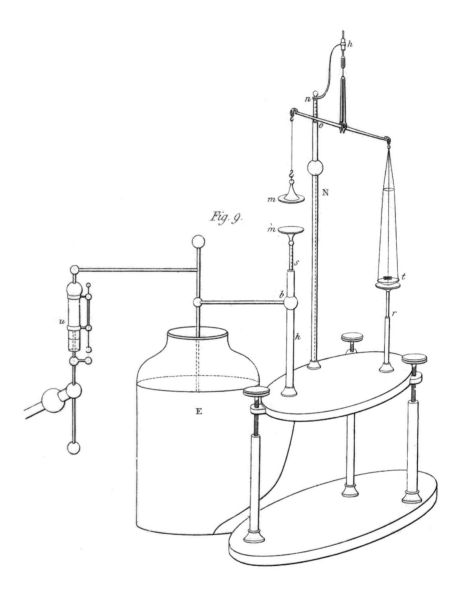

Fig. 9.

Harris 1834

conductor at the top of the glass pillar was electrified using a Leyden jar and the attraction of the scale-pan was counteracted by weights on the other arm of the balance.

Harris 1834, Fig. 9
C 311
Hackman 1978*a*, Pl. 23
Heilbron 1979, p. 476
Inventory No. 1929-96

L92 Harris's balance electrometer

Cox, W. C.

1840–74   Devonport

'Cox/Devonport'

H 420mm, **L** 350mm, **W** 190mm

This instrument is possibly slightly later than the previous item and was almost certainly acquired by King's College.

The instrument stands on an oval mahogany base on three levelling screws. At the back is the support for the balance which is a tall brass pillar with a brass arm. On either side of the instrument are pillars for use with the balance. The left-hand one is a brass pillar with a brass top which would have supported the scale-pan containing weights when the instrument was in use. The other has a glass stem and a brass upper section which can be raised or lowered; the top is boxwood covered in gilt brass. A scale running around the brass section is marked 0 to 25 to give the height of the pillar in order to perform experiments on Coulomb's inverse square law. In addition there is a small pillar with a screw and lignum vitae top at the front of the instrument which is not described by Harris. The balance is now missing.

Harris 1834, Fig. 9
C 312
Hackman 1978*a*, Pl. 23
Heilbron 1979, p. 476

Inventory No. 1927-1190

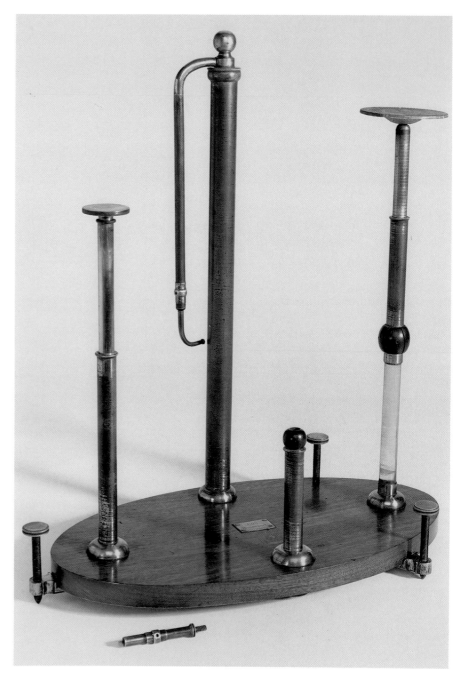

L93 Harris's electroscope

Watkins and Hill

1834–49   London

'Watkins & Hill, Charing Cross'

H 510mm, **D** 130mm, **D** 105mm (scale), **DE** 150mm

Again the instrument probably came into the collection via King's College.

Like Henly's electrometer, it uses the repulsion of like charges to cause divergence. An elliptical brass ring has two brass rods set vertically in opposite directions attached to its extremities. A brass ball on a brass rod extends backwards from the ring. Pivoted across

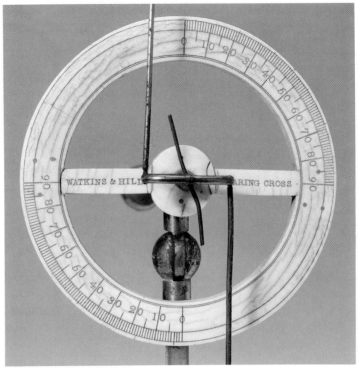

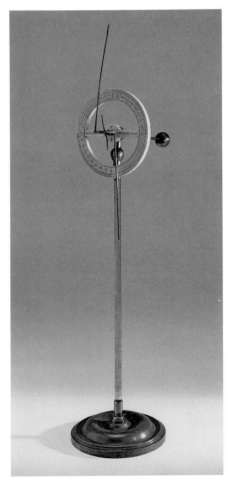

the short diameter of the ring is an axis carrying two short brass pins. At one time these pins would have held straw reeds to form indices which terminated in pith balls, but these are now lost. A circular ivory scale on which are marked two quadrants from 0° to 90°, corresponding to the loci of movement of the straw reeds, is situated behind these indices. The horizontal brass rod is mounted on a brass ball and socket joint and can be moved to a certain extent in a vertical circle. The instrument is insulated on a glass stem on a mahogany base.

When the brass ball behind the scale was electrified the straw indices would have been repelled by the brass rods. Harris referred to the instrument as an 'electroscope' rather than an 'electrometer', since the force of repulsion diminishes as the divergence increases.

Harris 1834, pp. 214–15, Pl. 2, Fig. 1
Noad 1859, pp. 26–7
C 308

Inventory No. 1927-1192

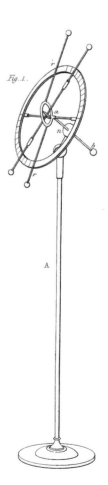

Harris 1834

L94 Coulomb torsion balance

second half nineteenth century

D 240mm (base), **H** 530mm, **D** 25mm

This item was added to the collection while it was at King's College. The instrument is now far from complete; the internal parts are missing. It consists of a glass cylinder on which is a horizontal paper scale marked from 0 to 30. The name Baird and Tatlock can just be seen but there is evidence of previous scales so this is probably not the original. The glass tube which held the filament is present but has become detached from the cylinder. The top is a flat mahogany disc with a brass pointer which can be moved over a circular paper scale marked in degrees.

Coulomb invented this type of electrometer in 1785. A filament consisting of a waxed silk thread was suspended from the top of the instrument. It carried a cross-piece, one end of which held a pith ball and the other a paper disc dipped in turpentine to reduce oscillations and provide a counter-balance. A second pith ball gilded with brass was held rigid by a clamp above a hole in the top of the glass cylinder so that it was at the same height as the first. If this ball was charged and allowed to touch the first they repelled each other and the filament twisted. By applying torsion on the filament by turning the pointer over the top scale the degree to which the balls could be made to approach each other could be observed. Coulomb's results confirmed the inverse square law for electrical repulsion.

Coulomb 1785*a*
Coulomb 1785*b*, pp. 601–3
Noad 1859, p. 32, Fig. 11
Heilbron 1979, pp. 470–3

Inventory No. 1929-117

L95 Apparatus for Richmann's experiment on electric charge

Newman, J.

1828–57   London

'*I NEWMAN/REGENT ST/LONDON*'

L 430mm, **W** 430mm, **DE** 20mm, **L** 170mm (tinfoil), **W** 170mm (panel)

This appears to be the main part of the apparatus used in Richmann's experiment, which demonstrates Franklin's discovery that the two charges disposed on the inner and outer surfaces of a Leyden jar are equal. It is a large pane of framed and varnished glass which has a central section which is coated on both sides with tinfoil. Either side could have been earthed if required. The pane of glass would have been supported vertically in a stand. Two pith balls would have been hung so as to touch each side of the coated glass. One side was then charged and the other

earthed so that one ball rose. The contact with earth was then broken and both balls were seen to rise equally. Various similar experiments could be observed with this apparatus.

Noad 1859, pp. 134–5, Fig. 93
Dictionary of Scientific Biography 1970–80, under Richmann, G. W.

Inventory No. 1927-1669

L96 Armstrong's boiler

after 1842
'PROVED|FOR THE|LONDON| PORTABLE GAS|COMPANY'
L 965mm (boiler), **D** 305mm (boiler), **L** 800mm (base), **W** 600mm (base), **L** 380mm (legs), **D** 70mm (legs), **L** 650mm (nozzle)

This item must have come into the collection via King's College since it was not invented until 1843. The London Portable Gas Company was active between 1828 and 1834.

It consists of a cylindrical boiler of black-painted wrought iron which has an opening with a screw thread at one end. It rests on a rolled iron stand in which the fire was housed. The stand has a tray at the bottom, two grates for coke, two doors, and a circular hole at one end which was probably for viewing. It stood on four glass insulating legs, only two of which remain. The whole is on a pine base with four handles. In addition there is an iron nozzle and stopcock, a connector, and a piece of iron piping.

The boiler closely resembles Armstrong's second design as proposed in 1843. In 1840 he noticed that, for a variety of steam engines, sparks could be produced from an insulated conductor placed in the vicinity of the steam jet, or more easily from the boiler itself. In a series of experiments he observed positive and negative charged steam by adding various chemicals to the water, discovered that the charge was caused by friction as the steam left the nozzle, and was led to believe in the one-fluid theory of electricity. By building larger boilers and improving the friction of the nozzle, he developed a hydroelectric machine which was the most powerful of its day.

Armstrong 1840
Armstrong 1842
Armstrong 1843, p. 2
C 278

Inventory No. 1927-1702

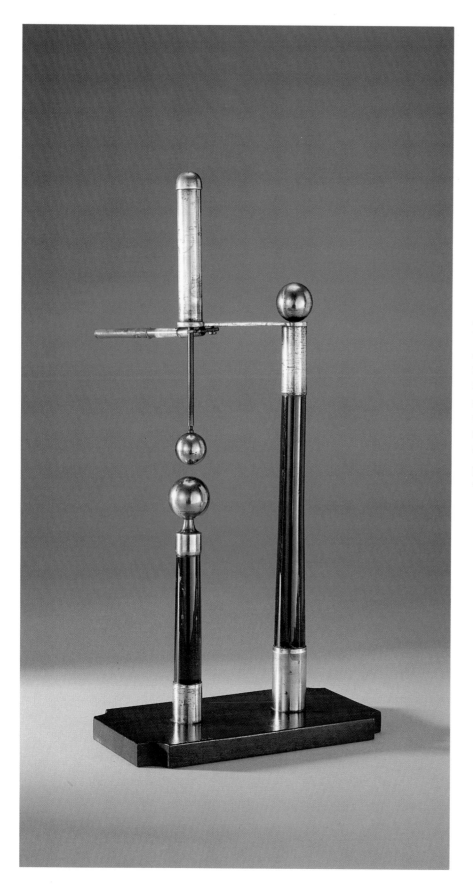

L97 Adjustable spark gap

first half nineteenth century
H 440mm, **L** 225mm, **W** 115mm

This piece of apparatus could have come into the collection in the 1780s when the Prince of Wales bought some electrical apparatus from George Adams the Younger. However, there is no particular reference to a spark gap and this instrument is probably of a later date.

The instrument consists essentially of two brass balls. The lower one is supported on an insulated blue glass pillar; the upper one can be held at the required height above it by a catch. The catch and upper ball are attached to a brass bar from a second taller pillar, also of blue glass. The catch has an insulated glass handle partially covered in sealing wax. The catch can be released to let the upper ball make contact with the lower one. Either ball could be earthed while the other was charged.

Inventory No. 1927-1444

L98 Discharge gap

first half nineteenth century
L 485mm, **W** 165mm, **H** 405mm

A piece of mahogany of rectangular cross-section is supported on glass pillars at each end. In its centre is a vertical hole through which passes a brass rod. The rod has a brass ball at each end. Directly beneath the lower ball a brass rod extends upwards from the mahogany base, also ending in a brass ball. The space between these balls constitutes the discharge gap; the distance can be adjusted. The lower rod can be earthed through the base.

Discharge gaps were used to prevent an excessive build-up of charge.

C 303

Inventory No. 1927-1704

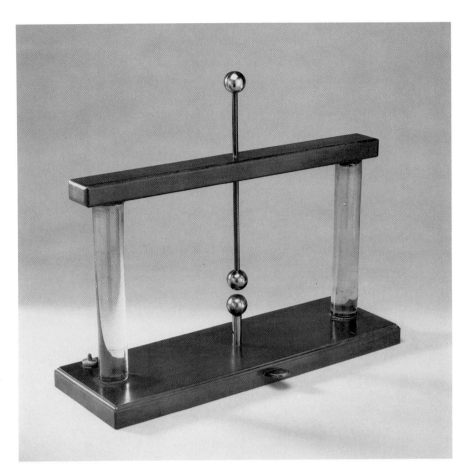

L99 Battery of Leyden jars

Newman, J.

1828–57   London

ON IVORY PLAQUE
'I NEWMAN/122 REGENT ST/ LONDON'

H 470mm, **L** 580mm, **W** 580mm,
H 310mm (jars), **D** 160mm (jars)

There are nine Leyden jars arranged in three rows of three; one jar is now broken. They have lead foil on their outer surfaces and tinfoil on their inner surfaces. Three of the jars have brass cages to collect or distribute the charge, while the others are connected to the brass charging balls by rigid rods extending to their bases. Three, six, or nine jars can be used at one time. The jars are housed in a mahogany box.

Adams 1787*a*, pp. 147–9, Fig. 65
C 289

Inventory No. 1929-126

L100 Leyden jar

first half nineteenth century
H 240mm, **D** 93mm

The tinfoil coating on the Leyden jar is in very poor condition. There is a brass ball conductor on a rod which passes through a mahogany cover. There is some brass wire and a piece of chain inside the jar.

C 287

Inventory No. 1927-1263

L101 Pair of Leyden jars

first half nineteenth century
H 260mm (both), **D** 90mm (both)

These Leyden jars are very similar and probably formed a pair. They have glass tubes down the centre which are coated with tinfoil. A brass wire runs from the brass ball conductor at the top through the glass tube to the base. They both have mahogany lids. One of the jars has lost most of its tinfoil.

C 286

Inventory Nos 1927-1266, -1270.

L102 Leyden jar

first half nineteenth century

H 250mm, **D** 90mm

The jar has a chain extending from the brass ball conductor through the mahogany lid to the base of the interior. Most of the tinfoil has come off both the surfaces.

C 286

Inventory No. 1927-1264

L103 Leyden jar

first half nineteenth century

H 220mm, **D** 65mm

This is a small Leyden jar with tinfoil on the inside of the glass and black-painted tinfoil on the outside. There is a brass hook attached to the sealing wax bung at the top. Four wires are twisted together inside the jar to form a cage with which to transmit the charge.

C 287

Inventory No. 1927-1265

L104 Leyden jar

first half nineteenth century

H 345mm, **D** 150mm

The Leyden jar has a tinfoil coating both internally and externally. There is a brass ball conductor on a brass rod which extends through the mahogany lid and joins eight wires which are twisted together to form a cage.

C 285

Inventory No. 1927-1271

L105 Leyden jar

first half nineteenth century

H 310mm, **D** 260mm

This is a large Leyden jar with painted lead foil on the outside and tinfoil on the inside. Eight pieces of brass wire form a cage in the jar and are twisted together at the top. A brass screw protrudes from the mahogany cap. There is also a thin piece of blackened brass wire which extends into the jar from a small hole in the cap.

C 287

Inventory No. 1927-1272

L106 Glass jar containing lead shot and conductor

first half nineteenth century

H 150mm (jar), **H** 240mm (variable), **D** 75mm (jar)

This is a glass jar about two-thirds full of lead shot. A brass conductor consisting of a rod with a spherical top fits through a cork bung and extends into the shot. It was probably not made for this purpose since it has a screw thread at the base.

C 287

Inventory No. 1927-1267

L107 Aurora tube on stand

first half nineteenth century

H 470mm, **D** 50mm, **D** 130mm (base)

This differs slightly from the more usual aurora tubes, such as E207, in that the conductors are not single spikes on balls but three-pronged forks. The luminous discharge would emerge in three strands. The lower brass cap can be unscrewed from the mahogany base and a threaded flange would allow an attachment to an air pump.

Watson 1752
Adams 1787a, p. 119, Fig. 49
C 293

Inventory No. 1927-1440

L108 Aurora globe

first half nineteenth century

H 450mm, **D** 170mm (base), **D** 160mm (globe)

The aurora globe has features in common with the carbon arc lamp L122 and is probably by the same maker. The spherical globe can be evacuated by means of the tap in the base. The lower brass ball conductor is fixed to the base while the upper one is adjustable. There is an insulated glass handle at the top. Globes and flasks were popular variations on aurora tubes.

Watson 1752
Noad 1859, p. 80
C 291

See also E207, L122

Inventory No. 1927-1274

L109 Fulminating pane

West, Francis

second quarter nineteenth century   London

L 138mm, **H** 210mm, **DE** 10mm

The fulminating pane consists of a rectangular piece of glass in a mahogany frame which has a brass ball conductor at one side and a hook from which a chain could be attached for an earth at the other side. A tinfoil line connects these two points; gaps in the tinfoil indicate where a spark would be seen if the instrument were charged. The gaps are arranged to give the word WEST, suggesting that it was made by Francis West of Fleet Street, London.

   This type of fulminating pane was popular for a long period between the late eighteenth and mid-nineteenth centuries.

Adams 1787*a*, pp. 91–2, Fig. 32   Calvert 1971, Nos 438, 439, 440
Noad 1859, p. 82
C 297   **Inventory No.** 1927-1185

L110 Fulminating pane

first half nineteenth century

H 455mm, **W** 90mm, **D** 150mm (stand), **D** 35mm (ball)

The rectangular pane is mounted vertically in a mahogany stand. There is a brass ball at the top for charging and tinfoil 'spangles' in a zigzag pattern running down to the base. The glass is painted in broad stripes horizontally so that the spark could be viewed through various colours. This type of pane had become available in the first half of the nineteenth century.

Sturgeon 1842, p. 85
C 300   **Inventory No.** 1927-1188

L111 Instrument to observe atmospheric electricity?

first quarter nineteenth century

L 320mm, **W** 120mm, **H** 780mm

Chaldecott suggests that this instrument may have been used to observe atmospheric electricity.

A tall mahogany pillar is situated on one side of the mahogany base. Above the pillar is a small glass rod with a conical plug of sealing wax which supports a tinplate cone. A brass wire is attached to the base of the cone and runs down to join the top of a short glass and mahogany pillar on the other side of the base. A brass ball about halfway down the wire originally held another brass wire horizontally. This wire had a small brass ball at each end; it is now missing.

C 302

Inventory No. 1927-1763

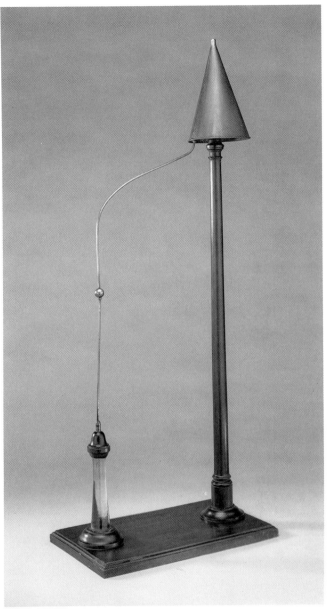

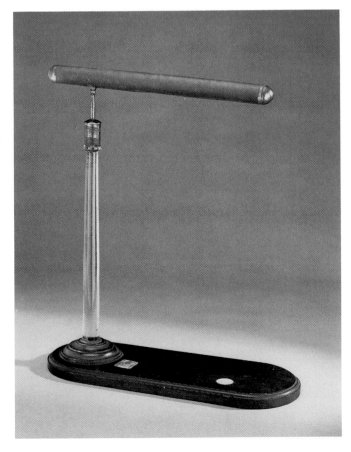

L112 Cylindrical conductor

first half nineteenth century

L 360mm (conductor), **D** 30mm (conductor), **L** 405mm, **W** 160mm, **H** 465mm

The hollow brass cylindrical conductor is coated in painted tinfoil and has brass hemispherical ends. It stands on a glass insulating rod on an oval mahogany base.

Inventory No. 1927-1443

L113 Conductor

mid-nineteenth century

H 245mm, **D** 110mm (base), **D** 30mm (ball)

The conductor is simply a brass ball mounted on a brass rod which screws into a mahogany base.

Inventory No. 1927-1261

L114 Electrostatic apparatus

Newman, J.

1828–57   London

ON BASE
'I NEWMAN/REGENT ST/ LONDON'

L 300mm (base), **W** 330mm (base), **H** 260mm, **H** 240mm (rods), **L** 115mm (wires)

The apparatus consists of a mahogany board 1 foot square with nine upright glass insulating rods positioned in three rows of three. The rods have brass collars at the top and steel pointers on which are pivoted brass wires, each $4\frac{1}{2}$ inches long. Each wire had a pith ball at either end, although these have mostly disintegrated or been lost. It is not certain what principle the apparatus demonstrated, but presumably the brass collars were charged and the wires aligned themselves to minimize the electrostatic potential.

Inventory No. 1927-1447

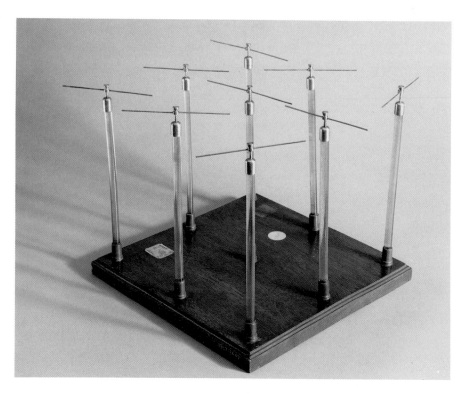

L115 Volta's crown of cups

first quarter nineteenth century

D 252mm (tray), **D** 255mm, **H** 130mm

The instrument consists of nine glasses which can be arranged in a horseshoe shape on a black-painted tinplate tray. Forked conductors, with one side zinc and the other copper, fit between each pair of glasses but do not touch inside the glasses. Single strips of zinc and copper are placed in the end glasses to form terminals. The presence of a tenth conductor indicates that there were originally ten glasses. When the glasses were filled with water, or preferably brine, a potential difference was set up between the terminals.

The 'couronne des tasses' was one of the experiments described in Volta's important paper on electric piles in 1800.

Volta 1800, Pl. 17, Fig. 1
C 313

Inventory No. 1927-1255

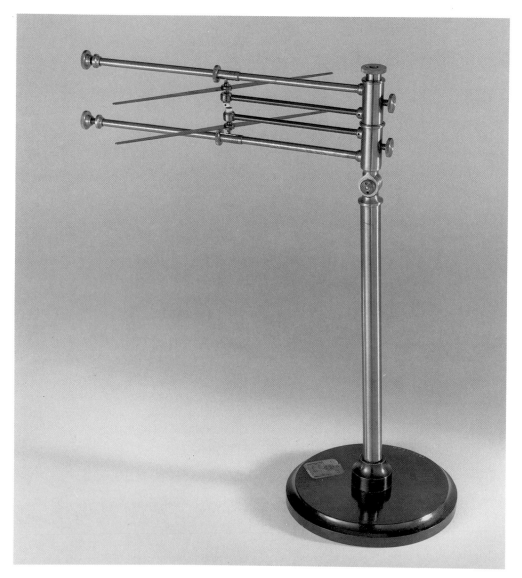

L116 Double 'oersted'

second quarter nineteenth century

H 395mm, **L** 283mm, **L** 215mm (needles), **D** 165mm (base)

This apparatus for showing Oersted's discovery may have come into the collection via King's College. It consists of a pair of steel astatic needles, i.e. a pair of needles with opposite poles aligned placed one above the other. This arrangement enables the magnetic field of the Earth to be discounted. The needles are able to rotate around an axis which is in two parts, each ending in an ivory knob. Horizontal brass rods above the upper needle and below the lower needle carry the electric current from terminals at their ends to terminals on the central brass pillar. Passing the current in the same direction simultaneously along both rods has the effect of turning both needles in the same direction, and since they are astatic the effect is enhanced. The apparatus is on a brass pillar on a painted boxwood base.

Oersted discovered the magnetic effect of an electric current in 1819. This instrument is a simple demonstration model of the type of instrument developed by Oersted, Ampère, and others around 1821.

Oersted 1821
Ampère 1826
Oersted 1928
Blondel 1982
Hendry 1986, pp. 55–71

Inventory No. 1927-1260

L117 Electric cell

second half nineteenth century

L 333mm, **H** 290mm, **W** 100mm, **D** 260mm (wheel)

This instrument came into the collection via King's College; it appears to have been made partly in a laboratory. Three concentric discs, the outer ones copper and the inner one zinc, rotate in a pine trough lined with sealing wax. The outer ones are connected at various points around their circumferences. An inner axle to the zinc disc is insulated from an outer axle to the copper discs. Both axles carry small brass discs which rotate in small troughs which would have been filled with mercury. Brass wires from each of the small troughs form the electrodes.

The trough would have been filled with an electrolytic solution and the large discs would have been turned using the boxwood handle. This would have the effect of preventing polarization. A potential difference would have been set up between the discs which would have been communicated to the mercury troughs via the axles and hence to the electrodes.

Inventory No. 1927-1521

L118 Needle galvanometer

Newman, J.

1834–57   London

'*I. NEWMAN/122/REGENT STREET/LONDON*

L 230mm, **W** 175mm, **H** 315mm

Like the majority of instruments using electric current, this item almost certainly came into the collection via King's College. It consists of a mahogany base board on three levelling screws on which is pasted a compass card. Above the card is a coil of wire. A brass rod suspends a pair of astatic needles, one of which can be rotated inside the coil and one above it. The thread on which they hang passes through an opening in the top of the coil. The glass dome protecting the instrument is now missing. The instrument is an early form of galvanometer using the coil attributed to Cumming and the astatic needle which was first used with a galvanometer by Nobili. The coil amplifies the effect of a current on a magnetic needle and the astatic pair negates the Earth's magnetic field.

Cumming 1822*a*
Cumming 1822*b*
Nobili 1834
C 315
Stock and Vaughan 1983

Inventory No. 1927-1275

L119 Saxton's magneto-electric machine

Newman, J.

1833–57   London

ON IRON HORSESHOE
'I. NEWMAN' and *'REGENT ST./LONDON'*
L 550mm, **W** 310mm, **H** 510mm

This machine is similar but not identical to that described by Saxton in 1834. It consists of seven painted steel horseshoe magnets fixed together horizontally on a mahogany platform on a mahogany base. The armature is constructed of two soft iron pieces connected together with an iron cross-piece and wound with insulated copper wire. The armature can be rotated by turning the walnut handle on the vertical mahogany wheel which, by means of a cord and small pulley, turns an axle running the length of the instrument. The end of the axle extends beyond the armature and carries a small brass disc. One end of the wire is connected to this, while the other is connected to the axle which passes through the disc and is insulated from it by an ivory ring.

This model is now incomplete; only the pillar which supported the platform holding the 'flood cups' survives. The pillar can be moved radially on a slider to enable the distance between the machine and this platform to be varied.

The machine was used in various ways. A diamond-shaped bar could have been attached to the axle and aligned to dip into a small cup of mercury. The disc also dipped continuously into a cup of mercury, and as the cups were connected the circuit was shorted whenever the bar touched the mercury. The large back e.m.f. produced sparks. Another experiment was the melting of platinum wire. A circuit was formed by connecting insulated wire to a brass sliding contact acting on the disc, connecting it to a piece of platinum wire in a glass tube, and then making contact with the axle. Alternat-

ing current could have been produced from terminals in the platform which were connected by sliding brass contacts with the disc and axle respectively. Probably the most common use was in the administration of electric shocks. Noad describes two alternative armatures but there is only evidence of one in this example.

The design represents an early form of magneto-electric machine: a machine converting mechanical energy into electrical energy using a magnetic field. The first was made by Pixii in 1832. This design was used by Saxton to decompose water, make sparks jump between mercury and a wheel, ignite wire, and give shocks to the mouth and tongue. Saxton's original machine was exhibited at the 1833 meeting of the British Association and subsequently at the Adelaide Gallery in London. When a description of Clarke's machine appeared in 1836, Saxton accused him of piracy. The two machines were the same in principle.

Noad 1859

Saxton 1834
Saxton 1836
Noad 1859, pp. 697–700, Figs 372–378
C 316   **Inventory No.** 1927-1277

NINETEENTH CENTURY MATERIAL

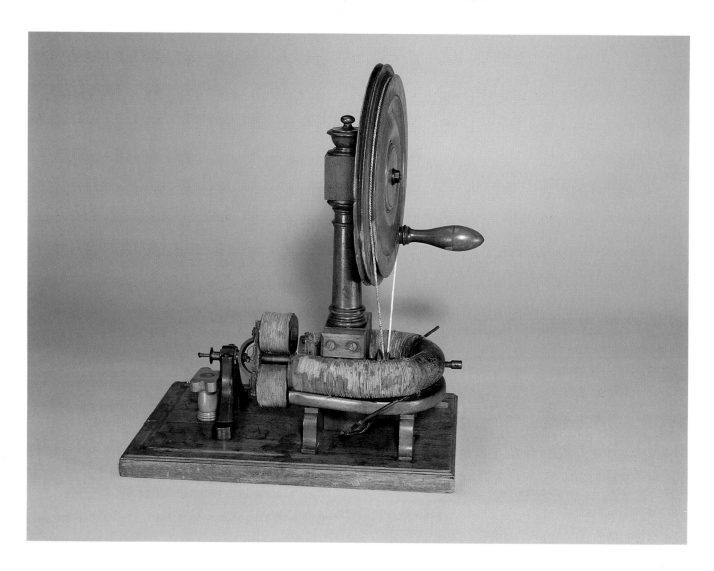

L120 Clarke's magneto-electric machine

Newman, J.

1836–57   London

L 465mm, **W** 295mm, **H** 550mm

Like Saxton's machine, this is not identical to the instrument described by Clarke in 1836 which has the magnet vertical. In principle it is the same as the Saxton machine, the only significant difference being in the composition of the magnet itself, a fact not mentioned by Clarke.

The magnet is a cylindrical piece of soft iron bent into a horseshoe and bound with thick insulated copper wire. The armature consists of two pieces of soft iron also bound with insulated copper wire, as in Saxton's machine. A walnut handle on a mahogany multiplying wheel turns the axle which turns the armature. The projection at the end of the

armature still carries the two-pronged 'wheel' or diamond-shaped bar which would have dipped into a cup containing mercury supported on the small boxwood pillar. As with Saxton's machine, the pillar on a slider remains but the cup does not.

Clarke's machine in this form requires a source of electric current to energize the electromagnet.

Saxton 1834
Clarke 1836
Saxton 1836
Sturgeon 1836–43, Vol. I, pp. 145–55
Noad 1859, p. 697
C 317

Inventory No. 1927-1788

L121 Induction coil with interrupter

Newman, J.

c. 1840–57　　　　　　　　　　　　　London

ON IVORY PLAQUE
'J. NEWMAN/122 REGENT STR/LONDON'
L 290mm, **W** 125mm, **H** 140mm

Like much of the material by Newman, this item came into the collection via King's College. It is a fairly early example of an induction coil, having a number of features in common with Sturgeon's and Bachhoffner's machines. It appears to have been altered at some subsequent date.

The instrument consists of a primary and secondary coil wound round a mahogany bobbin. The primary coil is of thick copper wire and the secondary is of thin copper wire. A silk cover conceals the winding. There are two terminals at each end of the instrument. One is connected to a brass stand which carries a brass spring. The spring makes contact with a brass rod which is connected to the primary coil and back to another terminal. The current also passes through thick wire wound around an electromagnet which is placed under the spring. When this operates, a steel keeper at the end of the spring is drawn down to the magnet and the circuit is broken. The electromagnet then releases the spring which again makes contact and so the process repeats itself. The maker's name is on an ivory plaque in the base of the instrument.

Sturgeon 1836–43, Vol. I, pp. 470–84 in particular p. 478, pp. 496–7 letter from G. H. Bachhoffner

Inventory No. 1927-1449

L122 Carbon arc lamp

second quarter nineteenth century
D 120mm (globe), **H** 400mm, **D** 125mm (base)

This is an early form of electric lamp. An ellipsoidal glass globe can be partially evacuated by means of a tap in the base. A carbon pole piece is fixed on a rod to the base and extends into the globe. A similar pole piece was mounted on the upper steel rod which runs through an air-tight collar to the ring handle at the top; however, this is now missing. If the top and bottom of the instrument were connected to the terminals of a large battery and the pieces of carbon were first in contact but then drawn apart, an intense arc of light would be seen between the poles.

Davy is usually credited with making the first arc lamp at the Royal Institution in 1802, but although his apparatus was essentially similar his intentions were more complex than the mere production of light.

Royal Institution 1802, p. 214, Fig. 2
C 314
Bowers 1982, p. 64

Inventory No. 1927-1179

L123 Lorentz 'instantaneous light machine'

after 1806

L 240mm, **W** 240mm, **H** 360mm

This item is identical to that described in Lorentz's patent specification of 1807. It is an oil lamp which is lit by an electric spark acting on a stream of hydrogen gas passing over the wick.

The large mahogany chest has three internal parts: an electrophorus in the base and two chambers above which are used to produce hydrogen at high pressure. The chambers are lined with painted tinplate. The upper chamber has a vessel into which zinc and hydrochloric acid can be put to generate the hydrogen. It has a cock and a pipe to the lower chamber and a connection with the electrophorus in the base which can be adjusted to give a spark gap. The front of the instrument carries the brass oil lamp which has a cylindrical oil bath with a wick protruding at the top. A brass cap protects the wick when not in use. Behind this, attached to the mahogany chest, is a brass representation of the sun with a small hole in the centre. The hydrogen escapes through this hole when the tap at the side of the instrument is turned.

To prepare the instrument for use the lower chamber is filled with water. The hydrogen is generated and passed down into the lower chamber through a pipe. It displaces the water which rises into the upper chamber, but is kept at considerable pressure by the action of gravity. The electrophorus is then charged. When the tap is turned the hydrogen is released and set alight by the spark. The resulting flame through the centre of the sun lights the lamp. The instrument would operate for many weeks with one charging.

Lorentz was not the first to devise this type of lamp lighter; Adams the Younger mentions Volta, Ingenhouz, and others. Anderson gives a full bibliography and cites Erhmann who des-

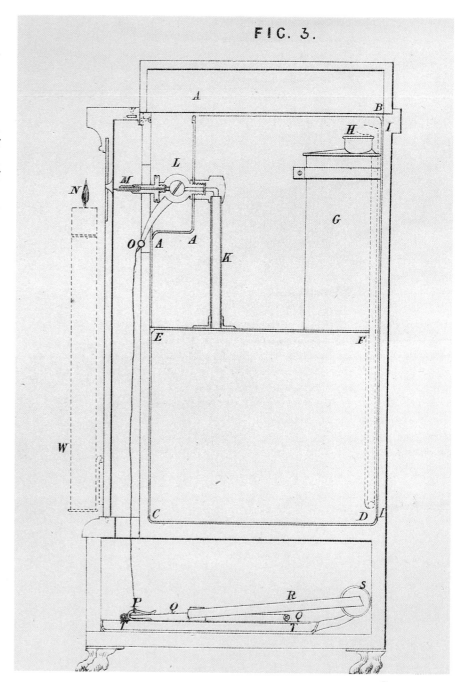

FIC. 3.

Lorentz 1807

cribed a machine on the same principles in 1780. A modification of Lorentz's lamp by Mayer is in the Playfair Collection. In 1823 these machines were superseded by the more convenient Dobereiner lamp.

Erhmann 1780

Adams 1799*b*, Vol. 2, pp. 98–101, Pl. 7, Fig. 6
Lorentz 1807
Dobereiner 1823
C 306
Anderson 1978, Item 64, pp.148–9, Fig. 57

Inventory No. 1927-1181

MAGNETISM

L124 Pair of bar magnets in box

Newman, J.

1828–57   London

STAMPED ON KEEPER AND ON CENTRAL WOOD
'*I. Newman Regent St London*'

L 510mm (box), **W** 123mm (box), **D** 30mm (box), **L** 435mm (magnets), **W** 38mm (magnets), **DE** 8mm (magnets)

This pair of magnets is one of several items by Newman, which could have been purchased by King's College before, or during, the time when the King George III collection was housed there. There are two steel bar magnets nearly 18 inches long, separated by a piece of mahogany, stamped 'I. Newman Regent St. London'. One of the keepers is also stamped but the other is not original. The magnets are in a mahogany box with a sliding lid. Newman was at 122 Regent Street between 1828 and 1857.

C 265

Inventory No. 1929-103

L125 Set of six magnets with keepers on board

Sandersons Brothers and Co.

after 1837   Sheffield

ON ONE MAGNET
'*Sandersons Brothers & Co Cast Steel Sheffield*'

L 555mm, **W** 132mm, **H** 35mm, **L** 507mm (magnets), **W** 45mm (magnets), **T** 7mm (magnets)

This item probably came into the collection via King's College. The steel magnets are arranged in two stacks of three with keepers across their ends. One magnet has the maker's name stamped on one side; each has a dot for the north pole. A pine board separates the stacks.

C 265

Inventory No. 1929-104

L125   L126

L126 Set of six magnets with keepers on board

mid-nineteenth century

L 420mm, **W** 115mm, **DE** 40mm, **L** 354mm (magnets), **W** 35mm (magnets), **T** 5mm (magnets)

This item probably came into the collection via King's College. It is quite likely that it was made by Sanderson Brothers and Co. since it is so similar to L125.

The six magnets are arranged in two stacks of three with keepers across their ends. Each has a small brass plug for the north pole and a small hole for the south pole. They also have a screw hole in the centre for suspension or mounting. The pine board separates the stacks.

C 265

Inventory No. 1929-105

L127 Two magnetic needles

mid-nineteenth century

H 225mm (1), **L** 310mm (1), **H** 200mm (2), **L** 265mm (2)

The needles are similar but do not constitute a pair. They are of steel with brass caps in the centre which pivot on steel spikes.

Inventory No. 1929-99

WEAPONS

L128 Sword tester

c. 1800

L 1120mm, **W** 90mm, **H** 135mm

The object consists of a mahogany model sword which is pivoted through its handle by a brass rod attached to mahogany mounts. Halfway along the length of the blade there is a brass joint which can be set at a range of angles by a butterfly nut, altering the effective curvature of the blade. It is thought to be a testing device for blades of various curvatures. It may be associated with L131.

C 340

Inventory No. 1927-1473

L129 Spring gun

Robert, H.

first half nineteenth century   Paris

HENRI/ROBERT/PARIS'

L 240mm, **W** 120mm, **H** 30mm,
D 10mm (balls)

The apparatus consists of a mahogany board with a piece running lengthways near the upper surface which can be pulled back against a spring along a channel. The piece can be held on a brass catch and a small white limestone ball can be placed in the channel. When the catch is released the ball is shot out along the channel. The actual spring mechanism is hidden. When not in use the balls are held in a small chamber in the side of the board. The brass cover for this chamber is stamped with the maker's name.

Henri Robert was a well-known maker of sundials, astronomical instruments, and chronometers in the first half of the nineteenth century. This instrument is not typical of his work.

C 344

Inventory No. 1927-1459

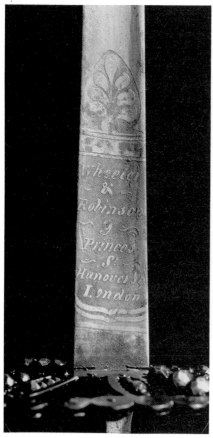

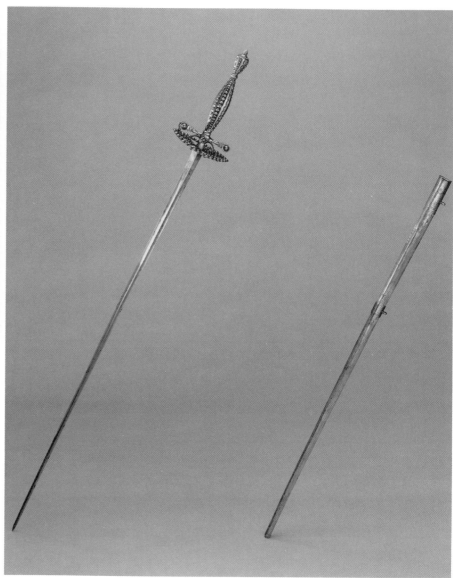

L130 Rapier and sheath

Wheeler and Robinson

c. 1810   London

BLADE TOP
'*Wheeler/&/Robinson/9/Princes/St/Hanover Sq/London*'

L 940mm, **W** 120mm (max), **DE** 70mm (max)

Although the blade is English, the handle is likely to be
French or Italian. This is a type of dress sword which would
have been worn at Court. The sheath appears to be similar to
papier mâché but is now broken.

C 338

Inventory No. 1927-1469

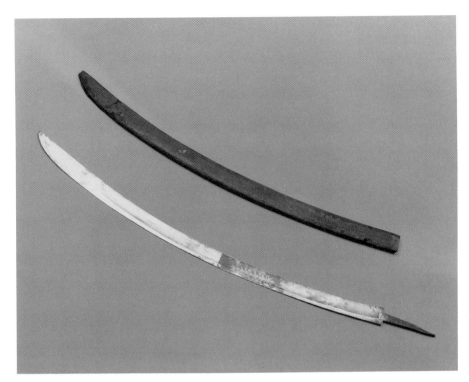

L131 Sabre blade and sheath

T.G.

c. 1803 England

ENGRAVED ON BLADE
'Warranted' *'GR'*
STAMPED ON TOP
Royal Crest 'T.G'
STAMPED ON TOP
'RRR'

L 950mm, **W** 45mm (sheath), **T** 10mm (sheath), **W** 33mm (blade), **T** 5mm (blade)

This is an English officer's sabre from about 1803. T.G was the blademaker. The blade is decorated with a royal coat of arms, a crown, the initials G.R, and some floral designs.

C 339

Inventory No. 1927-1470

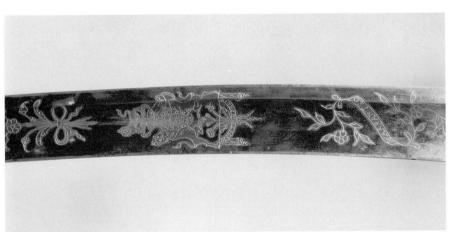

L132 Percussion gun

c. 1810

L 980mm, **W** 30mm, **DE** 120mm (max)

This is a sporting percussion gun which was probably converted from a flintlock in about 1820.

C 342

Inventory No. 1927-1468

PHOTOGRAPHY

L133 Eighteen pieces of photographic paper

1839–41

L 330mm, **W** 240mm **H** 40mm (box)

The contents are as follows:

1. Fifteen negative photograms made by the photogenic drawing process: all mounted on nine sheets of stiff paper; size approximately 205mm × 267mm. The series appears to be part of a trial of 'fixing' agents.
   - (i) Two leaf impressions with handwritten inscription on mount:
     - (a) 'Hypo-sulphite May 1839'
     - (b) 'Hypo-sulphite May 1839' 'Ammonia Aug 1840'.
   - (ii) Two leaf impressions with handwritten inscription on mount:
     - (a) 'Hypo-sulphite May 1839'
     - (b) 'Hypo-sulphite May 1839'.
   - (iii) One lace impression (faded) with handwritten inscription on mount: 'ammonia Aug. 1840'.
   - (iv) One photogenic drawing (image effaced) with handwritten inscription on mount: 'ammonia Aug 1840'.
   - (v) One photogenic drawing, grasses (? very faded) with handwritten inscription on mount: 'ammonia Aug 1840'.
   - (vi) Three photogenic drawings from engravings (faded) with handwritten inscriptions on mount: 'Ammonia Aug 1840'.
   - (vii) Two photogenic drawings (faded) with handwritten inscription on mount:
     - (a) feathers
     - (b) copy of engraving 'ammonia then Hyposulphite Aug. 1840'.
   - (viii) One photogenic drawing; copy of engraving (faded) with handwritten inscription on mount: 'Ammonia, then Hyposulphite Aug 1840'
   - (ix) Two photogenic drawings (faded) with handwritten inscriptions on mount:
     - (a) copy of engraving 'Iodide of Potash Aug 1840',
     - (b) lace impression 'Ammonia then Iodide of Potash Aug 1840'
2. One unmounted photogenic drawing; lace impression (faded) 220mm × 150mm.
3. One photogenic drawing; image effaced. Inscribed in pencil 'London by night' 192mm × 118mm.
4. Three photogenic drawings: images effaced.
5. One strip black lace suitable for making photogenic drawings. Approximately 190mm × 55mm.
6. One envelope, dark blue, containing pre-sensitized photogenic drawing paper, bears printed label headed 'HELLOGRAPHIC PAPER' followed by detailed instructions for producing and fixing photogenic drawings.
7. Envelope, dark green, containing pre-sensitized photogenic drawing paper. Bears printed sticker 'J C Newtons Haliographic Drawing'.
8. One handwritten note of instructions for producing photosensitive paper.
9. One sheet writing paper folded to contain fragments of thin mica sheet. One other thin sheet of mica, cracked and flaking.
10. One engraving on stiff paper (300mm × 213mm). Printed inscription 'FOSSILS, ENGRAVED ON A DAGUERREOTYPE PLATE, by the process of LL Boscowen Ibbitson Esq. With the apparatus at the Polytechnic Institution, Regent Street'.

The announcement of two practicable photographic processes in 1839 aroused tremendous interest and speculation in both scientific and artistic communities. Accounts of Talbot's work were reported in such fashionable journals as the *Athenaeum* and the *Literary Gazette*. Detailed practical hints soon appeared in British popular scientific journals such as *The Magazine of Science*, *The Mechanics Magazine*, and *The Mirror*. Materials and apparatus to make photogenic drawings were being advertised for sale as early as the summer of 1839. Unsurprisingly, in those early days both processes were imperfect and it required considerable skill, patience, and luck to produce satisfactory images. Even when a picture was produced, would-be photographers experienced problems in preserving their images and there was an extensive debate about suitable 'fixing' agents. It was also soon learned that paper was not the ideal base support for negative materials and experiments were made with thin sheets of mica as well as glass. Despite all the problems, many people set to work making copies of lace, leaves, and prints. Examples of Talbot's work were shown to Queen Victoria and were greatly admired by her and her circle.

Thomas 1964
Eder 1972
Arnold 1977
Ward and Stevenson 1986

Inventory No. 1927-1674

L134 Iodine sensitizing box for daguerreotype plates

c. 1841

L 260mm, **W** 240mm, **H** 100mm

The box would have been used to sensitize daguerreotype plates prior to use. There is a thick glass trough at the base which held the heated sensitizing chemicals. The top of the trough has a sliding glass cover which enables the plates to be exposed for a limited period, usually up to 2 minutes. Above this is a mahogany box which holds the frames. The smallest is $3\frac{1}{4}$ inches by $4\frac{1}{4}$ inches, or quarter-plate, a size favoured for daguerreotype portraits at the time. The plate holder L134 appears to belong with this object. There is a note inside the box signed by an I. J. Hurse which is probably later.

A typical procedure involved pouring an iodine solution into the trough, sufficient to give a depth of about $\frac{1}{4}$ inch. A polished daguerreotype plate was then placed in the appropriate frame and exposed to iodine vapour. The plate was then ready for use in the camera.

Anon 1845
Thornthwaite 1845
C 334

Inventory No. 1927-1508, -1496

L135 Mercury developing box for daguerreotype plates

1840–5

H 340mm, **L** 215mm, **W** 115mm

The base of the box is iron with a central basin for holding mercury. In the front is a circular glass observation window protected by a sliding panel to enable the operator to monitor the progress of development. The top of the box consists of a hinged door through which can be inserted vertically a plate holder for plates up to 6 inches × 8 inches in size. This latter feature is unusual in that the plate size is non-standard and most other contemporary developing boxes were designed to incline the plate at an angle of 45° to the vertical. It is possibly an adaptation of an earlier piece of apparatus.

The plate was placed in the box after exposure in the camera. Heat was applied to mercury in the basin by means of a spirit lamp. Early instructions for the daguerreotype process recommended heating to about 90°F. Many similar developing boxes were equipped with a thermometer, but where this was not present it was suggested that the cup be heated until the mercury was pleasantly warm to the finger. The progress of development was observed through the glass

window. When the plate was judged to be properly developed it was removed and fixed, usually with 'hypo'.

Thornthwaite 1843   C 332
Anon 1845   Thomas 1969, p. 47, cat. No. 314, 315
Inventory No. 1927-1502

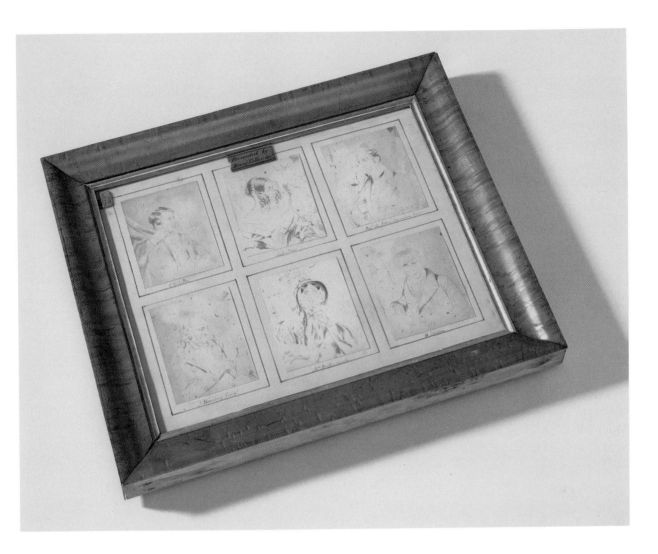

L136 Six calotype photographs in frame

Collen, Henry

1841–3

L 400mm (frame), **W** 340mm (frame)

These are six hand-retouched calotype portraits mounted in two rows of three on card within a glazed frame. Each portrait, approximately 77mm × 98mm is identified with a handwritten caption thus: 'C Geisler', 'Miss Ives', 'Revd Sir John Seymour Bart', (top row); 'J Henning Seuplr', 'Mrs Haldimand', 'Henry Collen'. (bottom row). A 'stick-on' label on the protective glass bears a handwritten inscription; 'Presented by Henry Collen'.

The photographic images have almost completely faded leaving the hand-retouching characteristic of Collen's calotypes prominently displayed. This small collection of Collen's work is believed to have been passed by a tutor to the royal family.

Henry Collen was a miniature painter of some repute. According to one authority, at least 100 of his paintings were exhibited at the Royal Academy of Arts during the period 1820–72. He was miniature painter to the Duchess of Kent and the Princess Victoria in 1835. In 1841 he became the first person licensed to practise Talbot's calotype process on a commercial basis. Although his portraits were at first well received, he failed to master the technical problems of calotype photography completely and the business eventually foundered around the middle of 1844.

The Chemist, Vol III, 1842, p. 122
C 335
Thomas 1964, pp. 24–5
Arnold 1977, pp. 138–40
Schaaf 1982, p. 365, Fig. 12

Inventory No. 1927-1670

L137 Daguerreotype portrait and negative image on copper

c. 1841

L 142mm (frame), **W** 102mm (frame)

The two images are mounted side by side behind a card mask in a passe-partout frame. They appear to depict the same gentlemen but the copper image is a mirror image of the daguerreotype.

No part of the image on copper shows obvious signs of relief but it is possibly an early attempt to produce a photographically-derived plate suitable for printing on paper.

Following the announcement of photography, several attempts were made to produce printing plates from daguerreotypes. Processes towards this end were described by Berres in Vienna, Donné and Figcau in Paris, and Grove in London. Part of all these techniques involved etching the unchanged silver of a daguerreotype with nitric or hydrochloric acid. The results were unsatisfactory.

Walker 1841
Eder 1978, pp. 577–8

Inventory No. 1927-1680

L138 Faded photograph on paper in frame

c. 1840

L 215mm, **W** 180mm

The paper bears the watermark 'R Turner Chafford Mills 1840', a papermaker known to be favoured by associates of W. H. F. Talbot.

The image is now totally effaced but it is reasonable to assume that it would have been made by Talbot's photogenic drawing processes.

C 672
Arnold 1977, p. 126 (Footnote)

Inventory No. 1927-1675

L139 Faded image in frame

c. 1840

L 400mm, **W** 280mm

The object consists of a positive and negative image of a flower pattern in pink and white on glass. The glass is cracked.

Inventory No. 1927-1676

L140 Box camera with focusing screen

1840–5

L 530mm (base), **W** 365mm (base), **H** 315mm

This is a mahogany sliding box camera for plates or paper of dimensions 12 inches × 10 inches. The camera is of the general pattern most commonly favoured by photographers during the 1840s and 1850s. It consists of a front box into which a slightly smaller rear box slides, both on a common base board. There is a horizontal sliding front panel which is pierced by a centring hole but not cut to take a lens. The rear box is cut with vertical grooves to contain the ground glass focusing screen; this would be replaced by the plate holder when the photographer was ready to capture his image. The camera is not jointed in the manner normally favoured by instrument-makers and may not be light-tight. It is possible that it was a speculative instrument which was not used for photography.

Coe 1978, pp. 20–1

Inventory No. 1927-1636

L141 Single dark slide, mahogany

1840–60

W 290mm, **L** 350mm

This piece of apparatus is standard for the period and goes with the camera above. The plate is held between two pieces of mahogany, one of which can be slid out of the frame to expose it to the light.

Inventory No. 1927-1665

L142 Double dark slide for 12 inch × 10 inch plates

c. 1851

L 365mm, **H** 300mm, **W** 25mm

The mahogany double dark slide would have been used as a holder for two glass plates or paper. 'Slide' or 'back' were the terms widely used to describe the flat lightproof container carrying the sensitized plate or paper in a camera. The double forms of slide sold during the 1850s were designed for use with Gustave Le Gray's waxed paper negative process, a modification of Talbot's calotype process.

Le Gray 1851
Willats 1851

Inventory No. 1927-1664

L143 Double dark slide

c. 1851

L 258mm, **W** 240mm

This dark slide is very similar to the above but would take paper of dimensions 8 inches × 7 inches, a non-standard size.

Thomas 1969, Item 15

Inventory No. 1927-1500

L144 Three electrotype engravings of the Parthenon in a frame

c. 1840

'Presented to the Parthenon Club by . . .'

L 505mm (frame), **W** 300mm (frame)

The three electrotype images are mounted individually on card and arranged vertically in a glazed wood frame. All three cards bear the inscription 'G Barclay Sp' in small italics immediately bottom right of the image. The central card only bears the inscription 'Presented to the Parthenon Club by' in large italics below the image and 'ELECTROTYPE' in capitals to the bottom left of the image.

The application of technology to the reproduction and multiplication of illustrations was one of the great preoccupations of the first half of the nineteenth century. One of the more important innovations was electrotyping which was widely announced in 1839, the same year as photography. Details of electrotyping appeared in journals and early instruction manuals alongside accounts of photographic practice.

Electrotyping involved the electrolytic deposition of metal from solution on to an engraved metal plate to form a matrix or mould from which prints could be pulled. The technique only became practicable when a steady current such as that produced by the Daniell cell became available.

The Magazine of Science, Vol. 1 1839–40, pp. 247, 393, 405, 411
Walker 1841
C 335
Wakeman 1973, pp. 74–6

Inventory No. 1927-1671

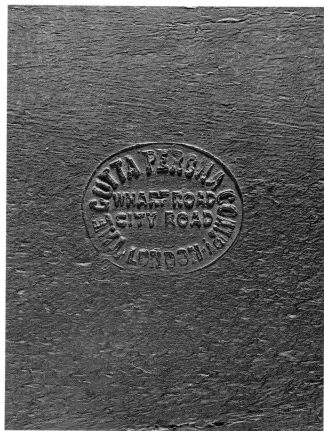

L145 Wet plate sensitizing bath

c. 1851   London

ON BOX
*'THE GUTTA PERCHA COMPY LONDON/
WHARF ROAD/CITY ROAD'*

L 312mm, **W** 265mm, **H** 22mm

The item is a narrow rectangular vessel with an open top made of gutta percha and bearing a manufacturer's trade mark 'The Gutta Percha Compy London Wharf Road City Road'. The same address for the company appears in the *London Post Office Directory* for 1851. Similar gutta percha vessels appear in most photographic dealer's catalogues of the period, sometimes described as dipping troughs. It is now in a poor condition.

Glass plates coated with collodion were sensitized with silver nitrate solution within a bath. A glass 'dipper' with a turned end was used to support and lower the plate into the solution. The bath is designed to take plates up to 12 inches × 10 inches in size.

Gutta percha was favoured as a material for the construction of photographic vessels on account of its lightness and flexibility. There was some fear that it reacted unfavourably with photographic chemicals and its use was widely debated in journals of the 1850s. The correspondents in these journals were usually anonymous.

Horne, Thornthwaite, and Wood, c. 1852
Photo Journal, Vol. IV, 21 January 1858, p. 139
Photo News, Vol. I, 1858, pp. 21, 120, 156
Photo News, Vol. III, 1858
C 333

Inventory No. 1927-1501

L146 Photographic printing frame

c. 1851

L 275mm, **H** 265mm, **W** 40mm

The item is a mahogany frame for printing on paper from glass negatives of maximum dimensions 8 inches × 9 inches. The frame is of unorthodox design, having the appearance of a dark slide, and is a non-standard size. It is fitted with a sliding front panel behind which is a sheet of plate glass allowing light to reach a narrow compartment containing negative and printing paper. The rear of the frame has an external hinged prop allowing it to be supported at an angle of 45°. Access to the compartment is through a leather hinged flap at the top of the frame. The inner walls of the compartment are stained with silver salts.

Inventory No. 1927-1498

MISCELLANEOUS

L147 Model palanquin

early nineteenth century

L 1570mm, **W** 350mm, **H** 370mm

The model palanquin is made of mahogany which has been painted green. Strength is given to the stucture by the addition of thick brass wires which are attached to the corners of the body and the carrying poles. The body has a sliding door on each side and two glass windows at the front and back, one of which is cracked. The base is made of straw matting on which is placed an upholstered cushion and small pillow. The pillow and cushion have been re-covered at a later date in imitation of the upholstery lining the internal walls. There is also a small mahogany drawer. The model stands on four feet carved with a classical motif when not in use. Its origins are European rather than Far Eastern, although palanquin models were not common in England until later in the century.

Singer *et al.* 1956, Vol. 1, p. 705

Inventory No. 1927-1934

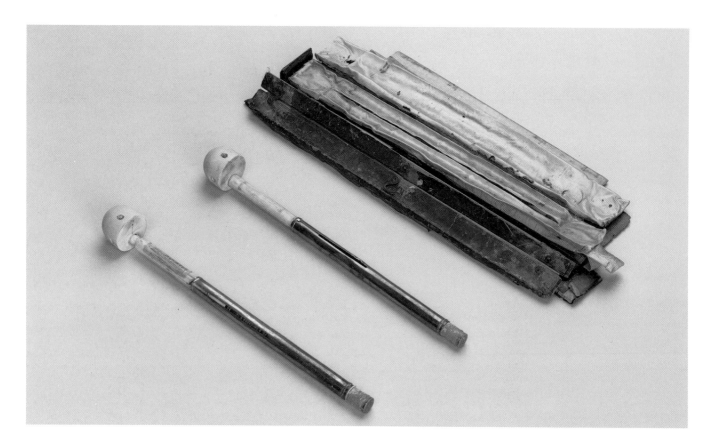

L148 Double instrument

Gilbert, G.

first quarter nineteenth century

ENGRAVED
Royal Coat of Arms
'G.R.'
'Invented by George Gilbert/L.H.V'
ON END OF EACH PIECE—IN INK
'B'

D 35mm (bases), **H** 30mm (bases), **L** 305mm, **D** 14mm (tubes)

The purpose of this instrument is not known. It has been dated as early nineteenth century because that is consistent with the materials of the case. There is no evidence that George Gilbert had any connection with the Gilbert family of instrument-makers at Tower Hill.

The instrument consists of a pair of identical pieces which have rounded ivory bases which screw into ivory tubes. A silver jacket surrounds each tube which can be pushed down by a calfskin-covered piece of wood acting on a spring. The silver jackets have a slit into which a silver pin attached to the ivory tube fits. However, there is no scale. The silver is engraved with the royal coat of arms, 'G.R', and 'invented by George Gilbert L.H.V.'. The tops of both instruments are marked with a 'B' in ink. It is now in a very poor condition.

Brown 1979, pp. 83–4

Inventory No. 1927-1537

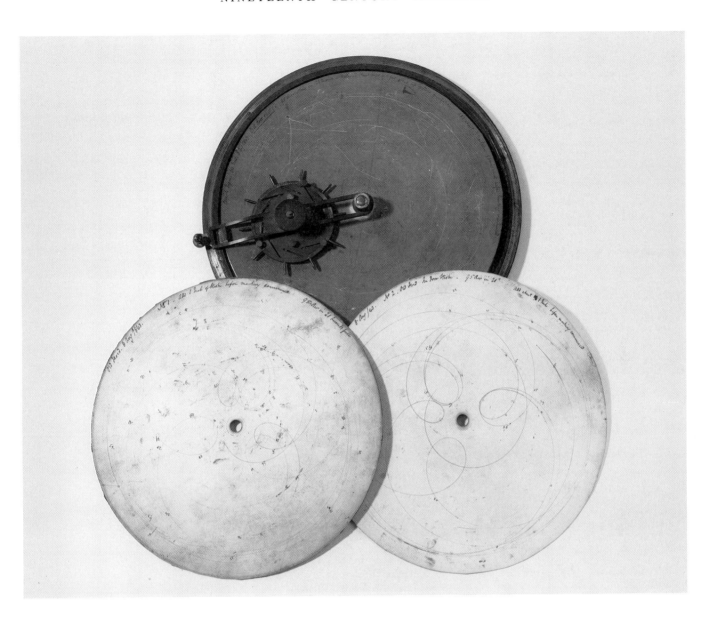

L149 Three indicator diagrams in circular frame, with recording attachment

before 1844

'Conical Engine Old . . . 8 Aug 1843 . . .'

D 350mm, **DE** 40mm

The recording attachment makes pencil marks on circular sheets of cardboard which are attached to the brass disc. There are 1000 divisions of the disc around its circumference. The device was used to test the performance of engines in some way, the number of revolutions being counted in a period of time. One indicator diagram has the following written around the circumference: 'Conical Engine Old. . . 8 Aug 1843. . . Outdoor. . .95 revolutions in 22 seconds water in standpipe 90 feet above top of stone foundation-add about 1/3 revolution of pencil before marking commenced'.

Inventory No. 1927-1524

L150 Davy lamp

c.1820

H 200mm, **D** 50mm

The lamp consists of a wick inserted in oil in a cylindrical brass container which forms the base. A gauze cylinder screws above the base and a gauze cap fits over the top. The mesh of the gauze is about 30 per inch or 900 per square inch. Three brass wires from the base link to form a hook to carry the instrument.

Davy invented this type of lamp in 1815. The gauze was used to prevent explosions with 'fire-damp' or methane gas by absorbing the heat from the flame. This is a relatively early example.

Davy 1816*a*
Davy 1816*b*
Davy 1816*c*, pp. 115–19
Hardwick and O'Shea 1916, pp. 548–724
C 76

Inventory No. 1927-1134

L151 Clockwork roasting jack

first half nineteenth century

INSCRIPTION
'*WARRANTED 1.*'

H 305mm, **L** 105mm, **W** 105mm

This is possibly one of the model jacks listed in the Queen's Catalogue, but it is more likely to be of later date. It is a bottle-type roasting jack commonly found in small households in the nineteenth century which has been doctored to reveal the mechanism. The spring is wound by a key or winch which is now missing. The spring turns a steel escapement wheel which turns the spit through several revolutions in one direction and then in the other. Lack of weight on the spit acts as a brake on the mechanism. The instrument is made of brass and steel and the top piece is of enamelled tinplate.

In 1773 Merlin filed a patent for 'a spring jack having a reflector to increase the heat and therefore save fuel', but it is not known if the application of clockwork to a roasting jack was his innovation. In 1790 Freemantle patented a jack which has similar features to this model but does not entirely correspond.

Q. C., Item 46, Two models of jacks
Merlin 1773
Freemantle 1790
Whittington 1794
Wright 1964, p. 51
Davidson 1986, p. 57

Inventory No. 1927-1458

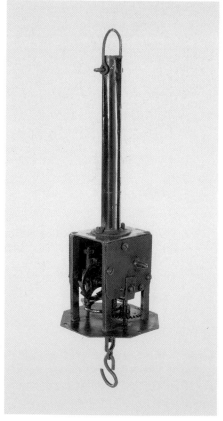

L152 Cone with carrier on needle pivot with glass rod in case

first quarter nineteenth century

H 40mm (cone)

The brass cone is painted red, the glass rod has a triangular cross-section, and the box is covered in red leather. The purpose of the instrument is not known. It has been dated from the materials of the box.

Inventory No. 1927-1546

L153 Minton vase

1830–40

H 305mm

The handle is broken and a flower is loose. Although this item has previously been described as a Rockingham vase, it has a serial number '1539' which indicates that it is Minton dating from the 1830s. Minton were making vases such as this at that time.

Inventory No. 1927-1777

PART 5
Less significant items

J1 Two hyperbolic curves, mounted on stands

first half nineteenth century

PAINTED—BASE
'hyperbola'

No. 1: **L** 695mm, **H** 475mm, **W** 160mm
No. 2: **L** 670mm, **H** 320mm, **W** 160mm

These are curved mahogany planks on supports.

Inventory No. 1927-1226

J2 Parallelepiped

nineteenth century

PAINTED
'87 oz'

L 265mm (side), **L** 210mm (opposite side), **H** 70mm

This is a solid mahogany parallelepiped which appears to have been used for centre of gravity experiments. The central section is painted black on the top and bottom surface. '87 oz' is painted in white on one black section.

C 247

Inventory No. 1927-1646

J3 Grooved wheel on shaft

second half eighteenth century

L 252mm, **D** 100mm (wheel), **D** 18mm (axle)

The boxwood wheel is mounted on a steel axle. The axle can be clamped to a table or similar surface using the brass butterfly nut.

Inventory No. 1927-1829

J4 Peg board and ten pegs

nineteenth century

L 208mm, **W** 208mm, **H** 40mm

This is a square pine board with eight peg holes in two rows of four. Four cuts in the base on each side correspond to the holes and suggest that weights were hung on threads tied to the pegs. Six pegs form a set and are made of mahogany, two more are slightly more ornate and also made of mahogany, and a further two are rather rough and made of pine.

Inventory No. 1927-1523

J5 Set of eight rectangular weights

second half eighteenth century or first half nineteenth century

L 30mm (total), **L** 15mm (weight), **W** 15mm (weight), **H** 10mm (weight)

Each weight has two hooks.

C 62

Inventory No. 1927-1862

J6 Set of cylindrical weights

last quarter eighteenth century

H 70mm ('25'), **H** 50mm ('15'), **D** 18mm (both)

Only two weights can now be traced. One of the existing weights has '25' engraved on the top and the other '15'. Both have one brass hook.

C 62

Inventory No. 1927-1768

J7 Five cylindrical weights, each with one hook

second half eighteenth century

'1/8', '1/8', '1/4', '1/4', '1/2'

L 20mm (smallest), **L** 40mm (largest), **D** 11mm

These weights may belong with E56 and E57, the mechanical powers apparatus.

C 62

Inventory No. 1927-1864

J8 Tapering pulley block with three wheels

second half eighteenth century

L 232mm, **D** 48mm (large pulley), **D** 40mm (middle pulley), **D** 32mm (small pulley)

The brass tapering pulley block contains three wheels giving a ratio of 4:5:6.

C 30

Inventory No. 1927-1214

J9 Two frames, each with two coaxial pulleys with cords

last quarter eighteenth century or first half nineteenth century

H 90mm (each), **D** 47mm (wheel)

These pulleys are heavier than the pulleys known to have been in Demainbray's or Adams's collections. They consist of iron frames, one with a solid hook and one with a wire hook, and solid brass wheels.

C 30

Inventory No. 1927-1215

J10 Detached finial with gilded decoration

first half eighteenth century

H 145mm, **D** 90mm (max), **D** 9mm (min)

The gilded lead finial has been detached from a larger piece of apparatus and was held in place by a rosewood dowel. A segment of the finial has been sawn away to allow for a flush fitting of this ornament.

Inventory No. 1927-1773

J11 Pair of pulley blocks

last quarter eighteenth century or first quarter nineteenth century

L 135mm (both), **D** 40mm (both)

The pulley wheels in each block are enclosed in a moulded brass pillar of square cross-section. They have circular end plates with six holes through which pass the cords. The other ends have large iron hooks which have corroded.

C 30

Inventory No. 1927-1826

J12 Tapering open-ended box with pulley wheel

second half eighteenth century or first half nineteenth century

L 126mm (box), **W** 102mm (box), **D** 47mm (pulley), **T** 14mm (pulley)

The open-ended mahogany box was mounted on a larger

piece of apparatus. This is suggested by the unpolished back of the box and the holes left by securing wood screws.

Inventory No. 1927-1887

J13 Nine single pulleys

early nineteenth century

D 130mm, 125mm, 90mm, 90mm, 65mm, 65mm, 43mm, 30mm, 24mm, **L** 180mm (largest), **L** 50mm (smallest)

This is a set or part of a set of solid brass pulley wheels in frames. Some have two iron hooks, some one, and some none.

C 30

Inventory No. 1927-1206

J14 Rod

second half eighteenth century or first half nineteenth century

L 415mm, **H** 30mm, **W** 12mm

The is made of steel with rounded pinheads of brass at each end. There is a hollow brass tube in the centre, presumably for fitting on to a stand.

Inventory No. 1927-1697

J15 Tripod stand with telescopic pillar and collar cross-piece

mid-eighteenth century

H 1100mm (max), **H** 810mm (min), **D** 620mm

The telescopic pillar has a thumb screw clamp while the collar cross-piece has a knurled knob. One leg is broken.

Inventory No. 1927-1836

J16 Tripod stand

second half eighteenth century

H 730mm, **D** 620mm

The stand is identical to J15 except that the telescopic pillar and collar cross-piece are missing.

Inventory No. 1927-1837

J17 Stand

second half eighteenth century or first half nineteenth century

H 283mm, **D** 208mm (base)

Chabrol sketched a stand with a round base in use with Demainbray's balance in his lecture notes of 1753. However, the sketches are far too crude to identify any particular object. The mahogany stand would have taken a cylindrical rod of smaller diameter than itself and fixed it by means of a screw at one side. The screw is now missing. The stand is weighted with a ring of lead in the base.

Chabrol 1753, pp. 17, 19, 109

Inventory No. 1929-127

J18 Adjustable table top on tripod

first half nineteenth century

W 600mm, **L** 600mm, **H** 770mm (max), **H** 460mm (min)

The adjustable pillar alters the height of the table top and is clamped with a thumb screw. The table is mahogany and the screw is of steel.

Inventory No. 1929-122

J19 Model of a winding headgear?

second half eighteenth century or first half nineteenth century

H 760mm, **L** 340mm, **W** 275mm, **D** 62mm (pulley), **W** 62mm (pulley)

This model is so incomplete that it is now impossible to identify. There was probably a mechanism on top of the structure which operated through the rectangular hole in the upper platform. This platform is supported on four legs and there is a lower platform near the base. Part of the way up the structure an iron axle is fixed horizontally to two cross-bars. It carries a thick mahogany pulley. With the exception of the axle the model is made entirely of mahogany. Some loose pieces are with it but it is far from certain that they belong.

Inventory No. 1927-1660

J20 Table

second half eighteenth century or first half nineteenth century

L 430mm, **W** 430mm, **H** 910mm

The table is made of mahogany with brass levelling screws.

Inventory No. 1929-123

J21 Tripod stand with cover for lamp

first half nineteenth century

D 165mm (cover), **D** 220mm (tripod), **H** 275mm (tripod)

It was previously believed that the tripod stand belonged with the Davy lamp L150, but there is no evidence for this. The top of the mahogany tripod stand has a brass plate with a recessed circular hole. There is a circular cover which has a ventilation plate with holes and another aperture.

C 76

Inventory No. 1927-1667

J22 Square plate held in elliptical picture frame

last quarter eighteenth century or first quarter nineteenth century

LABEL—BACK
'5 16/1/40'

L 115mm (plate), **W** 109mm (plate), **T** 1mm (plate)

The steel plate has been held in place with nails and gummed paper. The plate appears to be coated but this is obscured by rusting.

Inventory No. 1927-1767

J23 Polished mahogany ball

second half eighteenth century or first half nineteenth century

D 47mm

Inventory No. 1927-1882

J24 Six ivory balls

mid-eighteenth century

D 22mm (five balls), **D** 19mm (one ball)

It is difficult to attribute these balls with any certainty. Since Adams's apparatus tends to be complete, they are quite likely to have been Demainbray's. They have been used in experiments on pendulums and percussion since they have holes where threads have been attached.

Inventory No. 1927-1565

J25 Ivory ruler scale with support rings at each end

second half eighteenth century

SCALE
 '*2*', '*3*', '*4*'

L 77mm, **W** 10mm (max), **W** 2mm (min), **H** 23mm

The scale has three irregularly spaced crossed lines marked 2, 3, and 4. The support rings suggest that the scale was clamped to a glass tube.   **Inventory No.** 1927-1776/1

J26 Ivory ruler with support rings at each end

second half eighteenth century

SCALE
 0 ($\frac{1}{10}$) 6 inches

L 161mm, **W** 12mm (max), **W** 3mm (min), **H** 21mm

The scale has a 6 inch inscribed scale with tenth of an inch divisions. The support ring suggests that the scale was clamped to a glass tube.

Inventory No. 1927-1776/2

J27 Ivory ruler with support rings at each end

second half eighteenth century

SCALE
 0 ($\frac{1}{10}$) 6 inches

L 161mm, **W** 12mm (max), **W** 3mm (min), **H** 21mm

The scale has a 6 inch inscribed scale with tenth of an inch divisions. The support ring suggests that the scale was clamped to a glass tube.

Inventory No. 1927-1776/3

J28 Scale

early nineteenth century

SCALE
 5 ($\frac{1}{10}$) 32

L 710mm, **W** 20mm, **DE** 3mm

This is a wooden scale for an instrument, probably a barometer; there are holes for attachment and the scale goes up to 32 inches. Each tenth is marked.

Inventory No. 1927-1798

J29 Elliptical funnel-shaped hollow block

mid-nineteenth century

H 63mm, **D** 145mm, **DE** 62mm, **D** 154mm (max), **D** 144mm (min), **H** 62mm

Like L54 and J30, the block was probably used for acoustical experiments. The larger aperture appears to have been covered with a membrane. The other opening has three equidistant screw holes on its edge. The block is made of ash.

Inventory No. 1927-1485

J30 Cone-shaped framework for membrane

mid-nineteenth century

D 120mm (max), **D** 70mm (min), **H** 145mm, **T** 1mm

The framework and membrane would have been used for acoustical experiments. The framework is flexible and has been constructed from cardboard which has then been glued. Remains of the membrane are visible on the narrow end of the cone.

Inventory No. 1927-1487

J31 Oil lamp

early nineteenth century

H 80mm, **D** 33mm

The small oil lamp consists simply of a glass bottle with a brass funnel through which the wick passes.

Inventory No. 1927-1597

J32 Apparatus to show the expansion of liquids

second half nineteenth century

H 630mm, **D** 110mm (bulb), **W** 45mm (scale)

This appears to be teaching apparatus and most probably came from King's College. There is a spherical thick glass bulb at the base into which descends a narrow glass tube. The bulb is stopped with cork. The glass tube was sealed at the top with a smaller bulb but this is now broken. A paper scale on mahogany is crudely attached to the tube but there are no marks visible. There is evidence of a coloured liquid having been used for experiments in which the instrument probably served as a thermoscope.

Inventory No. 1927-1625

J33 Expansion of liquids apparatus with tripod stand

mid-nineteenth century

H 625mm **D** 35mm (thermometer)

The apparatus consists of a sealed brass cylinder with a threaded neck. Onto this is screwed a thermometer with stand. The apparatus is supported on a brass circular tripod which allows the cylinder to be heated from underneath. The thermometer has a graduated scale but it is unclear as to the units used. The thermometer is broken.

See also L68

Inventory No. 1927-1627

J34 Mercury-in-glass thermometer

last quarter eighteenth century or first quarter nineteenth century

H 420mm (back), **H** 390mm (tube), **W** 55mm, **DE** 10mm

It is not possible to say whether the pine back board relates to the broken thermometer tube.

C 91

See also L68

Inventory No. 1927-1708

J35 Calorimeter

nineteenth century

H 150mm, **D** 70mm

The copper calorimeter has a copper wire handle.

C 86

Inventory No. 1927-1534

J36 Calorimeter

first half nineteenth century

H 74mm, **D** 61mm

The interior of the copper vessel is tinplated.

C 86

Inventory No. 1927-1535

J37 Rectangular electrical conductor on glass rod

nineteenth century

H 455mm **L** 115mm **W** 120mm

The conductor is made of a rectangular piece of pine covered in tinfoil on a glass stand.

Inventory No. 1927-1446

J38 Two ornamental tripod stands with cross-beam conductors

first half nineteenth century

D 420mm (base), **H** 1410m, **L** 460mm (conductor)

The position of the cross-beam conductor bars can be adjusted both vertically and horizontally.

Inventory No. 1927-1227

J39 Four cylindrical conductors with attached rod or wire

nineteenth century

L 130–70mm, **D** 40–37mm

Two of the cylindrical conductors consist of a copper sheet nailed to a hollow wooden former. A cotton-covered wire has been soldered to each of the sheets. The other two conductors are rolled copper sheets with rods soldered to their edges.

Inventory No. 1927-1759

J40 Three glass insulating rods

first half nineteenth century

L 95mm, **D** 27mm

The glass rods have tapered ends and threaded brass bases. They were probably used for electrical experiments.

Inventory No. 1927-1527

J41 Two coils of insulated wire and a tightly wound ring of wire

second half nineteenth century

D 33mm (1 + 2), **L** 280mm (1 + 2), **D** 43mm (3), **T** 4mm (3)

The two coils of insulated wire form a loose spring which probably passed over a former. The third smaller ring of wire appears to be an induction loop as a second wire has been wound around the ring.

Inventory No. 1927-1775

J42 Insulating cup

second half eighteenth century or first half nineteenth century

D 52mm, **H** 30mm

The insulating cup is made of thick glass. It would probably have been placed under a chair leg for use in electrical experiments.

Inventory No. 1927-1564

J43 Rectangular frame with circular aperture

mid-nineteenth century

L 252mm, **W** 210mm, **D** 97mm (aperture), **DE** 18mm

The frame has a circular aperture with a threaded ring. The threaded ring is attached to the framework by wood screws. It would have housed a lens such as can be found in a camera or lantern projector.

Inventory No. 1927-1497

J44 Rotatable lens holder on pillar with circular base

second half nineteenth century

D 130mm (base), **D** 69mm (lens holder), **T** 12mm (lens holder), **H** 520mm

The lens holder can be rotated. The circular iron base has been cast and is attached to the brass pillar of square cross-section by a rivet. The lens holder is threaded onto the pillar.

Inventory No. 1927-1479

J45 Mirror in mount on swivel fitting

mid-nineteenth century

L 260mm, **D** 115mm

The mirror may have been part of an optical bench. It is in a brass swivel frame and can be turned in any direction.

C 218

Inventory No. 1927-1429

J46 Two telescopes

INSCRIPTION—TUBE
1 lens, 2 lens & 6 lens

No. 1: **L** 420mm (closed), **D** 83mm (max), **D** 60mm (min)
No. 2: **L** 510mm (closed), **D** 51mm (max), **D** 39mm (min)

Both telescopes are incomplete; one has a missing draw tube while the other has no lenses. The first telescope has an experimental optical arrangement which is indicated by the inscriptions on the tube. This instrument has been lined with playing cards.

Inventory No. 1927-1418

J47 Body tube of naval telescope

c. 1800
L 705mm (closed), **D** 90mm

This is the body of an octagonal naval telescope. There is an achromatic objective but the eyepiece and draw tube are missing.

Inventory No. 1927-1422

J48 Terrestrial telescope

last quarter eighteenth century
L 190mm (closed), **D** 55mm

The telescope has two brass draw tubes, a shagreen-covered body, and silver mounts and caps. The eyepiece lenses are missing. The objective appears to be achromatic.

Inventory No. 1927-1423

J49 Two dark red filters with cylindrical mount, in box

first half nineteenth century
D 20mm, **H** 14mm, **D** 80mm (box), **H** 16mm (box)

The lid for the pine box is broken, and the spacing piece for the filters is missing.

C 207
Inventory No. 1927-1554

J50 Eyepiece lens and draw tube

eighteenth century
L 100mm, **D** 36mm

The inside of the mahogany tube is painted black and the lens has been stopped down with paper to one-third of its diameter.

C 207
Inventory No. 1927-1555

J51 Segment from a draw-tube telescope

first half eighteenth century
L 53mm, **D** 38mm

The segment of tube is possibly from the eyepiece end of the telescope. It is covered in green shagreen.

Inventory No. 1927-1561

J52 Telescope part

second half eighteenth century
D 60mm, **L** 90mm

There are no lenses in this object which consists of a boxwood lens mount in a green vellum tube with a brass band at one end.

Inventory No. 1927-1560

J53 Eyepiece lens and cap from a telescope

second half eighteenth century or first half nineteenth century
D 23mm (max), **D** 17mm (min), **H** 15mm

The brass eyepiece cap only contains the eye lens and has a slotted shutter dust cap. It is threaded with decorative milling.

C 207
Inventory No. 1927-1603

J54 Eyepiece tube from a transit telescope

second half nineteenth century
D 35mm (max), **D** 17mm (min), **L** 55mm

The brass tube still has the securing points for the three vertical and one horizontal cross hairs. A broken short length of wire is in place. The eye lens is missing.

Inventory No. 1927-1572

J55 Eyepiece cap with shutter

second half eighteenth century or first half nineteenth century

D 55mm, **H** 25mm

The brass eyepiece cap is probably from a telescope. Nail holes in the edge of the lens cap suggest a cardboard or vellum draw tube.

Inventory No. 1927-1547

J56 Dark red solar filter in mount

first half nineteenth century

D 37mm, **T** 9mm

The filter is in a brass mount. The edge of the filter cap has been milled.

C 207

Inventory No. 1927-1541

J57 Oval pivoted mirror from a scioptric ball

second half eighteenth century

H 94mm, **D** 66mm

The ring and hook arm are decorative and are probably from a scioptric ball.

C 218

Inventory No. 1927-1570

J58 Eyepiece with draw tube

first half nineteenth century

D 23mm, **L** 203mm

The large spacing of the lenses suggests that this might be a terrestrial eyepiece for a telescope. There is also a locating pin on the draw tube.

Inventory No. 1927-1566

J59 Dark red solar filter in mount

first half nineteenth century

D 42mm, **T** 10mm

The filter is in a brass mount. The edge of the filter cap has been milled.

C 207

Inventory No. 1927-1542

J60 Oval plane mirror in cylindrical mount

first half nineteenth century

L 25mm, **D** 21mm

The mirror is mounted diagonally to provide a convenient viewing position for use with a telescope eyepiece. It could also be used in conjunction with a quadrant or transit telescope. It is made of speculum metal and the mount is brass. Four screws allow for adjustment.

Inventory No. 1927-1550

J61 Collection of thirteen microscope slides

1850–65 (Topping slides), eighteenth century (others)

ON THREE OF FIVE SINGLE SLIDES
'Prepared by/C.M. Topping/4 New Winchester St/Penton ville/London'

L 110mm (eight slides), **W** 24mm (eight slides), **L** 72mm (five slides), **W** 20mm (five slides)

The slides are of two types. Those in ebony and ivory mounts are earlier than the glass slides which have printed paper labels.

C 193

Inventory No. 1927-1563

J62 Telescope eyepiece

second half eighteenth century

H 78mm, **D** 60mm

No lens remains. The complete instrument may have been a telescope or a prospect glass. The one draw tube is covered in shagreen. The mounts are horn and the lens cap is silver.

C 178

Inventory No. 1927-1529

J63 Drawer with handle

second half eighteenth century or first half nineteenth century

SCRATCHED
'MICROMETER SCALES, LUCERNAL MICRO-SCOPE'

L 170mm, **W** 165mm, **H** 27mm

The mahogany drawer is warped and contains shaped inserts for parts of a microscope or a similar scientific instrument. The drawer is of the right proportion to have come from the base of a Culpeper microscope.

Inventory No. 1927-1598

J64 Spectacle case

eighteenth century

L 140mm, **W** 41mm, **H** 27mm

The case is a flattened papier mâché cylinder which is rounded at each end. The case fits together as two equal interleaving halves.

Inventory No. 1927-1596

J65 Scioptric ball? in mount

first half nineteenth century

D 205mm, **L** 115mm, **H** 90mm

Although the instrument has all the features of a scioptric ball, the narrow aperture makes this identification uncertain.

A small boxwood ball has a brass tube protruding at each side. It is mounted in a mahogany disc which can move between two fixed mahogany discs linked together by brass strips. There are signs that the device was part of something larger or attached to a wall or shutter.

Inventory No. 1927-1764

J66 Convex lens in circular mount

D 112mm

The mahogany circular mount has two locating holes on opposite edges. These would have allowed the lens to swivel on the pillar mount that is now missing. The focal length of the lens is 1630mm or about 5 feet 6 inches.

C 207

Inventory No. 1927-1580

J67 Convex lens in slide holder

last quarter eighteenth century or nineteenth century

L 80mm, **W** 73mm, **T** 1mm, **D** 62mm

The slide mount has been made from the pages of a book, the text of which is still visible. The focal length is 1370mm or about 4 feet 6 inches.

C 207

Inventory No. 1927-1581

J68 Convex lens

last quarter eighteenth century or nineteenth century

D 51mm, **T** 1mm

The lens has been roughly bevelled and has a slightly green tinge. The focal length is approximately 1310 mm or about 4 feet 3 inches.

C 207

Inventory No. 1927-1589

J69 Convex lens

last quarter eighteenth century or nineteenth century
D 40mm, **T** 1.5mm

The edge of the lens has a double bevel; the glass is colourless and free of bubbles. The focal length of the lens is aproximately 880mm or nearly 3 feet.

C 207

Inventory No. 1927-1590

J70 Two convex lenses

last quarter eighteenth century or first half nineteenth century

INSCRIPTION
15 feet focus

D 99mm (1), **T** 5mm (1), **D** 103mm (2), **T** 3mm (2)

The edges of the two lenses have been bevelled and the glass has a distinct green tinge. Both lenses have a focal length of about 5m or 15 feet.

C 207

Inventory No. 1927-1567

J71 Three convex lenses and one glass disc in box

last quarter eighteenth century or nineteenth century

LENS 1
'24 inch 36 inch'
LENS 2
'24 inch 36 inch'

L 280mm (box), **W** 75mm (box), **H** 23mm (box), **D** 60mm (1), **D** 63mm (2), **D** 50mm (3), **D** 61mm (4)

The three convex lenses all have bevelled edges. The fourth piece of glass is roughly edged and appears to be an unworked lens blank. Three of the lenses have a green tinge while the last has a brown tinge. The focal lengths are as follows: 1, 1370mm or about 4 feet 6 inches; 2, 1360mm or about 4 feet 6 inches; 3, 1850mm or about 6 feet.

C 207

Inventory No. 1927-1605

J72 Circular convex lens

last quarter eighteenth century or first half nineteenth century

D 70mm, **DE** 3mm

The glass has a slightly green tinge with bevelled edges. The lens has a focal length of 920mm or 3 feet.

C 207

Inventory No. 1927-1562

J73 Plano-convex lens

last quarter eighteenth century or nineteenth century
D 104mm, **T** 4mm

The glass has a green tinge and has bevelled edges. The focal length is greater than 15m or 50 feet.

Inventory No. 1927-1575

J74 Two lenses, one concave and one convex

last quarter eighteenth century or nineteenth century
'focus 15 feet concave [vieny]'
'70 inch's'

D 106mm (1), **T** 5mm (1), **D** 88mm (2), **T** 4mm (2)

The first lens is colourless with a bevelled edge, while the second lens has a slight green tinge with a rough cut edge. The focal lengths are of the order of the inscriptions.

C 207

Inventory No. 1927-1576

J75 Convex lens

last quarter eighteenth century or nineteenth century
'66in 5/10'
'66in 5.10'

D 78mm, **T** 3mm

The glass has a slight green tinge and roughly bevelled edges The focal length is approximately that on the inscription.

C 207

Inventory No. 1927-1577

J76 Two glass blanks and a lens

last quarter eighteenth century or nineteenth century
D 36mm, **T** 2mm (1), **T** 2mm (2), **T** 1.5mm (3)

All three items have bevelled edges; one has a slight green tinge. Two of the pieces may be glass blanks as they appear to be flat, while the other has a focal length of 2 feet.

C 207

Inventory No. 1927-1591

J77 Three lenses

last quarter eighteenth century or nineteenth century
D 78mm, **DE** 2mm (1), **DE** 3mm (2), **DE** 1.75mm (3)

Two lenses are slightly convex. One lens has bevelled edges and a strong green tinge. Another has a slight brown tinge and is full of air bubbles. Their focal lengths are approximately 16 feet and and 6 foot 6 inches respectively. The three may have made up an achromatic triplet.

C 207

Inventory No. 1927-1588

J78 Box containing three lenses, one mirror, one strip, and five diaphragms

nineteenth century
H 25mm (box), **D** 165mm (max), **D** 95mm (min)

The concave circular mirror is probably a condenser from a microscope, while the diaphragm could have been used on an optical bench.

C 207

Inventory No. 1927-1574

J79 Square lens holder

nineteenth century
D 63mm (lens holder), **L** 75mm (framework), **W** 76mm (framework), **H** 31mm (framework)

The lens holder frame could be slotted into a larger piece of optical apparatus such as an optical bench.

Inventory No. 1927-1587

J80 Three lenses in mounts

last quarter eighteenth century
D 96mm (lenses), **D** 124mm (mounts), **DE** 30mm

The three lenses are all of crown glass, 4 inches in diameter, with long focal lengths. They are each set in pine frames.

C 207   **Inventory No.** 1927-1582, -1583, -1584

J81 Three biconcave lenses and one disc of glass

nineteenth century
D 26mm **T** 4mm (1), **T** 2mm (2), **T** 2.5mm (3), **D** 13mm (4), **T** 1.5mm (4)

Each lens has a ground annulus.

C 207   **Inventory No.** 1927-1594

J82 Two circular discs

last quarter eighteenth century or nineteenth century
D 28mm, **T** 1.3mm (1), **T** 1mm (2)

The glass discs appear to be flat.

C 207   **Inventory No.** 1927-1593

J83 Nine lenses

second half eighteenth century or nineteenth century
D 12mm (approx)

C 207   **Inventory No.** 1927-1610

J84 Three lenses

last quarter eighteenth century or nineteenth century
D 34mm, **T** 1mm (1), **T** 1mm (2), **T** 1mm (3)

The lenses have rough edges; their focal lengths are approximately 20 inches, 32 inches, and 3 foot 6 inches.

C 207   **Inventory No.** 1927-1592

J85 Six lenses

second half eighteenth century or first half nineteenth century

D 9mm (two lenses), **D** 10mm (two lenses), **D** 12mm (two lenses)

Three lenses are unfinished with rough edges. The others are convex with double bevelled edges. The focal lengths are approximately 8 inches.

C 207

Inventory No. 1927-1611

J86 Green glass lens in frame

mid-nineteenth century

D 220mm, **D** 130mm (aperture), **DE** 32mm

This is a plano-convex lens of long focal length mounted in a brass ring set in a square mahogany frame. The frame is now broken. It was probably attached to a larger object.

C 207

Inventory No. 1927-1519

J87 Six lenses with three perforated cones in a box

second half eighteenth century or first half nineteenth century

D 3–6mm (lenses)

The lenses are either for components of microscope objectives or for use in eyepieces. The cones are made of brass.

C 207

Inventory No. 1927-1612

J88 Plano-convex lens mounted in flange

mid-nineteenth century

D 50mm, **T** 11mm

The flange has been constructed by soldering a metal strip to a round plate. The lens is highly convex and could only be used as a condenser lens for either a lantern or a microscope. The two flange cut-outs are circular and have different radii. These are identical to those found on J104 which suggests that they are related.

C 207

See also J104

Inventory No. 1927-1551

J89 Twelve lens mounts

nineteenth century

Various dimensions

These mounts are in very poor condition.

Inventory No. 1927-1543

J90 Box containing 23 circular lens blanks and a plano-concave lens

second half eighteenth century or first half nineteenth century

D 40mm (blanks ($\times 21$)), **T** 2mm (blanks ($\times 21$)), **D** 32mm (blanks ($\times 2$)), **T** 2mm (blanks ($\times 2$)), **D** 35mm (lens), **T** 3mm (lens)

Most of the lens blanks appear to have been cast as circular discs; the rest have been roughly cut. The concave lens has bevelled edges and has a ground annulus on one side. The box is made of cardboard.

C 207

Inventory No. 1927-1540

J91 Eight lenses

second half eighteenth century or first half nineteenth century

D 14–19mm, **T** 1.5–3.5mm

Three of the lenses have bevelled edges, the rest have roughly finished edges. The focal lengths range between 6 inches and 13 inches.

C 207

Inventory No. 1927-1609

J92 Box containing thirteen pieces of glass

first half nineteenth century

INSCRIPTION ON A PIECE OF GLASS
'*Thomas Plate*'

various dimensions

The box contains one piece of ground glass and one piece of plate glass, with the remainder consisting of square panes except for two broken circular pieces.

Inventory No. 1927-1538

J93 Plano-convex lens in cylindrical mount

mid-nineteenth century

D 28mm, **H** 17mm

The lens appears to be the eye lens from a larger eyepiece. It has a focal length of around $\frac{4}{10}$ inch.

C 207

Inventory No. 1927-1559

J94 Box containing miscellaneous pieces of glass

nineteenth century

HANDWRITTEN—SIDE OF BOX
'*Box containing Lens*'

L 200mm, **W** 50mm, **H** 50mm

The box contains six strips which may be microscope slides, four panes, three rods, and five tubes.

Inventory No. 1927-1606

J95 Cardboard box containing optical parts

first half nineteenth century

'*24 inch, 36 inch*'

L 230mm (box), **W** 130mm (box), **H** 40mm (box)

The box contains a piece of mirror, a lens in a paper mount, a lens of approximately 36 inches focal length, a plano-concave lens, a lens with a hole in the middle, and four pieces of glass.

C 207

Inventory No. 1927-1604

J96 Glass lens in threaded mounting cell

first half nineteenth century

D 63mm, **DE** 20mm

The brass threaded mount is part of a larger optical instrument. The edges of the brass mount have decorative milling. The lens has a focal length of approximately 16 feet. The glass is cracked.

C 207

Inventory No. 1927-1558

J97 Lens holder, eye stop ring, and beading

second half eighteenth century or nineteenth century

D 70mm (lens holder), **DE** 13mm (lens holder), **D** 33mm (eye stop), **DE** 5mm (eye stop), **W** 9mm (beading)

The rim of the boxwood lens holder exactly fits the objective lens cell of the larger of the two vellum telescopes J46. The brass eye stop ring fits the eyepiece lens of the other vellum telescope J46.

See also J46

Inventory No. 1927-1586

J98 Six lenses

nineteenth century

D 27mm (two lenses), **D** 23mm (two lenses), **D** 20mm (two lenses)

Three of the lenses are blanks and have not been fully ground; they have very rough edges. Two other lenses have bevelled edges and have a concave–convex figure. The last lens is biconvex with rough edges and contains many air bubbles. Its focal length is 900mm or about 3 feet.

C 207

Inventory No. 1927-1608

J99 Two convex lenses in circular box

nineteenth century

'24 inch'

D 63mm, **T** 2.5mm, **D** 68mm (box), **H** 11mm (box)

The lenses have a slight green tinge with rough unbevelled edges. One lens has a paper annulus.

C 207

Inventory No. 1927-1601

J100 Box containing 29 pieces of glass plate and three convex lenses

nineteenth century

L 50mm (29), **W** 50mm (29), **D** 22mm (lenses), **T** 2.5mm (lenses)

Twenty of the square plate pieces have been partly ground; the other nine pieces are untouched. The three lenses are incomplete and still have pitch attached to them.

Inventory No. 1927-1599

J101 Three concave lenses and one convex lens in box

nineteenth century

D 67mm, 75mm, 34mm, 50mm, **H** 14mm (box), **L** 196mm (box), **W** 105mm (box)

Three of the four lenses have bevelled edges while the remaining one is very roughly finished. The convex lens has a slightly brown tinge and has a focal length of about 4m or 13 feet.

C 207

Inventory No. 1927-1602

J102 Box containing two concave lenses and one convex lens

second half eighteenth century or first half nineteenth century

D 36mm (box), **H** 30mm (box), **D** 28mm (lenses)

The convex lens has a focal length of 45mm or $1\frac{1}{2}$ inches.

C 207

Inventory No. 1927-1556

J103 Microscope objective

second quarter eighteenth century

L 30mm, **D** 35mm

The lens is mounted in boxwood and has a short green vellum tube attached.

Inventory No. 1927-1552

J104 Plano-convex lens mounted in flange with hook and rod

mid-nineteenth century

L 140mm (rod and hook), **D** 6mm (rod and hook), **D** 50mm (flange), **T** 20mm (flange)

This item is almost identical to J88.

C 207

Inventory No. 1927-1553

J105 Two square plates

second half eighteenth century or first half nineteenth century

L 80mm, **W** 80mm, **T** 4mm

Each mahogany plate has a notch cut in its edge midway along with a nail offset from the centre.

Inventory No. 1927-1239

J106 Lens in frame, in box

nineteenth century

D 175mm

The lens is in a brass frame in a mahogany box. It is badly cracked. It would have had a very long focal length.

C 207

Inventory No. 1927-1499

J107 Two tapering open-ended boxes

first half nineteenth century

No 1: **L** 650mm, **W** 60mm, **DE** 95mm (max), **DE** 65mm (min)

No 2: **L** 360mm, **W** 60mm, **DE** 95mm (min), **DE** 120mm (max)

The insides of the mahogany boxes are covered with matt black paint which suggests an optical use.

Inventory No. 1927-1517

J108 Tapering open-ended box with detachable panel

first half nineteenth century

L 550mm, **DE** 55mm, **W** 155mm (max), **W** 85mm (min)

Again, the inside of the mahogany box is covered with matt black paint which suggests an optical use.

Inventory No. 1927-1518

J109 Two tapering pillars with rectangular plates

second half eighteenth century or first half nineteenth century

D 90mm (max), **D** 30mm (min), **L** 340mm

The pillars are threaded at each end with two nuts and a rectangular plate. The broader end of the pillars has been attached to a board from a piece of apparatus. The rectangular plates would act as washers so that other apparatus could be attached to the top of the pillars.

See also E164, E165

Inventory No. 1927-1490

J110 A pair of circular stands with fulcra

first half nineteenth century

H 230mm, **D** 150mm (base)

The stands are identical and are made of turned mahogany while the fulcra are brass. The stands have rosewood turnkeys to adjust the height.

Inventory No. 1927-1477

J111 Circular pillar stand with fulcrum

first half nineteenth century

D 57mm (base) **H** 210mm (base), **W** 15mm (fulcrum), **L** 11mm (fulcrum), **H** 15mm (fulcrum)

The brass pillar supports a steel fulcrum. Both the base and the column have been turned. The column has an octagonal section which is threaded at both ends.

Inventory No. 1927-1476

J112 Adjustable clamp with square-section sleeve pieces

first half nineteenth century

L 150mm, **W** 42mm, **H** 120mm

The adjustable swivel clamp has two square-section sleeve pieces. The sleeve pieces would be attached to a stand which would allow for height adjustment of the clamp. The jaws of the clamp are circular; each of the adjustments has a clamping bolt.

Inventory No. 1927-1405

J113 Circular plate in frame with central hole

first half nineteenth century

D 115mm

The glass plate is in a brass framework with a central hole. One side of the plate is coated with black pitch, part of which has worn away. On the edge of the framework there is a circular channel which suggests that the plate formed a lid for a glass cylinder or similar vessel.

C 218

Inventory No. 1927-1426

J114 Glass rod

nineteenth century

D 12mm

The tube is made of thick glass with a brass cap at one end; it is now broken.

Inventory No. 1927-1178/2

J115 Gas cylinder

second half nineteenth century

D 120mm, **L** 680mm

The cylinder is made of iron with an octagonal cast iron base. The control valve wheel handle is missing.

Inventory No. 1927-1638

J116 Gas cylinder

second half nineteenth century

D 130mm, **L** 710mm

The valve on top of the cylinder is controlled with a wheel-shaped handle. The cast iron cylinder is mounted in a trunnion stand.

Inventory No. 1927-1637

J117 Tapering buckets with suspension wire

last quarter eighteenth century or first quarter nineteenth century

H 12mm (1), **D** 20mm (1) **H** 25mm (2), **D** 26mm (2)

The buckets are constructed of soldered silver sheet.

Inventory No. 1927-1895

J118 Cylindrical vessel with nozzle and filler cap

first half nineteenth century

H 100mm, **D** 65mm

The vessel is brass and the bolt is steel. A threaded turned bolt with gasket forms a sealing cap which allows the vessel to be filled.

Inventory No. 1927-1536

J119 Parallel wires held in a rectangular frame

nineteenth century

L 262mm, **W** 209mm, **T** 14mm

Each of the 28 parallel copper wires is 8 inches long and are spaced $\frac{1}{4}$ inch apart. It is not known to what purpose this apparatus was put. The copper wire is not under tension but has been distorted.

Inventory No. 1927-1492

J120 Glass cylinder with stopper

nineteenth century

INSCRIPTION
'8,4'

D 18mm , **L** 61mm , **H** 28mm (stopper), **D** 15mm (stopper)

One end of the tube has been ground flat, but it makes a poor seal with the stopper. The other end of the tube has been cut or broken off as the edges are rough.

C 129

Inventory No. 1927-1600

J121 Steel turnings in a glass jar

Beatson, Clark and Company 1918

after 1917 (jar)

'bottle patent No 117403'

H 120mm, **D** 70mm

This could be exhibiting the advantages of a new glass stopper invented by Beatson, Clark and Company in 1918. If so it is the latest known addition to the collection. Possibly the steel turnings were kept in the jar to see if any corrosion took place. Since the jar has been damaged some corrosion has unavoidably occurred. Alternatively the turnings could simply have been put in the bottle when the collection was transferred to the Museum.

Inventory No. 1927-1533

J122 Two thick-walled glass tubes with a scale

second half nineteenth century

SCALE—HANDWRITTEN IN INK
1830 (10) 2040

L 480mm, **W** 50mm, **DE** 25mm

The instrument appears to be some form of barometer or manometer, or possibly a hydrometer for dense liquids. It consists of two glass tubes with their bases in a bung. One has a sealed bulb at the top while the other is broken. The scales on the paper behind the tubes are identical and run from 1830 to 2040 in tens downwards.

Inventory No. 1927-1735

J123 Three test tubes

second half nineteenth century

L 52mm (1), **D** 9mm (1), **L** 39mm (2), **D** 10mm (2), **L** 39mm (3), **D** 8mm (3)

Inventory No. 1927-1607

J124 Box with two handles

nineteenth century

L 335mm, **W** 95mm, **DE** 80mm

This appears to be a very shallow mahogany box or drawer in two pieces. There is a rosewood handle attached to each side. Additional hinges are missing.

Inventory No. 1927-1885

J125 Plank with slot

second half eighteenth century or nineteenth century

L 1380mm, **W** 150mm, **L** 1030mm (slot)

Inventory No. 1927-1845

J126 Box mounted on slotted beam which carries three stands

nineteenth century

SCALE IN INCHES
1–29

H 260mm (stands), **L** 810mm (stands), **W** 150mm (stands), **H** 160mm (box and beam), **L** 200mm (box and beam), **W** 150mm (box and beam)

This piece of apparatus was designed to allow for adjustment amongst the three stands in both length and height. It may have been attached to a larger stand.

Inventory No. 1927-1844

J127 Rectangular framework with two brackets

second half eighteenth century

L 865mm, **W** 270mm, **T** 95mm

The two brackets have two holes each for wooden dowels.

Inventory No. 1927-1835

J128 Board with centrally mounted collar

second half eighteenth century or first half nineteenth century

L 760mm, **H** 100mm, **T** 12mm

The collar is secured to the board with mahogany screws and could be held on a shaft by a grub-screw.

Inventory No. 1927-1838

J129 Two T-section beams with clamps

PAINTED—FACES
'V' 'A' 'B' 'C' 'D' 'E' 'F'
'. . . V . . . F G H . . R'

L 780mm (beam), **W** 60mm (beam), **T** 30mm (beam)

The clamp mechanism on the beam bracket could have been used to secure it to a table top. It does not fit the Philosophical Table as its top is too thin. The section of the beam fits the clamps from J130. In addition the knurled securing bolts on each object match, further suggesting an association.

Inventory No. 1927-1832

J130 Two clamps with knurled retaining bolts

L 180mm, **W** 65mm, **H** 80mm

Used with J129

Inventory No. 1927-1890

J131 Plank with wedge-shaped end

second half eighteenth century or first half nineteenth century

L 1103mm, **T** 10mm, **W** 59mm

The ends of the mahogany plank are bevelled but only on one side. In addition there are three locating holes.

Inventory No. 1927-1834

J132 Fulcrum on pillar with circular base

last quarter eighteenth century or first quarter nineteenth century

D 45mm (base), **H** 76mm, **L** 47mm (fulcrum), **W** 12mm (fulcrum)

The pillar is threaded into the base. The fulcrum has one slotted pivot; the other is a straight hole. It is entirely made of brass.

Inventory No. 1927-1869

J133 Rectangular base with four circular holes

second half eighteenth century or first half nineteenth century

L 1370mm, **W** 153mm, **H** 22mm

The holes are equally spaced pairs at each end of the mahogany base. A faded discoloration shows where four circular pillars were mounted.

Inventory No. 1927-1833

J134 Wedge

second half eighteenth century or first half nineteenth century

H 70mm, **L** 410mm, **W** 162mm

The mahogany wedge is in the form of a semi-parabola.

Inventory No. 1927-1840

J135 Cylindrical block with four equidistant holes

second half eighteenth century or first half nineteenth century

D 295mm, **H** 95mm

Inventory No. 1927-1823

J136 Two rectangular blocks and seven loose screws

nineteenth century

L 153mm, **W** 47mm, **H** 23mm

Each mahogany block has seven or eight wood screw holes of which four are countersunk. The blocks appear to be detached from a larger piece of apparatus.

Inventory No. 1927-1881

J137 Beam clamp with two milled nuts

nineteenth century

H 22mm, **L** 270mm, **W** 28mm

This clamp was probably used in conjunction with a larger piece of apparatus. The beam is mahogany and the nuts are brass.

Inventory No. 1927-1888

J138 Three units with slots

second half eighteenth century

W 106mm, **T** 17mm, **H** 58mm

The blocks are made of mahogany. They may be part of the Adams mechanics material but there is no direct evidence.

Inventory No. 1927-1886

J139 Mahogany block with inclined ends

second half eighteenth century or nineteenth century

L 230mm, **W** 85mm, **H** 24mm

There are two brass studs, diagonally across from each other on the underside of the block.

Inventory No. 1927-1889

J140 Triangular block with butterfly screw and two washers

nineteenth century

H 100mm, **W** 110mm, **L** 23mm

The butterfly screw is made from a mahogany screw soldered to a triangular piece of brass.

Inventory No. 1927-1892

J141 Unit with slot and saw-cut

second half eighteenth century or first half nineteenth century

W 122mm, **T** 12mm, **H** 77mm

The vertical keyway slot could support a similarly shaped beam, while the saw-cut which is a right angles could hold a piece of card. The unit is made of mahogany.

Inventory No. 1927-1893

J142 Small brass spanner

second half eighteenth century or first half nineteenth century

L 45mm

Inventory No. 1927-2046

J143 Small cruciform key and tool

second half eighteenth century or first half nineteenth century

L 52mm, **W** 34mm, **T** 5mm

The brass key, in addition to its winding function, contains three square-section ends. These could be used as internal spanners.

Inventory No. 1927-2045

J144 Turned rod with a worm gear at one end and a rectangular plug at the other end

second half nineteenth century

L 250mm, **D** 30mm (max), **D** 11mm (min)

The bronze object has been cast and then turned on a lathe. The mounting points at each end are clearly visible.

Inventory No. 1927-1772

J145 Apparatus with diaphragm on circular block

mid-nineteenth century

D 130mm, **H** 90mm

The apparatus is almost certainly optical in nature and could have been used to adjust the alignment of a lens or prism in an optical demonstration. The diaphragm is made of cardboard, the adjusting mechanism is of boxwood, and the base is of mahogany.

Inventory No. 1927-1522

J146 Graduated semicircle scale on iron base

mid-nineteenth century

'*KCL*'

D 100mm (max), **D** 35mm (min), **H** 115mm

The silvered semicircular scale is marked from 0° to 90° in each direction. The movable scale marker is attached to a collar on which a piece of wire or rod has broken off.

Inventory No. 1927-2057

J147 Three tapering T-section beams with brackets

second half eighteenth century or first half nineteenth century

L 1395mm, **W** 65mm (max), **W** 37mm (min), **H** 39mm (max), **H** 25mm (min)

One end of the mahogany beams is surmounted with an offset circular bracket while the other end has a flat plate. This T-shaped plate has a threaded hole which houses a thumbscrew bolt in two of the three beams.

Inventory No. 1927-1894

J148 Parts of a curve

mid-nineteenth century

PAINTED
    '*Involute of a Circle. Rad: 1 3/4 In:*'

H 22mm (1), **W** 55mm (1), **L** 650mm (1), **H** 22mm (2), **W** 55mm (2), **L** 440mm (2)

The curve was probably part of the set of figures including L1; it is now in a poor condition. It is made of mahogany.

C 246

Inventory No. 1929-128

J149 Release mechanism on support arm

nineteenth century

L 148mm, **W** 20mm, **DE** 20mm

The brass release mechanism is on a mahogany arm. It is probably from a larger instrument.

Inventory No. 1927-1698

J150 Tube with two-pronged collar connection, tap, and nozzle

second half eighteenth century or first half nineteenth century

L 530mm, **D** 45mm (max)

The brass nozzle is threaded onto the tap which is soldered onto the tube.

Inventory No. 1927-1770

J151 Board with bracket

nineteenth century

L 177mm, **W** 85mm, **H** 115mm

This is simply a mahogany board with a brass bracket which has a circular hole near the top screwed to one end. There are signs that another bracket was originally attached to the other end.

Inventory No. 1927-1796

J152 Four tapering glass rods

L 355mm (one), **D** 23–37mm

Three of the four rods are broken and three have a green tinge while the other is colourless. This last rod has been lacquered and superficially looks brown. The tips of the rods are slightly bevelled, suggesting that they might be part of an insulating stool or table.

Inventory No. 1927-1699

J153 Stands for magnets

first half nineteenth century
1: **L** 1170mm, **W** 250mm, **H** 110mm
2: **L** 1140mm, **W** 240mm, **H** 130mm

The rectangular mahogany stands have two locating lips for each magnet.

Inventory No. 1927-1652, -1653

J154 Rod with screw nut

second half eighteenth century or first half nineteenth century
L 550mm, **D** 9mm, **D** 25mm (nut)

This is simply a copper rod painted black with a brass nut screwed to one end.

Inventory No. 1927-1694

J155 Rectangular clamp with milled thumbscrew

nineteenth century
H 35mm, **W** 17mm (max), **W** 10mm (min), **L** 29mm

The brass clamp could have been used to hold a part in place on a larger apparatus or demonstration.

Inventory No. 1927-1769

J156 Threaded pillar with butterfly screw

mid-eighteenth century
L 90mm (two screws), **H** 83mm, **D** 15mm (max), **D** 8mm (min)

The brass pillar appears to have been mounted.

Inventory No. 1927-1765/2

J157 Threaded pillar with two cross-pieces

mid-eighteenth century
H 113mm, **D** 15mm, **L** 30mm (cross-piece)

The brass pillar appears to have been mounted.

Inventory No. 1927-1765/1

J158 Tray with 25 thin metal discs, a ceramic cone, and a lens

mid-nineteenth century
L 270mm (tray), **W** 195mm (tray), **DE** 43mm (tray), **D** 18mm (lens mount), **D** 53mm (concave lens), **D** 15mm (brass ring), **D** 50mm (discs)

Used with L73

Inventory No. 1927-1595

J159 Bell with striker

second half eighteenth century or first half nineteenth century
L 260mm, **H** 410mm, **D** 145mm (bell)

The brass bell is suspended from a mahogany frame which is rotatable on the weighted mahogany base. It was probably used for experiments on central forces.

Inventory No. 1927-1350

J160 Pump plate

nineteenth century
SCALE
9–0–9

D 215mm, **H** 330mm

This brass pump plate has a ground glass top and a manometer.

Inventory No. 1927-1349

J161 Glass cylinder

nineteenth century

L 205mm, **D** 160mm

This glass cylinder is ground on the outside and may be a shade for a lamp.

C 129

Inventory No. 1927-1354

J162 Glass cylinder

nineteenth century

L 203mm, **D** 157mm

C 129

See J161

Inventory No. 1927-1355

J163 Glass sphere

nineteenth century

L 420mm, **D** 250mm

This glass sphere has brass collars which have been repaired or replaced at some time. The condition is good although the tap is not original.

C 129

Inventory No. 1927-1356

J164 Bell jar

nineteenth century

H 282mm, **D** 140mm

C 129

Inventory No. 1927-1369

J165 Glass flask

nineteenth century

H 245mm

This round-bottomed flask has a brass collar threaded on the outside which allows the flask to screw into a large piece of apparatus.

C 129

Inventory No. 1927-1385

J166 Two tubes

nineteenth century

L 455mm, **D** 30mm

These tubes, made of thick glass, each have a brass cap.

C 129

Inventory No. 1927-1392

J167 Glass tube with stand

nineteenth century

H 545mm (stand), **L** 460mm (tube), **D** 40mm (tube), **L** 170mm (brass core)

This glass tube fits over a cylindrical brass core. It has a stand with a mahogany base with brass upright and two arms.

C 129

Inventory No. 1927-1393

J168 Glass cylinder

nineteenth century

L 264mm, **D** 48mm

This glass cylinder has a brass base with a milled screw in it.
C 129

Inventory No. 1927-1396

J169 Pump plate

nineteenth century

H 205mm, **D** 100mm

This brass pump plate has a nozzle screwed on its upper side. The pump plate is screwed on a brass connector with a tap which in turn screws into a wooden stand.

Inventory No. 1927-1401

J170 Brass disc

nineteenth century

D 115mm, **W** 5mm

Inventory No. 1927-1402

J171 Pump plate with metal plate and two sunk cups

Newman, J.

1775–1857 (possibly composite)

STAMPED ON TAP
'I. NEWMAN'

D 120mm, **H** 95mm

This metal pump plate is lead weighted and has two sunk cups. The tap attached may not belong to it.

Inventory No. 1927-1406

J172 Tube

nineteenth century

L 480mm, **D** 15mm

This brass tube has a brass collar and brass spring and a threaded wooden cap at one end.

Inventory No. 1927-1409

J173 Manometer

nineteenth century

L 180mm (overall), **W** 28mm (overall)

This mercury manometer sits in a glass tube which has a brass cap.

Inventory No. 1927-1450

J174 Glass tube

nineteenth century

L 830mm

This glass tube has one closed end. The other end had a ground glass stopcock which has been broken off.

Inventory No. 1927-1690

J175 U-tube

nineteenth century

STAMPED ON ONE SCREW
'KCL'
scale
'19–0–20'

H 505mm

This glass U-tube is supported by an iron stand with three brass levelling screws.

Inventory No. 1927-1940

J176 Nine plugs

nineteenth century

These brass plugs are accessories for an air pump.

Inventory No. 1927-2050

J177 Pipe

nineteenth century
L 950mm, **D** 12mm

This brass pipe was for hydrostatic demonstrations involving a jet of water or for showing the path of projectiles.

Inventory No. 1927-1408

J178 Jar

nineteenth century
H 330mm

This is a cylindrical glass jar for unknown purposes.

Inventory No. 1927-1389

J179 Four cups in stand

nineteenth century
H 580mm (stand), **L** 65mm (cup)

These brass cups are painted on the inside and all have a brass tap screwed in their ends. They fit on a mahogany stand which has provision for five cups. They may be for hydrostatic demonstrations.

Inventory No. 1927-1747

J180 Four nozzles

nineteenth century
L 280mm (max)

These nozzles can be screwed into the plate of an air pump.

Inventory No. 1927-2051

J181 Three taps

eighteenth or nineteenth century
L 40mm

Two of these taps are made of brass and the other of steel. They are for pneumatics experiments.

Inventory No. 1927-2049

J182 Six screws

nineteenth century
L 45mm

These steel screws have brass butterfly heads and washers.

Inventory No. 1927-2047

J183 Spanner

second half eighteenth century
L 125mm, **T** 11mm

This brass spanner is double ended and is probably for an air pump.

Adams 1799a, Vol. 1, Pl. 1, Fig. 1, Item f

Inventory No. 1927-2044

J184 Two keys

nineteenth century
L 80mm

Both these keys are made of brass; one has a wooden handle. It is likely they are for use with an air-pump or a similar piece of equipment.

Inventory No. 1927-2043

J185 Crank

nineteenth century

L 140mm

The wooden handle on this crank can be adjusted to vary its length. The brass shaft has three square holes.

Inventory No. 1927-2042

J186 Three spanners

eighteenth or nineteenth century

L 140mm, 170mm, 115mm

Two of these three pinhole spanners are made of brass and one of steel. All three have wooden handles. They are for use with an air pump.

Inventory No. 1927-2041

J187 Wheel

second half eighteenth century or first half nineteenth century

D 87mm, **DE** 18mm

This wheel is made of boxwood with a groove in the circumference. It has a brass disc on each side.

Inventory No. 1927-1849

J188 Flask on stand

nineteenth century

D 190mm, **H** 245mm

This glass flask is mounted on a mahogany support which in turn is supported on a mahogany base by a short brass pillar. The flask and support can rotate on this pillar.

Inventory No. 1927-1879

J189 Spanner

eighteenth century

L 230mm

Inventory No. 1927-1693

J190 Bell jar

second half eighteenth century or nineteenth century

H 240, **D** 127mm

This glass bell jar has a ground folded-over rim and a ground neck with stopper.

C 129

Inventory No. 1927-1363

J191 Weighted cylinder

D 60mm, **H** 65mm

This cylinder is made of boxwood.

Inventory No. 1927-1345

J192 Steel pivot

second half eighteenth century or first half nineteenth century

L 80mm, **D** 5mm

Inventory No. 1927-1336

Bibliography

Abraham Sharp (1963). *Abraham Sharp: mathematician and astronomer 1643–1743*. Exhibition Catalogue, Bolling Hall Museum, Bradford.

Adams, G. (1746). *A catalogue of mathematical, philosophical and optical instruments*, bound with *Micrographia illustrata*. George Adams, London.

—— (1748). *The description and use of a new sea quadrant*. John Hart, London.

—— (1761a). *A description of the pneumatic apparatus*. MS.

—— (1761b). *Pneumatics scrapbook*. MS.

—— (1761–2). *Principles of mechanics—plates*. MS.

—— (1762). *A description of an apparatus for explaining the principles of mechanicks*. MS.

—— (1766). *A treatise describing and explaining the construction and use of new celestial and terrestrial globes*. London.

—— (1777). *A catalogue of mathematical, philosophical and optical instruments*, bound with *A treatise describing the construction and use of new celestial and terrestrial globes* (4th edn). London.

Adams, G., Jr (1787a). *Catalogue of mathematical and philosophical instruments* bound with *An essay on electricity with an essay on magnetism*. London.

—— (1787b). *An essay on electricity, with an essay on magnetism* (3rd edn). London.

—— (1787c). *Essays on the microscope*.

—— (1791). *Geometrical and graphical essays*. George Adams, London.

—— (1794). *Lectures on natural and experimental philosophy*, 5 Vols. London.

—— (1797). *Geometrical and graphical essays* (2nd edn).

—— (1798). *Essays on the Microscope*, bound with *A catalogue of optical, mathematical and philosophical instruments* (3rd edn). W. and S. Jones, London.

—— (1799a). *An essay on electricity* (5th edn). London.

—— (1799b). *Lectures on natural and experimental philosophy etc.*, 5 Vols (2nd edn). W. and S. Jones, London.

Albert, W. (1972). *The turnpike road system in England 1663–1840*. Cambridge University Press.

Alexander, H. G. (1956). *The Leibniz–Clarke correspondence, together with extracts from Newton's Principia and Opticks*. Manchester University Press.

Algarotti, F. (1739). *Newtonismo per il dame*. London.

Allan, D. G. C. (1974). The Society of Arts and Government, 1754–1800. Public encouragement of arts, manufactures, and commerce in eighteenth century England. *Eighteenth century studies*, **7**, 434–52.

—— (1979). *William Shipley, founder of the Royal Society of Arts: a biography with documents* (revised edn). Scolar Press, London.

—— (1989). The Laudable Association of Anti-Gallicans. *Royal Society of Arts Journal*, **137**, 623–8.

Altick, R. D. (1978). *The shows of London*. Harvard University Press, Cambridge, MA.

Amontons, G. (1688). Sur un nouveau barometre. *Histoire de L'Académie Royale des Sciences*, **2**, 39.

Ampère, A. M. (1826). *Theorie des phénomènas électro-dynamiques*, Paris.

Anderson, P. J. (1893). *Officers and graduates of University and King's College Aberdeen MVD – MDCCCLX*. Aberdeen.

Anderson, R. G. W. (1978). *The Playfair Collection and the teaching of chemistry at the University of Edinburgh 1713–1858*. Royal Scottish Museum, Edinburgh.

Anon (1845). *Practical hints on the daguerreotype*. London.

Appleby, J. H. (1990). Erasmus King: eighteenth-century experimental philosopher. *Annals of science*, **47**, 375–92.

Arago, F. (1811). Sur une modification remarquable qu'éprouvent les rayons lumineux dans leur passage à travers certains corps diaphanes, et sur quelques autres nouveaux phénomènes d'optique. *Mémoires de la classe des sciences mathématiques*, 93–134.

—— (1854–62). *Oeuvres de François Arago*, Vol. 7 (ed. J. A. Barral). Paris.

Armitage, A. (1966). *Edmond Halley*. Nelson, London.

Armstrong, W. G. (1840). On the electricity of effluent steam. *Philosophical magazine*, **17**, 452–7.

—— (1842). On the cause of the electricity of effluent steam. *Philosophical magazine*, **20**, 5–8.

—— (1843). On the efficacy of steam as a means of producing electricity and on a curious action of a jet of steam upon a ball. *Philosophical magazine*, **22**, 1–5

Ashworth, P. (1990). Davis of Windsor. *Windlesora (Journal of the Windsor local history publications group)*, **9**.

Atkinson, A. D. (1952). William Derham, FRS (1657–1735). *Annals of science*, **8**, 368–92.

Austin, J. F. and McConnell, A. (1980*a*). James Six FRS. Two hundred years of the Six's self-registering thermometer. *Notes and records of the Royal Society of London*, **35**, 49–65.

—— (1980*b*). *J. Six. The construction and use of a thermometer*. Nimbus, London.

Bailey, A. M. (1776). *The advancement of arts, manufactures, and commerce* (revised edn). Alexander Bailey, London.

Bailey, W. (1786). *Description of the machines in the repository of the Society of Arts*. London.

Baker, H. (1742). *The microscope made easy*. London.

—— (1751). Baker correspondence in the John Rylands University Library of Manchester. MS English 19.

—— (1753). *Employment for the microscope*. London.

Baltrusaitis, J. (1977). *Anamorphic art* (trans. Strachan). Chadwyck-Healey, Cambridge.

Banfield, E. (1985). *Barometers: stick or cistern tube*. Baros, Trowbridge.

Barrière, P. (1951). *L'académie de Bordeaux. Centre de culture internationale au XVIIIe siècle (1712–1792)*. Bière, Bordeaux.

Barrow, I. (1675). *Archimedes opera methodo nova illustrata succincte demonstrata per Is Barrow*.

Beatson, Clark and Co. (1918). Improvements in or relating to glass bottle stoppers. British Patent 117403.

Beckett, J. V. (1981). *Coal and tobacco: the Lowthers and the economic development of West Cumberland, 1660–1760*. Cambridge University Press.

Bedini, S. (1966). Lens making for scientific instrumentation in the seventeenth century. *Applied optics*, **5**, 687–94.

Bélidor, B. (1737). *Architecture Hydraulique*, 2 Vols. Paris.

Bennett, A. (1787). Description of a new electrometer. *Philosophical transactions of the Royal Society of London*, **77**, 26–34.

Bennett, J. A. (1984). *Catalogue 5. Spectroscopes, prisms and gratings*. Whipple Museum of the History of Science, Cambridge.

—— (1987). *The divided circle*. Phaidon Christie's, Oxford.

Benton, W. A. (1940–8). *An illustrated record of the objects and pictures in the Avery Historical Museum relating to the history of weighing*, Vol. 23. Avery, Birmingham.

Bernoulli, J. (1771). *Lettres astronomiques*. Jean Bernoulli, Berlin.

Berry, H. F. (1915). *A history of the Royal Dublin Society*. Longmans Green, London.

Bevis, J. (1769). Observation of the last transit of Venus, and of the eclipse of the Sun the next day; made at the house of Joshua Kirby, Esq. at Kew. *Philosophical transactions of the Royal Society of London*, **59**, 181–91.

Bion, N. (1723). *Traité de la construction et des principaux usages des instruments mathematique*. La Haye.

—— (1758). *The construction and principal uses of mathematical instruments* (trans. E. Stone). London.

Birch, T. (1744). *The life of the honourable Robert Boyle*. Millar, London.

Black, J. (1987). *The English press in the eighteenth century*. University of Pennsylvania, Philadelphia, PA.

Blackmore, H. L. (1983). *Firearms accessories in Hugh Pollard's history of firearms*. Country Life Books, London.

Blanchard, R. (1955). *The Englishman. A political journal by Richard Steele*. Clarendon Press, Oxford.

Blondel, C. (1982). *A.-M. Ampère et la création de l'électrodynamique (1820–1827)*. Bibliotheque Nationale, Paris.

Bolle, B. (1982). *Barometers*. Argos, Watford.

Bonelli, M. L. (1968). *Il Museo di Storia della Scienza a Firenze*. Cassa di Risparmio, Florence.

Boss, V. (1972). *Newton and Russia. The early influence, 1698–1796*. Harvard University Press, Cambridge, MA.

Bouguer, P. (1748). De la mesure des diametres des plus grandes planetes: Description d'un nouvel instrument qu'on peut nommer héliomètre, propre à les determiner observations sur le soleil. *Memoires de l'Académie Royales des Sciences*, **61**, 11–34.

Bourne, J. (1846). *Treatise on the steam engine*. London.

Bowers, B. (1975). *Sir Charles Wheatstone FRS, 1802–1875*. Her Majesty's Stationery Office, London.

—— (1982). *A history of electric light and power*. Peter Peregrinus, Stevenage, for the Science Museum.

Boyle, R. (1660). *New experiments physico-mechanicall*. Oxford.

—— (1725). *The philosophical works of the honourable Robert Boyle* (ed. P. Shaw). Innys, London.

Bracegirdle, B. (1978). *A history of microtechnique*. Heinemann, London.

—— (1986). The first automatic microtome. *Proceedings of the Royal Microscopical Society*, **21**, 145–7.

Bradbury, S. (1967). *The evolution of the microscope*. Pergamon, Oxford.

Bradley, R. (1739). *New improvements of planting and gardening* (7th edn). London.

Braun, I. A. (1767). De admirando frigore artificiali quo mercurius seu hydrargyos est congelatus dissertatio. *Novi commentarii academiae scientiarum imperialis petropoli*, **11**, 27–34.

Brewer, J. (1976). *Party ideology and popular politics at the accession of George III*. Cambridge University Press.

—— (1989). *The sinews of power: war, money and the English state, 1688–1783*. Unwin Hyman, London.

Brewster, D. (1849). Description of a new stereoscope. *Report of the 19th Meeting of the British Association for the Advancement of Science, Birmingham*, pp. 6–7.

—— (1851). Description of several new and simple stereoscopes for exhibiting, as solids, one or more representations of them on a plane. *Transactions of the Royal Scottish Society of Arts*, **3**, 247–59.

—— (1856). *The stereoscope*. London.

Britten, F. J. (1982). *Britten's old clocks and watches* (9th edn) (ed. G. H. Baillie, C. Ilbert and C. Clutton). Methuen, London.

Brooke, J. (1972). *King George III*. Constable, London.

Brown, J. (1661). *The description and use of a joynt rule*. J. Brown and H. Sutton, London.

—— (1671). *The description and use of the triangular quadrant*. John Wingfield, London.

Brown, Joyce (1979). *Mathematical instrument-makers in the Grocers' Company 1688–1800*. Science Museum, London.

Brown, O. (1982). *Catalogue 2. Balances and weights*. Whipple Museum of the History of Science, Cambridge.

Bryden, D. J. (1972a). *Scottish scientific instrument-makers 1600–1900*. Royal Scottish Museum, Edinburgh.

—— (1972b). Three Edinburgh microscope makers: John Finlayson, William Robertson and John Clark. *Book of the Old Edinburgh Club*, **33** (3), 165–76.

—— (1988). *Catalogue 6. Sundials and related instruments*. Whipple Museum of the History of Science, Cambridge.

Buckland, J. S. (1984). Thomas Savery, his steam engine workshop of 1702. *Transactions of the Newcomen Society*, **56**, 1–20.

Butler, J. and Hoskin, M. (1987). The archives of Armagh Observatory. *Journal for the history of astronomy*, **18**, 295–307.

Bynum, W. F. and Porter, R. (ed.) (1985). *William Hunter and the eighteenth-century medical world*. Cambridge University Press.

Calvert, H. R. (1971). *Scientific trade cards in the Science Museum Collection*. Her Majesty's Stationery Office, London.

Canton, J. (1751). A method of making artificial magnets without the use of natural ones. *Philosophical transactions of the Royal Society of London*, **47**, 31–8.

—— (1753–4). Electrical experiments with an attempt to account for their several phenomena, together with some observation on thunder-clouds. *Philosophical transactions of the Royal Society of London*, **48**, 350–8.

Cantor, G. N. (1983). *Optics after Newton: theories of light in Britain and Ireland, 1704–1840*. Manchester University Press.

Cassels, J. W. S. (1979). The Spitalfields Mathematical Society. *Bulletin of the London Mathematical Society*, **11**, 241–58.

Cavallo, T. (1777). *A complete treatise of electricity in theory and practice*. London.

—— (1782). *A complete treatise on electricity* (2nd edn). London.

—— (1786–95). *A complete treatise on electricity* (3rd edn). London.

—— (1798). *An essay on the medicinal properties of factitious airs*.

Chabrol, Le père François (1753). *Abrège des leçons de phisique expérimentalle, données en 1753, par le Docteur Maimbray: avec des remarques critiques*. MS.

Chaldecott, J. A. (1951). *Handbook of the King George III collection of scientific instruments*. Her Majesty's Stationery Office, London.

—— (1975). Josiah Wedgwood (1730–95) scientist. *British journal for the history of science*, **8**, 1–16.

Chambers, W. (1763). *Plans, elevations, sections and perspective views of the gardens and buildings at Kew, Surry*. W. Chambers, London.

Chew, V. K. (1968). *Physics for princes*. Her Majesty's Stationery Office, London.

C., J. (1694). A paper about magnetism or concerning the changing and fixing the polarity of a piece of iron. *Philosophical transactions of the Royal Society of London*, **18**, 257–62.

Clarke, E. M. (1836). Description of E. M. Clarke's magnetic-electrical machine. *Philosophical magazine*, **9**, 262–6.

Clarke, T. N., Morrison-Low, A. D., and Simpson, A. D. C. (1989). *Brass and glass: scientific instrument making workshops in Scotland as illustrated by instruments from the Arthur Frank Collection at the Royal Museum of Scotland*. National Museums of Scotland, Edinburgh.

Clay, R. S. and Court, T. H. (1932). *History of the microscope*. Griffin, London.

Coe, B. (1978). *Cameras from Daguerreotypes to instant pictures*. Marshall Cavendish, London.

Collins, J. (1710). *The description and use of 4 several quadrants*. London.

Colton, J. (1974). Kent's Hermitage for Queen Caroline at Richmond. *Architectura*, **4**, 181–91.

Conyers, J. (1678). Extract of a letter from Mr John Conyers of his improvement of Sir Samuel Moreland's speaking trumpet & C. *Philosophical transactions of the Royal Society of London*, **12**, 1027–9.

Cooke, Troughton and Simms Ltd. (1959). *At the sign of the orrery: the origins of the firm of Cooke, Troughton & Simms, Material collected by E. Wilfred Taylor and J. Simms Wilson*, York.

Cope, Z. (1953). *William Cheselden 1688–1752*. Livingstone, Edinburgh.

Corner, B. C. and Booth, C. C. (1971). *Chain of friendship: selected letters of Dr John Forthergill of London 1735–1780*. Belknap, Cambridge, MA.

Costabel, P. (1948). Le paradoxe de Mariotte. *Archives internationales d'histoire des sciences*, **2**, 446–7.

Coulomb, C. A. (1785a). Description d'une boussole dont l'aiguille est suspendue par un fil de soie. *Mémoires de l'Académie Royale*, **99**, 560–8.

—— (1785b). Mémoires sur l'electricite et le magnétisme. *Mémoires de l'Académie Royale*, **99**, 569–638.

Court, T. H. and von Rohr, M. (1929–30). Chronological list of the more important members of the Worshipful Company of Spectaclemakers. *Transactions of the Optical Society*, **31** (2), 71–90.

Cowper, M. (1865). *Diary of Mary Countess Cowper, 1714–20*. Murray, London.

Cranfield, G. A. (1962). *The development of the provincial newspaper 1700–1760*. Clarendon Press, Oxford.

Crawforth, M. A. (1984). *Handbook of old weighing instruments*.

International Society of Antique Scale Collectors, Chicago.

Crisp, H. (1925). *The Crisp Collection of antique microscopes. Sale catalogue from Steven's Auction Rooms.* London, 17 February 1925.

Crommelin, C. A. (1951). *Descriptive catalogue of the physical instruments of the 18th century.* Communication 81, National Museum of the History of Science, Leiden.

Cudworth, W. (1889). *The Life and correspondence of Abraham Sharp.* London.

Cumberland (1765). Inventory of household-furniture of His Late Royal Highness The Duke of Cumberland at Windsor Great Lodge. Royal Archives CP, Vol. 1. Royal Library, Windsor Castle. MS.

Cumming, J. (1822a). On the application of magnetism as a measure of electricity. *Transactions of the Cambridge Philosophical Society,* **1**, 281–6.

—— (1822b). On the connection of galvanism and magnetism. *Transactions of the Cambridge Philosophical Society,* **1**, 269–79.

Curll, E. (1736). *The rarities of Richmond.*

Cutcliffe, S. H. (ed.) (1984). *Science and technology in the eighteenth century.* The Institute, Bethlehem, PA.

Cuthbertson, J. (1807). *Practical electricity and galvanism.* London.

Daniell, J. F. (1830). On a new register pyrometer, for measuring the expansions of solids, and determining the higher degrees of temperature upon the common thermometric scale. *Philosophical transactions of the Royal Society of London,* **120**, 257–86.

Daumas, M. (1972). *Scientific instruments of the 17th and 18th centuries and their makers* (trans. M. Holbrook). Batsford, London.

Davenport, S. (1737). *A description of the new-invented table air-pump.* Stephen Davenport, London.

Davidson, C. (1986). *A woman's work is never done* (2nd edn). Chatto & Windus, London.

Davy, H. (1816a). An account of an invention for giving light in explosive mixtures of fire-damp in coal mines by consuming the fire-damp. *Philosophical transactions of the Royal Society of London,* **106**, 23–4.

—— (1816b). Further experiments on the combustion of explosive mixture confined by wire-gauze with some observations on flame. *Philosophical transactions of the Royal Society of London,* **106**, 115–19.

—— (1816c). On the fire-damp of coal mines, and on methods of lighting the mines so as to prevent explosion. *Philosophical transactions of the Royal Society of London,* **106**, 1–22.

de Magellan, J. H. (1779). *Description et usages des nouveaux barometres, pour mesurer la hauteur des montagnes et la profondeur des mines* W. Richardson, London.

—— (1780). *Description d'une machine nouvelle de dynamique, inventée par Mr G. Atwood . . . dans une lettre adressée à monsieur A. Volta . . . par J. H. de Magellan.* W. Richardson, London.

—— (1783). *Description of a glass-apparatus* (3rd edn).

de Servière, G. (1719). *Recueil d'ouvrages curieux de mathématique et de mecanique ou description du cabinet de Monsieur Grollier de Servière.* Lyon.

de Virville, A. F. (1723). *Construction d'un nouveau barometre.*

Dear, P. (1985). Totius in verba: rhetoric and authority in the early Royal Society. *Isis,* **76**, 145–61.

Dekker, E. (1986). *The Leiden sphere.* Museum Boerhaave, Leiden.

Delamain, R. (1632). *The making description and use of a small portable instrument in the form of a mid trapezia called a horizontall quadrant.* R. Hawkings, London.

Della Porta, G. B. (1658). *Natural magick.* London.

Demainbray, S. C. T. (1750–2). *A short account of a course of natural and experimental philosophy: consisting of 40 lectures.*

—— (1753). *Programme, ou idée generale d'un cours de physique experimentale, en trente-quatre lecons.*

—— (1754a). *Programme, ou idée generale d'un cours de physique experimentale, en trente-quatre lecons.*

—— (1754b) *The syllabus of a course of natural and experimental philosophy.*

—— (1761). Letter from Demainbray to Bute, 29 January 1761. MS.

Derham, W. (1714). *The artificial clock maker.* London.

Desaguliers, J. T. (1713). *A catalogue of the experiments in Mr Desaguliers's course.* London.

—— (1717). *Physico-mechanical lectures. Or, an account of what is explain'd and demonstrated in the course of mechanical and experimental philosophy given by J. T. D. Desaguliers,* Bridger, and Vream, London.

—— (1728a). An account of a machine for measuring any depth in the sea with great expedition and certainty. *Philosophical transactions of the Royal Society of London,* **35**, 559–62.

—— (1728b). *The Newtonian system of the world the best model of government: an allegorical poem.* London.

—— (1734). *A course of experimental philosophy.* London.

—— (1738). An account of some magnetical experiments. *Philosophical transactions of the Royal Society of London,* **40**, 385.

—— (1744). *A course of experimental philosophy* (2nd edn), 2 Vols. London.

Dickson, P. G. M. (1967). *The financial revolution in England. A study in the development of public credit 1688–1756.* Macmillan, London.

Diderot, D. (1767). *Recueil de planches sur les sciences, les arts libéraux et les arts méchaniques.* Paris.

Dietrich, Baron d'Holbach (1752). *L'art de la verriere de Neri, Merret et Kunckel.* Paris.

Disney, A. N., Hill, C. F., and Watson-Baker, W. E. (1928). *Catalogue of the Royal Microscopical Society Collection.* Microscopical Society, London.

DNB (1885–1901). *Dictionary of national biography from earliest times to 1900.* Spottiswoode, London.

Dobereiner, J. W. (1823). Neu entdeckte merkwürdige Eigenschaften des suboxyds des Platins. *Annalen der Physik*, **2**, 269–73.

Dolland, J. (1753*a*). On an improvement of retracting telescopes by increasing the number of eye-glasses. *Philosophical transactions of the Royal Society of London*, **48**, 103.

—— (1753*b*). A description of a contrivance for measuring small angles by Mr John Dolland, communicated by Mr J. Short. *Philosophical transactions of the Royal Society of London*, **48**, 178–81.

—— (1754). An explanation of an instrument for measuring small angles by Mr John Dollond Part 2. *Philosophical transactions of the Royal Society of London*, **48**, 551–64.

—— (1758). Object glasses for telescopes. British Patent 721.

Dollond, P. (1765). An account of an improvement made by Mr Peter Dollond in his new telescopes. *Philosophical transactions of the Royal Society of London*, **55**, 54–6.

Dossie, R. (1768). *Memoirs of agriculture, and other oeconomical arts*, Vol. 1. J. Nourse, London.

DSB (1970–80) *Dictionary of Scientific Biography* (ed. C. C. Gillispie). Charles Scribener, New York.

du Quet, M. (1706). *Machines approvées par l'Académie Royale des Sciences de Paris* (ed. Gallon). Paris.

Dubost, M. (1747). Description d'un moulin. In *Machines approvés par l'Académie Royale des Sciences de Paris*, Vol. 7 (ed. Gallon). Paris. (1777).

Dutens, L. (1806). *Memoirs of a traveller now in retirement*. Phillips, London.

Eder, J. M. (1978). *History of photography*. Dover, New York.

Edinburgh encyclopaedia (1830). Blackwood.

Ehrmann, F. L. (1780). *Description et usage de quelques lampes à air inflammable*. Strasburg.

Eisenstein, E. L. (1979). *The printing press as an agent of change*. Cambridge University Press.

Emerson, R. L. (1979). The Philosophical Society of Edinburgh, 1737–1747. *British journal of the history of science*, **12**, 154–91.

Emerson, W. (1758). *The principles of mechanics explaining and demonstrating the general laws of motion* (2nd edn). London.

Encyclopaedia Britannica (7th edn) (1842). Edinburgh.

Engelsman, S. B. (1983). Leeuwenhoek's microscopes. In *Beads of glass: Leeuwenhoek and the early microscope* (ed. B. Bracegirdle), pp. 28–40. Museum Boerhaave, Leiden, and Science Museum, London.

Erlam, H. D. (1954). Alexander Monro, primus. *University of Edinburgh journal*, **17**, 77–105.

'Espinasse, M. (1956). *Robert Hooke*. Heinemann, London.

Euler, L. (1795). Letters of Euler to a German princess (trans. H. Hunter). London.

Evans, G. M. (1988). Early paddle wheels. *Marine propulsion international*, (July/August) 20–2.

Fairclough, O. (1975). Joseph Finney and the clock and watch makers of eighteenth century Liverpool. Unpublished M.A. Thesis, University of Keele.

Farey, J. (1827). *A treatise on the steam engine*. London.

Farrell, M. (1981). *William Whiston*. Arno Press, New York.

Faujas de Saint Fond, B. (1907). *A journey through England and Scotland to the Hebrides in 1784*. Hopkins, Glasgow.

Fauque, D. (1983). Les origines de l'héliomètre. *Revue d'histoire des sciences*, **36** (2), 153–70.

Fawcett, T. (1972). Popular science in 18th-century Norwich. *History today*, **22**, 590–5.

Ferguson, J. (1757). *Astronomy explained upon Sir Isaac Newton's principles* (2nd edn). London.

—— (1764*a*). *Lectures on select subjects*. London.

—— (1764*b*). The description of a new and safe crane which has four different powers; invented by Mr James Ferguson F.R.S. *Philosophical transactions of the Royal Society of London*, **54**, 24–8.

Force, J. E. (1985). *William Whiston: honest Newtonian*. Cambridge University Press.

Foster, S. (1659). *Miscellanies: or Mathematical Lucubrations of Mr Samuel Foster* (ed. John Twydsden). London.

Fox, R. H. (1919). *Dr John Fothergill and his friends*. Macmillan, London.

Franklin, B. (1965). *The papers of Benjamin Franklin, Vol. 8., April 1, 1758 through December 31, 1759* (ed. L. W. Labaree). Yale University Press, New Haven, CT.

—— (1986*a*). *The Autobiography and other writings* (ed. K. Silverman). Penguin, New York.

—— (1986*b*). *The papers of Benjamin Franklin, Vol. 25, October 1, 1777 through February 28, 1778* (ed. W. B. Willcox). Yale University Press, New Haven, CT.

Freemantle, W. (1790). *Vertical weight and spring roasting jack*. British Patent 1737, 24 March 1790.

Gallon (ed.) (1777). *Machines approvées par l'Académie Royale des Sciences de Paris*. Paris.

Gardiner, W. (1737). *Practical surveying improved*. London.

Gascoigne, J. (1984). Politics, patronage and Newtonianism: the Cambridge example. *Historical journal*, **27**, 1–24.

—— (1989). *Cambridge in the age of the Enlightenment: science, religion and politics from the Restoration to the French Revolution*. Cambridge University Press.

George, M. D. (1966). *London life in the eighteenth century*. Penguin, Harmondsworth.

Gibbs, F. W. (1951*a*). Peter Shaw and the revival of chemistry. *Annals of science*, **7** (3), 211–37.

—— (1951*b*). Robert Dossie (1717–1777) and the Society of Arts. *Annals of science*, **7**, 149–72.

—— (1952). William Lewis, MB, FRS (1708–1781). *Annals of science*, **8**, 122–51.

Gilbert, W. (1600). *De magnete*. London.

Goldsmith, O. (1776). *A survey of experimental philosophy*. London.

Golinski, J. V. (1983). Peter Shaw: chemistry and communication in Augustan England. *Ambix*, **30**, 19–29.

—— (1988). Utility and audience in eighteenth-century chemistry: case studies of William Cullen and Joseph

Priestley. *British journal of the history of science*, **21**, 1–31.
—— (1989). A Noble spectacle: phosphorus and the public culture of science in the early Royal Society. *Isis*, **80**, 11–39.

Goodison, N. (1977). *English barometers 1680–1860* (revised edn). Antique Collectors Club, Woodbridge.

Gouk, P. M. (1982). Acoustics in the early Royal Society 1660–1680. *Notes and records of the Royal Society of London*, **36**, 155–75.

Gould, R. T. (1923). *The marine chronometer*. Potter, London.

Gowing, R. (1983). *Roger Cotes: natural philosopher*. Cambridge University Press.

Gregory, J. (1663). *Optica promota*.

Griffiss, W. (1755). *A short account of a course of mechanical and experimental philosophy and astronomy*. London.

Grove, G. (1954). *Grove's dictionary of music and musicians*. Macmillan, London.

Guerlac, H. (1963). Francis Hauksbee, expérimentateur au profit de Newton. *Archives internationales d'histoire des sciences*, **16**, 113–28.
—— (1964). Sir Isaac and the ingenious Mr Hauksbee. In *Mélanges Alexandre Koyré* (ed. I. B. Cohen and R. Taton), Vol. 1, pp. 228–53. Hermann, Paris.
—— (1981). *Newton on the continent*. Cornell University Press, Ithaca, NY.

Guggisberg, F. G. (1900). *'The Shop': The story of the Royal Military Academy*. Cassell, London.

Gunther, A. E. (1984). *An introduction to the life of the Rev. Thomas Birch DD FRS 1705–1766*. Halesworth Press, Halesworth.

Gunther, R. W. T. (1920–67). *Early science in Oxford*. Privately printed, Oxford.

Hackmann, W. D. (1978a). Eighteenth century electrostatic measuring devices. *Annali dell' Istituto e Museo di storia della Scienza di Firenze*, **3**, 3–58.
—— (1978b). *Electricity from glass: the history of the frictional electrical machine 1600–1850*. Sijthoff & Noordhoff, Leiden.

Hales, S. (1727). *Vegatable staticks*. London.
—— (1728). An account of a machine for measuring any depth in the sea, with great expedition and certainty. *Philosophical transactions of the Royal Society of London*, **35**, 559–62.
—— (1731). *Statistical essays* (2nd edn). W. Innys, London.
—— (1768). *A Treatise on ventilators*. Manby, London.

Halley E. (1686). An historical account of the trade winds and monsoons. *Philosophical transactions of the Royal Society of London*, **16**, 153–68.
—— (1716). The art of living under water: or, a discourse concerning the means of furnishing air at the bottom of the sea. *Philosophical transactions of the Royal Society of London*, **29**, 492–9.

Hambly, M. (1988). *Drawing instruments 1580–1980*. Sotheby's, London.

Hamill, J. (1986). *The craft: a history of English freemasonry*. Crucible, Wellingborough.

Hammond, J. H. (1981). *The camera obscura*. Adam Hilger, Bristol.

Hanna, B. T. (1982). Polinière and the teaching of experimental physics at Paris 1700–1730. In *Eighteenth century studies presented to Arthur M. Wilson* (ed. P. Gay). Russell and Russell, New York.

Hans, N. A. (1951). *New trends in education in the eighteenth century*. Routledge and Kegan Paul, London.

Hardwick, F. W. and O'Shea, L. T. (1916). Notes on the history of the safety lamp. *Transactions of the Institution of Mining Engineers*, **51** (5–6), 548–72.

Harley, R. D. (1970). *Artists' pigments 1600–1835*. Butterworths, London.

Harris, G. (1847). *The life of Lord Chancellor Hardwicke*. Moxon, London.

Harris, J. (1704). *Lexicon technicum*. London.
—— (1719). *Astronomical dialogues between a gentleman and a lady*. London.

Harris, M. (1987). *London newspapers in the age of Walpole*. Associated University Press, London.

Harris, W. S. (1834). On Some Elementary Laws of Electricity. *Philosophical transactions of the Royal Society of London*, **124**, 213–45.

Hase, W. and Bate, R. B. (1824). *Description of the patent improved treadmill*. Norwich.

Hauksbee, F. (1709). *Physico-mechanical experiments*. London.
—— (1719). *Physico-mechanical experiments on various subjects* (2nd edn). Senex and Taylor, London.
—— (1970). *Physico-mechanical experiments on various subjects*. Johnson Reprint Corporation, New York.

Hauksbee, F., Jr (1714). *A course of mechanical, optical, hydrostatical and pneumatical experiments to be performed by Francis Hauksbee; and the explanatory lectures read by William Whiston*. London.
—— (1743). *A further account of the effects of Mr Hauksbee's alternative medicine, as applied in the cure of the venerial disease*. J. Roberts, London.

Hearnshaw, F. J. C. (1929). *The centenary history of King's College London 1828–1928*. Harrap, London.

Heath, T. L. (1897). *The works of Archimedes*. Cambridge University Press.

Heathcote, N. H. de V. (1955). Franklin's introduction to electricity. *Isis*, **46**, 29–35.

Heilbron, J. L. (1979). *Electricity in the 17th and 18th centuries. A study of early modern physics*. University of California, Berkeley, CA.
—— (1983). *Physics at the Royal Society during Newton's presidency*. William Andrews Clark Memorial Library, University of California at Los Angeles.

Heinke, F. W. and Davis, W. G. (1873). *History of diving from the earliest times to the present date* (4th edn). London.

Helsham, R. (1739). *A course of lectures in natural philosophy*. London.

—— (1743). *A course of lectures in natural philosophy* (2nd edn). London.

Henderson, E. (1870). *Life of James Ferguson, FRS, in a brief autobiographical account* (2nd edn). Edinburgh.

Hendry, J. (1986). *James Clerk Maxwell and the theory of the electromagnetic field.* Adam Hilger, Bristol.

Henly, W. (1778). Observations and experiments tending to confirm Dr. Ingenhousz's theory of the electrophorus; and show the impermeability of glass to electric fluid. *Philosophical transactions of the Royal Society of London,* **48**, 1049–55.

Herschel, Mrs J. (1876). *Memoirs and correspondence of Caroline Herschel.* Murray, London.

Hewson, J. B. (1983). *A history of the practice of navigation* (2nd edn). Brown, Ferguson, Glasgow.

Hill, J. (1770). *The construction of timber from its early growth; explained by the microscope, and proved from experiments, in a great variety of kinds.* London.

Hitchins, H. L. and May, W. E. (1955). *From lodestone to gyrocompass* (2nd edn). Hutchinson, London.

Hodgson, J. (1706). *The theory of navigation demonstrated.* London.

—— (1725). *A system of the mathematics.* Page, London.

Holmes, G. (1982). *Augustan England: professions, state and society, 1680–1730.* Allen and Unwin, London.

Holmes, J. (1793). *A short narrative of the genius, life, and works of the Late Mr John Smeaton, F.R.S.* London.

Hooke, R. (1674). *Animadversions of the first part of the Machina Coelestis of Johannes Hevelius . . . together with an explanation of some instruments made by Robert Hooke.* John Martyn, London. Reprinted in Gunther (1920–67), Vol. 8.

—— (1676). *A description of helioscopes and some other instruments made by Robert Hooke.* John Martyn, London. Reprinted in Gunther (1920–67), Vol. 8.

—— (1679). *Lectiones Cutlerianae.* London. Reprinted in Gunther (1920–67), Vol. 8.

Hopkins, W. (1835). On aerial vibrations in cylindrical tubes *Transactions of the Cambridge Philosophical Society,* **5**, 231–70.

Horne, J. (1765). *The petition of an Englishman.* Sumpter, London.

Horne, J., Thornthwaite, W. H., and Wood (c. 1852). *A catalogue of photographic apparatus and chemical preparations.* London.

Horrebow, P. (1740–1). *Operum mathematico-physicorum,* 3 Vols. Copenhagen.

Horrins, J. (1835). *Memoirs of a trait in the character of George III.* Edwards, London.

Houben, G. M. M. (1982). *The weighing of money.* G. M. M. Houben, Zwolle.

Howse, D. (1989). *Nevil Maskelyne: the seaman's astronomer.* Cambridge University Press.

Hughes, E. (1951). The early journal of Thomas Wright of Durham. *Annals of Science,* **7** (1), 1–24.

Hume, K. J. (1980). *A history of engineering metrology.* Mechanical Engineering Publications, London.

Hunt, F. V. (1978). *Origins of acoustics.* Yale University Press, New Haven, CT.

Hunter, M. (1982). *The Royal Society and its fellows 1660–1700. The morphology of an early scientific institution.* Monograph 4, British Society for the History of the Science, Chalfont St Giles.

—— (1989). *Establishing the new science. The experience of the early Royal Society.* Boydell, Woodbridge.

Hunter, W. (1781). Essay on a new method of applying the screw by William Hunter, Surgeon. *Philosophical transactions of the Royal Society of London,* **71**, 58–66.

Hurst, W. R. (1928). *An outline of the career of John Theophilus Desaguliers MA LLD FRS.* London.

Hutchison, S. C. (1968). *The history of the Royal Academy 1768–1968.* Chapman and Hall, London.

Hutton, C. (1795). *A mathematical and philosophical dictionary.* Johnson and Robinson, London.

Huygens, C. (1888). *Oeuvres complètes 1890–1950,* Vol. 8. Haarlem.

Inkster, I. (1977). Science and society in the metropolis: a preliminary examination of the social and institutional context of the Askesian Society of London, 1796–1807. *Annals of science,* **34**, 1–32.

—— (1979). London science and the Seditious Meetings Act of 1817. *British journal of the history of science,* **12**, 192–6.

—— (1980). The public lecture as an instrument of science education for adults—the case of Great Britain, c.1750–1850. *Paedagogica historica,* **20**, 80–107.

Inkster, I. and Morrell, J. (1983). *Metropolis and province: science in British culture, 1780–1850.* Hutchinson, London.

Jacob, M. C. (1974). Early Newtonianism. *History of science,* **12**, 142–6.

—— (1988). *The cultural meaning of the scientific revolution.* Knopf, New York.

Jamnitzer, W. (1568). *Perspectiva corporum regularium.* Nuremberg.

Jesse, J. H. (1867). *Memoirs of the life and reign of George III* (2nd edn).

JFI (1830). *Journal of the Franklin Institute,* **10**, 226, Pl. 2.

Johnson, S. (1925). *Lives of the English poets.* London.

Kay, R. (1968). *The diary of Richard Kay, 1716–1751, a Lancashire Doctor* (ed. W. Brockbank and F. Kenworthy). Chetham Society, Manchester.

Keill, J. (1776). *Introduction to natural philosophy or lectures on physics etc. read in University of Oxford.* Glasgow.

Kidwell, P. A. (1988). *Josiah Lyman's protracting trigonometer,* Rittenhouse. Vol. 3, No. 1.

King, E. (1750). *Catalogue of the experiments made by Mr King, in his course of natural philosophy.* London?

King, H. C. (1955). *The history of the telescope.* Griffin, London.

King, H. C. and Millburn, J. R. (1978). *Geared to the stars.* Adam Hilger, Bristol.

King, R. (1985). *Royal Kew.* Constable, London.

Kisch, B. (1965). *Scales and weights: an historical outline.* Yale University Press, New Haven, CT.

Knight, G. (1744). An account of some magnetical experiments. *Philosophical transactions of the Royal Society of London,* **43**, 161–6.

—— (1746). An account of some magnetical experiments. *Philosophical transactions of the Royal Society of London,* **44**, 656–72.

—— (1766). Patent Specification No. 850.

Knowles Middleton, W. E. (1964). *The history of the barometer.* Johns Hopkins, Baltimore.

Kunckel, J. (1679). *Ars vitrana experimentalis oder Vollkommene Glasmacker-Kunst.* Frankfurt and Leipzig.

Lalande, J. le F. (1792). *Astronomie* (3rd edn). Paris.

—— (1980). *Journal d'un voyage en Angleterre 1763* (ed. H. Monod-Cassidy). Voltaire Foundation, Oxford.

Lane, T. (1767). Description of an electrometer invented by Mr Lane, with an account of some experiments made by him with it. *Philosophical transactions of the Royal Society of London,* **57**, 451–60.

Law, A. (1966). Teachers in Edinburgh in the eighteenth century. *Book of the Old Edinburgh Club,* **32**, 108–57.

Law, R. J. (1965). *The steam engine.* Her Majesty's Stationery Office, London.

—— (1970–2a). Henry Hindley of York 1701–1771, part I. *Antiquarian horology,* **7**, 205–21.

—— (1970–2b). Henry Hindley of York 1701–1771, part II. *Antiquarian horology,* **7**, 682–701.

Lawrence, C. J. (1988). Alexander Monro primus and the Edinburgh manner of anatomy. *Bulletin of the history of medicine,* **62**, 193–214.

Lawrence, S. C. (1988). Entrepreneurs and private enterprise: the development of medical lecturing in London, 1775–1820. *Bulletin of the history of medicine,* **62**, 171–92.

Le Gray, G. (1851). *Plain directions for obtaining photographic pictures,* Part 2. T. and R. Willats, London.

Letsome, S. and Nicholl, J. (1739). *A defense of natural and revealed religion being a collection of the sermons preached at the lecture founded by the Honourable Robert Boyle, Esq, from the year 1691 to the year 1732.* London.

Leupold, J. (1713). *Anamorphosis mechanica nova.* Leipzig.

Leybourn, W. (1672). *Panorganon.* William Birch, London.

—— (1682). *Dialing, plain, concave, convex, projective, refractive.* Awnsham Churchill, London.

—— (1694). *Recreations of divers kinds.* London.

Lindqvist, S. (1984). *Technology on trial: the introduction of steam power technology into Sweden 1715–1736.* Almqvist and Wiksell, Stockholm.

Lindsay, E. M. (1969). The astronomical instruments of H.M. King George III presented to Armagh Observatory. *Irish astronomical journal,* **9**, 57–68.

Linebaugh, P. (1977). The Tyburn riot against the surgeons. In *Albion's fatal tree: crime and society in eighteenth-century England* (D. Hay, P. Linebaugh, J. G. Rule, E. P. Thompson, and C. Winslow), pp. 65–117. Penguin, Harmondsworth.

Lipsius, J. (1596). *Poliorceticon.* Antwerp.

Loftis, J. (1950–1). Richard Steele's censorium. *The Huntingdon Library quarterly,* **14**, 43–66.

Long, R. (1742–64). *Astronomy,* 5 Vols. Cambridge.

Lorentz, R. (1807). Producing light and fire instantaneously, British Patent 3007.

Lubbock, C. A. (1933). *The Herschel chronicle.* Cambridge University Press.

Lyle, D. (1762). *The art of shorthand improved.* London.

Lysons, D. (1811). *The environs of London* (2nd edn). London.

McClellan, J. E. (1985). *Science reorganized: scientific societies in the eighteenth century.* Columbia University Press, New York.

McConnell, A. (1982). *No sea too deep.* Adam Hilger, Bristol.

Mackay, C. (1956). *Extraordinary popular delusions and the madness of crowds.* Harrap, London.

McKendrick, N., Brewer, J., and Plumb, J. H. (1982). *The birth of a consumer society: the commercialization of eighteenth-century England.* Europa, London.

Maclaurin, C. (1801). *A treatise on fluxions* (2nd edn). London.

MacLeod, C. (1988). *Inventing the industrial revolution.* Cambridge University Press.

Maddison, R. E. W. (1959). The portraiture of the Honourable Robert Boyle FRS. *Annals of science,* **15**, 141–214.

—— (1969). *The life of the Honourable Robert Boyle FRS.* Taylor and Francis, London.

Mahoney, M. S. (1980). *Christiaan Huygens: The measurement of time and of longitude at sea.* In *Studies on Christian Huygens* (ed. H. J. M. Bos *et al.*). Swets and Zeitlinger B.V., Lisse.

Maitland, W. (1772). *The history of London from its foundation to the present time,* 2 Vols. Wilkie, London.

Maluf, R. B. (1986). *Jean Antoine Nollet and experimental natural philosophy in eighteenth-century France.* University Microfilms International, Ann Arbor, MI.

Manuel, F. E. (1968). *A portrait of Isaac Newton.* Belknap, Cambridge, MA.

Manwaring, E. W. (1925). *Italian landscape in 18th century England.* Oxford University Press, New York.

Manzini, C. A. (1660). *L'occhiale all 'occhio.* Bologna.

Mariotte, E. (1718). *The motion of water and other fluids. Being a treatise of hydrostaticks.* London.

Martin, B. (1740). *A catalogue of philosophical, mathematical and optical instruments.*

—— (1746). *The description and use of a new, portable, table air-pump and condensing engine.* Benjamin Martin, London.

—— (1759). *New elements of optics.* London.

Maskelyne, N. (1771). Description of a method of measuring difference of right ascension and declination with Dollond's micrometer together with other new applications of the same. *Philosophical transactions of the Royal Society of London,* **61**, 536–46.

May, W. E. (1973). *A history of marine navigation.* Foulis, Henley-on-Thames.

Merlin, J. (1773). Spring jack having a reflector to increase the heat and therefore save fuel. British Patent 1032, 29 January 1773.

Meyer, G. D. (1955). *The scientific lady in England 1650–1760*. University of California Press, Berkeley, CA.

Michel, H. (1947). *Traité de l'astrolabe*. Gauthier-Villars, Paris.

Michell, J. (1750). *A treatise of artificial magnets*. Cambridge.

Millburn, J. R. (1976). *Benjamin Martin: author, instrument-maker, and 'country showman'*. Noordhoff, Leiden.

—— (1981). Demonstrating the motion of comets. *Space education*, **1**, 2, 55–8.

—— (1983). The London evening courses of Benjamin Martin and James Ferguson, eighteenth-century lecturers on experimental philosophy. *Annals of science*, **40**, 437–55.

—— (1985). James Ferguson's lecture tour of the English Midlands 1771. *Annals of science*, **42**, 397–415.

—— (1986). *Retailer of the sciences: Benjamin Martin's scientific instrument catalogues, 1756–1782*. Vade-Mecum, London.

—— (1988). The Office of Ordnance and the instrument-making trade in the mid-eighteenth century. *Annals of science*, **45**, 221–93.

—— (1993). *The Adams family of Fleet Street: instrument makers to King George III*. Vade-Mecum, London.

Millburn, J. R. and King, H. C. (1988). *Wheelwright of the heavens. The life and work of James Ferguson, FRS*. Vade-Mecum, London.

Miller, D. P. (1988). 'My favourite studdys': Lord Bute as naturalist. In *Lord Bute: essays in re-interpretation* (ed. K. W. Schweizer), pp. 213–31. Leicester University Press.

—— (1989). 'Into the Valley of Darkness': reflections on the Royal Society in the eighteenth century. *History of science*, **27**, 155–66.

Morland, S. (1671). *Tuba Stentoria-Phonica*. London.

Murray, D. (1930). *Chapters in the history of bookkeeping, accountancy and commercial arithmetic*. Jackson, Wylie, Glasgow.

Musson, A. E. and Robinson, E. (1969). *Science and technology in the industrial revolution*. Manchester University Press.

Nairne, E. (1777a). An account of some experiments, made with an air-pump on Mr Smeaton's principle; together with some experiments with a common air-pump. *Philosophical transactions of the Royal Society of London*, **67**, 614–37.

—— (1777b). An account of some further experiments made with the same air-pumps on Mr Smeaton's principle, the results of which were different from the former. *Philosophical transactions of the Royal Society of London*, **67**, 637–48.

Newton, I. (1729). *Mathematical principles of natural philosophy* (trans. A. Motte). London.

—— (1967). *The correspondence of Isaac Newton 1694–1709*, Vol. 4 (ed. J. F. Scott). Royal Society, Cambridge.

—— (1981). Demonstrating the motion of comets. *Space Education*, **1**, 2, 55–8.

Niceron, J. F. (1638). *Perspective curieuse* (1st edn). Paris.

Nichols, J. (1790). *Bibliotheca Topographica Britannica*. London.

—— (1966). *Literary anecdotes of the eighteenth century*. Kraus Reprint, New York.

Nightingale, T. (1790). Machine for calendering, glazing and dressing [various materials]. British Patent 1783.

NMM (1970). *An inventory of the navigation amd astronomy collections in the National Maritime Museum, Greenwich*. National Maritime Museum, Greenwich.

Noad, H. M. (1859). *A manual of electricity* (4th edn). London.

Nobili, L. (1834). Galvanometio. *Memoire e observazioni edite ed inedite del Cavaliere Leopoldo Nobili colla descrizione ed analisi de'suoi apparati ed istrumenti*, **2**, 27.

Nollet, J. A. (1740a). Sur les instruments qui sont propres aux expériences de l'air. *Historie de l'Académie Royale des Sciences: mémoires*, **53**, 385–432.

—— (1740b). Sur les instruments qui sont propres aux expériences de l'air. Seconde partie. *Histoire de l'Académie Royale des Sciences: mémoires*, **53**, 567–85.

—— (1741). Sur les instruments qui sont propres aux expériences de l'air. Troisième partie. *Histoire de l'Académie Royale des Sciences: mémoires*, **54**, 338–62.

—— (1749a). *Recherches sur les causes particulières des phénomenes électriques*. Paris.

—— (1749b). Vaisseau de verre. *Mémories de l'Académie Royale des Sciences*, **62**, 461–6.

—— (1754). *Leçons de physique*. Amsterdam and Leipzig.

Nooth, J. M. (1775). The description of an apparatus for impregnating water with fixed air, and of the manner of conducting that process. *Philosophical transactions of the Royal Society of London*, **65**, 59–66.

O'Dea, W. T. (1958). *A short history of lighting*, London.

Oersted, H. C. (1821). Consideration au l'électromagne-tisme. *Bibliotheca Universalle*, **18**, 11.

—— (1928). Experimenta circa effectum conflictus electrici in acum magneticam. *Isis*, **10**, 435–44.

Paine, T. (1794). *The age of reason*, Part 1. Paris.

Pardies, I. (1672). Letter to Oldenburg, 21 May 1672 (Vol. 9).

Parent, A. R. (1703). *Recherches de physique et de mathématique*. Paris.

Partington, J. R. (1962). *A history of chemistry*, Vol. 3. Macmillan, London.

Pascal, B. (1663). *Traites de l'equilibre des liqueurs*. Paris.

Paulson, R. (1979). *Popular and polite art in the age of Hogarth and Fielding*. University of Notre Dame Press.

Peachey, G. C. (1924). A memoir of William and John Hunter. G. C. Peachey, Plymouth.

Pemberton, H. (1728). *A view of Sir Isaac Newton's philosophy*. Palmer, London.

Perronet, J. R. (1783). *Description des projets et la construction des Ponts de Neuilli, de Mantes, d'Orléans*, Vol. 2. Paris.

Philip & Sons Ltd. (1987a). *The crystal sphere*. London.

—— (1987b). *Sale catalogue* (April). London.

Phillips, H. (1964). *Mid-Georgian London*. Collins, London.

Pinchbeck, C. (1762). Contrivance to improve the walking wheel crane. *Transactions of the Society for the Encouragement of Arts*, **4**, 183–5.

Porter, R. (1989). *Health for sale*. Manchester University Press.

Porter, R., Schaffer, S., Bennett, J., and Brown, O. (1985). *Science and profit in 18th century London*. Whipple Museum of the History of Science, Cambridge.

Price, D. J. (1952). The early observatory instruments of Trinity College, Cambridge. *Annals of science*, **8** (1), 1–12.

Priestley, J. (1767). *The history and present state of electricity* (1st edn). London.

—— (1769). *The history and present state of electricity* (2nd edn). London.

—— (1772). An account of a new electrometer continued by Mr William Henly and of several electrical experiments made by him, in a letter from Dr Priestley to Dr Franklin. *Philosophical transactions of the Royal Society of London*, **62**, 359–64.

—— (1775). *The history and present state of electricity* (3rd edn). London.

—— (1775–7). *Experiments and observations on different kinds of air* (2nd edn), 3 Vols. London.

Pulteney, R. (1790). *Historical and biographical sketches . . . botany*. London.

Q.C. *A catalogue of the apparatus of the philosophical instruments, Her Majesty has deposited in the Royal Observatory at Richmond with an account of the presents, any sundry persons made to Her Majesty's collection*. MS.

Rackstrow, B. (1748). *Miscellaneous observations . . . experiments on electricity*. London.

Record (1912). *Record of the Royal Society of London* (3rd edn). Royal Society of London.

Redwood, J. (1976). *Reason, ridicule and religion: the age of Enlightenment in England, 1660–1750*. Thames and Hudson, London.

Rees's Cyclopaedia (1819). Longman *et al.*

Reynolds, T. S. (1983). *Stronger than a hundred men*. Johns Hopkins, Baltimore, MD.

Richmond (1770). *A list of some curious mathematical bodys, figures etc. from the collection of the Great Mr Boyle*. MS.

Rigaud, G. (1882). Dr Demainbray and the King's Observatory at Kew. *Observatory*, **5** (66), 279–85.

Rigaud, S. P. (1831). Some account of James Stirling FRS. *Edinburgh journal of science*, **5**, 191–6.

Ritchie, W. (1826). On a new photometer, founded on the principles of Bouguer. *Transactions of the Royal Society of Edinburgh*, **10**, 443–5.

Robert, H. (1840). Pour un chronometre adjudicateur remplacasent les bougies dans les ventes. French Patent 7662, 1st series.

Robertson, A. (1829). Construction of paddles for propelling ships, boats or vessels on water. British Patent 5749, 7 January 1829.

Robertson, J. (1805). *The elements of navigation* (7th edn), Vol. 2. London.

Robinson, E. (1962). The profession of civil engineer in the eighteenth century: a portrait of Thomas Yeoman, FRS, 1704(?)–1781. *Annals of science*, **18**, 195–215.

Robinson, F. J. G. (1970). A philosophic war: an episode in eighteenth-century scientific lecturing in North-East England. *Architectural and Archaeological Society of Durham and Northumberland*, **2**, 101–9.

—— (1972). Trends in education in Northern England during the eighteenth century: a biographical study. Unpublished Ph.D. Thesis, University of Newcastle-upon-Tyne.

Roche, D. (1978). *Le siècle des lumières en province. Académies et académiciens provinciaux, 1680–1789*. Mouton, Paris.

Rogers, N. (1978). Popular protest in early Hanoverian London. *Past and present*, **79**, 70–100.

Ronan, C. A. (1969). *Edmond Halley: genius in eclipse*. Macdonald, London.

Rousseau, G. S. (1982). *Tobias Smollett: essays of two decades*. Clark, Edinburgh.

Rousseau, G. S. and Porter, R. S. (ed.) (1980). *The ferment of knowledge: studies in the historiography of eighteenth-century science*. Cambridge University Press.

Rowbottom, M. E. (1965). The teaching of experimental philosophy in England, 1700–1730. *Actes du XIIe Congrès International d'Histoires des Sciences*, Vol. 4, pp. 46–53. Blanchard, Paris.

—— (1968). John Theophilus Desaguliers (1683–1744). *Proceedings of the Huguenot Society of London*, **21**, 196–218.

Royal Institution (1802). *Journals of the Royal Institution*, Vol. 2. Royal Institution, London.

Rudé, G. (1971). *Hanoverian London*. Secker and Warburg, London.

—— (1983). *Wilkes and liberty: a social study* (new edn). Lawrence and Wishart, London.

Rutherforth, T. (1748). *A system of natural philosophy*, 2 Vols. Cambridge.

Saladin, G. A. (1986). *Il teatro di filosofia sperimentale di Giovani Poleni*. Edizioni Lint, Trieste.

Saul, E. (1735). *An historical and philosophical account of the barometer* (2nd edn). London.

Saverien, A. (1753). *Dictionaire universel de mathématique et de physique*. Paris.

Savery, S. (1753). A new way of measuring the difference between the apparent diameter of the Sun at the times of the Earth's perihelion and aphelion. With a micrometer placed in a telescope invented for that purpose. *Philosophical transactions of the Royal Society of London*, **48**, 167–78.

Savery, T. (1699). An account of Tho Savery's engine for raising water by the help of fire. *Philosophical transactions of the Royal Society of London*, **21**, 228.

—— (1702). *The miner's friend*. London.

Saxton, J. (1834). Description of a revolving keeper magnet for producing electrical currents. *Journal of the Franklin Institute*, **17**, 155–6.

—— (1836). J. Saxton on his magneto-electrical machine with remarks on Mr E. M. Clarke's paper in the preceding number. *Philosophical magazine*, **9**, 360–5.

Schaaf, L. (1982). Henry Collen and the Treaty of Nanking. *History of photography*, **6**, 353–66.

Schaffer, S. (1983). Natural philosophy and public spectacle in the eighteenth century. *History of science*, **21**, 1–43.

—— (1986). Scientific discoveries and the end of natural philosophy. *Social studies of science*, **16**, 387–420.

Scott, R. H. (1885). The history of the Kew Observatory. *Proceedings of the Royal Society of London*, **39**, 37–86.

Scriblerus (1731). *Whistoneutes*. London.

Secord, J. A. (1985). Newton in the nursery. *History of science*, **23**, 127–51.

Senex, J. (1738). A contrivance to make the poles of the diurnal motion in a celestial globe pass round the poles of the ecliptic. *Philosophical transactions of the Royal Society of London*, **40**, 203–4.

Setchell, J. R. M. (1970a). The friendship of John Smeaton FRS with Henry Hindley, instrument and clockmaker of York. *Notes and records of the Royal Society of London*, **25** (1), 79–86.

—— (1970b). Further information on the telescopes of Hindley of York. *Notes and records of the Royal Society of London*, **25** (2), 189–92.

's Gravesande, W. J. (1720). *Mathematical elements of physicks, proved by experiments: being an introduction to Sir Isaac Newton's philosophy* (1st edn) (trans. J. Keill). G. Strahan, London.

—— (1747). *Mathematical elements of natural philosophy* (6th edn), 2 vols. London.

Shapin, S. (1974). The audience for science in eighteenth century Edinburgh. *History of science*, **12**, 95–121.

—— (1988a). Robert Boyle and mathematics: reality, representation, and experimental practice. *Science in context*, **2**, 23–58.

—— (1988b). The house of experiment in seventeenth-century England. *Isis*, **79**, 373–404.

Shapin, S. and Schaffer, S. (1985). *Leviathan and the air-pump*. Princeton University Press.

Shapin, S. and Thackray, A. (1974). Prosopography as a research tool in history of science: the British scientific community 1700–1900. *History of science*, **12**, 1–28.

Sheppard, T. and Musham, J. F. (1923). *Money scales and weights*. Spink, London.

Short, J. (1752). An account of a horizontal top invented by Mr Serson. *Philosophical transactions of the Royal Society of London*, **47**, 352–3.

Sicca, C. M. (1986). Like a shallow cave by nature made: William Kent's 'natural' architecture at Richmond. *Architectura*, **16**, 68–82.

Simcock, A. V. (1984). *The Ashmolean Museum and Oxford science 1683–1983*. Museum for the History of Science, Oxford.

Singer, C. Holmyard, E. J. Hall, A. R. and Williams, T. I. (1956). *A history of technology*, Vol. 7. Clarendon Press, Oxford.

Six, J. (1782). An account of an improved thermometer. *Philosophical transactions of the Royal Society of London*, **72**, 72–81.

—— (1794). *The construction and use of a thermometer*. Maidstone.

Smeaton, J. (1751). A letter from Mr J. Smeaton to Mr John Ellicott, FRS, concerning some improvements made by himself in the air-pump. *Philosophical transactions of the Royal Society of London*, **47**, 415–28.

—— (1752). A description of a new tackle or combination of pullies. *Philosophical transactions of the Royal Society of London*, **47**, 494–7.

—— (1759). An experimental enquiry concerning the natural powers of water and wind to turn mills and other machines depending on a circular motion. *Philosophical transactions of the Royal Society of London*, **51**, 100–74.

—— (1786). Observations on the graduation of astronomical instruments, with an explanation of the method invented by the late Mr Henry Hindley, of York. *Philosophical transactions of the Royal Society of London*, **76**, 1–47.

Smith, J. (1688). *A compleat discourse on the nature, use, and right managing . . . Baroscope*. Watts, London.

Smith, R. (1738). *A compleat system of opticks*. Cambridge.

Smollett, T. (1771). *The expedition of Humphry Clinker*. London.

Snow, H. W. (1834). On some elementary laws of electricity. *Philosophical transactions of the Royal Society of London*, **24**, 213–46.

Sprat, T. (1959). *History of the Royal Society*. Routledge, Kegan, Paul, London.

Steele, R. and Gillmore, J. (1718). *An account of the fish-pool*. London.

Stephens, A. (1813). *Memoirs of John Horne Tooke*. London.

Stewart, L. (1981). Samuel Clarke, Newtonianism, and the factions of post-revolutionary England. *Journal of history of ideas*, **42** (1), 53–72.

—— (1986a). Public lectures and private patronage in Newtonian England. *Isis*, **77**, 47–58.

—— (1986b). The selling of Newton: science and technology in early eighteenth-century England. *Journal of British studies*, **25**, 178–92.

Stock, J. T. and Vaughan, D. (1983). *The development of instruments to measure electric current*. Science Museum, London.

Sturgeon, W. (1836–43). *Annals of electricity, magnetism and chemistry and guardian of experimental science*, 10 Vols, London.

—— (1842). *Lectures on electricity*. London.

Sutherland, V. L. S. and Mitchell, L. G. (ed.) (1986). *The eighteenth century*. The history of the University of Oxford, Vol. V, Clarendon Press, Oxford.

Taylor, B. (1721). (Extract of a letter from Dr Brook Taylor FRS to Sir Hans Sloan dated 25 June 1714.) Giving an account of some experiments relating to magnetism. *Philosophical transactions of the Royal Society of London*, **31**, 204–8.

Taylor, E. G. R. (1954). *The mathematical practitioners of Tudor and Stuart England 1485–1714*. Cambridge University Press.

—— (1966). *The mathematical practitioners of Hanoverian England 1714–1840*. Cambridge University Press.

Taylor, E. W. and Wilson, J. S. (1944). *At the sign of the orrery*. Troughton and Simms Ltd, York.

Thackray, A. (1965). The business of experimental philosophy. *Actes du XII* Congrès International d'Histoire des Sciences*, Vol. 3B, pp.155–9. Blanchard, Paris.

—— (1974). Natural knowledge in cultural context: the Manchester model. *American historical review*, **79**, 672–709.

Thiout, A. (1741). *Traité d'horologerie*. Paris.

Thomas, D. B. (1964). *The first negatives*. Her Majesty's Stationery Office, London.

—— (1969). *The Science Museum Photography Collection*. Her Majesty's Stationery Office, London.

Thornthwaite, W. H. (1843). *Photographic manipulation*. Edward Palmer, London.

—— (1845). *A guide to photography*. London.

Thurkow, G. (1978). The dotchin. *Equilibrium*, autumn, 55–9.

Turner, A. J. (1981). William Oughtred, Richard Delamain and the horizontal instrument in seventeenth century England. *Annali dell' Istituto e Museo di Storia della Scienza di Firenze*, **6**, 99–124.

—— (1985). *The time museum: astrolabes. Astrolabe related instruments*. Time Museum, Rockford.

—— (1987). *Early scientific instruments. Europe 1400–1800*. Philip Wilson for Sotheby's Publications, London.

Turner, G. L'E. (1966). Decorative tooling on 17th and 18th century microscopes and telescopes. *Physis*, **8**, 99–128.

—— (1967a). The auction sales of the Earl of Bute's instruments, 1793. *Annals of science*, **23**, 213–42.

—— (1967b). A portrait of James Short, FRS, attributable to Benjamin Wilson, FRS. *Notes and records of the Royal Society of London*, **22**, 105–12.

—— (1969). James Short FRS and his contribution to the construction of reflecting telescopes. *Notes and records of the Royal Society of London*, **24**, 91–108.

—— (1973). Descriptive catalogue of Van Marum's scientific instruments. *Martinus Van Marum: life and work*, Vol. 4, Part 2. Noordhoff, Leiden.

—— (1974a). Henry Baker FRS, founder of the Bakerian Lecture. *Notes and records of the Royal Society of London*, **29**, 53–79.

—— (1974b). The Portuguese agent: J. H. de Magellan. *Antiquarian horology*, **9**, 74–6.

—— (ed.) (1976). *The patronage of science in the nineteenth century*. Noordhoff, Leiden.

—— (1978). Apparatus of science in the eighteenth century. *Revista da Universidade de Coimbra*, **26**, 15–31.

—— (1989). *The great age of the microscope: the Collection of the Royal Microscopical Society through 150 years*. Adam Hilger, Bristol.

Turner, G. L'E. and Levere, T. H. (1973). *Van Marum's scientific instruments in Teyler's Museum*, Vol. 4 (ed. E. Lefebvre and J. G. de Bruijn). Noordhoff, Leiden.

Tweedie, C. (1922). *James Stirling: a sketch of his life and works along with his scientific correspondence*. Clarendon Press, Oxford.

Underwood, E. A. (1977). *Boerhaave's men at Leyden and after*. Edinburgh University Press.

Universal magazine of knowledge and pleasure, September 1755.

van der Krogt, P. (1984). *Old globes in the Netherlands: a catalogue of terrestrial and celestial globes made prior to 1850 and preserved in Dutch collections* (trans. W. ten Haken). HES Uitgevers, Utrecht.

van Musschenbroek, J. (1739). *Description de novelles sortes de machines pneumatiques*. Luchtmans, Leiden.

van Musschenbroek, P. (1725). De Viribus Magneticis. *Philosophical transactions of the Royal Society of London*, **33**, 370–8.

—— (1739). *Essai de physique*, Leiden.

—— (1744). *The elements of natural philosophy . . . students in universities* (trans. J. Colson). Nourse, London.

—— (1762). *Introductio ad philosophiam naturalem*. Lugduni Batavorum.

Volta, A. (1782). Of the method of rendering very sensible the weakest natural or artificial electricity. *Philosophical transactions of the Royal Society of London*, **72**, 7–33.

—— (1800). On the electricity excited by the mere contact of conducting substances of different kinds. *Philosophical transactions of the Royal Society of London*, **90**, 403–31.

von Uffenbach, Z. C. (1934). *London in 1710 from the travels of Zacharias Conrad von Uffenbach* (trans. and ed. W. H. Quarrell and M. Mare). Faber, London.

Vream, W. (1717). *Description of the air pump*. London.

Wakeman, G. (1973). *Victorian book illustration*. David and Charles, Newton Abbot.

Walker, C. V. (1841). *Electrotype manipulation* (3rd edn), Part 2 (bound with Fisher, G. T. *Photogenic manipulation*). George Knight and Sons, London.

Walker, R. B. (1973). Advertising in London newspapers, 1650–1750. *Business history*, **15**, 112–30.

Walker, R. J. B. (1979). *Old Westminster Bridge*. David and Charles, Newton Abbot.

Wallis, J. (1677). Concerning a new musical discovery. *Philosophical transactions of the Royal Society of London*, **12**, 839.

Ward, A. (1989). 'Sr Joshua and His Gang': William Blake the Royal Academy. *Huntingdon Library quarterly*, **52**, 75–95.

Ward, C. (1837). Musical drums. British Patent 7505.

Ward, F. A. B. (1981). *A catalogue of European scientific instruments in the Department of Medieval and later Antiquities.* British Museum, London.

Ward, J. (1740). *The lives of the professors of Gresham College.* London.

Ward, J. and Stevenson, S. (1986). *Printed light.* Her Majesty's Stationery Office, Edinburgh.

Warner, D. J. (1967). Celestial technology. *Smithsonian journal of history*, **2** (3), 35–48.

—— (1987). Air pumps in American education. *Physics teacher*, February, 82–5.

Watkins, J. (1831). *The life and times of 'England's Patriot King' William the Fourth.* London.

Watson, W. (1752). An account of the phaenomena of electricity in vacuo with some observations thereupon. *Philosophical transactions of the Royal Society of London*, **47**, 362–76.

Weatherill, L. (1988). *Consumer behaviour and material culture in Britain 1660–1760.* Routledge, London.

Webb, M. I. (1954). *Michael Rysbrack Sculptor.* Country Life, London.

Wedgwood, J. (1782). An attempt to make a thermometer for measuring the higher degrees of heat. *Philosophical transactions of the Royal Society of London*, **72**, 305–26.

—— (1784). An attempt to compare and connect the thermometer for strong fire. *Philosophical transactions of the Royal Society of London*, **74**, 358–83.

—— (1786). Additional observations on making a thermometer for measuring the higher degrees of heat. *Philosophical transactions of the Royal Society of London*, **76**, 390–408.

Wedgwood (1978). *Josiah Wedgwood: the arts and sciences united. Catalogue of the exhibition held at the Science Museum.* Josiah Wedgwood and Sons, Stafford.

Weld, C. R. (1848). *A history of the Royal Society.* Parker, London.

Werkmeister, L. (1967). *A newspaper history of England 1792–93.* University of Nebraska Press, Lincoln, NB.

Westfall, R. S. (1980). *Never at rest: a biography of Isaac Newton.* Cambridge University Press.

Wheatland, D. P. (1968). *The apparatus of science at Harvard 1765–1800.* Harvard University Press.

Wheatstone, C. (1823). New Experiments on Sound. *Annals of philosophy*, **6**, 2, pp. 81–90.

—— (1827). Experiments on audition. *Quarterly journal of science*, **3**, 67–72.

—— (1828). On the resonances or reciprocated vibrations of columns of air. *Quarterly journal of science*, **3**, 175–83.

—— (1838). Contributions to the physiology of vision.

Philosophical transactions of the Royal Society of London, **128**, 371–94.

Whipple, R. S. (1926). An old catalogue and what it tells us of the scientific instruments and curios collected by Queen Charlotte and King George III. In *Proceedings of the optical convention*, Part 2, pp. 502–28. London.

Whiston, W. (1716). *An account of a surprizing meteor . . . March the 6th, 1715/16 at night* (2nd edn). Senex, Taylor, London.

White, G. H. (1959). *The complete peerage.* St Catherine's Press, London.

Whittington, W. (1794) Machine for roasting meat or other food. British Patent 1986.

Willats, T. and R. (1851). *Catalogue of photographic apparatus.* T and R. Willats, London.

Williams, W. J. (1928). Masonic personalia, 1723–39 (Transactions of Quatuor Coronati Lodge). *Ars Quatuor Coronatorum*, **40**, 30–42.

Wilson, B. (1778). New experiments and observations on the nature and use of conductors. *Philosophical transactions of the Royal Society of London*, **68**, 245–313.

Wilson, M. I. (1984). *William Kent: architect, designer, painter, gardener, 1685–1748.* Routledge, Kegan, Paul, London.

Wing, D. A. (1900). The Pool family of Easton, Massachusetts, *Rittenhouse*, **4**, 4.

Wing, V. and Leybourn, W. (1649). *Urania practica or practical astronomie.* R. Leybourn, London.

Wollaston, W. H. (1813). On a method of freezing at a distance. *Philosophical transactions of the Royal Society of London*, **103**, 71–4.

Wood, H. T. (1913). *A history of the Royal Society of Arts.* John Murray, London.

Woodford, T. (1756). *A treatise containing the description of a new and curious quadrant made and finished by the masterly hand of the excellent mechanic John Rowley.* London.

Woolf, H. (1959). *The transits of Venus. A study of eighteenth-century science.* Princeton University Press, Princeton, NJ.

Woolrich, A. P. (1986). *Mechanical arts and merchandise* Archeologische Pers, Eindhoven.

Wright, L. (1964). *Home fires burning.* Routledge, Kegan, Paul, London.

Wright, T. (1720). *A description of an astronomical instrument, being the orrery reduc'd.* T. Wright, London.

Wynter, H., Ltd (1976). *A catalogue of scientific instruments,* Vol. 3 (1). London.

Wynter, H. and Turner, A. (1975). *Scientific instruments.* Studio Vista, London.

Young, T. (1807). *A course of lectures on natural philosophy and the mechanical arts.* Joseph Johnson, London.

Zuck, D. (1978). Dr Nooth and his apparatus. *British journal of anaesthesia*, **50,** 393–405.

Inventory number index

List relating museum inventory numbers to catalogue entries

1884-79	E45	1927-1130	D102	-1164,		
1889-39	E36	1927-1131	D33	-1165		
1925-136	E156	1927-1132	E22	1927-1166	D133
1927-1101	M11	1927-1134	L150	1927-1167	D134
1927-1102	M48	1927-1135	E17	1927-1168	E153
1927-1103	M22	1927-1136	L52	1927-1169	E182
1927-1104	D99	1927-1137	D143	1927-1170	E181
1927-1105	M4	1927-1138	L14	1927-1171	E180
1927-1106/pt 1	M47	1927-1139	D136	1927-1172	E183
1927-1106/pt 2	D147	1927-1140	L82	1927-1173	E223
1927-1107	D15	1927-1141	D62	1927-1175	E57
1927-1108	E58	1927-1142	E195	1927-1176	E207
1927-1109	D113	1927-1143	E189	1927-1177	E196
1927-1110	M26	1927-1144	E213	1927-1178/pt 1	E205
1927-1111	M49	1927-1145	E216	1927-1178/pt 2	J114
1927-1112	D24	1927-1146	E19	1927-1179	L122
1927-1113	M102	1927-1147	E50	1927-1180	E70
1927-1114	M100	1927-1148	E49	1927-1181	L123
1927-1115	M67	1927-1149	E52	1927-1182	E197
1927-1116	M63	1927-1150	E14	1927-1183	L89
1927-1117	L72	1927-1151	E7	1927-1184	D26
1927-1118	M77	1927-1152	D60	1927-1185	L109
1927-1119	M13	1927-1153	D53	1927-1186	E190
1927-1120	D47	1927-1154	L79	1927-1188	L110
1927-1121	M20	1927-1155	L80	1927-1189	E56
1927-1122	M94	1927-1156	D142	1927-1190	L92
1927-1123	M65	1927-1157	E219	1927-1191	L85
1927-1124	M66	1927-1158	D73	1927-1192	L93
1927-1125	M52	1927-1159	D54	1927-1193	E204
1927-1126	M51	1927-1160	D55	1927-1195	D141
1927-1127	M70	1927-1161	E154	1927-1196	E44
1927-1128	M72	1927-1162,	E152	1927-1197	D16
1927-1129	D2	-1163,		1927-1198	E75

1927-1199	D117	1927-1238	D101	1927-1284	L34
1927-1200	M73	1927-1239	J105	1927-1285	D35
1927-1201	M89	1927-1240	D1	1927-1286	D36
1927-1202	E15	1927-1241	M32	1927-1287	E76
1927-1203	E16	1927-1242	M17	1927-1288	D41
1927-1204	D115	1927-1243	M46	1927-1289	L51
1927-1205	M16	1927-1244	D46	1927-1290	L44
1927-1206	J13	1927-1245	E193	1927-1291	P9
1927-1207/pt 1	M21	1927-1246	L84	1927-1292	P4
1927-1207/pt 2	D98	1927-1247	E174	1927-1293	P16
1927-1208	L22	1927-1248	L75	1927-1294	P48
1927-1209	L41	1927-1249	D132	1927-1295	E77
1927-1210	E5	1927-1250	D131	1927-1296	E78
1927-1211	D3	1927-1251	D37	1927-1297	L50
1927-1212	D5	1927-1252	D38	1927-1298/pt 1	P45
1927-1213	D6	1927-1253	L13	1927-1298/pt 2	P46
1927-1214	J8	1927-1254	D27	1927-1299, -1300	E217
1927-1215	J9	1927-1255	L115	1927-1301, -1302	D42
1927-1216	M38	1927-1256	L78		
1927-1217	M19	1927-1257	E164	1927-1303	P28
1927-1218	E6	1927-1258	E165	1927-1304	D122
1927-1219	M12	1927-1259	D18	1927-1305	E80
1927-1220	M81	1927-1260	L116	1927-1306	E81
1927-1221/pt 1	M5	1927-1261	L113	1927-1307	E82
1927-1221/pt 2	M84	1927-1262	E206	1927-1308	P15
1927-1222/pt 1	M82	1927-1263	L100	1927-1309	P29
1927-1222/pt 2	M83	1927-1264	L102	1927-1310	E83
1927-1222/pt 3	M85	1927-1265	L103	1927-1311/pts 1 and 2	P33
1927-1222/pt 4	M87	1927-1266	L101		
1927-1222/pt 5	M88	1927-1267	L106	1927-1312	L43
1927-1223	M68	1927-1268	E200	1927-1313	P6
1927-1224	L12	1927-1269	E201	1927-1314/pt 1	P41
1927-1225	L18	1927-1270	L101	1927-1314/pt 2	P43
1927-1226	J1	1927-1271	L104	1927-1315	P2
1927-1227	J38	1927-1272	L105	1927-1316	P34
1927-1228	L15	1927-1273	E199	1927-1317	P62
1927-1229	M90	1927-1274	L108	1927-1318	P38
1927-1230	M44	1927-1275	L118	1927-1319	P39
1927-1231	M45	1927-1276	E191	1927-1320	P40
1927-1232/pt 1	M33	1927-1277	L119	1927-1321	E84
1927-1232/pt 2	M61	1927-1278	E71	1927-1322	P19
1927-1233	M42	1927-1279, -1280	E177	1927-1323	P23
1927-1234	M8			1927-1324	P42
1927-1235	D103	1927-1281	E178	1927-1325	M1
1927-1236	D4	1927-1282	D68	1927-1326/pt 1	P13
1927-1237	M25	1927-1283	D40		

1927-1326/pt 2	P14	1927-1371	E109	1927-1417	E187
1927-1326/pt 3	P22	1927-1372	E91	1927-1418	J46
1927-1327	E86	1927-1373	E92	1927-1419	D72
1927-1328	P49	1927-1374	E94	1927-1420	E184
1927-1329	P61	1927-1375	P27	1927-1421	E185
1927-1330	P12	1927-1376	E106	1927-1422	J47
1927-1331	E83	1927-1377	E96	1927-1423	J48
1927-1332	P5	1927-1378	E95	1927-1424	D140
1927-1333	P35	1927-1379	E97	1927-1425	E176
1927-1334	P36	1927-1380	E108	1927-1426	J113
1927-1335	P37	1927-1381	P47	1927-1427	D144
1927-1336	J193	1927-1382	L48	1927-1428	E175
1927-1337	P11	1927-1383	E75	1927-1429	J45
1927-1338	P8	1927-1384	D34	1927-1430	L76
1927-1339	P44	1927-1385	J166	1927-1431	D69
1927-1340	P51	1927-1386	D145	1927-1432	D75
1927-1341	P52	1927-1387	D31	1927-1433	D76
1927-1342	P58	1927-1388	L42	1927-1434	D70
1927-1343	P54	1927-1389	J181	1927-1435	D57
1927-1344	L32	1927-1390	P60	1927-1436	D74
1927-1345	J195	1927-1391	E98	1927-1437	D71
1927-1346	P24	1927-1392	J167	1927-1438	D61
1927-1347	P26	1927-1393	J168	1927-1439	E208
1927-1348	E87	1927-1394	D48	1927-1440	L107
1927-1349	J161	1927-1395	D120	1927-1441	E211
1927-1350	J159	1927-1396	J169	1927-1442	E212
1927-1351	E88	1927-1397	P50	1927-1443	L112
1927-1352	E93	1927-1398	D49	1927-1444	L97
1927-1353	D125	1927-1399	D125	1927-1445	E202
1927-1354	J162	1927-1400	E90	1927-1446	J37
1927-1355	J163	1927-1401	J170	1927-1447	L114
1927-1356	J164	1927-1402	J171	1927-1448	E214
1927-1357	E89	1927-1403	D32	1927-1449	L121
1927-1359	E92	1927-1404	E99	1927-1450	J174
1927-1360,	E91	1927-1405	J112	1927-1451	E11
-1361,		1927-1406	J172	1927-1452	D43
-1362		1927-1407	D30	1927-1453	D100
1927-1363	J194	1927-1408	J180	1927-1454	L33
1927-1364	L49	1927-1409	J173	1927-1455	D12
1927-1365	L47	1927-1410	E192	1927-1456	E13
1927-1366	L46	1927-1411	E101	1927-1457	D11
1927-1367	P30	1927-1412	D119	1927-1458	L151
1927-1368	L45	1927-1413	E37	1927-1459	L129
1927-1368	E92	1927-1414	D118	1927-1460	E161
1927-1369	J165	1927-1415	D139	1927-1461	E159
1927-1370	E79	1927-1416	E186	1927-1462	E160

1927-1607	J123	1927-1652,	J154	1927-1697	J14
1927-1608	J98	-1653		1927-1698	J150
1927-1609	J91	1927-1654	E35	1927-1699	J153
1927-1610	J83	1927-1655	D8	1927-1700	E38
1927-1611	J85	1927-1656	L28	1927-1701	E39
1927-1612	J87	1927-1657	L26	1927-1702	L96
1927-1614	D116	1927-1658	L27	1927-1703	E198
1927-1615	L20	1927-1659	E34	1927-1704	L98
1927-1616	L19	1927-1660	J19	1927-1705	L23
1927-1617	E209	1927-1661	E169	1927-1706	L71
1927-1618	L30	1927-1663	E21	1927-1707	E116
1927-1618	D28	1927-1664	L142	1927-1708	J34
1927-1619	D51	1927-1665	L141	1927-1709	E120
1927-1620	D50	1927-1666	M111	1927-1710	E142
1927-1621	E61	1927-1667	J21	1927-1711	E148
1927-1622	D44	1927-1668	E66	1927-1712	E143
1927-1623	E102	1927-1669	L95	1927-1713	E144
1927-1624/pt 1	P1	1927-1670	L136	1927-1714	E125
1927-1624/pt 3	E107	1927-1671	L144	1927-1715	E145
1927-1625	J32	1927-1672,	L8	1927-1716	E147
1927-1626	L68	-1673		1927-1717	E117
1927-1627	J33	1927-1674	L133	1927-1718	E126
1927-1628	E103	1927-1675	L138	1927-1719	E123
1927-1629	E105	1927-1676	L139	1927-1720	E134
1927-1630	D7	1927-1677	P18	1927-1722	E121
1927-1631	L31	1927-1678	P17	1927-1723	E130
1927-1632	E63	1927-1679	E210	1927-1724	E139
1927-1633	D114	1927-1680	L137	1927-1725	E132
1927-1634	D22	1927-1681	D93	1927-1726	E138
1927-1635	D9	1927-1682	D97	1927-1727	E140
1927-1636	L140	1927-1683	D90	1927-1728	E137
1927-1637	J116	1927-1684	D92	1927-1729	E127
1927-1638	J115	1927-1685	D91	1927-1730	E128
1927-1639	M29	1927-1686	D94	1927-1731	E136
1927-1640	L77	1927-1687	D95	1927-1732	E135
1927-1641	D17	1927-1688	D96	1927-1733	L70
1927-1642	D10	1927-1689/pt 1	P31	1927-1734	E141
1927-1643	L1	1927-1689/pt 2	P7	1927-1735	J122
1927-1644	L7	1927-1690	J175	1927-1735	J192
1927-1645	D77	1927-1691	E158	1927-1737	E122
1927-1646	J2	1927-1692	E188	1927-1738	E118
1927-1647	D85	1927-1693	J192	1927-1739	E119
1927-1648	D87	1927-1694	J176	1927-1740	E131
1927-1649	D84	1927-1694	J155	1927-1741	E133
1927-1650	D78	1927-1695	M69	1927-1742	E124
1927-1651	D79	1927-1696	L17	1927-1743	E129

1927-1744	E146	1927-1788	L120	1927-1833	J133
1927-1745	E115	1927-1789	M103	1927-1834	J131
1927-1746	L35	1927-1790	M104	1927-1835	J127
1927-1747	J182	1927-1791	M105	1927-1836	J15
1927-1748	E104	1927-1792	M28	1927-1837	J16
1927-1749	D29	1927-1793	M30	1927-1838	J128
1927-1750	E70	1927-1794	M27	1927-1839	E163
1927-1751	D80	1927-1795	M76	1927-1840	J134
1927-1752	D89	1927-1796	J152	1927-1841	D106
1927-1753	D83	1927-1797	M86	1927-1842	L4
1927-1754	D86	1927-1798	J28	1927-1843	L3
1927-1755	D88	1927-1799	M39	1927-1844	J126
1927-1756	D82	1927-1800/pt 1	M9	1927-1845	J125
1927-1757	D81	1927-1800/pt 2	M10	1927-1846	L6
1927-1758	D146	1927-1801	M71	1927-1847	L2
1927-1759	J39	1927-1802	M43	1927-1848	L5
1927-1760	L36	1927-1804	M60	1927-1849	J190
1927-1761	L37	1927-1805	M62	1927-1850	M35
1927-1762	L38	1927-1806,	M37	1927-1851	M50
1927-1763	L111	-1807		1927-1852	M80
1927-1764	J65	1927-1808	E53	1927-1853	M64
1927-1765/pt 1	J158	1927-1809	M109	1927-1854	M78
1927-1765/pt 2	J157	1927-1810	P20	1927-1855	D23
1927-1766	P32	1927-1811	P21	1927-1856	M34
1927-1767	J22	1927-1812	E149	1927-1857	M40
1927-1768	J6	1927-1813	L69	1927-1858	M92
1927-1769	J156	1927-1814	E69	1927-1859/pt 1	M93
1927-1770	J151	1927-1815	E68	1927-1859/pt 2	M99
1927-1771	D121	1927-1816	E67	1927-1860/pt 1	M41
1927-1772	J145	1927-1817	E113	1927-1860/pt 2	M101
1927-1773	J10	1927-1818	D21	1927-1861	M107
1927-1774	E8	1927-1819	M59	1927-1862	J5
1927-1775	J41	1927-1820	M55	1927-1863	M14
1927-1776/pt 1	J25	1927-1821	M57	1927-1864	J7
1927-1776/pt 2	J26	1927-1822	L21	1927-1865	M106
1927-1776/pt 3	J27	1927-1823	J135	1927-1866	M36
1927-1777	L153	1927-1824	M23	1927-1867	M87
1927-1778	E60	1927-1825	M24	1927-1868	M79
1927-1780	E3	1927-1826	J11	1927-1869	J132
1927-1781	E2	1927-1827	M56	1927-1870	M31
1927-1782	E150	1927-1828	M58	1927-1871	E7
1927-1783	E9	1927-1829	J3	1927-1872	M95
1927-1784	L16	1927-1829	J177	1927-1873	D105
1927-1785	E30	1927-1830	M54	1927-1874	L25
1927-1786	E10	1927-1831	D104	1927-1875	M3
1927-1787	L9	1927-1832	J129	1927-1876	L24

1927-1877	E55	1927-1927	M114	1927-2065/pt 5	E229
1927-1878	L29	1927-1928	D14	1927-2065/pt 6	E230
1927-1879	J191	1927-1929	D13	1927-2065/pt 7	E231
1927-1880	D25	1927-1930	E62	1927-2065/pt 8	E232
1927-1881	J136	1927-1931	D109	1927-2065/pt 9	E233
1927-1882	J23	1927-1932	D19	1927-2065/pt 10	E234
1927-1885	J124	1927-1933	D112	1927-2065/pt 11	E235
1927-1886	J138	1927-1934	L147	1927-2065/pt 12	E236
1927-1887	J12	1927-1935	D20	1927-2065/pt 13	E237
1927-1888	J137	1927-1936	D110	1927-2065/pt 14	E238
1927-1889	J139	1927-1937	D107	1927-2065/pt 15	E239
1927-1890	J130	1927-1938	D108	1927-2065/pt 16	E240
1927-1891	M110	1927-1939	D111	1927-2065/pt 17	E241
1927-1892	J140	1927-1940	J178	1927-2065/pt 18	E242
1927-1893	J141	1927-1941	L39	1927-2065/pt 19	E243
1927-1894	J148	1927-1942	E65	1927-2065/pt 20	E244
1927-1895	J117	1927-2020	L56	1927-2065/pt 21	E245
1927-1896	M15	1927-2021	D123	1927-2065/pt 22	E246
1927-1897	E59	1927-2037	P55	1927-2065/pt 23	E247
1927-1898	E23	1927-2038	P57	1927-2065/pt 24	E248
1927-1899	P59	1927-2039	M113	1927-2065/pt 25	E249
1927-1900	M96	1927-2040	E12	1927-2065/pt 26	E250
1927-1902	M98	1927-2041	J189	1927-2065/pt 27	E251
1927-1903	M97	1927-2042	J188	1927-2065/pt 28	E252
1927-1904	E194	1927-2043	J187	1927-2065/pt 29	E253
1927-1905	L90	1927-2044	J186	1927-2065/pt 30	E254
1927-1906	P53	1927-2045	J143	1927-2065/pt 31	E255
1927-1909	E112	1927-2046	J142	1927-2065/pt 32	E256
1927-1910	E114	1927-2047	J185	1927-2065/pt 33	E257
1927-1911	D124	1927-2048	P63	1927-2065/pt 34	E258
1927-1912/pt 1	M18	1927-2049	J184	1927-2065/pt 35	E259
1927-1912/pt 2	M53	1927-2050	J179	1927-2065/pt 36	E260
1927-1913	E33	1927-2051	J183	1927-2065/pt 37	E261
1927-1914	E29	1927-2052	J144	1927-2065/pt 38	E263
1927-1915	E28	1927-2052	E85	1927-2065/pt 39	E262
1927-1916	E31	1927-2053	P25	1927-1883, -1884	M112
1927-1917	M74	1927-2054	P3	1927-1582, -1583, -1584	J80
1927-1918	E24	1927-2055	E179		
1927-1919	E32	1927-2056	E18		
1927-1920	E27	1927-2057	J147	1928-790	D65
1927-1921	E26	1927-2058	P56	1928-803	D137
1927-1922	E51	1927-2063	E64	1928-819	D67
1927-1923	E25	1927-2065/pt 1	E225	1928-829	D63
1927-1924	E1	1927-2065/pt 2	E226	1928-830	D64
1927-1925	E4	1927-2065/pt 3	E227	1928-851	D66
1927-1926	L73	1927-2065/pt 4	E228		

1929-93	L87	1929-106	D130	1929-120	D59
1929-94	L86	1929-107	D129	1929-121	L11
1929-95	L88	1929-108	E74	1929-122	J18
1929-96	L91	1929-109	D128	1929-123	J20
1929-97	L55	1929-111	E72	1929-124	E54
1929-98	M7	1929-112	D126	1929-125	D39
1929-99	L127	1929-113	M6	1929-126	L99
1929-100	E73	1929-114	M75	1929-127	J17
1929-101	D52	1929-115	P10	1929-128	J149
1929-102	D127	1929-116	M2	1929-129	D148
1929-103	L124	1929-117	L94	1929-130	E215
1929-104	L125	1929-118	L83	1938-706	D138
1929-105	L126	1929-119	D58	1949-116	E157

Index

Where an index entry refers to an illustration, page numbers are printed in **bold type**. A page number followed by an n indicates that the reference is to a footnote on that page, and a page number followed by tab indicates that the reference is to a table on that page.